Handbook of Normative Multiagent Systems

Handbook of Normative Multiagent Systems

Edited by
Amit Chopra
Leendert van der Torre
Harko Verhagen
Serena Villata

© Individual authors and College Publications 2018. All rights reserved.

ISBN 978-1-84890-285-5

College Publications
Scientific Director: Dov Gabbay
Managing Director: Jane Spurr

http://www.collegepublications.co.uk

Cover produced by Laraine Welch
Printed by Lightning Source, Milton Keynes, UK

All rights reserved. No part of this publication may be reproduced, stored in a retrieval system or transmitted in any form, or by any means, electronic, mechanical, photocopying, recording or otherwise without prior permission, in writing, from the publisher.

CONTENTS

PREFACE — vii

LIST OF CONTRIBUTORS — xiii

PART I INTRODUCTION — 1

HARKO VERHAGEN, MARTIN NEUMANN, MUNINDAR P. SINGH
Normative Multi-Agent Systems: Foundations and History — 3

PART II MODELING — 27

NATASHA ALECHINA, MEHDI DASTANI, BRIAN LOGAN
Norm Specification and Verification in Multiagent Systems — 29

VIVIANE T. SILVA, WAMBERTO W. VASCONCELOS, JESSICA S. SANTOS, JEAN O. ZAHN, MAIRON BELCHIOR
Modeling Normative Conflicts in Multiagent Systems — 57

CHRISTOPHER K. FRANTZ, GABRIELLA PIGOZZI
Modeling Norm Dynamics in Multiagent Systems — 73

NICOLETTA FORNARA, TINA BALKE-VISSER
Modeling Organizations and Institutions in MAS — 143

ROB CHRISTIAANSE
Modeling Norms Embedded in Society: Ethics and Sensitive Design — 171

PART III ENGINEERING — 207

MATTEO BALDONI, CRISTINA BAROGLIO, AMIT K. CHOPRA, AKIN GÜNAY
Interaction Protocols — 209

MATTEO BALDONI, CRISTINA BAROGLIO, OLIVIER BOISSIER, JOMI F. HÜBNER, ROBERTO MICALIZIO
Norm-Aware and Norm-Oriented Programming — 231

PART IV LOGICAL ANALYSIS 249

GABRIELLA PIGOZZI, LEENDERT VAN DER TORRE
Multiagent Deontic Logic and its Challenges from a Normative Systems Perspective 251

XAVIER PARENT, LEENDERT VAN DER TORRE
Detachment in Normative Systems: Examples, Inference Patterns, Properties 305

CELIA DA COSTA PEREIRA, BEISHUI LIAO,
ALESSANDRA MALERBA, ANTONINO ROTOLO,
ANDREA G. B. TETTAMANZI, LEENDERT VAN DER TORRE,
SERENA VILLATA
Handling Norms in Multiagent Systems by Means of Formal Argumentation 345

EMILIANO LORINI
Logics for Games, Emotions and Institutions 375

Preface

This Handbook of Normative Multiagent Systems is a community effort aimed at providing a comprehensive and up-to-date view of the state-of-the-art and current trends in the lively research field of normative multiagent systems (NorMAS).

Workshops on normative multiagent systems have been organised (nearly) yearly since 2005. At a meeting held during the Dagstuhl NorMAS 2015 Seminar, the participants decided to produce a handbook, extending the roadmap drawn up after a Dagstuhl Seminar in 2012. The presentations and lively discussions at the Seminar opened the way to the actual implementation of the handbook. Very positive feedback, many suggestions, and general support collected during the year after the Dagstuhl Seminar boosted the process. The handbook website was set up (www.normativemas.org), and an intense exchange of ideas at large was started, in order to shape the whole initiative and the contents of the handbook. First versions of the chapters were produced and submitted in the first part of 2016. Each chapter was examined by two reviewers, whose feedback has been crucial to improve the quality of the final versions of the chapters produced by the authors by the summer of 2017. Some of the chapters, suitably modified to stand as journal papers, have appeared in two special issue of the IfCoLog Journal of Logics and their Applications (Volume 4, Issue 9) published in November 2017, and (Volume 5, Issue 2) published in April 2018. Authors' proofreading finally completed the long effort resulting in the present handbook, whose organisation and contents are briefly summarised below.

The volume is organised into four parts: A. Introduction, B. Modeling formalisms, C. Engineering, D. Logical Analysis.

The introduction in Part A consists of a single chapter on the foundations and the history of the field. Norms are widely used to represent ethical, legal and social aspects of multiagent systems, and normative multiagent systems provide a promising model for human and artificial agent coordination because they integrate norms and individual intelligence. Apart from the history of norms in computer science in general and NorMAS in particular, the chapter's main focus is on social norms as used in the social sciences where norms are considered one of the main origins of social order. The main theoretical approaches are presented including game theory as a model of choice and societal effects of choice.

The aim of the part B of this handbook is to describe the major achievements the NorMAS research field attained in *modeling* normative multiagent systems, as well as the main challenges that still remain open in this regard. Many modelling issues arise when there is the need to define a normative multiagent system for a specific usage scenario, e.g., how to specify the norms and how to verify that the current state of affairs actually satisfies such norms, how to model the fact that norms may change and this change may impact on other norms (leading to potential conflicts) and on the verification procedures, how to detect and subsequently manage possible violations of the norms, how to model organisations and institutions which play a fundamental role in NorMAS. In addition to these questions, where NorMAS is informed and inspired by different disciplines such as computer science but also sociology, psychology and cognitive science, we realise that NorMAS also need to model how norms emerge in (agent) societies, and how such social norms can be studied through agent-based simulation models. This section covers some of these aspects and answers some of the questions proposed above, discussing the solutions presented in the literature and the open issues. We describe now with some more details the content of each of the chapters of this section.

The first chapter in this part formally analyses the issue of modeling norm specification and verification in multiagent systems (MAS). In particular, violation conditions of regulative norms may correspond to conditions on states, actions, or arbitrary temporal patterns. They may be specified semantically or expressed syntactically in a suitable temporal logic, or in a programming language. Verification problems for norms or rather for normative systems involve verifying consistency of norms, verifying whether violation conditions hold, and finally verifying whether a system where norms are enforced satisfies some system objective.

The second chapter in this part has the goal to present the elements used by several approaches to represent a norm and the techniques found in the literature for the detection and resolution of conflicts between norms in multiagent systems. The techniques used to detect normative conflicts are classified in two main groups *(i)* approaches that deal with normative conflicts at design time, and *(ii)* approaches that deal with normative conflicts at runtime. The approaches used to resolve normative conflicts are divided in two kinds: *(i)* norm prioritization, and *(ii)* norm update. The norm prioritisation strategy priorities one of the norms in conflict by overriding the other under particular circumstances. In the norm update strategy, one of the norms in conflict is updated in order to eliminate the conflict.

In the third paper, how to model norm dynamics in multiagent systems is discussed. More precisely, all existing life cycle models are reviewed looking at normative processes from a holistic perspective, which include the introduction of individual life cycle models and their contextualisation with specific contributions that exemplify life cycle processes. They provide a comprehensive contemporary overview of individual contributions to the area of NorMAS and the systematic comparison of the discussed life cycle models. Based on this analysis, they also propose a refined life cycle model that resolves terminological ambiguities and ontological inconsistencies of the existing models, while reflecting the contemporary view on norm formation and emergence.

The fourth chapter analyses possible solutions to model organisations and institutions in multiagent systems. Institutions and organisations are two concepts within the MAS community that are commonly referred to when the question arises on how to ensure that an (open) MAS exhibits some desired properties, while the agents interacting in that MAS have some degree of autonomy at the same time. The authors give a brief introduction to the two concepts as its related ideas, outlining research done in the area of NorMAS and giving pointers on current challenges for modelling institutions and organisations.

The fifth chapter aims at discussing research directions towards the modelling of those norms that are embedded in the society, with a particular attention to ethics and sensitive design. After elaborating on the notions of decision rights, responsibility and accountability, the original design question is rephrased into seven key questions formulated in a principled way using the procedure to classify and analyse an ethical system applied to the code of conduct of Nike. Also the issue of defining the notion of a model as the start and result of the design process is addressed in this chapter.

Part C is concerned with the engineering of normative multiagent systems. This part is inspired from the fact that a founding basis for the field of multi-agent systems was the novel *social* abstractions it offered for engineering *open* systems of autonomous agents. For example, instead of conceptualizing a software system merely as a collection of components, it conceptualized the system as an *organization* of agents who interactions were subject to normative expectations, that is, *norms*. Whereas computing has historically focused on mechanisms that regiment interactions, work in multiagent systems has focused on norm-based architectures that regulate via norms. The chapter titled *Interaction Protocols* and *Norm-Aware and Norm-Oriented Programming* expand on the foregoing themes. To specify an architecture is to specify the interactions among its components, in our case, agents.

The chapter on interaction protocols discusses approaches for specifying interaction protocols. It discusses two types of protocols, for constraining message ordering and occurrence in multiagent systems and for capturing the normative *meanings* of messages, the latter being higher-level social constraints. In fact, a focus on meaning is a specialty of multiagent systems research. The chapter dwells at length on *commitment protocols*, an exemplar of meaning-based approaches.

The chapter on programming discusses approaches for specifying and implementing *computational organizations* of agents. It discusses how interaction protocols and norms figure in the specification of agents and the reasoning they perform. Significant emphasis is placed on the *goals* of agents and how the *environment* and norms feature in the achievement of those goals. The discussion is made concrete with examples from *JaCaMo*, a platform for developing and running multiagent systems.

Part D is concerned with logically analysing normative multiagent systems. Given the profound importance of norms in multiagent systems, it is fundamental to understand, e.g., which norms are valid in certain environments, how to interpret them, and to determine the deontic conclusions of such norms. For a complete review on deontic logic and logic-based normative systems, we refer the reader to the Handbook of deontic logic and normative systems [Gabbay et al., 2013]. The point of introducing formal definitions in this chapter is just to have a reference for the interested reader.

The first chapter in this part discusses how deontic logic—the field of logic that is concerned with normative concepts such as obligation, permission, and prohibition—can be used for reasoning about normative multiagent systems, It gives an overview of several challenges studied in deontic logic, with an emphasis on challenges involving agents. It starts with traditional modal deontic logic using preferences to address the challenge of contrary-to-duty reasoning, and STIT theory addressing the challenges of nondeterministic actions, moral luck and procrastination. Then it turns to alternative norm-based deontic logics detaching obligations from norms to address the challenge of Jorgensen's dilemma, including the question how to derive obligations from a normative system when agents cannot assume that other agents comply with their norms. Also some traditional challenges are discussed from the viewpoint of normative systems. Normative multiagent systems need to combine normative reasoning with agent interaction, and thus raise the challenge to relate the logic of normative systems to game theory. For a full coverage of related work the reader is refer to the handbook of deontic logic and normative systems

The second chapter of part D addresses the most basic question to ask a normative system: which obligations, permissions and institutional facts can be detached from a set of rules or conditional norms in the context of the rest of the system. Consider a driver who is surprised by a child crossing the street. A car is approaching from the other lane. Moral code and criminal law tells him to evade the child. Traffic law prohibits him to change lanes, risking a frontal confrontation with the truck. Moreover, there are norms to predict his own children in the back of the car. Which decision the driver should take, depends on the details of the situation. This is the detachment problem. In the near future, such detachment problems should not only be solved by human drivers, but also by driverless cars. Benchmark examples are used in the chapter to compare ways to reason with normative systems. An overview of several benchmark examples of normative reasoning and deontic logic is given: Van Fraassen's paradox, Forrester's paradox, Prakken and Sergot's cottage regulations, Jeffrey's disarmament example, Chisholm's paradox, Makinson's Moebius strip, and Horty's priority examples. Inference patterns are used to compare different ways to reason with normative systems, and more abstract properties are defined to compare different ways to reason with normative systems. The ten introduced properties can be used also as requirements for the further development of formal methods for normative systems and deontic logic.

The third chapter in this part discusses three examples from the literature of handling norms by means of formal argumentation, illustrating that formal argumentation is used to enrich and analyse normative multiagent systems in various ways. First, the authors discuss how existing ways to resolve conflicts among norms using priorities can be represented in formal argumentation. Based on such representation results, formal argumentation can be used to explain the detachment of obligations and permissions from hierarchical normative systems in a new way. Second, they discuss how formal argumentation can be used as a general theory for developing new approaches for normative reasoning, using a dynamic ASPIC-based legal argumentation theory. Third, they show how argumentation can be used to reason about other challenges in the area of normative multiagent systems as well, by discussing a model for arguing about legal interpretation. The aim to discuss these three examples is to inspire new applications of formal argumentation to the challenges of normative reasoning in multiagent systems.

The fourth chapter in this part discusses the formal analysis of cognitive and institutional concepts. A conceptual framework is introduced that clarifies the relationship between intention and action and the role of intention in practical reasoning; explains how moral attitudes such as standards, ideals and moral values influence decision-making; explains how preferences are formed on the basis of desires and moral values; clarifies the distinction between the concept of goal and the concept of preference; elucidates how mental attitudes including beliefs, desires and intentions trigger emotional responses, and how emotions retroactively influence decision-making and mental attitudes by triggering belief revision, desire change and intention reconsideration. Then, the authors explain how game theory and logic have been used in order to develop formal models of such cognitive phenomena. They put special emphasis on a specific branch of game theory, called epistemic game theory, and on a specific family of logics, so-called agent logics.

Altogether we believe that the chapters included in this volume achieve two goals. On the one hand, they provide enough introductory material so that a newcomer can get acquainted with the essentials of the field. On the other hand, they cover more advanced issues so that anyone interested in the field may have a comprehensive, though of course not exhaustive, reference on the state-of-the-art and get useful insights for future developments.

We are pleased to conclude with some dutiful expressions of gratitude. We thankfully acknowledge the contribution of all the authors and the reviewers who made this volume possible, and the help of all the colleagues who provided comments, suggestions, critiques, and encouragements during the development of the initiative. Chopra's participation in this effort was partially supported by EPSRC grant EP/N027965/1 (Turtles). Last but not least, special thanks go to College Publications and in particular to Jane Spurr for her invaluable continued support.

Amit Chopra
Leon van der Torre
Harko Verhagen
Serena Villata

BIBLIOGRAPHY

[Gabbay et al., 2013] D. Gabbay, J. Horty, and X. Parent. *Handbook of Deontic Logic and Normative Systems*. College Publications, 2013.

List of Contributors

Natascha Alechina, University of Nottingham, Nottingham, NG8 1BB United Kingdom

Matteo Baldoni, Università degli Studi di Torino Dipartimento di Informatica, via Pessinetto 12, I10149, Torino (TO) Italy

Tina Balke-Visser, Vanderlande, Systems Group, The Netherlands

Cristina Baroglio, Università degli Studi di Torino, Dipartimento di Informatica, via Pessinetto 12, I10149, Torino (TO) Italy

Mairon Belchior, Universidade Federal Fluminense, Av. Gal. Milton Tavares de Souza, s/n, Boa Viagem - Niterói/RJ - Brazil, CEP: 24210-346

Olivier Boissier, Mines Saint-Etienne, 158 cours Fauriel, 42023 Saint-Etienne, France

Amit K. Chopra, School of Computing and Communications, Lancaster University, Infolab21 Lancaster LA1 4WA, United Kingdom

Rob M.J. Christiaanse Faculty of Technology, Policy and Management, Delft University of Technology, Delft, The Netherlands and EFCO BV, Amsterdam

Célia da Costa Pereira, Université Cte dAzur, I3S, CNRS, Templiers 1 , 930, Route des Colles - BP145—, 06903 Sophia Antipolis Cedex, France

Mehdi Dastani, Utrecht University, Princetonplein 5, 3584 CC Utrecht, The Netherlands

Nicoletta Fornara, Universit della Svizzera italiana, Faculty of Communication Sciences, Lugano, Switzerland

Christopher K. Frantz, Norwegian University of Science and Technology (NTNU), Department of Computer Science (IDI), 2815 Gjøvik, Norway

Akin Günay, School of Computing and Communications, Lancaster University, Infolab21 Lancaster LA1 4WA , United Kingdom

Jomi F. Hübner, Federal University of Santa Catarina, Department of Automation and Systems Engineering, PO Box 476, Florianópolis, SC, 88040-900 Brazil

Bweishui Liao, Center for the Study of Language and Cognition, Xixi Campus, Zhejiang University, Hangzhou, 310028 China

Brian Logan, University of Nottingham, Nottingham, NG8 1BB United Kingdom

Emiliano Lorini, CNRS, IRIT, Toulouse University, 118 route de Narbonne, 31062 Toulouse, France

Alessandra Malerba, CIRSFID, University of Bologna, Italy

Roberto Micalizio, Università degli Studi di Torino, Dipartimento di Informatica, via Pessinetto 12, I10149, Torino (TO) Italy

Martin Neumann, Johannes-Gutenberg University Mainz, Institute for Sociology, Sociology of technology and innovation and simulation methodologies, Johannes Welder Weg 20, 55128 Mainz , Germany

Xavier Parent, University of Luxembourg, Faculté des Sciences, de la Technologie et de la Communication, Maison du Nombre, 6 Avenue de la Fonte, L-4364 Esch-sur-Alzette, Luxemburg

Gabriella Pigozzi, Université Paris-Dauphine, PSL Research University, CNRS, LAMSADE, 75016 Paris, France

Antonio Rotolo, CIRSFID, University of Bologna, Italy

Jessica S. Santos, IBM Research, Av. Pasteurs 138/146, Urca - Rio de Janeiro/RJ, Brazil, CEP: 22.290-903 and Universidade Federal Fluminense, Av. Gal. Milton Tavares de Souza, s/n , Boa Viagem - Niterói/RJ - Brazil , CEP: 24210-346

Minundar P. Singh, Department of Computer Science, North Carolina State University, USA

Andrea G.B. Tettamanzi, Université Cte dAzur, I3S, Inria, CNRS, Templiers 1 , 930 Route des Colles - BP145, 06903 Sophia Antipolis Cedex, France

Viviane Torres da Silva, IBM Research, Av. Pasteurs 138/146, Urca - Rio de Janeiro/RJ, Brazil CEP: 22.290-903

Leendert van der Torre, University of Luxembourg, Faculté des Sciences, de la Technologie et de la Communication, Maison du Nombre, 6 Avenue de la Fonte, L-4364 Esch-sur-Alzette, Luxemburg

Wamberto Vasconcelos, Department of Computing Science, University of Aberdeen, Meston Walk, Old Aberdeen, AB24 3UE, Aberdeen, United Kingdom

Harko Verhagen, Department of Computer and Systems Sciences, Stockholm University, 164 07 Kista, Sweden

Jean O. Zahn, Universidade Federal Fluminense, Av. Gal. Milton Tavares de Souza, s/n , Boa Viagem - Niterói/RJ - Brazil, CEP: 24210-346

PART I

INTRODUCTION

1
Normative Multiagent Systems: Foundations and History

HARKO VERHAGEN, MARTIN NEUMANN, MUNINDAR P. SINGH

ABSTRACT. This chapter provides a brief history of the field of normative multiagent systems, including highlights of the main intellectual themes in this field; how those themes have played out over the years; a summary of some major challenges and how well those challenges have been addressed; and some promising directions for future research. Human Interaction, Social Theory, Norms as the origins of social order; Historic overview over development of social science with respect to research on norms; Short sketch of relevant theories. Game theory will be also treated here as a theoretical point of view on society.

1 Introduction

This chapter provides a quick overview of the history of normative multiagent systems and its foundations. We will start by delving into the latter and start from social science as the area in which norms and their effects and dynamics have been modeled and measured. The study of norms, in its modern form, extends back approximately a hundred years to the beginning of the twentieth century. In the 1920s, as the fields of sociology and social psychology were becoming generally accepted as part of academia, the study of norms became a central theme. In anthropology, Margaret Mead famously studied norms in Samoan society, leading to the idea that norms are different in different societies while in social psychology Muzafer Sherif conducted experimental studies of the adoption of beliefs in conformance to others [Sherif, 1937], especially authority figures, and subsequently on group formation [Sherif *et al.*, 1955]. Sherif's work related the social sphere to the psychological sphere. The close association of norms and sanctions was recognized from the outset, for example, in the work of Radcliffe-Brown [Radcliffe-Brown, 1934]. The centrality of norms in sociology peaked in the 1950s with the work of Talcott Parsons [1949, 1951] that defined the functionalist melody of the day, a view in which different institutions contribute and strife towards a state of balance and social equilibrium. As a counterpart to this, and the development in the 1960s of anti-authoritarian attitudes, the influence of microsociology, trying to understand social interactions in small groups using qualitative approaches and an individualistic-constructionist view on social processes took over, moving norms to the background and on the whole in large denying the existence of supraindividual concepts. To quote Margaret Thatcher "There is no such thing as society" [Thatcher, 1987]. Balancing the macrosociological and microsociological apporach has been the aim of for instance Giddens [1984] and his structuration theory and the work of Coleman [1990] and his (in)famous Coleman boat focussing on the connections between microlevel interactions and macrolevel entities such as norms. Lately, there has been a resurgence of attention on norms and allied concepts under the rubric of investigations of prosocial behavior, for example, [Simpson and Willer, 2008] and [Therborn, 2002]. For obvious reasons, norms have been of importance in the field of criminology as well, for example, the discussion in Gibbs [1965] on the definition and classification of norms.

The relationship of norms to communication and the modeling of societies is natural, and quite germane to the study of multiagent systems. Norms provide the framework within which members of the society interact. That is, the norms of a society provide the rules of encounter under which that society's members interact and in particular communicate. Thinkers, notably including Searle [1995], associate norms with how our social reality is constructed. A major distinction is between *constitutive norms* and *regulative norms*. Constitutive norms determine what counts as what in the specific society. For example, raising your hand may count as placing a bid in an open outcry auction but may count as hailing a taxi on a busy street. Regulative norms determine when one may act in a certain manner. For example, you may not whistle in an open outcry auction or even whistle shrilly if walking on the sidewalk, but you may accompany your taxi hailing hand wave with a loud whistle. Paul Grice [1975], the philosopher of language, proposed what are now called the Gricean maxims of communication. These maxims correspond to norms of communication. We would expect that they would be followed in ordinary conversation though they may be flouted for dramatic effect, as in humor, irony, or sarcasm.

The study of norms bifurcated into two main branches. The first branch, like the original conception, is concerned with community standards. This body of work has followed with mechanisms that lead to the emergence of norms and their robustness against norm-violating behaviors of (some) members of society. We classify the works by Robert Axelrod [1986] on the iterated prisoners dilemma and by David Lewis on conventions in coordination games in this branch. From the field of social ontology the game theoretical approach to norms is developed by amongst others Cristina Bicchieri [2005] following on the work of Edna Ullmann-Margalit [1977].

The second branch sprung out of studies in deontic logic. Georg von Wright introduced deontic logic as a distinct field of study in the 1950s. Deontic logic provides constructs for dealing with what is permitted, forbidden, or obligatory. Von Wright's studies led him to the idea of understanding norms as they relate to individual action [von Wright, 1963]. This branch has led to an intense study of a variety of norm types, including commitments, prohibitions, and authorizations.

2 Social norms: history and core concepts

Social norms can be found in nearly every interaction between humans. Whether they team up for solving problems or just shake hands, it seems nearly impossible that interaction is unaffected by social prescriptions. In the interactions social norms are at the interface between the individual and society. On the one hand, norms are enacted by individual habits and behavior. On the other hand they characterize and differentiate large groups and societies, for instance by norms to shake hands or to bow for greeting. This two sided effect is characteristic for social norms: Interaction of individual agents shapes social structure while on the other hand agency is shaped by social structure [Conte *et al.*, 2013]. Norms are macro-level patterns generated by interactions of individual actors. However, the transmission of social norms and cultural values to the individual shapes the individual in a way that has even been denoted as a 'second birth' [Claessens, 1972]. For this reason the study of social norms can be approached from a social or individual perspective. However, any attempt has to keep in mind that at least implicitly and in very principle investigations of the relation between individual and society imply two opposing angles of basic assumptions [Geulen, 1991]. One can propose either a harmony or antagonism between the individual and society. The assumption of a harmony is advocated by philosophers such as Aristotle, Leibniz or Hegel. Representatives of the assumption of an antagonism are, for example, Hobbes or Rousseau. These two philosophers stand exemplary for two again opposing consequences of the assumption of a principle antagonism, namely arguing for the need for social order as Hobbes did in postulating the Leviathan or advocating individual freedom as Rousseau did [Neumann, 2013]. These two basic attitudes can also be found in social science theories on social norms.

As social norms mediate between the individual and society it is no wonder that their investigation is closely related to the emergence of the scientific disciplines of both sociology and psychology. Faced with changing labor relations in the industrial revolution Emil Durkheim posed the question how social integration can go along with growing individual autonomy. In his book on the division of labor [Durkheim, 1893] he found the answer to this research question in the study of social norms and their change. Durkheim regarded norms as a central ingredient for securing social order. Durkheim can be regarded as the paradigmatic example of a sociological approach to social norms. The paradigmatic example for a psychological approach to the question of how the individual is shaped by the social world can be found a few decades later in the work of Sigmund Freud, namely his postulation of id, ego and super ego in which the super ego represents society [Freud, 1953]. However, while Durkheim and Freud are paradigmatic for the different approaches to norms from the perspective of society and the perspective of the individual, they are both also paradigmatic examples for research implying a principle antagonism between individual and society: Both agree society acts as an alien force on the individual. However, while Durkheim searched for means to preserve social order in a rapidly changing world, Freud's theory has at least been used for attempts to release the individual from overarching demands of the super ego.

However, what actually are social norms? Examples are easily found, such as tipping in a restaurant, shaking hands or respect of property. Obviously these examples also show that norms strongly vary between different cultures and societies and also with regard to their salience within a society [Conte et al., 2013]. As unproblematic as it is finding examples, it becomes more difficult to provide a clear cut definition. Over decades several classifications in attempt of a definition have been provided (for example, [Gibbs, 1965], [Therborn, 2002], [Interis, 2011]). For example, it is not agreed if and how norms relate to values which are justified by moral standards [Gibbs, 1965], if they are unintended results of interaction [Bicchieri and Muldoon, 2011] or whether they need to be based in deontic beliefs [Bicchieri, 2005] that even may be codified in legal norms, as for example in Weber's definition of law: "An order will be called law if it is externally guaranteed by the probability that coercion (physical or psychological), to bring about conformity or avenge violation, will be applied by a staff of people holding themselves specially ready for that purpose." [Weber, 1921]. The recourse to law brings us to the point of norm enforcement. It is not agreed whether a right or even obligation of norm enforcement [Axelrod, 1986] is indispensable for an attempt of a definition. Nevertheless we refer here to an attempt provided by [Interis, 2011] as a definition which is broadly enough to cover at least most of the cases discussed in the literature so far. Obviously, this may change in course of time. This definition can be regarded as a baseline that can be extended in more specific accounts:

Norms are a regularity of intentional behavior within a certain population [Interis, 2011].

The notion of 'intentional' behavior remains silent whether the intention is based on a deontic belief [Bicchieri, 2005], i.e. if people follow norms because they think it is obliged to do so. Certainly following an obligation is intentional behavior. However, intentional behavior need not be based on obligations. It is a broader concept. Thus norms entail in a minimal account the following components:

1. An individual component: a belief.

2. A social component: the intentional behavior is a regular pattern within a group.

These characteristics entail that norms imply both a psychological and a social component. A norm has a psychological component as it is an individual belief. However, it has also a social component as it is a regularity of behavior in a group. This reflects that norms are a central element in the relation between structure and agency, making norms fundamental in the two way dynamics of creating social order.

Note that in the absence of a deontic belief that people *ought* to behave in a certain way norm enforcement can only be achieved as an unintended consequence of for example mutual adjustment. For instance people may develop certain pathways in the grass of a park that might lead to a snack bar or the toilet. Once many people took a certain path the grass might vanish in a way that looks like a path and so other people follow the route as well. Ellickson 2001 describes an example of the spontaneous emergence of property rights among indigenous American societies in Labrador. Insofar the sociological approach to the theory of norms is weaker than legal theories or normative systems based on deontic logic which cover important aspects of this handbook. This resembles the idea of harmony between individual and society. Approaches that advocate an antagonism between individual and society typically emphasize that norms describe prescribed (or prohibited) behavior, i.e. the need for punishing norm deviation for securing salience of norm abiding behavior. This can certainly be found as for example the prohibition to smoke in restaurants, or the prescription to wear safety belts during the take-off of an airplane. Enforcement of prescriptions or prohibitions may range from laws, enforced by specialized staff, as in the example of Weber's definition cited above to informal signs of disapproval. In sum, a more rigid concept of norms entails a further element:

1. A deontic: a conduct is obliged or prohibited

Including a deontic belief brings sociological research on norms closer to fields of normative reasoning, such as deontic logic or legal theory, examined in other chapters of this handbook. However, including a deontic belief further increases the number of necessary ingredients for specifying normative behavior. Refraining from smoking in restaurants or wearing safety belts during take-off are behavior regulations in specific circumstances: namely take-off of an airplane or sitting in a restaurant. These prescriptions or prohibitions are conditional to specific situations. These situations describe the circumstances in which the norm becomes valid [Hechter and Opp, 2001]. If the conditions are not fulfilled the norm is not active. For instance, I may smoke outside the restaurant.

1. Situational conditions: specify the scope of application of norms.

This more rigid concept of norms beyond mere behavior regularities enables a differentiation between conventions, norms and values. Conventions are different from norms as they may be regularities of behavior and even intentional behavior without a deontic belief. Values on the other hand may not be restricted to specific circumstances but of universal validity.

2.1 Social norms: theoretical approaches

So far the background and definition of norms that can be agreed on among different sociological theories has been provided. In the next section a brief overview of the main theoretical approaches in sociological research will be provided, highlighting the different contributions of these different theoretical approaches to the explanations of social norms. Very broadly the different sociological theories can be characterized by different distinct research questions that are in the focus of research in these diverging traditions. Certainly this is only a tendency. All theories provide answers to a bundle of questions and reducing them to only one particular question is certainly a simplification. The same holds already for the classification of particular research to certain theoretical accounts. Nevertheless it might help the reader as a kind of guide through the thicket of numerous theories and problems. Given this warning, very broadly the main research questions can be described by three distinct questions [Neumann, 2008].

1. First thing to note is that norms have an effect on society. As social interaction is regulated by social norms also the appearance of the whole society is shaped by norms. This was the research question that already puzzled Emil Durkheim. Likewise, for instance, most of the research in legal theory is less concerned with the causal mechanisms, for instance in parliamentary processes, that bring certain laws into being. Rather legal research addresses the question of what effects can be expected by certain legislative regulations. This research question can be characterized as asking for the *functional effects* of social norms.

2. The individual, psychological component calls for the question how actors are transformed in such a way that they factually follow norms. How and by what mechanisms is agency shaped by social structure? This can be characterized as the problem of *transformation* of human agency.

3. The social component of a regularity of behavior within a group implies the question how the spreading or decay of norm-abiding behavior within a population can be explained. This can be characterized as the problem of *transmission* of norms.

The main theoretical approaches can be distinguished by approaching norms from a macro- or micro-social perspective. Again very broadly theories of norms can be summarized by the categories of role theory as a macro-social account and identity theories and rational choice accounts as micro-social approaches [Bicchieri and Muldoon, 2011]. The distinction between macro- and micro-social emphasized a different focus on individual habits and actions on the one hand and on the social level on the other hand.

Macro Theory

Historically, sociological theoretizing of norms has been elaborated first by macro-social approaches. Norms have been a central theoretical term in social systems theory of the second half of the 20th century. Within this account norms have been a central theoretical building block of so-called role theory. Going back to Durkheim, role theory has been elaborated in particular by Parsons [Parsons, 1949][Parsons, 1951]. The basic theoretical assumption of role theory is that individual behavior is guided by so-called role sets. These role sets are social prescriptions which are specified by a set of norms. Thus social norms provide the foundations of this theoretical account. In contrast to micro social theories, the explanation of how social norms come into existence is not the central focus of role theory, but rather norms are used for an explanation.

Such an explanation can be illustrated by taking a closer look at the so-called homo sociologicus [Dahrendorf, 1965]. The notion of a 'homo sociologicus' paraphrases key elements of role theory. Dahrendorf [Dahrendorf, 1965] considers the example of meeting a person at a cocktail party (comp. [Neumann, 2008]). By getting acquainted with a stranger one might learn that he is a married academic and is father of two children. His professional occupation is being teacher. He is a German citizen and has been refugee after the second World War. He might have said that he had some problems in fitting in in a catholic city in West-Germany as he is protestant. The talk might have continued but already this information might be sufficient for the feeling of getting familiar with this man. In more recent times it might be that our refugee is Muslim with an oriental destination. This information might give us certain expectations who he is and how he will behave. However, Dahrendorf emphasizes that all this information is about social facts. As a teacher our dialogue partner is faced with certain obligations and expectations. This leads to certain stereotype of what is a classical teacher. This is a social role. Moreover while Protestantism (or other religions) and professional occupations certainly influence personal habits and attitudes, they transcend the individual: The religion existed before this individual was born and will exist when he dies. Millions of other individuals are Protestants (or members of other religions). They even need not know each other, but nevertheless survey research reveals that religion is an explanatory factor for attitudes and value orientations for instance [Troitzsch, 2015]. Social structure provides a casting mold for individual behavior [Durkheim, 1895].

Thus role theory emphasizes the structural constraints imposed on individual behavior by social norms and thus assumes an antagonism between the individual and society. However, these constraints provide a certain reliability in what kind of behavior can be expected in certain situations. Role theory claims that this is a precondition that enables creation and maintenance of social relations. Unconstrained by social norms interaction would result in chaotic coexistence of articulations to which individuals could not make sense of. Parsons [Parsons, 1949] emphasized that the ends of individual actions are not arbitrary, but rather are prescribed by social norms. Thereby social norms regulate the proper function of the social system. Thus primary focus of role theory is on examining the *functional effects* of norms by enabling social integration. For instance, the role of the father is to educate his child. The role of the lecturer is crucial for the socialization of pupils. Thus, both roles are functionally relevant for the reproduction of the society [Neumann, 2008]. This was already Durkheim's research question when he investigated how social integration is possible in a functionally differentiated industrial society.

Role theory dominated sociological theory in the 1960s and 1970s. Criticized already in the 1960s as regarding individual actors as mere puppets ([Wrong, 1961], [Homans, 1964]), in the past decades its influence significantly decayed. Theoretical assumptions of role theory could not been empirically verified or had to be refined ([Bicchieri, 2005], [Bicchieri and Xiao, 2009]), and questions remained unanswered concerning the scope and limits of the force of social norms in determining individual behavior. Social systems theory and role theory attempts at a functionalist explanation of social structures: originating from biology a functionalist explanation asks what system components are needed for the functioning of the overall system. If a phenomenon P is found in a society there are reasons for the members of society to practice P. However, there are also effects of P for the society: Its *function* that might be different from the reasons to practice P. However, a feedback loop is postulated: decrease of P triggers social disintegration which triggers increase of P. Philosophy of science remains skeptical with regard to the explanatory status of a functionalist explanation as it fails to provide mechanisms which could provide causal explanations of the generation of social phenomena. It is criticized that the existence of a certain phenomenon P is explained by postulating that P is necessary for the function of the system. This is claimed as invalid because the explanation of P depends on the phenomenon to be explained ([Hempel, 1994], [Nagel, 1981], [Mayntz, 1961]).

A central question of this macro-social account was 'why are there phenomena such as social norms' but remained indecisive with regard to the question: 'how are these phenomena brought about'? Nevertheless, this macro-social account can be regarded as a starting point for further elaboration of sociological theory of norms. It remains a challenge to provide causal explanations of the functional effects of norms. More recent approaches attempt to fill the gap left open by this account. For instance, the question how and why individuals follow the structural constraints imposed on them addresses the question of *transformation* of actors. For this purpose the micro level of individual actors has to be taken into account.

Role theory answers this question by recourse to socialization theory. At this point the theory relies on findings from other scientific disciplines. Parsons and Shils [Pasons and Shils, 1951] provide and attempt by referring to Freud's psychoanalytic framework which had been state of the art in the 1950s. In the next section more recent accounts of identity theories will be highlighted.

Micro theories a: Identity theories

Identity theories address the question how the individual copes with categories provided by society in the process of generating attitudes and values ([Tajfel, 1970], [Tajfel, 1981]). For instance, individuals might wear certain clothing to express their religious conviction. Thus identity theories investigate the problem of the *transformation* of individual agency by social categories. Identity theories argue that the mechanism of norm internalization can be found in the development of a sense of identity ([Bosma and Kunnen, 2001], [Krappmann, 2006], [Neumann, 2013]). Central claim of identity theories is that identity formation is shaped by cultural patterns by being engaged in social relations [Granovetter, 1985]. While identity theories differentiate between personal and social identity, in this context the concept of social identity is essential. Social identity describes the part of the self-perception which is influenced by peer- and reference groups. Peer groups denote the social network in which the individual is directly embedded. For instance, youth groups may identify themselves by certain clothing. However, social identity may also transcend direct individual relations to a self-identification with large scale groups without personal acquaintanceship. For instance, individuals may identify themselves with their religious group or a certain political party. These groups are denoted as reference group [Neumann, 2013].

The roots of this socio-psychological research go back to Mead ([Mead, 1934]), Piaget ([Piaget, 1932], [Piaget, 1947]), and Kohlberg ([Kohlberg *et al.*, 1996]) who focus on the cognitive development by which humans become morally responsible agents. Mead's theory was groundbreaking by investigating the cognitive development of children as for instance, the capability of role-taking, i.e. to regard oneself from the perspective of other persons. In the same line Piaget distinguished a heteronomous stage of moral development in which norms are perceived as given by an external normative authority and an autonomous stage of moral development in which norms are perceived as product of free agreement. The moral development has been summarized by Kohlberg [Fittkau, 1993] as a three stage process from a pre-conventional to a post-conventional level during which juveniles develop the power of judgment [Neumann, 2010]. Thus childhood is particular relevant for the development of the cognitive complexity of humans that allows for internalization of social norms ([Fuhrer and Trautner, 2005], [Krappmann, 2006]). For instance, [Phinney, 1990] reports that only at the age of about four years children begin to stabilize ethnic and gender identities. Nevertheless, acquisition and change of social identities is a life-long process. For instance, immigrants might identify with the host society or individuals acquiring professional identities as, for example, professor or physician.

Central aspects of identities are the inner perception of oneself on the one hand and ascriptions of others on the other hand. Typical examples for ascriptions of others are gender or ethnicity. In some societies also belonging to certain classes may be imposed on the individual. However, individuals may integrate ascriptions of others into their own self-perception, for example by joining a working class movement. In this context the theory of self categorization becomes relevant [Turner *et al.*, 1987]. Self categorization investigates how ascriptions of others become part of the personal self perception, i.e. how social identities become personal identities. Of particular relevance for this process are the concepts of in-group and out-group [Haslam, 2001]. In-groups are those to which the individual feels associated. In-group identification enables collective action. For instance an individual may regard him or herself as part of a sports team whereas the other team in a match is regarded as the out-group. In sports this is even accentuated by different tricots. Regarding oneself as member of the team enables collective action such as playing football together. However, it has to be noted that the differentiation between in-groups and out-groups depends on the level of abstraction. For instance members of different teams in a sports game may regard themselves as belonging to the in-group of sportsmen in contrast to for example journalists reporting the match. Moreover, the categories of identification are not static, but always fluid and dependent on the situation. Outside of a sports game, the sportsperson may regard him or herself not as a member of a certain sports team but as member of a certain family together with their spouse husband and their children [Turner *et al.*, 1987]. The link between self-categorization and the theoretical problems imposed by sociological role theory is established by the fact that the categories of self-identification are the social roles described by the sociological account. Self categorization in terms of a certain social role triggers that an individual acts in accordance with the norms that define the particular role. If for instance an individual is primed with the social role of a banker the individual is more likely to act as prescribed by the role set of a banker than in the case when the individual is primed by the role set of family [Cohn *et al.*, 2014].

The question is left open however, why and how self-categorization according to social roles triggers certain behaviors. This question refers to motivational theories, as the theory of self-determination [Deci and Ryan, 2000]. Self-determination theory argues that socially embedded identity allows for action selection and thus individual freedom. This is highlighted by the theory of self-determination [Deci and Ryan, 2000]. Action determination can be intrinsically or extrinsically motivated. According to the theory of self-determination the identity of individuals contributes to the development of their intrinsic motivation. Whereas in principle obedience is extrinsically motivated because it is a social prescription (or prohibition), through the process of internalization norms it becomes part of personal identity. Thereby a transformation of the human actor transforms external regulation into self-determination [Deci and Ryan, 2000]. Internalization of extrinsic motivation represents the bridge between psychological integrity and social cohesion [Neumann, 2013]. Thus in contrast to the role theoretical perspective that the social constrains the individual, identity theories reveal the basic assumption of harmony between individual and society.

However, the question of the origins of the social categories or how certain categories spread in a population is out of the scope of social psychological theories. In fact, identities are always fluid and individuals committed to several group identities that might even be conflicting [Bicchieri and Muldoon, 2011]. The problem of how norms spread in populations is the problem of *transmission*. This is the focus of rational choice theories to be examined next.

Micro theories b: rational choice theories

Rational choice theories are theories of individual action selection according to the criterion of maximizing individual expected utility. Following the Scottish moral philosophy, most prominently represented by David Hume or Adam Smith, it is assumed that individuals seek happiness and try to avoid pain (Rawls 2000). As the starting point of an explanation is individual action, rational choice theories can be characterized as micro social theories. However, the preference ordering that gives rise to expected utilities is out of the scope of this theoretical account [Stigler and Becker, 1977]. This stands in contrast to identity theories which focus exactly on this question. Within this framework norms modulate action selection as deviation of norms conforming behaviour is threatened by sanctions. Becoming victim of sanctions reduces utility which is included in the calculation of expected utilities. As long as the danger of being observed and punished is greater than the benefits of norm deviation individuals follow a norm. Within this framework it is not necessary that individuals obey norms because they want or feel obliged to it. Norms are equated with normal behavior [Campenní *et al.*, 2009]. There is no need for individuals to know the norms [Neumann, 2008]. Norms are a behavioral regularity (compare with [Interis, 2011]) enforced by sanctions. This is a basic mechanism for the *transmission* of norms. In sum, rational choice is a theory of norm enforcement. The central theoretical term is the notion of sanctions [Coleman, 1989]. This reveals the basic assumption of an antagonism between individual and society.

The theoretical accounts discussed so far investigate the relation between an isolated individual and society as a whole, may it be from the perspective of society as in macro social accounts or from the perspective of the individual as in micro social accounts. However, the analytical framework of rational choice becomes particular fruitful by enlarging the perspective from action to interaction. *Game theory* investigates strategic interaction, i.e. situations in which the expected utility of the choice of action is dependent on actions of others. Individuals know the structure of the situation and take this into account in their choice of action [Gintis, 2000]. For analyzing the theoretically expected result the concept of a Nash equilibrium is particularly relevant [Nash, 1950]. This is a situation in which no player individually can change the course of action without losses of expected utility. Thus a Nash equilibrium stabilizes interaction. This framework is of particular relevance for the study of norms as it provides a setting for studying the coordination of individual actions. This is what the functional analysis of macro social theory revealed as the central function of norms. For instance, it is mutually beneficial to drive on a particular side of the road. If all drive on one side, no driver can decide individually to drive on the other side without damaging himself or herself. Thus a norm is reached: a regularity of intentional behavior. In such so-called coordination games achieving norm obedience is rather unproblematic and need not be enforced by an external sanctioning agency as it is in the individual self-interest of the involved actors to following the norm. Thus there is a harmony between individual interest and collective gains.

More problematic are social dilemmas in cooperation games. These games highlight the antagonism between individual and society. The most prominent example is the prisoner's dilemma. Social dilemmas characterize situations in which it is in the individual self-interest not to cooperate, but only if the other players choose to cooperate. In this case the defector gains the maximum benefit while the cooperator is cheated. However, if all defect, all loose. If the payoff is denoted as T, for 'temptation', the reward for cheating a cooperator, R for the reward of mutual cooperation, P for 'punishment', denoting the risk of mutual cheating and S, the sucker who cooperates and is cheated the structure in a dilemma situation is the following: $T > R > P > S$. A dilemma is characterized by a structure in which holds that $T > R$. In this case there is trade-off between individual self-interest and collective gains and thus a temptation to cheat. The dilemma is that the Nash equilibrium results in mutual defection, that is, cheating [DeLanda, 2011].

In such situations norms may provide an incentive to push the actors away from being trapped in mutual defection [Bicchieri, 2005]. Ullman-Margarit [Ullmann-Margalit, 1977] describes a dramatic example of two soldiers that have the duty of holding a certain position rear cover for their troop while being attacked by an enemy. Holding the position is an important back-up for their troop. If they fulfill their duty they have a chance to hold the position and survive. However, this is a life threatening situation. If one of the soldiers decides to escape, his or her chances for survival will increase but the comrade will certainly be killed. In the first instance this situation can be analyzed as a dilemma for the two soldiers: Certainly it holds that $T > R$. However, a third party of the troops is introduced which is the main sucker in this situation. It seems less obvious that it holds: $R > P$. Rationally it can be expected that both soldiers try to save their life individually and leave their troop helplessly surrendering to the enemies. It seems that $P > R$, which makes defection even more likely. In this case internalization of the military norm of honor might induce the courage to risk the personal life in the name of higher values, that is, that the two soldiers collectively arrive at a situation of R, that is, that they both bravely fight the enemy. This framework has been widely applied, in particular for analyzing the management of public goods [Ostrom, 1990]. However, here we arrive at an analysis which again describes function effects of norms as in the macro-social role theory. This does not clarify how norms are transmitted. This can be analyzed by adding a dynamic perspective to the analysis of strategic interaction. This is studied by so-called evolutionary n-person game theory.

Evolutionary game theory investigates the dynamics of strategies in repeated interactions. This is significantly different from one-shot games. For instance in repeated interactions of two actors (with memory) in a prisoner's dilemma mutual cooperation would become a Nash equilibrium. This can be examined by models that include a replicator dynamics that enables differential reproduction of different strategies. More successful strategies replicate whereas less successful ones die out. Success is measured according to utility values gained in previous interactions. This is denoted as their fitness. In this setting different strategies compete to become evolutionary dominant. Thus rationality is reconstructed as an evolutionary search process [Smith, 1982]. The dynamics can be investigated by agent-based simulation models of different complexity, for example interacting agents might have a memory or not. This facilitates also to include more than two agents. In every round two agents in a population are selected randomly to play against each. Successful agents replicate whereas unsuccessful once are deleted from the simulation. The simplest example is the evolution of tit-for-tat as dominant strategy in interactions of agents without memory against strategies of pure defection and pure cooperation [Axelrod, 1984]. A central criterion for evaluation of stability is not a Nash equilibrium but evolutionary stable strategies (ESS) [Smith and Price, 1973]. ESS includes that an equilibrium is robust against perturbations, as for example insertion of new strategies for example by mutation [Binmore, 1998]. For instance, in a simulation of the prisoner's dilemma first tit-for-tat may become dominant. Once tit-for-tat is dominant all agents cooperate in the long run. Then it may become less costly to avoid punishment at all, i.e. pure cooperation may insert the population (as a perturbation through mutation). However, once cooperation becomes dominant insertion of pure defection (again by mutation) can easily spread in the population. Thus tit-for-tat is not an ESS. Evolutionary n-person game theory enables the examination of the evolution of solidarity norms such as reciprocal altruism, and even indirect altruism in large anonymous societies [Bicchieri *et al.*, 2004]. For this reason, it is a key framework for identifying mechanisms for the transmission of norms that can be extended for studying further mechanisms such as signaling and reputation [Berger and Rauhut, 2015].

2.2 Summary on sociology and norms

As social norms are at the interface between the individual and society the roots of research originate both from sociology and psychology. In the work of Emil Durkheim and Sigmund Freud social norms can be found at the very beginning of the emergence of these scientific disciplines. Therefore a definition needs to entail an individual element of a belief and a social element that the belief is shared in a group, leading to a certain regularity of intentional behavior. More restrictive concepts of social norms specify a particular kind of belief, namely a deontic belief that certain behavior is prescribed or prohibited in certain situations.

The different theoretical approaches can broadly be characterized as macro social theories that examine norms as properties of whole societies or micro social theories that focus on individual actors and actions and their consequences for aggregated social groups. Mainly social norms have been investigated to examine social roles in social systems theory as a macro social account, and micro social accounts of identity theories and rational choice theories, including game theory and evolutionary game theory. In a rational choice account, game theory is of particular relevance for the study of norms as it provides a framework for investigating interactions. These different approaches focus on different aspects related to the scientific investigation of norms, namely their functional effects, normative transformation of human actors and transmission of norms. Sociological role theory investigates functional effects of social integration. Identity theories focus on the transformation of individual actors and the most prominent contribution of rational choice theory to the comprehension of norms is to provide mechanisms of the transmission of norms.

Note, that investigating the relation between individual and society implies taking a perspective as starting point for the investigation: in principle the relation between individual and society need to be perceived as either harmonious or antagonistic. However, it has to be taken into account that this is only a broad characterization of tendencies. This holds also for the classification of the different theories and their main contributions.

3 Norms, AI, and normative multiagent systems

There has not been a significant attention to norms as we understand them in the broad Artificial Intelligence community. One notable exception is the work of Shoham and Tennenholz on social laws and conventions (see, for example, [Shoham and Tennenholtz, 1995; Shoham and Tennenholtz, 1997]). The study of language has considered norms but usually in the spirit of considering fixed sets of norms, such as the Gricean maxims mentioned in the introduction. Works on norms have appeared in AI forums but those works are better classified as belonging to the normative multiagent systems (NorMAS) community—which makes sense because NorMAS is considered one of the themes of interest to AI.

3.1 NorMAS Community History

Normative multiagent systems research began to form in the mid 1990s just as the field of distributed artificial intelligence was gelling and becoming more clearly focused on the challenges of multiagent systems as opposed merely to the challenges of potentially interacting individual agents. An interest that set multiagent systems apart from some distributed artificial intelligence was the emphasis on system-level concerns and the idea that the agents are autonomous and may represent competing interests. This contrasted with the distributed artificial intelligence theme of distributed problem solving where one party allocated problems (or tasks) to different agents, thereby presuming the agents all served the same interest and hence limiting the agents' autonomy.

In multiagent systems, there was increasing interest in formulating how agents could come together to live and work together, whether cooperatively or competitively. One branch of research pursued abstractions based on game theory, where there was usually a clear statement of the options (strategies) and payoffs of the concerned agents. The other branch of research pursued abstractions inspired by human societies, including the social controls as effected through organizations, institutions, and norms.

Possibly the first broad event focused on normative multiagent systems was the Workshop on Norms, Obligations, and Conventions organized as part of the First International Conference on Multiagent Systems (ICMAS) in Kyoto in December 1996. The workshop led to an influential Artificial Intelligence and Law special issue. The NorMAS series of workshops started in 2005 at the AISB seminar series [Boella *et al.*, 2005] which resulted in a double special issue of the journal Computational and Mathematical Organization Theory [Boella *et al.*, 2006b], the introduction paper [Boella *et al.*, 2006a] of which is among the highest cited in the research area. At about the same time (2006), the related research community Coordination, Organization, Institutions and Norms in Agent Systems (COIN) started.

3.2 Defining NorMAS

One major part of nascent research community is the definition of its topic. This was indeed part of the work on the special issue of the inaugural workshop. In the introduction [Boella *et al.*, 2006a] of the special issue [Boella *et al.*, 2006b] the research area is defined as follows:"A normative multiagent system is a multiagent system with normative systems in which agents can decide whether to follow the explicitly represented norms, and the normative systems specify how and in which extent the agents can modify the norms." On the other hand, a new community is dynamic so just one year later a new definition was presented in the JAAMAS special issue [Boella et al., 2008b] (based on the 2007 Dagstuhl seminar on Normative Multiagent Systems): "A normative multiagent system is a multiagent system organized by means of mechanisms to represent, communicate, distribute, detect, create, modify, and enforce norms, and mechanisms to deliberate about norms and detect norm violation and fulfillment" [Boella et al., 2008a]. This second definition expresses a more dynamic and interactionist view on norms. The same paper and seminar also formulated a list of 10 challenges that would drive the community forward in its research endeavours for the following years—challenges for agent programmers to provide tools for:

1. agents supporting communities in their task of recognizing, creating, and communicating norms to agents

2. agents to simplify normative systems, recognize when norms have become redundant, and to remove norms

3. agents to enforce norms.

4. agents to preserve their autonomy

5. agents to construct organizations

6. agents to create intermediate concepts and normative ontology, for example to decide about normative gaps

7. agents to decide about norm conflicts

8. agents to voluntarily give up some norm autonomy by allowing automated norm processing in agent acting and decision making

9. conviviality.

10. legal responsibility of the agents and their principals

How to program normative systems is a challenge that increasingly comes up in the literature. See for instance the Special Issue on Foundations of Social Computing of the ACM Transactions on Internet Technology (TOIT) in which [Baldoni et al., 2014] conclude that explicit representation of social relationships is key to the realization of socio-technical systems in which regulations control interaction options. Chopra and Singh propose a programming model for normative systems wherein norms are given first-class status and used to characterize the correct behavior of agents [Chopra and Singh, 2016].

How to acquire norms is a nontrivial challenge and one that has been approached from multiple angles. The most obvious direction is to have agents interact and let the appropriate norms emerge. However, the semantics of the concept of norms may remain unclear using such a strategy. In this regard, Morales and colleagues [Morales et al., 2013; Morales et al., 2015] have proposed a series of approaches to learn a normative system from the repeated interactions of agents in varying settings. They show some norm languages that can be learned under reasonable assumptions. Mashayekhi and colleagues [Mashayekhi et al., 2016] consider the emergence of norms in a hybrid setting in which the technical elements can prevent the most egregious violations (called integrity violations, for example, a collision in the traffic domain) and can offer recommendations to the agents to avert anticipated risky situations (called conflicts). The agents acquire strategies indicating the establishment of a norm where the deviation from a few leads to their being punished. A second direction for norm acquisition is by relying upon existing documentation, in natural language, to determine what norms hold. The documentation could be formal and created before the fact to guide agent behavior, for example, in regulations and contracts. Based on a study of several hundred real-life business contracts, Gao and Singh [Gao and Singh, 2014] show how to identify norms of the major types in sentences in real-life contract.

Or, more interestingly, the natural language documents may themselves have been created through the natural "innocent" interactions of people trying to go about their business. In this regard, the work of Dam and colleagues [Avery et al., 2016; Dam et al., 2015] is interesting in mining norms from software repositories based on the actions of software engineers as they collaborate on a project. We would classify the work of Kalia and colleagues [Kalia et al., 2013] on mining commitments from chat and email messages exchanged by workers in (and clients of) an enterprise.

4 Future Directions

The study of normative multiagent systems, we would argue, is still much in its inception. As the world moves inexorably toward ubiquitous and increasingly advanced information and communications technologies, we are continually uncovering problems that require deeper and broader treatments of computational norms: how to represent them; reason about them; elicit, develop, or refine them; and implement them—and how to do at the level of the artificial, natural, or hybrid societies that these norms characterize. In this endeavour, consultation of the social science and philosophical roots of norms in the previous section will be key.

4.1 Sociotechnical Systems

A particularly promising direction is that of sociotechnical systems. Part of the importance of sociotechnical systems arises from emerging applications of information and communication technologies, which inevitably involve the interactions of multiple people and organizations (hence social) over complex sensing, actuating, computing, storage, and communication infrastructures (hence technical). The problems of security and privacy are inherently sociotechnical in nature because securing the social or technical elements in exclusion of the other type is simply inadequate. In this regard, Patkos and colleagues [Patkos *et al.*, 2015] present several perspectives on privacy.

Although the notion of sociotechnical systems as combining considerations of social and technical elements has been studied for years, it has usually been approached in one of two main ways. The classical works in Science and Society studies refer to social and technical in the large. For example, they may consider how work and human interactions and power structures in an organization change through the introduction of new technology, such as email. These studies are valuable in that they seek to understand when and how an organization or society may introduce a new technology and the repercussions thereof. However, they do not seek computational models as such because the notion of technology is treated as an environmental input to an organization, not as a technical resource to be computed about. At the other extreme, in the user interfaces research community, they refer to any technical resource, for example, a smart phone, as a technical element and anything involving a human, even a single human, as social. Here the idea is to understand user experience, including whether a technical element helps a user solve his or her problem and the kinds of errors a user may make. These works too are not computational and though they focus on social and technical in the small, they end up with an impoverished model of the social elements.

In contrast, the NorMAS take on sociotechnical systems is to characterize them as normative systems [Singh, 2013]. In other words, the social elements are represented computationally as norms and their reliance on and effect upon the technical elements is computationally characterized. In this manner, we can hope to achieve, not just claims of broad generality about social and technical elements, but model organizations and their prospective member agents, and reason about their properties. Kafalı and colleagues [Kafalı *et al.*, 2016] show how to model and construct a sociotechnical system with respect to some privacy-relevant requirements from stakeholders.

Adopting this conception of sociotechnical systems opens up possibilities for some research directions that are not only important in practice but bring up new conceptual and theoretical challenges. Christiaanse and colleagues [Christiaanse *et al.*, 2014] provide an extensive sociocognitive model that would support how agents interact within a sociotechnical system and to form a sociotechnical system. This work is based on the work on the WIT trinity by Noriega et al. [Noriega *et al.*, 2015], [Noriega *et al.*, 2017]. In the WIT trinity importance is given to three different stances to view a social-technical system from: the actual world in which the system is used (W), the Ideal system (I), and the Technological artifacts that implement it (T). Nardin and colleagues [Nardin *et al.*, 2016] study the nature and variety of sanctions in sociotechnical systems bringing in considerations of a sanctioning process geared toward achieving governance.

Chopra and colleagues [Chopra et al., 2014; Chopra and Singh, 2014] seek to expand requirements engineering to support sociotechnical systems explicitly. In previous works, for example, on Tropos [Bresciani et al., 2004], the social elements are considered only in the early stages of modeling but are replaced by a system actor, a technical element, that captures specific solutions to the elicited requirements but omits the social constructs. In contrast, Chopra and colleagues take a normative stance wherein the social elements are represented explicitly in a computational manner in the resulting specification. Each participant can contribute a system actor who can work with the applicable norms, and may comply or violate the norms as it sees fit.

BIBLIOGRAPHY

[Avery et al., 2016] Daniel Avery, Hoa Khanh Dam, Bastin Tony Roy Savarimuthu, and Aditya K. Ghose. Externalization of software behavior by the mining of norms. In *Proceedings of the 13th International Conference on Mining Software Repositories (MSR)*, pages 223–234, Austin, Texas, May 2016. ACM.

[Axelrod, 1984] Robert M. Axelrod. *The evolution of cooperation*. Basic Books (AZ), 1984.

[Axelrod, 1986] Robert Axelrod. An evolutionary approach to norms. *American Political Science Review*, 80(4):1095–1111, December 1986.

[Baldoni et al., 2014] Matteo Baldoni, Cristina Baroglio, and Federico Capuzzimati. A commitment-based infrastructure for programming socio-technical systems. *ACM Transactions on Internet Technology (TOIT)*, 14(4):23:1–23:23, December 2014.

[Berger and Rauhut, 2015] Roger Berger and PD Dr Heiko Rauhut. Reziprozität und Reputation. In *Handbuch Modellbildung und Simulation in den Sozialwissenschaften*, pages 715–742. Springer, 2015.

[Bicchieri and Muldoon, 2011] Cristina Bicchieri and Ryan Muldoon. Social norms. In *The Stanford Encyclopedia of Philosophy*. Stanford Univeristy, 2011.

[Bicchieri and Xiao, 2009] Cristina Bicchieri and Erte Xiao. Do the right thing: but only if others do so. *Journal of Behavioral Decision Making*, 22(2):191–208, 2009.

[Bicchieri et al., 2004] Cristina Bicchieri, John Duffy, and Gil Tolle. Trust among strangers. *Philosophy of Science*, 71(3):286–319, 2004.

[Bicchieri, 2005] Cristina Bicchieri. *The grammar of society: The nature and dynamics of social norms*. Cambridge University Press, 2005.

[Binmore, 1998] Ken G Binmore. *Game theory and the social contract: just playing*, volume 2. MIT press, 1998.

[Boella et al., 2005] Guido Boella, Leendert van der Torre, and Harko Verhagen, editors. *Proceedings of the Symposium on Normative Multi-Agent Systems*. The Society for the Study of Artificial Intelligence and the Simulation of Behaviour, 2005.

[Boella et al., 2006a] Guido Boella, Leendert van der Torre, and Harko Verhagen. Introduction normative multiagent systems. *Computational and Mathematical Organization Theory*, 12(2):71 – 80, 2006.

[Boella et al., 2006b] Guido Boella, Leendert van der Torre, and Harko Verhagen. Special issue on normative multiagent systems. *Computational and Mathematical Organization Theory*, 12(2 – 3), 2006.

[Boella et al., 2008a] Guido Boella, Leendert van der Torre, and Harko Verhagen. Introduction to the special issue on normative multiagent systems. *Journal of Autonomous Agents and Multi-Agent Systems*, 17(1):1 – 10, 2008.

[Boella et al., 2008b] Guido Boella, Leendert van der Torre, and Harko Verhagen. Special issue on normative multiagent systems. *Journal of Autonomous Agents and Multi-Agent Systems*, 17(1), 2008.

[Bosma and Kunnen, 2001] Harke A Bosma and E Saskia Kunnen. Determinants and mechanisms in ego identity development: A review and synthesis. *Developmental review*, 21(1):39–66, 2001.

[Bresciani et al., 2004] Paolo Bresciani, Anna Perini, Paolo Giorgini, Fausto Giunchiglia, and John Mylopoulos. Tropos: An agent-oriented software development methodology. *Journal of Autonomous Agents and Multi-Agent Systems (JAAMAS)*, 8(3):203–236, May 2004.

[Campenní et al., 2009] Marco Campenní, Giulia Andrighetto, Federico Cecconi, and Rosaria Conte. Normal= normative? the role of intelligent agents in norm innovation. *Mind & Society*, 8(2):153, 2009.

[Chopra and Singh, 2014] Amit K. Chopra and Munindar P. Singh. The thing itself speaks: Accountability as a foundation for requirements in sociotechnical systems. In *Proceedings of the IEEE International Workshop on Requirements Engineering and Law (RELAW)*, page 22, Karlskrona, Sweden, 2014. IEEE Computer Society.

[Chopra and Singh, 2016] Amit K. Chopra and Munindar P. Singh. From social machines to social protocols: Software engineering foundations for sociotechnical systems. In *Proceedings of the 25th International World Wide Web Conference*, pages 903–914, Montréal, April 2016. ACM.

[Chopra et al., 2014] Amit K. Chopra, Fabiano Dalpiaz, F. Başak Aydemir, Paolo Giorgini, John Mylopoulos, and Munindar P. Singh. Protos: Foundations for engineering innovative sociotechnical systems. In *Proceedings of the 18th IEEE International Requirements Engineering Conference (RE)*, pages 53–62, Karlskrona, Sweden, August 2014. IEEE Computer Society.

[Christiaanse et al., 2014] Rob Christiaanse, Aditya Ghose, Pablo Noriega, and Munindar P. Singh. Characterizing socio-cognitive technical systems. In *Proceedings of the European Conference on Social Intelligence (ECSI)*, volume 1283 of *CEUR*, pages 336–346, Barcelona, November 2014. CEUR-WS.org.

[Claessens, 1972] Dieter Claessens. *Familie und Wertsystem: eine Studie zur" zweiten soziokulturellen Geburt" des Menschen und der Belastbarkeit der" Kernfamilie"*. Duncker & Humblot, 1972.

[Cohn et al., 2014] Alain Cohn, Ernst Fehr, and Michel André Maréchal. Business culture and dishonesty in the banking industry. *Nature*, 516(7529):86–89, 2014.

[Coleman, 1989] Jules L Coleman. The rational choice approach to legal rules. *Chi.-Kent L. Rev.*, 65:177, 1989.

[Coleman, 1990] James Coleman. *Foundations of Social Theory*. Harvard university press, 1990.

[Conte et al., 2013] Rosaria Conte, Giulia Andrighetto, and Marco Campennì. *Minding norms: Mechanisms and dynamics of social order in agent societies*. Oxford University Press, 2013.

[Dahrendorf, 1965] Ralf Dahrendorf. *Homo sociologicus. Ein Versuch zur Geschichte, Bedeutung und Kritik der Kategorie der sozialen Rolle*. Westdeutscher Verlag, Cologne et Opladen, 1965.

[Dam et al., 2015] Hoa Khanh Dam, Bastin Tony Roy Savarimuthu, Daniel Avery, and Aditya K. Ghose. Mining software repositories for social norms. In *Proceedings of the 37th IEEE/ACM International Conference on Software Engineering (ICSE)*, pages 627–630, Florence, Italy, May 2015. IEEE Computer Society. New Ideas and Emerging Results Track.

[Deci and Ryan, 2000] Edward L Deci and Richard M Ryan. The" what" and" why" of goal pursuits: Human needs and the self-determination of behavior. *Psychological inquiry*, 11(4):227–268, 2000.

[DeLanda, 2011] Manuel DeLanda. *Philosophy and simulation: the emergence of synthetic reason*. Bloomsbury Publishing, 2011.

[Durkheim, 1893] Emile Durkheim. *De la division du travail social: étude sur l'organisation des sociétés supérieures*. F. Alcan, 1893.

[Durkheim, 1895] Emile Durkheim. *Les règles de la méthode sociologique*. F. Alcan, 1895.

[Ellickson, 2001] Robert C Ellickson. The evolution of social norms: a perspective from the legal academy. In Michael Hechter and Karl-Dieter Opp, editors, *Social norms*, pages 35–75. Russell Sage Foundation, 2001.

[Fittkau, 1993] B Fittkau. *Pädagogisch-psychologische Hilfen für Erziehung, Unterricht und Beratung. 2 Bde*. Aachen: Hahner, 1993.

[Freud, 1953] Sigmund Freud. *Abriss der Psychoanalyse: Das Unbehangen in der Kultur*. Fischer, 1953.

[Fuhrer and Trautner, 2005] Urs Fuhrer and Hanns Martin Trautner. Entwicklung von Identität. *Soziale, emotionale und Persönlichkeitsentwicklung*, pages 335–424, 2005.

[Gao and Singh, 2014] Xibin Gao and Munindar P. Singh. Extracting normative relationships from business contracts. In *Proceedings of the 13th International Conference on Autonomous Agents and MultiAgent Systems (AAMAS)*, pages 101–108, Paris, May 2014.

[Geulen, 1991] Dieter Geulen. Die historische Entwicklung sozialisationstheoretischer Ansätze. In Klaus Hurrelmann and Dieter Ulich, editors, *Neues Handbuch der Sozialisationsforschung*, pages 21–54. Beltz Verlag, 1991.

[Gibbs, 1965] Jack P Gibbs. Norms: The problem of definition and classification. *American Journal of Sociology*, 70(5):586–594, 1965.

[Giddens, 1984] Anthony Giddens. *The constitution of society*. University of California Press, 1984.

[Gintis, 2000] Herbert Gintis. *Game theory evolving: A problem-centered introduction to modeling strategic behavior*. Princeton university press, 2000.

[Granovetter, 1985] Mark Granovetter. Economic action and social structure: The problem of embeddedness. *American journal of sociology*, 91(3):481–510, 1985.

[Grice, 1975] H. Paul Grice. Logic and conversation. In Peter Cole and Jerry L. Morgan, editors, *Syntax and Semantics, Volume 3: Speech Acts*. Academic Press, New York, 1975. Reprinted in [Martinich, 1985].

[Haslam, 2001] S Alexander Haslam. *Psychology in organizations*. Sage, 2001.

[Hechter and Opp, 2001] Michael Hechter and Karl-Dieter Opp. *Social norms*. Russell Sage Foundation, 2001.

[Hempel, 1994] Carl G Hempel. The logic of functional analysis. *Readings in the philosophy of social science*, pages 349–75, 1994.

[Homans, 1964] George C Homans. Bringing men back in. *American Sociological Review*, pages 809–818, 1964.

[Interis, 2011] Matthew Interis. On norms: A typology with discussion. *American Journal of Economics and Sociology*, 70(2):424–438, 2011.

[Kafalı et al., 2016] Özgür Kafalı, Nirav Ajmeri, and Munindar P. Singh. Revani: Revising and verifying normative specifications for privacy. *IEEE Intelligent Systems*, 31(5):8–15, September 2016.

[Kalia et al., 2013] Anup K. Kalia, Hamid R. Motahari Nezhad, Claudio Bartolini, and Munindar P. Singh. Monitoring commitments in people-driven service engagements. In *Proceedings of the 10th IEEE International Conference on Services Computing (SCC)*, pages 160–167, Santa Clara, California, June 2013. IEEE Computer Society.

[Kohlberg et al., 1996] Lawrence Kohlberg, Wolfgang Althof, Fritz Oser, and Fritz Oser. *Die Psychologie der Moralentwicklung*. Suhrkamp, 1996.

[Krappmann, 2006] Lothar Krappmann. Spannungsfeld zwischen gesellschaftlicher Reproduktion und entstehender Handlungsfähigkeit. In *Theorien, Modelle und Methoden der Entwicklungspsychologie*, pages 369–408. Hogrefe, 2006.

[Martinich, 1985] Aloysius P. Martinich, editor. *The Philosophy of Language*. Oxford University Press, New York, 1985.

[Mashayekhi et al., 2016] Mehdi Mashayekhi, Hongying Du, George F. List, and Munindar P. Singh. Silk: A simulation study of regulating open normative multiagent systems. In *Proceedings of the 25th International Joint Conference on Artificial Intelligence (IJCAI)*, pages 373–379, New York, July 2016. IJCAI.

[Mayntz, 1961] Renate Mayntz. Kritische Bemerkungen zur funktionalistischen Schichtungstheorie. *Kölner Zeitschrift für Soziologie und Sozialpsychologie*, 13(5):10–28, 1961.

[Mead, 1934] George Herbert Mead. *Mind, self and society*, volume 111. Chicago University of Chicago Press, 1934.

[Morales et al., 2013] Javier Morales, Maite López-Sánchez, Juan Antonio Rodriguez-Aguilar, Michael Wooldridge, and Wamberto Vasconcelos. Automated synthesis of normative systems. In *Proceedings of the 12th International Conference on Autonomous Agents and MultiAgent Systems (AAMAS)*, pages 483–489, St. Paul, Minnesota, May 2013. IFAAMAS.

[Morales *et al.*, 2015] Javier Morales, Maite López-Sánchez, and Juan Antonio Rodriguez-Aguilar. Synthesising liberal normative systems. In *Proceedings of the 14th International Conference on Autonomous Agents and MultiAgent Systems (AAMAS)*, pages 433–441, Istanbul, May 2015. IFAAMAS.

[Nagel, 1981] Ernest Nagel. *The structure of science: Problems in the logic of scientific explanation*. Hackett Publishing Company, 1981.

[Nardin *et al.*, 2016] Luis G. Nardin, Tina Balke-Visser, Nirav Ajmeri, Anup K. Kalia, Jaime S. Sichman, and Munindar P. Singh. Classifying sanctions and designing a conceptual sanctioning process for socio-technical systems. *The Knowledge Engineering Review*, 31(2):142–166, March 2016.

[Nash, 1950] John F. Nash. Equilibrium points in n-person games. *Proceedings of the national academy of sciences*, 36(1):48–49, 1950.

[Neumann, 2008] Martin Neumann. Homo socionicus: a case study of simulation models of norms. *Journal of Artificial Societies and Social Simulation*, 11(4):6, 2008.

[Neumann, 2010] Martin Neumann. Norm internalisation in human and artificial intelligence. *Journal of Artificial Societies and Social Simulation*, 13(1):12, 2010.

[Neumann, 2013] Martin Neumann. Social constraint. In *Simulating Social Complexity*, pages 335–364. Springer, 2013.

[Noriega *et al.*, 2015] Pablo Noriega, Julian Padget, Harko Verhagen, and Mark D'Inverno. Towards a framework for socio-cognitive technical systems, 2015.

[Noriega *et al.*, 2017] Pablo Noriega, Jordi Sabater-Mir, Harko Verhagen, Julian Padget, and Mark d'Inverno. Identifying affordances for modelling second-order emergent phenomena with the WIT framework. In *Autonomous Agents and Multiagent Systems - AAMAS 2017 Workshops, Visionary Papers, São Paulo, Brazil, May 8-12, 2017, Revised Selected Papers*, pages 208–227, 2017.

[Ostrom, 1990] Elinor Ostrom. *Governing the commons: The evolution of institutions for collective action*. Cambridge, Cambridge University Press, 1990.

[Parsons, 1949] Talcott Parsons. *The Structure of Social Action*. New York: Free Press, 1949.

[Parsons, 1951] Talcott Parsons, editor. *The social system*. IL: Free Press of Glencoe, 1951.

[Pasons and Shils, 1951] Talcott Pasons and Edward A. Shils, editors. *Towards a general theory of action*. Harvard University Press, 1951.

[Patkos *et al.*, 2015] Theodore Patkos, Giorgos Flouris, Panagiotis Papadakos, Antonis Bikakis, Pompeu Casanovas, Jorge González-Conejero, Rebeca Varela Figueroa, Anthony Hunter, Gudjon Idir, George Ioannidis, Marta Kacprzyk-Murawska, Andrzej Nowak, Jeremy V. Pitt, Dimitris Plexousakis, Agnieszka Rychwalska, and Alexandru Stan. Privacy-by-norms: Privacy expectations in online interactions. In *Proceedings of the 3rd Workshop on Self-Adaptive and Self-Organising Socio-Technical Systems held at the International Conference on Self-Adaptive and Self-Organizing Systems*, pages 1–6, Cambridge, Massachusetts, September 2015. IEEE Computer Society.

[Phinney, 1990] Jean S Phinney. Ethnic identity in adolescents and adults: review of research. *Psychological bulletin*, 108(3):499, 1990.

[Piaget, 1932] Jean Piaget. *Le jugement moral chez l'enfant*. Paris, PUF, 1932.

[Piaget, 1947] Jean Piaget. *La psychologie de l'intelligence*, volume 249. Armand Colin, 1947.

[Radcliffe-Brown, 1934] Alfred R. Radcliffe-Brown. Social sanction. In Edwin R. A. Seligman, editor, *Encyclopaedia of the Social Sciences*, volume XIII, pages 531–534. Macmillan Publishers, 1934. Reprinted in *Structure and Function in Primitive Society*, chapter 11, pages 205–211, The Free Press, Glencoe, Illinois, 1952.

[Searle, 1995] John R. Searle. *The Construction of Social Reality*. Free Press, New York, 1995.

[Sherif *et al.*, 1955] Muzafer Sherif, B. Jack White, and O. J. Harvey. Status in experimentally produced groups. *American Journal of Sociology*, 60(4):370–379, January 1955.

[Sherif, 1937] Muzafer Sherif. An experimental approach to the study of attitudes. *Sociometry*, 1(1):90–98, July 1937.

[Shoham and Tennenholtz, 1995] Yoav Shoham and Moshe Tennenholtz. On social laws for artificial agent societies: Off-line design. *Artificial Intelligence*, 73((1-2)):231–252, 1995.

[Shoham and Tennenholtz, 1997] Yoav Shoham and Moshe Tennenholtz. On the emergence of social conventions: Modeling, analysis, and simulations. *Artificial Intelligence*, 94((1-2)):139–166, 1997.

[Simpson and Willer, 2008] Brent Simpson and Robb Willer. Altruism and indirect reciprocity: The interaction of person and situation in prosocial behavior. *Social Psychology Quarterly*, 71(1):37–52, March 2008.

[Singh, 2013] Munindar P. Singh. Norms as a basis for governing sociotechnical systems. *ACM Transactions on Intelligent Systems and Technology (TIST)*, 5(1):21:1–21:23, December 2013.

[Smith and Price, 1973] J Maynard Smith and George R Price. The logic of animal conflict. *Nature*, 246(5427):15–18, 1973.

[Smith, 1982] John Maynard Smith. *Evolution and the Theory of Games*. Cambridge university press, 1982.

[Stigler and Becker, 1977] George J Stigler and Gary S Becker. De gustibus non est disputandum. *The american economic review*, 67(2):76–90, 1977.

[Tajfel, 1970] Henri Tajfel. Experiments in intergroup discrimination. *Scientific American*, 223(5):96–103, 1970.

[Tajfel, 1981] Henri Tajfel. *Human groups and social categories: Studies in social psychology*. Cambridge University Press, 1981.

[Thatcher, 1987] Margaret Thatcher. Interview for woman's own 31 october 1987, pp. 810. www.margaretthatcher.org/document/106689, 1987. Accessed: 2018-03-25.

[Therborn, 2002] Göran Therborn. Back to norms! on the scope and dynamics of norms and normative action. *Current Sociology*, 50(6):863–880, 2002.

[Troitzsch, 2015] Klaus G Troitzsch. Extortion racket systems as targets for agent-based simulation models. comparing competing simulation models and emprical data. *Advances in Complex Systems*, 18(05n06):1550014, 2015.

[Turner et al., 1987] John C Turner, Michael A Hogg, Penelope J Oakes, Stephen D Reicher, and Margaret S Wetherell. *Rediscovering the social group: A self-categorization theory*. Basil Blackwell, 1987.

[Ullmann-Margalit, 1977] E Ullmann-Margalit. The emergence of norms., 1977.

[von Wright, 1963] Georg Henrik von Wright. *Norm and Action: A Logical Enquiry*. International Library of Philosophy and Scientific Method. Humanities Press, New York, 1963.

[Weber, 1921] Max Weber. *Wirtschaft und Gesellschaft*. Tübingen: JCB Mohr, 1921.

[Wrong, 1961] Dennis H Wrong. The oversocialized conception of man in modern sociology. *American sociological review*, pages 183–193, 1961.

PART II

MODELING

2
Norm Specification and Verification in Multiagent Systems

NATASHA ALECHINA, MEHDI DASTANI, BRIAN LOGAN

ABSTRACT. This article presents a high-level overview of the literature on norms and their uses in multiagent systems. We distinguish the main types of norms used in multiagent systems, and the ways in which the behaviour of a system can be modified through the enforcement of norms. We first review the formal approaches used to study norms and norm enforcement mechanisms. We then explain the syntax and semantics of the key specification languages used to represent norms, and briefly survey some programming frameworks that support the implementation of normative multiagent systems. Finally, we briefly review the key research questions and techniques in the important area of norm verification.

1 Introduction

Norms are generally conceived as standards of behaviour [Bicchieri, 2006; Elster, 2009]. In the norm literature (e.g., [Bicchieri, 2006; Bicchieri and Mercier, 2014; Elster, 2009]), various norm types have been distinguished based on the authorities that issue and enforce the norms. Examples of norm types are legal, social, moral and rational norms. Legal norms requires a legal body that issues norms and a corresponding executive body that enforces norms. For example, the legislative body of a state can issue traffic laws and the executive body of the state can enforce the traffic laws. Social norms often emerge through interaction within a community of individuals, who are subsequently in charge of enforcing the norm. For example, the amount of labour in a workplace can emerge as a social norm, after which those who work too hard or too little get criticised or even ignored/excluded from the workplace. Moral norms differ from legal and social norms as there is no authority or society required to issue and enforce moral norms. Moral norms are seen as a product of reasoning or internalisation of some external standards. Individuals follow their moral norms because of other internal reasons such as deliberation or emotions. Finally, rational norms includes prescriptive rational rules such as axioms of logics or equilibria in games. In general, norms are prescriptive in the sense that they prescribe which states, actions or behaviour to pursue or avoid.

In multiagent systems, norms are often used to ensure the overall objectives of the system. In order to organise a multiagent system in such a way that the standards of behaviour are actually followed by the agents, norms should be enforced by means of regimentation or sanctioning, e.g., [Jones and Sergot, 1993; Grossi, 2007; Dastani et al., 2013]. When regimenting norms, agents' behaviours leading to violations of norms are made impossible. Regimentation prevents agents from reaching a forbidden state or performing a forbidden action. Enforcing norms by regimentation decreases agent autonomy significantly. Norms can be regimented in various ways. For example, norms can be incorporated in the agent's decision making mechanisms so that all the agent's executions will be compliant with the norms. Norms can also be regimented externally by ignoring violating actions or undoing their effects. In the latter case, the enforcement mechanism is assumed to have control over the effects of the agents' actions, e.g., the enforcement mechanism can decide not to pass messages between some agents or to undo the effect of the agents' actions in the multiagent environment. Instead, norms enforcement can be based on the idea of responding after a violation of the norms has occurred. Such a response, which includes sanctions, aims to return the system to an optimal state. For sanction-based enforcement it is essential that the norm violating actions are observable by the system (e.g., fines can be issued in traffic systems only if the speed of cars can be observed). Sanction-based enforcement allows agents to violate norms and therefore contributes to the flexibility and autonomy of the agents' behaviour [Castelfranchi, 2004].

One of the key questions regarding norm enforcement in multiagent systems is whether the enforcement of a given set of norms can ensure some given desirable system properties. In particular, provided that a multiagent system does not satisfy some given desirable system properties, does the enforcement of a given set of norms modify the system in such a way that the desirable system properties are ensured. This problem is one of the versions of norm verification problem. Another related problem is to generate a set of norms that, when enforced in the system, ensures the desirable system properties. This latter problem is called norm synthesis problem. Both problems require a procedure to update a system with a set of norms. Such a procedure implements a norm enforcement mechanism. Another key question regarding norm enforcement is the expressive power of norms. In general, there is a trade-off between expressiveness of norms and the computational complexity of the verification, synthesis and update problems: the more expressive norms, the higher computational complexity of the problems.

In this chapter, we ignore the problem of norm synthesis and cover approaches to specification and verification of normative systems related to regulative norms, that is norms that can be violated. We survey various approaches to norm specification and cover different types of regulative norms such as state-, action-, and behaviour-based norms[1]. For verification, we only cover approaches using model-checking, because they are by far the more prevalent. However, there exists work using theorem proving for verification, for example [Governatori et al., 2013].

2 Background

In this section we introduce the necessary background on transition systems and temporal logics used in the specification and verification of norms. This includes background on temporal logics such as Linear-Time Temporal Logic LTL [Pnueli, 1977],

---

[1] Action-based norms is the term most widely used in the literature; sometimes we refer to those norms as *transition-based* to cover both norms specified in terms of actions and in terms of events. We will also sometimes refer to norms specified in terms of behaviours or temporal patterns as *path-based* norms.

Computation Tree Logic CTL and CTL^* [Clarke et al., 1986], Alternating-Time Temporal Logic ATL and ATL^* [Alur et al., 2002]. In the exposition of LTL, LTL + Past, CTL, and CTL^* below, we largely follow the notation in [Schnoebelen, 2003].

The logical languages we introduce below are defined relative to a set of propositional atoms Π, and talk about *state transition systems*, or transition systems for brevity. A transition system is a graph where states are vertices (decorated with propositional atoms) and transitions are edges. In a *labelled* transition system, edges are also decorated with labels, or action names.

Definition 1 (State Transition System). A *state transition system* is a tuple $M = (S, R, V)$, where S is a finite, non-empty set of states, $R \subseteq S \times S$ is a transition relation (for simplicity, we assume that R is a total relation, that is, some transition is possible in every state), and V is a propositional valuation $S \longrightarrow 2^\Pi$. A *pointed transition system* is a pair (M, s_I), where M is a transition system and $s_I \in S$ is the initial state. A *labelled* transition system is built using a set L of labels. It is a tuple $M = (S, \{R_a : a \in L\}, V)$, where each $R_a \subseteq S \times S$ is a transition relation.

A transition system can be used to describe the lifecycle of an agent, or a business process, or a system consisting of multiple interacting processes or agents. States correspond to configurations of the system at a moment in time. The transition relation corresponds to actions or events which change the state, and the valuation function assigns a set of atoms to a state (intuitively, the set of atoms which hold in that state).

Given a state transition system $M = (S, R, V)$, a *path* through M is a sequence $s_0, s_1, s_3, \ldots$ of states such that $s_i R s_{i+1}$ for $i = 0, 1, \ldots$. A *fullpath* is a maximal path and a *run* of M is a fullpath which starts from a state $s_I \in S$ designated as the initial state of M. We denote runs by $\rho, \rho', \ldots$, and the state at position i on ρ by $\rho[i]$. Intuitively, a path represents a finite history of events in the system, and a run corresponds to a complete infinite history or computation of the system. We denote the set of all runs in M by $\mathcal{P}(M)$.

For a state $s \in S$, the *tree* rooted at s is the infinite tree $T(s)$, obtained by unfolding M from s (the nodes of T are finite paths starting from s ordered by the prefix relation). $T(M) = T(s_I)$ is the *computation tree* of M. Note that branches of $T(M)$ are runs of M.

Linear Time Temporal Logic (LTL) The syntax of LTL is defined as follows:

$$\phi, \psi ::= p \mid \neg \phi \mid \phi \wedge \psi \mid \mathcal{X}\phi \mid \phi \mathcal{U} \psi$$

where $p \in \Pi$, $\neg$ stands for not, $\wedge$ for and, $\mathcal{X}$ means next state, and $\mathcal{U}$ stands for until. Other propositional connectives $\vee$ (or) and $\rightarrow$ (implies) are defined in a standard way. It is also possible to define $\mathcal{F}\phi$ (ϕ holds some time in the future) as $\top \mathcal{U} \phi$ and $\mathcal{G}\phi$ (always ϕ) as $\neg \mathcal{F} \neg \phi$.

The truth definition for formulas of LTL is given inductively with respect to a run $\rho \in \mathcal{P}(M)$ and a position i on ρ. We omit M, $\mathcal{P}(M)$, $T(M)$ etc. when it is clear from the context:

$\rho, i \models p$ iff $p \in V(\rho[i])$

$\rho, i \models \neg \phi$ iff $\rho, i \not\models \phi$

$\rho, i \models \phi \wedge \psi$ iff $\rho, i \models \phi$ and $\rho, i \models \psi$

$\rho, i \models \mathcal{X}\phi$ iff $\rho, i+1 \models \phi$

$\rho, i \models \phi \mathcal{U} \psi$ iff $\exists j \geq i$ such that $\rho, j \models \psi$ and $\forall k : i \leq k < j, \rho, k \models \phi$

A run ρ satisfies an LTL formula ϕ if $\rho, 0 \models \phi$. A transition system M satisfies an LTL formula ϕ, written as $M \models \phi$, if all runs in $\mathcal{P}(M)$ satisfy ϕ.

Extending LTL with Path Quantifiers The syntax of CTL^* is defined as follows:

$$\phi, \psi ::= p \mid \neg \phi \mid \phi \wedge \psi \mid \mathcal{X}\phi \mid \phi \mathcal{U} \psi \mid E\phi$$

(adding a quantifier over paths E, with the intended meaning 'there exists a continuation of the run satisfying ϕ). The universal quantifier $A\phi$ (on all runs) is defined as $\neg E \neg \phi$.

The truth definition for LTL is extended with

$\rho, i \models E\phi$ iff for some run $\rho' \in T(M)$ which is identical to ρ on the first i indices, $\rho', i \models \phi$.

A CTL^* formula ϕ is true in a transition system M, $M \models \phi$, iff $\rho, 0 \models \phi$ for all runs ρ in $T(M)$.

Note that any LTL formula is a CTL^* formula. A system M satisfies an LTL formula ϕ iff it satisfies a CTL^* formula $A\phi$. CTL^* is strictly more expressive than LTL. For example, it can express the existence of a choice point: there is a future where in the next state p holds, and a future where in the next state $\neg p$ holds, $E\mathcal{X}p \wedge E\mathcal{X}\neg p$.

Computation Tree Logic CTL is the fragment of CTL^* where every temporal modality ($\mathcal{U}$ or $\mathcal{X}$) must be under the immediate scope of a path quantifier (E or A). The semantics is inherited from CTL^*. Alternatively, the logic can be defined as follows, independently from CTL^*. The syntax is

$$\phi, \psi ::= p \mid \neg \phi \mid \phi \wedge \psi \mid E\mathcal{X}\phi \mid E(\phi \mathcal{U} \psi) \mid A(\phi \mathcal{U} \psi)$$

The semantics can be defined without reference to runs, only to states corresponding to positions on a run, as follows:

$s \models E\mathcal{X}\phi$ iff there is a branch of the tree $T(M)$ starting from s such that for the next state s' on that branch, $s' \models \phi$

$s \models E(\phi \mathcal{U} \psi)$ iff there is a branch ρ of the tree $T(M)$ with $\rho[i] = s$ such that there exists a state $s_j = \rho[j], j \geq i$, on that branch such that $s_j \models \psi$ and for all states $s_k = \rho[k]$ with $i \leq k < j$, $s_k \models \phi$

$s \models A(\phi \mathcal{U} \psi)$ iff for all branches ρ of the tree $T(M)$ with $\rho[i] = s$ there exists a state $s_j = \rho[j], j \geq i$, on that branch such that $s_j \models \psi$ and for all states $s_k = \rho[k]$ with $i \leq k < j$, $s_k \models \phi$

Linear Time Temporal logic with Past Although the expressive power of temporal logics does not change with the addition of past operators [Gabbay, 1987], it is convenient to consider temporal logics which talk not just about the future, but also about the past.

The syntax of $LTL + Past$ formulas is defined as follows:

$$p \in \Pi \mid \neg\phi \mid \phi \wedge \psi \mid \mathcal{X}\phi \mid \phi\mathcal{U}\psi \mid \mathcal{X}^{-1}\phi \mid \phi\mathcal{S}\psi$$

where $\mathcal{X}^{-1}$ means previous state, $\mathcal{S}$ stands for since (as in, ϕ has been true since ψ became true). The truth definition for formulas is given relative to $T(M)$, a run ρ and the state at position i on ρ:

$\rho, i \models \mathcal{X}^{-1}\phi$ iff $i > 0$ and $\rho, i-1 \models \phi$

$\rho, i \models \phi\mathcal{S}\psi$ iff $\exists j \leq i$ such that $\rho, j \models \psi$ and $\forall k : i \geq k > j, \rho, s_k \models \phi$

Alternating Time Temporal Logic (ATL) ATL formulas are interpreted on concurrent game structures.

Definition 2 (Concurrent Game Structure). A Concurrent Game Structure (CGS) is a tuple $M = (S, V, a, \delta)$ which is defined relative to a set of agents $\mathcal{A} = \{1, \ldots, n\}$ and a set of propositional variables Π, where:

- S is a non-empty set of states

- $V : S \to \wp(\Pi)$ is a function which assigns each state in S a subset of propositional variables

- $a : S \times \mathcal{A} \to \mathbb{N}$ is a function which indicates the number of available moves (actions) for each player $i \in \mathcal{A}$ at a state $s \in S$ such that $a(s, i) \geq 1$. At each state $s \in S$, we denote the set of joint moves available for all players in $\mathcal{A}$ by $A(s)$. That is

$$A(s) = \{1, \ldots, a(s, 1)\} \times \ldots \times \{1, \ldots, a(s, n)\}$$

- $\delta : S \times \mathbb{N}^{|\mathcal{A}|} \to S$ is a partial function where $\delta(s, m)$ is the next state from s if the players execute the move $m \in A(s)$.

The language of ATL is defined as follows:

$$p \in \Pi \mid \neg\phi \mid \phi \wedge \psi \mid \langle\!\langle C \rangle\!\rangle \mathcal{X}\phi \mid \langle\!\langle C \rangle\!\rangle G\phi \mid \langle\!\langle C \rangle\!\rangle \phi\mathcal{U}\psi$$

where $C \subseteq \mathcal{A}$. Intuitively, $\langle\!\langle C \rangle\!\rangle \gamma$ means 'the group of agents C has a strategy, all executions of which satisfy the formula γ, whatever the other agents in $\mathcal{A} \setminus C$ do'.

Definition 3 (Move). Given a CGS M and a state $s \in S$, a move (or joint action) for a coalition $C \subseteq \mathcal{A}$ is a tuple $\sigma_C = (\sigma_i)_{i \in C}$ such that $1 \leq \sigma_i \leq a(s, i)$.

By $A_C(s)$ we denote the set of all moves for C at state s. Given a move $m \in A(s)$, we denote by m_C the actions executed by C, $m_C = (m_i)_{i \in C}$. The set of all possible outcomes of a move $\sigma_C \in A_C(s)$ at state s is defined as follows:

$$out(s, \sigma_C) = \{s' \in S \mid \exists m \in A(s) : m_C = \sigma_C \wedge s' = \delta(s, m)\}$$

Definition 4 (Strategy). Given a CGS M, a strategy for a subset of players $C \subseteq \mathcal{A}$ is a mapping F_C which associates each finite path $s_I, \ldots, s$ to a move in $A_C(s)$.

A fullpath ρ is consistent with F_C iff for all $i \geq 0$, $\rho[i+1] \in out(\rho[i], F_C(\rho[0], \ldots, \rho[i]))$. We denote by $out(s, F_C)$ the set of all such fullpaths ρ starting from s, i.e. where $\rho[0] = s$. The truth definition, as for CTL, can be given relative to states in M:

- $s \models \langle\!\langle C \rangle\!\rangle \mathcal{X} \phi$ iff there exists a strategy F_C which such that for all $\rho \in out(s, F_C)$, $\rho[1] \models \phi$

- $s \models \langle\!\langle C \rangle\!\rangle \mathcal{G} \phi$ iff there exists a strategy F_C such that for all $\rho \in out(s, F_C)$, $\rho[i] \models \phi$ for all $i \geq 0$

- $s \models \langle\!\langle C \rangle\!\rangle \phi \mathcal{U} \psi$ iff there exists a strategy F_C such that for all $\rho \in out(s, F_C)$, there exists $i \geq 0$ such that $\rho[i] \models \psi$ and $\rho[j] \models \phi$ for all $j \in \{0, \ldots, i-1\}$

3 Norm Specification

In this section, we discuss how norms can be stated precisely and what it means for a norm to be violated. Many different approaches to specifying norms can be found in the literature. For example, some authors specify norms semantically, with respect to some formal model of the system (e.g., given a specification of the system, we can state that certain actions which are possible under this specification are forbidden by a norm), while others specify norms syntactically, as expressions of a formal language.[2] Alternatively, norms may be specified directly in terms of programming constructs. The specification may also depend on how the norms are enforced (regimentation or sanctioning), whether the subject of the norm is a single agent or a group of agents, etc. We therefore base our classification on whether a particular approach to norms specifies norms and their violation in terms of states (Section 3.1), in terms of actions or transitions (Section 3.2), or in terms of paths or behaviours (Section 3.3). We show how norms specified in terms of transitions and paths can be (re)expressed in temporal logic, allowing different approaches to specifying norms (and their violation) to be precisely compared. In Section 3.4 we address recent arguments that 'real life' norms cannot be expressed in temporal logics. Finally, in Section 3.5, we classify proposals for norm programming frameworks in the literature in terms of whether they can express state, transition or behaviour norms.

3.1 State-based Norms

Norms can be specified in terms of (properties of) states. For example, in [Alchourrón and Bulygin, 1981; Dastani et al., 2013] norms are specified by means of a set of violating states (the set of norm compliant states is the complement of the set of violating

---

[2]The distinction is often somewhat blurred, as specification of the system is also usually done in some formal language.

states). A state-based norm may prohibit or require states, e.g., a car is prohibited to park at a certain location or a car is obliged to have insurance. State-based norms may apply to a single agent or to a group of agents, for example, it may be prohibited for more than eight people to be in an elevator at the same time. It is generally assumed that norms may conflict, e.g., a car may be prohibited to park at a certain location while it is obliged to load a cargo at the same location. In order to specify state-based norms, various proposals have been put forward. In the rest of this section, we survey some of these proposals.

Counts-as Norms

Specifying norms directly in terms of states can sometimes be cumbersome. 'Counts-as' rules allow specification of norms in terms of properties (sets of states).

Regulative norms, also called deontic norms, can be seen as statements classifying system's states as complying or violating. Counts-as rules, together with a specific 'violation' atom $Viol$, can be used to classify system states. The so-called "counts-as reduction" of deontic norms builds on the tradition of the reductionistic approach in deontic logic started with the work of Anderson [Anderson, 1957; Anderson, 1958b; Anderson, 1958a] and Kanger [Kanger, 1971]. The idea of such reductionist approach is that the statement "ϕ is obligatory" in interpreted as the statement "$\neg\phi$ necessarily implies a violation" (i.e., $\neg\phi$ is prohibited), represented by the counts-as rule $\neg\phi \Rightarrow Viol$. Conditional deontic norms of the form "if C, ϕ is prohibited" can be represented by counts-as rules of the forms "ϕ counts-as $Viol$ in context C".

In contrast to regulative norms, constitutive norms establish a social institution by creating and classifying new facts, called institutional facts [Searle, 1995]. Institutional facts build on brute and institutional facts, and define new institutional facts. Following Searle [Searle, 1995], constitutive norms create and classify institutional facts by statements of the form "ϕ counts as ψ in context C", where ϕ is a brute or institutional fact and ψ is an institutional fact. In this way, a constitutive norm can be seen as defining institutional facts. Counts-as rules are used to represent constitutive norms [Boella and van der Torre, 2004; Aldewereld et al., 2009]. Note that the $Viol$ atom used in regulative norms can be seen as an institutional fact with the special interpretation indicating that some facts are considered as violating states.

Representing deontic norms using counts-as statements, one can consider ϕ as denoting *brute* facts (system's states), while the $Viol$ atom denotes an institutional fact. In this way, norms impose *institutional* descriptions upon the brute ones, e.g. "ϕ is a violation state". In the case of constitutive norms, one can consider ϕ as denoting brute or institutional fact, while ψ denotes institutional facts. Thus, a constitutive norm defines which brute or institutional fact can be considered as institutional fact. Counts-as statements could be complex and exhibit rich logical structure as shown, for instance, in [Grossi, 2007]. Counts-as rules are often used with an additional context condition that specifies the applicability of the counts-as rule. For example, $\phi \Rightarrow Viol$ in ψ indicates that in the context denoted by ψ, the brute fact ϕ is considered as a violation and thus prohibited. In [Boella and van der Torre, 2004], regulative and constitutive norms are modelled as agents' goals and beliefs, respectively, which are in turn specified by rules in input/output logic. They show how counts-as relations, which represent norms, can formally be specified in input/output logic.

Counts-as rules are also used to represent norms and sanctions in an organisational setting. For example, in [Dastani et al., 2013] counts-as rules are used to specify a normative artefact that is responsible for the control and coordination of software agents, and in [Bulling and Dastani, 2011] counts-as rules are used to determine a game theoretic mechanism that enforces certain socially preferred outcomes. To specify regimenting and sanctioning norms in normative artefact, special violation atoms `viol`$_\perp$ and `viol`$_i$ are introduced, respectively. These special atoms constitute the consequent of counts-as rules to represent obligations and prohibitions. For example, the counts-as rule { `book(a)`, `late(a)` } ⇒ {`viol`$_1$} represents the library norm which states that it is forbidden to being late in returning book a. Note that this prohibition is a sanctioning norm as it uses `viol`$_1$ atom (instead of `viol`$_\perp$). For each violation atom `viol`$_i$ a counts-as rule can be used to represent how to sanction such a violation. For example, the counts-as rule {$viol_1$} ⇒ {$fined$} indicates that a sanctioning fine should incur in response to the violation `viol`$_1$. A normative artefact controls and coordinates the agents' activities by determining the effect of the agents' actions in their environment. An artefact is assumed to observe the agents' actions, to evaluate them with respect to a given set of norms, and to determine the effects of these actions. The realising effects can be ignoring the action effect in case a regimenting norm is violated, or adding sanctions to the resulting states in case a sanctioning norm is violated. The decisions as to which norms are violated and which sanctions should be imposed are determined by taking the closure of the environment state, where the agents' actions are performed, under the sets of counts-as rules representing norms and sanctions.

Norms as Defeasible Rules

Norms are often conflicting and require formalisms to capture and cope with conflicts. One possible formalism to represent conflicting norms is by means of defeasible rules. In BOID [Broersen et al., 2002], defeasible rules are used to represent an agent's mental and motivational attitudes. In this framework, norms are considered as constituting an agent's motivational attitude that is used in the agent's deliberation process to determine the agent's behaviour. An agent can have conflicting motivational attitudes, e.g., an agent's obligation may conflict with other obligations or even with the agent's desires or intentions. In BOID, mental and motivational attitudes are represented by defeasible rules of the form $a \xhookrightarrow{x} b$, where $x \in \{B, O, I, D\}$ denotes possible mental attitudes such as beliefs, obligations, intentions and desires. A rule of the form $a \xhookrightarrow{O} b$ is interpreted as "if a is derived as a goal, then the agent is obliged that b is a goal". The goal generation operation, within the BOID deliberation process, applies defeasible rules iteratively and on the basis of a given order on rules to derive maximally consistent set of goals. It should be noted that norms in BOID are restricted to obligations, which are considered as a motivational attitude of an agent. Obligations in BOID are propositional properties (certain states are obligatory), similar to beliefs, desires, and intentions. Violation conditions of obligations in BOID can therefore be expressed in propositional logic. Since agents are allowed to have conflicting obligations and some obligations are not included in the set of maximally consist set of goals, an agent may comply with some and violate other norms. The following set of defeasible rules specifies an example of a BOID agent, who intends to attend a conference, is obliged to have a cheap room close to a conference site, but believes there are no cheap hotels nearby the conference site:

$$\text{cheap\_room} \stackrel{B}{\hookrightarrow} \neg\text{close\_to\_conf\_site}$$
$$\text{close\_to\_conf\_site} \stackrel{B}{\hookrightarrow} \neg\text{cheap\_room}$$
$$\top \stackrel{I}{\hookrightarrow} \text{go\_to\_conference}$$
$$\text{go\_to\_conference} \stackrel{O}{\hookrightarrow} \text{cheap\_room}$$
$$\text{go\_to\_conference} \stackrel{O}{\hookrightarrow} \text{close\_to\_conf\_site}$$

3.2 Action-based Norms

As with state-based approaches to specifying norms, norms specified in terms of transitions (e.g., actions, events), can be specified directly as a set of (prohibited) transitions [Ågotnes *et al.*, 2007; Ågotnes *et al.*, 2010; Knobbout and Dastani, 2012] (with compliant transitions defined as the complement of the set of violating transitions). Norms specified in terms of transitions may apply to an action by a single agent or an action performed by a group of agents, for example, it may be prohibited that more than 3 school children enter a shop together [Aldewereld *et al.*, 2013].

Action or event-based norms are used in frameworks for specifying *institutions* or agent societies, see e.g., [Cliffe *et al.*, 2007].

3.3 Behaviour-based Norms

As with norms specified in terms of states and transitions, norms specified in terms of paths or temporal patterns of behaviour can be specified directly as the set of violating runs (with compliant runs defined as the complement of the violating runs) [Alechina *et al.*, 2015; Bulling *et al.*, 2013]. However, when the number of traces is infinite, alternative approaches are necessary. As with state and action-based approaches, norms specified in terms of behaviours may apply to a single agent or to a group of agents.

Conditional Norms with Deadlines and Sanctions

Conditional norms with deadlines and sanctions were introduced in [Dastani *et al.*, 2009]. Conditional norms are triggered (detached) in certain states of the environment and have a temporal dimension specified by a deadline. The satisfaction or violation of a detached norm depends on whether the behaviour of the agent(s) brings about a specified state of the environment before a state in which the deadline condition is true. Norms can be enforced by means of sanctions or they can be regimented by disabling actions in specific states.

Definition 5 (Norms). Let $cond, \phi, d$ be boolean combinations of propositional variables from Π and $san \in \Pi$. A *conditional obligation* is represented by a tuple $(cond, O(\phi), d, san)$ and a *conditional prohibition* is represented by a tuple $(cond, P(\phi), d,$

san). A norm set N is a set of conditional obligations and conditional prohibitions.

Conditional norms are evaluated on runs of the physical transition system. A conditional norm $n = (cond, Y(\phi), s, san)$, where Y is O or P, is *detached* in a state satisfying its condition *cond*. Detached norms persist as long as they are not obeyed or violated, even if the triggering condition of the corresponding conditional norm does not hold any longer. A detached obligation $(cond, O(\phi), d, san)$ is *obeyed* if no state satisfying d is encountered before execution reaches a state satisfying ϕ, and *violated* if a state satisfying d is encountered before execution reaches a state satisfying ϕ. Conversely, a detached prohibition $(cond, P(\phi), d, san)$ is obeyed if no state satisfying ϕ is encountered before execution reaches a state satisfying d, and violated if a state satisfying ϕ is encountered before execution reaches a state satisfying d. If a detached norm is violated in a state s, the sanction corresponding to the norm is applied (becomes true) in s.

We say that a detached norm is annulled in a state s' immediately after a state s in which the norm is obeyed or violated, unless the same norm is detached again in s'. Note that given a state s in a transition system, we cannot say whether a norm is violated in s; to determine that, we need to know the path taken to reach s (e.g., whether any norms were detached in the past), and there may be more than one path to s. This is the reason why conditional norms are evaluated on runs of the system rather than in states.

Violation conditions of conditional norms can be expressed in temporal logic LTL $+Past$ as follows.

Definition 6 (Norm Violation). A state $\rho[i]$ violates a conditional obligation $(cond, O(\phi), d, san)$ on run ρ in $T(M)$ iff

$$T(M), \rho, i \models d \wedge \neg\phi \wedge ((X^{-1}(\neg\phi \wedge \neg d) \, \mathcal{S} \, (cond \wedge \neg\phi \wedge \neg d)) \vee cond)$$

$\rho[i]$ violates a conditional prohibition $(cond, P(\phi), d, san)$ iff

$$T(M), \rho, i \models \phi \wedge \neg d \wedge ((X^{-1}(\neg\phi \wedge \neg d) \, \mathcal{S} \, (cond \wedge \neg\phi \wedge \neg d)) \vee cond)$$

Note that whether $\rho[i]$ violates a norm is determined by the prefix of ρ ending in $\rho[i]$, and is not dependent on the future of $\rho[i]$.

Expressing Norms by Temporal Logic Formulas

Norm violation conditions can be expressed directly by a formula of some temporal logic. Instead of specifying e.g., conditional obligations and prohibition and then expressing their violation conditions in temporal logic, we can say that all states or all runs satisfying a temporal logic formula ϕ are prohibited. For group norms, ATL can be used to express norm violation conditions.

Norms as Team Plans

A history may also result from or an obligation to achieve some state or to carry out some actions by a group of agents [Grossi et al., 2004]. For example, an obligation on hospital staff may require two nurses to be on duty during a particular shift. In [Grossi et al., 2004], such obligations are expressed in Propositional Dynamic Logic

(PDL) (see, for example, [Harel, 1984]).[3] In [Alechina et al., 2014], similar group obligations are specified in LTL with $done(a, i)$ atoms, where $done(a, i)$ stands for 'action a has just been performed by agent i'.

Norms as Defeasible Rules

In [Governatori et al., 2013], defeasible rules are used to specify conflicting norms, in particular, contrary-to-duty and permissive norms. In its basic form, a contrary-to-duty norm consists of a primary norm and a secondary norm, which comes into effect when the primary norm is violated. An example of a basic contrary-to-duty norm is the obligation of a customer to pay an invoice within 7 days, and if the customer does not pay the invoice within 7 days, then the customer should pay the invoice plus 5 percent interest within 15 days. In a general case, a contrary-to-duty norm consists of a sequence of norms such that when the first norm in the sequence is violated, then the second norm is in force, but if the first two norms are violated, then the third norm is in force, etc. An example of the general case of contrary-to-duty norm is the obligation of a customer to pay an invoice within 7 days, and if the customer does not pay the invoice within 7 days, then the customer should pay the invoice plus 5 percent interest within 15 days, and if the customer does not pay the invoice plus 5 percent interest within 15 days, then the customer should pay the invoice with 10 percent interest within one month. Permissive norms are exceptions to obligations and prohibitions, and an explicit permissive norm is seen as an explicit derogation of an obligation or a prohibition. For example, a general prohibition regarding the use of private protected personal data can be derogated with a permission in the sense that the permission makes an exception to the general prohibition. A contrary-to-duty or permissive norm is specified by a defeasible rule, indexed by an obligation or permission, where the consequent consists of an ordered sequence of obligations or permissions. A contrary-to-duty norm has the general form $a \Rightarrow_O b \otimes c$ and is read as "in case a holds, then b obliged, but if the obligation b is not fulfilled, then the obligation c is activated and in force". The example contrary-to-duty norm above can be represented as follows:

$$invoice \Rightarrow_O payin7days \otimes pay + 7\%in15days \otimes pay + 10\%in30days$$

A permissive norm has the general form $a \Rightarrow_P b \odot c$. Such a rule can be used to represent permissions of the type "in situation a, the subject is entitled, in the order of preference, to option b or option c". However, the reading of permissive norms is slightly different from the reading of contrary-to-duty norms since permissions cannot be violated, i.e., we cannot read the permissive norm by means of "if permission b is violated". In [Governatori et al., 2013] it is argued that in the case of a permissive norm, one can proceed in the chain from b to c whenever $O\neg b$ holds. The preference operator $\odot$ establishes a preference order among permissions, and in case the opposite obligation is in force, another permission holds. In the next section, we show that the violation conditions of contrary-to-duty norms can also be expressed in temporal logic.

---

[3]PDL is yet another logic for describing labelled transition systems, which we did not cover in Section 2 in the interests of brevity.

3.4 Expressibility of Norms in Temporal Logic

In the previous sections, we have shown that the main classes of norms described in the literature can be naturally treated as conditions on runs or histories. Such conditions can be specified in a suitable temporal logic, for example, Linear Time Temporal Logic (LTL). Depending on the goals of the specification and verification process, we can either use LTL to define the set of runs which obey the norm, or to define the set of runs which violate the norm (one is simply a negation of another).

Recently, doubts were raised in [Governatori, 2015] regarding suitability of LTL and other temporal logics for expressing 'real life' norms. Basically, what the argument in [Governatori, 2015] really shows is that a translation of deontic notions such as obligations and permissions into temporal logic which interprets 'obligatory' as 'always true' and 'permitted' as 'eventually true' does not work, as could be expected. However, [Governatori, 2015] is now often cited as an argument against using temporal logic for specifying norms in general. We would like to revisit the example which is considered paradoxical when specified in LTL in [Governatori, 2015], and show that it is possible to exactly specify the set of conditions on runs which satisfy the norms from the example using standard LTL.

The example is as follows (we compress it slightly without changing the meaning, and use the same variable names for propositions):

1. collection of personal information (A) is forbidden unless authorised by the court (C)

2. The destruction of personal information collected illegally before accessing it (B) excuses the illegal collection

3. collection of medical information (D) is forbidden unless collection of personal information is permitted

As pointed out in [Governatori, 2015], this classifies possible situations as compliant and non-compliant as follows:

- situations satisfying C are compliant

- situations not satisfying C, where A happens but B happens as well, are weakly compliant (or correspond to a small violation; in the setting of conditional norms, this would deserve a small sanction)

- situations where C is false, where A happens and B does not, are violations

- situations not satisfying C where D happens are violations

- situations not satisfying C but also not satisfying A and D are compliant

The classification above is not very precise, since A, B, C, and D are treated as state properties which are true or false at the same time. Later in [Governatori, 2015] a temporal relation between A and B is introduced: if C is false and A happens, then B should happen some time after that to compensate for the violation of A[4].

Hence it is very easy to classify runs into compliant or violating in LTL:

- Fully compliant runs:

$$\mathcal{G}(C \vee (\neg C \wedge \neg A \wedge \neg D))$$

    (everywhere, either there is a court authorisation, or there is no collection of personal or medical information)

- Weakly compliant runs:

$$\mathcal{F}(\neg C \wedge A) \wedge \mathcal{G}(\neg C \wedge A \rightarrow \mathcal{F}B) \wedge \mathcal{G}(\neg C \rightarrow \neg D)$$

    (there is at least one violation of prohibition on collection of personal information, but each such violation is compensated by B in the future; there are no violations of prohibition on collecting medical information)

Finally, violations are specified as follows:

- Violating runs:

$$\mathcal{F}(\neg C \wedge (D \vee (A \wedge \neg \mathcal{F}B)))$$

Note that the three formulas above define a partition of all possible runs. Clearly, there is nothing paradoxical in this specification of the set of norms.

For the sake of completeness we reproduce here the formalisation of the same set of norms in [Governatori, 2015] and analyse where the paradoxical results come from. The set of norms is formalised in [Governatori, 2015] as follows:

N1 $\neg C \rightarrow (\neg A \otimes B)$

N2 $C \rightarrow \mathcal{F}A$

N3 $\mathcal{G} \neg A \rightarrow \mathcal{G} \neg D$

N4 $\mathcal{F}A \rightarrow \mathcal{F}D$

N1 is intended to say that B compensates for a violation $\neg C \wedge A$. It uses a connective $\otimes$ which was introduced for expressing contrary to duty obligations. The truth definition for $\otimes$ as given in [Governatori, 2015] is
$TS, \sigma \models \phi \otimes \psi$ iff $\forall i \geq 0$, $TS, \sigma_i \models \phi$ or $\exists j, k : 0 \leq j \leq k$, $TS, \sigma_j \models \neg \phi$ and $TS, \sigma_k \models \psi$, where TS is a transition system, and σ a run in TS. This makes $\otimes$ equivalent to

$$\mathcal{G}\phi \vee \mathcal{F}(\neg \phi \wedge \mathcal{F}\psi)$$

---

[4]It would have perhaps been better not to treat B as a state property, but as a property of a run, 'data not accessed until destroyed', which is expressible as $\neg Read \mathcal{U} Destroyed$, but we will stick with the formalisation in [Governatori, 2015] to make comparison easier. Another issue is that instead of requiring B to happen 'eventually', in real life there would be some time limit on when it should happen (such as in the next state).

This condition is similar to our characterisation of weakly compliant runs, although it is stated as a property which should be true for all runs. The condition **N2** is one of the really problematic ones. It aims to say that if C holds, then A is permitted; 'permitted' is identified with 'will eventually happen'. It is quite clear that permission of A cannot be expressed as 'A will eventually happen'; the two have completely different meanings. This does not mean that LTL cannot be used for specifying norms, it just means that this particular way of specifying norms in LTL is inappropriate. **N4** is problematic in the same way: instead of saying that an occurrence of D is not a violation under the same conditions as when an occurrence of A is not a violation, it says that if A is going to happen then D is going to happen – which is again a completely different meaning. **N3** attempts to say that if A is prohibited then D is prohibited. However, instead it implies that if A happens (the antecedent $\mathcal{G}\neg A$ is false) then it does not matter whether D happens (the implication is still true). Given this formalisation, which is inappropriate in multiple ways, [Governatori, 2015] produces an example run where **N1–N4** are true and the prohibition on collecting medical information is violated. The run consists of just two states t_1, t_2:

$$t_1 \models \neg C \wedge A \wedge D$$

$$t_2 \models B$$

which is a weakly compliant run as far as violating prohibition of A is concerned, but a non-compliant run as far as violating the prohibition of D is concerned. With our LTL specification of the set of norms it is classified as a violating run since $\mathcal{F}(\neg C \wedge D)$ holds on it. It does satisfy **N1–N4**, but clearly the problem is with **N1–N4** rather than with the intrinsic difficulty of classifying norm violating patterns in temporal logic.

3.5 Programming Norms

Another approach to specifying norms is directly in terms of programming constructs. The specification of norms can either be endogenous, i.e., form part of the programs of the (norm-compliant) agents comprising the MAS, or exogenous, i.e., form part of the program of some form of organisational framework or middleware. In these approaches, what it means for a norm to be violated is ultimately reducible to the operational semantics of the program, framework or middleware which operationalises the normative programming constructs and defines all norm-compliant executions of the normative MAS. In this section we briefly survey some of the main approaches in the literature and classify them in terms of whether they can express state, transition or history based norms.

An approach that integrates norms in a BDI-based agent programming architecture is proposed in [Meneguzzi and Luck, 2009]. This extends the AgentSpeak(L) architecture with a mechanism that allows agents to behave in accordance with a set of non-conflicting path-based norms. The agents can adopt obligations and prohibitions with deadlines, after which plans are selected to fulfil the obligations or existing plans are suppressed to avoid violating prohibitions.

There has also been considerable work on normative programming frameworks and middleware to support the development of normative multi-agent organisations, and such frameworks are often designed to inter-operate with existing BDI-based agent programming languages. The AMELI [Esteva *et al.*, 2004a] middleware is based on the ISLANDER formal framework [Esteva *et al.*, 2002]. ISLANDER is a modelling language for specifying institutions in terms of institutional rules and norms. AMELI facilitates agent participation within the institutional environment and supports regimentation of norms relating to agents' communication actions. AMELI is thus restricted to expressing (a particular type of) transition-based norms. Other approaches, e.g., [Esteva *et al.*, 2004a; Garcia-Camino *et al.*, 2005; Silva, 2008; García-Camino *et al.*, 2009], support more general action-based norms, and prescribe actions that should or should not be performed by agents. $\mathcal{S}$-$\mathcal{M}$OISE$^+$ provides support for normative MAS based on the MOISE organisational model. In $\mathcal{M}$OISE$^+$, a deontic specification states a role's permissions and obligations for missions (sets of goals). An organisational manager agent ensures that agent actions (e.g., committing to a mission) do not violate organisational constraints, including norms. However, while $\mathcal{S}$-$\mathcal{M}$OISE$^+$ provides an API which allows agents to discover their obligations, violation of obligations is not monitored by the organisational manager. JaCaMo is similar to $\mathcal{S}$-$\mathcal{M}$OISE$^+$. In JaCaMo, the organisational infrastructure of a multiagent system consists of organisational artefacts and agents that together are responsible for the management and enactment of the organisation. An organisational artefact employs a normative program which in turn implements a $\mathcal{M}$OISE$^+$ specification. Other frameworks such as ORA4MAS [Hübner *et al.*, 2010a] provide support for both norm regimentation and enforcement, however monitoring and enforcement must be explicitly coded in organisational artefacts.

Other norm-based programming languages have been proposed that use high-level norms to represent what the agents should establish or should avoid, in terms of a declarative description of a system state, rather than specifying which actions actions should or should not be performed. One such language is the Organisation Oriented Programming Language (2OPL) for the implementation of normative organisations [Tinnemeier *et al.*, 2009; Dastani *et al.*, 2009]. In this approach, an organisation is viewed as a software entity that exogenously coordinates the interaction between agents and their shared environment. 2OPL provides programming constructs to specify 1) the initial state of an organisation, 2) the effects of agents' actions in the shared environment, and 3) the applicable norms and sanctions. In 2OPL norms can be either enforced by means of sanctions or regimented. The interpreter of 2OPL is based on a cyclic control process. At each cycle, the observable actions of the individual agents (i.e., communication and environment actions) are monitored, the effects of the actions are determined, and norms and sanction are imposed if necessary. An advantage of 2OPL approach is its complete operational semantics such that normative organisation programs can be formally analysed by means of verification techniques (see, e.g., [Astefanoaei *et al.*, 2009; Dastani *et al.*, 2013; Alechina *et al.*, 2013]). A number of normative programming languages have recently been proposed that are similar in spirit to the 2OPL language. The normative language of the THOMAS multi-agent architecture [Criado *et al.*, 2010] supports conditional norms with deadlines, sanctions and rewards. Conditions refer to actions (and optionally states). Norms are enforced rather than regulated, and sanctions may be applied by agents rather than the organization. The normative infrastructure does not restrict interactions between agents. A rule-based system implemented in Jess maintains a fact base representing the organisational state, detects norm activation and monitors violations. NPL/NOPL [Hübner *et al.*, 2011] allows the expression of norms with conditions, obligations and deadlines, and norms may be regimented or enforced. Sanctions are represented as an obligation that an agent apply the sanction to the agent that violated the norm. A translation of $\mathcal{M}$OISE$^+$ specifications into NOPL programs is described in [Hübner *et al.*, 2010b].

4 Norm Verification

Verification of norms involves a variety of questions, answers to which all rely on the specification of norms. These questions include:

- Is a given set of norms consistent [Esteva *et al.*, 2004b]? If not, compute a maximal consistent subset of this set [Alechina *et al.*, 2012].

- Given a transition system and a set of norms, are any of the norms violated [Alechina *et al.*, 2013]?

- A variant of the question above, called runtime norm monitoring, see for example [Alechina *et al.*, 2015]: given a current finite run of the system (in other words, given a finite history of the system so far), are any norms violated or about to be violated?

- Verification of the effect of applying the norms [Ågotnes *et al.*, 2010; Alechina *et al.*, 2013]: given a transition system M and a set of norms N, after the norms are enforced on M, does some system objective ϕ hold in the resulting transition system? The result of applying N to M is called an implementation of a normative systems on M in [Ågotnes *et al.*, 2010] and is called a normative update of M with N in [Alechina *et al.*, 2013].

4.1 Norm Consistency

In this section we consider the problem of whether a set of norms are consistent. This is the focus of, for example, [Esteva *et al.*, 2004b]. The authors of [Esteva *et al.*, 2004b] consider two kinds of norms in electronic institutions, integrity norms which prohibit some actions after some condition occurs, and obligations, which make some actions obligatory after some condition occurs. Both are expressed in first order logic. In order to verify that an electronic institution is norm consistent, a 'dialogue' (essentially, a record of interactions between agents) must be found where there are no violations of integrity norms and there are no pending obligations. The verification problem is decidable when the domain of the ontology describing the institution is finite, so the norms can be propositionalised and the problem of checking consistency reduced to theorem proving in propositional logic.

The problem of finding a maximal consistent set of obligations arises in approaches such as BOID [Broersen *et al.*, 2001] and in the decision making mechanism of the normative programming language N2APL [Alechina *et al.*, 2012]. Essentially, the problem of finding a maximally consistent set of norms (or norms and goals) arises when a rational agent needs to decide which course of action to commit to (since it cannot commit to an inconsistent set, and at the same time may wish to obey as many norms as possible while achieving as many goals as possible). Under certain assumptions, in N2APL this problem is solvable in polynomial time (it is reduced to checking whether a certain set of plans with durations and deadlines can be scheduled in the available time).

4.2 Norm Compliance

In this section, we consider the problem: 'given a structure M and a set of norms N, are there any norm violations in M?'. If the set of norms is given semantically, we simply check whether any of the semantic conditions hold in M (are there any violating states or transitions; note that a set of violating runs even if given semantically needs to be represented in a finite way, e.g. by an automaton or by a regular expression). This problem arises as part of the problem of normative update in [Alechina *et al.*, 2013] (before sanctions could be applied, all norm violations need to be found).

If the set of norms is specified syntactically, and the set of formulas N' describes violation conditions of N, we have a model-checking problem [Clarke *et al.*, 1986; Alur *et al.*, 2002; Baier and Katoen, 2007]: for each formula ϕ in N', does M satisfy ϕ?

The model-checking problem is, given a transition system M and a formula ϕ, does $M \models \phi$ hold? The model-checking problem for different temporal logics has different complexity. The model-checking problem for CTL can be solved in time $O(|M| \times |\phi|)$; it is PTIME-complete. The model-checking problem for LTL is PSPACE-complete. It can also be solved in time $2^{O(|\phi|)} \times O(|M|)$, that is, exponential in the formula and linear in the size of the transition system, which corresponds to a more practical model-checking method than the PSPACE algorithm. The model-checking problem for CTL^* is PSPACE-complete. It can also be solved in time $2^{O(|\phi|)} \times O(|M|^2)$. The model-checking problem for ATL is PTIME-complete.

4.3 Runtime Norm Verification

According to Bauer et al. [Bauer *et al.*, 2011], runtime verification deals with those verification techniques that allow checking whether an execution of a system under scrutiny satisfies or violates a given correctness property. The problem of runtime verification is often formulated as follows: given a system to be checked and a correctness property, check whether an execution of the system satisfies or violates the correctness property. The process of runtime verification consists of various stages such as monitor synthesis, system instrumentation, and execution analysis. In the first stage, the correctness property is used to generate a monitor, which is basically a decision procedure for the property. In the second stage, relevant events of the system are fed into the monitor, and finally in the third stage, the system execution is analysed to decide whether the correctness property is satisfied or violated.

There are a variety of different formalisms that are proposed in the literature to specify and develop monitors that encode the correctness properties. These proposals varies from runtime verification specific formalisms to general purpose formalisms. Some RV-domain specific formalisms, as listed in [Bauer *et al.*, 2011], are language oriented formalisms such as extended regular expressions [Sen and Roşu, 2003], tracematches by the ApectJ team [Allan *et al.*, 2005], query-oriented languages such as PQL [Martin *et al.*, 2005], and rule-based approaches [Barringer *et al.*, 2007]. More generic and general purpose formalisms to specify and develop monitors are various fragments of linear temporal logic [Giannakopoulou and Havelund, 2001; Havelund and Rosu, 2004; Stolz and Bodden, 2006], various types of automata such as security automata [Schneider, 2000] for encoding safety properties and edit automata [Ligatti *et al.*, 2009] for encoding non-safety properties, and aspect-oriented programming such as AspectJ that can be used to develop monitors.

Assuming that correctness properties are closely related to norms in the sense that both are properties that system executions can satisfy or violate, techniques from runtime verification can be used to check norm violations at runtime. For runtime norm verification, the monitor synthesis stage is most relevant as it encodes a norm to a monitor that is subsequently used to decide violation/satisfaction of the norm at runtime. In the following, we present some of the general formalisms from runtime verification literature that can be used to encode norms for runtime norm monitoring purposes.

Runtime Verification for LTL-based norms

As mentioned above, various fragments of LTL are used to specify norms. In standard LTL, a formula specifies a property of infinite runs. However, following [Bauer *et al.*, 2011], the goal of runtime verification is to check properties given finite prefixes of infinite runs. Given that norms are specified as LTL formula, runtime norm verification should check whether finite prefixes of infinite runs are compliant or violate norms. For runtime verification of LTL properties, [Bauer *et al.*, 2011] proposes a three valued semantics. Adopting this semantics for norms, a finite run can be norm compliant, norm violating, or inconclusive in the sense that the norm cannot be said to be satisfied or violated. In general, given a finite run r, a norm n is violated if there is no continuation of r that satisfies n, satisfied if all possible continuations of r satisfy n, and inconclusive otherwise. Formally, let Π be a set of atomic propositions, $\Sigma = 2^\Pi$ be an finite alphabet, Σ^ω be the set of all infinite words (runs), and Σ^* be the set of all finite words (runs), $r\sigma \in \Sigma^\omega$ be an infinite run starting with finite prefix $r \in \Sigma^*$ followed by infinite run $\sigma \in \Sigma^\omega$, and $\models_{LTL}$ be the standard LTL satisfaction relation.

r satisfies n	if $\forall \sigma \in \Sigma^\omega : r\sigma \models_{LTL} n$
r violates n	if $\forall \sigma \in \Sigma^\omega : r\sigma \not\models_{LTL} n$
r is inconclusive wrt n	otherwise

Given that arbitrary LTL formula can be evaluated on finite runs, [Bauer *et al.*, 2011] describe the construction of a (deterministic) finite state machine that can read a finite run and determine whether it satisfies, violates, or is inconclusive with respect to a LTL property.

As shown in [Bauer *et al.*, 2011], the size of the resulting monitor is double exponential in the size of $|\phi|$ and the cause of this is related to the construction of the Büchi automata and the construction of the product automaton.

One interesting characteristic of their construction is that the satisfaction and violations of properties can be decided as early as possible. This feature is particularly important for adopting this approach for runtime monitoring of norms. It should be noted that the adoption of this approach for runtime verification of norms is limited to the detection of norm violations and cannot deal with norm enforcement to regiment or sanction violations (see e.g., [Alechina *et al.*, 2013; Alechina *et al.*, 2015]).

Runtime Enforcement for Safety-Progress Properties

Norms can be enforced on a system by means of regimentation or sanctions. In the first case, the violation of norms are prevented by either ignoring/undoing the violating actions or by halting/blocking the execution of the system. In case of sanctioning, norm violation are allowed but compensated by intervening in the system run. In the context of norm enforcement, the specification of norms is not only for the monitoring purposes, but also for the intervention. The specification of a norm should therefore include a regiment/sanction modality, and in the case of sanctioning, also the sanction that should be imposed upon the norm violation. In the field of runtime verification, mechanisms are devised to enforce properties at runtime [Schneider, 2000; Ligatti *et al.*, 2005; Ligatti *et al.*, 2009]. Examples of these mechanisms are truncation, suppression, insertion, and edit automata. These automata are known under the general term security automata that are designed to enforce security properties. The properties that security automata can enforce are specified with respect to the general Safety-Progress classification of properties [Chang *et al.*, 1993].

A truncation automaton is defined with respect to a system and can be seen as a sequence recogniser and is designed to halt the system run when the system attempts to invoke a forbidden operation. Such an automaton is defined as a finite or countably infinite state machine and with respect to a set of actions of the system under scrutiny. The transition function of a truncation automaton is a partial function and indicates whether to accept the current operation of the system under scrutiny and move to a new state or to halt the target program. Formally, a truncation automaton is tuple (Q, q_0, δ) defined with respect to a system with action set $\mathcal{A}$, where Q is the set of possible states of the automaton, $q_0 \in Q$ is the initial state of the automaton, and $\delta : \mathcal{A} \times Q \to Q$ is a partial transition function that determines which system actions to be accepted. The operational semantics of the truncation automaton is defined by the following transition rules:

$$(a; \sigma, q) \xrightarrow{a} (\sigma, q') \quad \text{if } \delta(a, q) = q' \quad \text{(STEP)}$$
$$(\sigma, q) \dotarrow (\cdot, q) \quad \text{otherwise} \quad \text{(STOP)}$$

The transition rule STEP accepts the current system action a allowing the system run to proceed and the transition rule STOP halts the system run. In [Falcone *et al.*, 2011], it is shown that the class of properties that truncation automaton can enforce is the class of safety properties of the form $\mathcal{G}\phi$, where ϕ is a past formula.

Edit automata are more powerful and extend truncation automata. An edit automaton is defined as a finite or countably infinite state machine and with respect to a set of actions of the system under scrutiny. The transition function of an edit automaton includes the partial transition function of the truncation automaton, but add two new partial transition functions with disjoint domains. One partial transition function indicates whether or not an operation of the system under scrutiny should be suppressed or accepted, while the second one specifies the insertion of a finite sequence of operations to be inserted in the run of the system under scrutiny. An edit automaton can thus allow, suppress, or halt the execution of the system under scrutiny or even insert finite sequences of operations in the execution of the system. Formally, an edit automaton is tuple $(Q, q_0, \delta, \gamma, \omega)$ defined with respect to a system with action set $\mathcal{A}$, where Q, $q_0 \in Q$ and δ are defined as with the truncation automata, $\gamma : \mathcal{A} \times Q \rightarrow \vec{\mathcal{A}} \times Q$ is a partial function that specifies the insertion of a finite sequence of actions into the system run, and $\omega : \mathcal{A} \times Q \rightarrow \{-, +\}$ is a partial function that specifies whether system actions should be accepted or suppressed. In edit automata it is assumed that δ and ω have the same domain, while δ and γ have disjoint domains. The operational semantics of the edit automaton is defined by the following transition rules:

$$(a; \sigma, q) \xrightarrow{a} (\sigma, q') \quad \text{if } \delta(a, q) = q' \text{ and } \omega(a, q) = + \quad \text{(STEPA)}$$
$$(a; \sigma, q) \xrightarrow{\cdot} (\sigma, q') \quad \text{if } \delta(a, q) = q' \text{ and } \omega(a, q) = - \quad \text{(STEPS)}$$
$$(a; \sigma, q) \xrightarrow{\tau} (a; \sigma, q') \quad \text{if } \gamma(a, q) = (\tau, q') \text{ and } \omega(a, q) = - \quad \text{(INS)}$$
$$(\sigma, q) \xrightarrow{\cdot} (\cdot, q) \quad \text{otherwise} \quad \text{(STOP)}$$

The transition rule STEPA accepts the current system action a allowing the system run to proceed, the transition rule STEPS suppresses the current system action a but allows the system run to proceed, the transition rule INS adds a finite sequence of actions τ and allows the system run to proceed, and finally the STEP transition rule halts the system run. In [Falcone et al., 2011], it is shown that the class of properties that edit automaton can enforce is the class of response properties of the form $\mathcal{GF}\phi$, where ϕ is a past formula.

Runtime Norms Verification with Aspect Oriented Programming

Aspect-oriented programming is an extension to object-oriented programming and allows software developers to create software systems that can grow to meet changing requirements. This is done by supporting dynamic modifications of software systems without changing their static object-oriented model. The dynamics modifications can be realised by including some new code to satisfy changing requirements in a separate single location rather than incorporating it at various locations in the existing software. Aspect oriented programming also allows to extend software systems to satisfy new requirements even if the code of the software system is not available. The key concepts of aspect oriented programming are point-cut and advice. An object-oriented program exposes joint points which can be selected by pointcuts. Such join points are points of execution in the software application where an intervention has to be realised in order to meet the new requirements. For example, a point-cut may refer to the point just before or after the execution thread enters or exits a method of some object. An advice is the new code that should be added to the existing object-oriented model to ensure the new requirements. This additional code implements the intervention in the execution of the existing object-oriented software application. A simple example of an advice is the logging code that a developer wants to apply just before or after the execution thread enter or exits a method of some object. A point-cut together with a corresponding advice is then called an aspect. The most mature aspect oriented programming language is AspectJ, which is an extension of Java.

Following the parallel with runtime verification, one may use aspect oriented programming to specify norms and enforce them during the execution of a software system. In the object oriented programming paradigm, state-based norms can be specified in terms of some state variables, action-based norms can be specified in terms of method calls, and behavioural norms can be implemented by creating some additional data structures. The enforcement of norms can be realised by means of point-cuts referring to the execution points where the value of some state variables change, when some method is called, or a combination thereof with some additional data structures. In particular, a norm can be specified by means of some point-cuts with corresponding queries on values of state-variables or arguments of method calls. The enforcement of a norm by means of regimentation or sanctioning can be modelled as an aspect that combines some point-cut and an advice. In particular, the violation of norms and possible sanctions can be implemented in the advice of an aspect.

For example, consider a behavioural/temporal norm that is specified in terms of a condition, an obligation/prohibition and a deadline. Such a norm specifies that some state should be reached or prevented (or some actions should be performed or avoided), as soon as the specified condition is met and before the deadline is reached. The violation of a norm applies an intervention procedure that in case of norm regimentation halts the software execution or in case of norm sanctioning imposes a sanction. A conditional norm can be implemented by means of a number of related point-cuts and advice pairs, in particular, one pair for the condition of the norm, one for the content of the norm (obligation or prohibition), and one for the deadline of the norm.

When the condition pointcut of the norm is reached, then the condition advice checks whether the norm should be instantiated and detached. If so, then a detached norm (obligation or prohibition) is created and stored in a specially designed data structure called detached norm list. Suppose the norm is an obligation. If the obligation pointcut is reached, then the obligation advice will check whether the stored detached obligation in the detached norm list is fulfilled and can be removed from the list. Moreover, if the pointcut of the deadline is reached, then the advice of the deadline pointcut will not proceed the call if the obligation is regimented, and otherwise executes the sanction that corresponds to the obligation. Suppose the norm is a prohibition. If the pointcut of the prohibition is reached, then its advice checks whether a detached prohibition exists. If so, then the prohibition is considered as violated. In case the prohibition is regimented, the advice will not proceed the pointcut's method call. Otherwise, if the prohibition is sanctioned, then the sanction will be executed. If the deadline pointcut is reached and the norm is a prohibition, then the detached prohibition is considered as fulfilled and removed from the detached norm list.

4.4 Normative Update

In this section, we consider the following *normative update problem*: 'given a structure and a set of norms, does the structure where this set of norms is enforced, satisfy a certain system objective?'. Formally, it can be characterised as follows. Given a model M of a computational system (e.g., a transition system) that does not satisfy an overall desirable system property ϕ (e.g., a LTL formula), decide whether the enforcement of a set of norms N (e.g., conditional norm with deadline) on M, denoted as $M \upharpoonright N$, satisfies the desirable system level property ϕ. N can be either a regimentation norm or a sanctioning norm with corresponding sanction that are imposed on violations. Thus, given $M \not\models \phi$, decide whether $M \upharpoonright N \models \phi$. This problem has been studied for state-based, action-based, and behaviour-based norms.

In [Ågotnes et al., 2010], norms are specified semantically as a set of prohibited transitions (edges). A set of norms implemented on a transition system M results in a new transition system, $M \upharpoonright N$, which is M with all prohibited edges removed. In [Ågotnes et al., 2010], the 'reasonableness assumption' is made, which is that $M \upharpoonright N$ is always non-empty: no set of norms will disable all possible transitions in the system. The most basic question to ask is whether $M \upharpoonright N$ satisfies some design objective ϕ. The authors state that the same approach would work with state-based norms, where a norm correspond to a set of prohibited states. In the latter case, $M \upharpoonright N$ is M with all the states prohibited by N removed.

In [Alechina et al., 2013], the notion of applying a set of norms to a transition system is studied for conditional norms with sanctions and deadlines. [Alechina et al., 2013] state that their approach subsumes that of [Ågotnes et al., 2010]. They distinguish two kinds of norms: regimenting and sanctioning norms. *Regimenting norms* can be used to ensure that certain state or behaviours never occur. If a norm labels a state with the distinguished sanction atom $san_\perp$, then the run containing this state is removed from the set of runs of the system by the normative update. *Sanctioning norms*, on the other hand, can be used to penalise rather than eliminate certain execution paths. An undesirable state (from the point of view of the system designer) may or may not be achievable by an agent (or agents) depending on the resources the agent is able or willing to commit to achieving it.

The *normative update* of a physical transition system is defined by applying sanctioning and regimenting norms to the computation tree of the system. The application of sanctioning norms changes the valuation of the violating states (sanction atoms are added), while the application of regimenting norms removes branches of the tree where one of the states violates the regimenting norm.

Definition 7 (Normative Update). Let $M = (S, R, V, s_I)$ be a finite transition system, $T(M)$ be the computational tree of M, and N a finite set of conditional obligations and prohibitions. The normative update of $T(M)$ with N, denoted as $T^N(M)$, is obtained from $T(M)$ as follows:

- for every state s of $T(M)$, if s violates a sanctioning norm $n \in N$, then the sanction atom of n is added to the valuation of s

- all branches which contain a state violating a regimenting norm $n \in N$ are removed from $T(N)$.

Observe that each node s' of $T^N(M)$ contains sanction atoms of all norms violated in s'. Observe also that runs which contain a state satisfying the distinguished sanction atom $san_\perp$ are removed from $T(M)$. As in [Ågotnes et al., 2010], [Alechina et al., 2013] also assume that $T^N(M)$ is non-empty, i.e., that regimentation does not remove all possible paths from the system.

To formulate system objectives, [Alechina et al., 2013] introduce two logics, variants of CTL and ATL, where path quantifiers are annotated with multisets of sanctions incurred on a path. This allows them to express properties like 'when norms are enforced, the agent(s) are unable to bring about a state satisfying (a bad property) ϕ without incurring at least sanctions Z', where Z is a multiset of sanctions. For both logics, the problem of checking whether the normative update satisfies a property is PSPACE-complete.

In [Knobbout et al., 2016], two variants of a dynamic modal logic are proposed to characterise and reason about norm dynamics. The first variant of the logic is devised for updates with state-based norm, while the second variant is devised for updates with action-based norms. The proposed logics come with corresponding sound and complete proof systems. The logics are devised to represent norm updates and to reason about the effect of such updates on a system specification. The logics provide update operators for adding norms to a system specification and to reason about the effects of such updates on the system specification. A target system is modelled as a labelled transition system that determines the effect of actions on the system states. Motivated by the Anderson's reduction [Anderson, 1958a], violation atoms are used to label the norm violating states.

The first type of norms are state-based and have the form $(\phi, +v, Act_R)$ or $(\phi, -v, Act_R)$, where ϕ is the norm condition, $+v$ and $-v$ are the norm effect, and Act_R is a set of repair actions. Adding such a norm to a system updates the system in such a way that for every ϕ state the violation atom v holds until a repair action from Act_R occurs. The idea of the repair action is that its occurrence repairs the system violation. Adding a norm of the form $(\phi, -v, Act_R)$ has a similar effect except that the proposition v stops to hold until the norm effect is repaired. An example of such a norm is $(station, +v, \{buy\_sub\})$, which represents the norm that being in a (train) station causes violation v unless this effect is repaired by buying a subscription. Updating a system with a norm duplicates its states to create two types of states: states in which norm effects are active and states in which norm effects are repaired. The transition relation is then modified in such a way that any violating action ends up in a state where the norm effect is active, and that any repair action ends up in states where the norm effect is repaired. The second type of norms are action-based of the form $Act_T, \phi, +v, Act_r$ and $Act_T, \phi, -v, Act_r$, where the new Act_T component is the set of actions that trigger a norm in the sense that after the triggering actions for every ϕ state the violation atom v holds until a repair action from Act_R occurs. Similar reading is used for norms of the form $Act_T, \phi, -v, Act_r$. An example of this second norm type is $(unchecked, station, +v, \{leave\})$, which represents the norm that unchecked entrance of a (train) station causes violation atom v unless this effect is repaired by buying a subscription.

5 Summary

Violation conditions of regulative norms may correspond to conditions on states, actions, or arbitrary temporal patterns. They may be specified semantically or expressed syntactically in a suitable temporal logic, or in a programming language. Verification problems for norms or rather for normative systems involve verifying consistency of norms, verifying whether violation conditions hold, and finally verifying whether a system where norms are enforced satisfies some system objective. We summarise the material covered in this chapter in a table below.

Specification	Verification Problems
State-based	Consistency; Compliance; Update
Action-based	Consistency; Compliance; Run-time monitoring; Run-time enforcement; Update
Behaviour-based	Consistency; Compliance; Run-time monitoring; Run-time Enforcement; Update

Table 1. Summary of specification and verification of norms

BIBLIOGRAPHY

[Ågotnes *et al.*, 2007] Thomas Ågotnes, Wiebe van der Hoek, and Michael Wooldridge. Normative system games. In *Proceedings of the 6th International Joint Conference on Autonomous Agents and Multiagent Systems (AAMAS '07)*, pages 1–8, New York, NY, USA, 2007. ACM.

[Ågotnes *et al.*, 2010] Thomas Ågotnes, Wiebe van der Hoek, and Michael Wooldridge. Robust normative systems and a logic of norm compliance. *Logic Journal of the IGPL*, 18(1):4–30, 2010.

[Alchourrón and Bulygin, 1981] Carlos E. Alchourrón and Eugenio Bulygin. The expressive conception of norms. In Risto Hilpinen, editor, *New Studies in Deontic Logic*, pages 95–124. OUP Oxford, 1981.

[Aldewereld *et al.*, 2009] Huib Aldewereld, Sergio Álvarez-Napagao, Frank Dignum, and Javier Vázquez-Salceda. Engineering social reality with inheritance relations. In Huib Aldewereld, Virginia Dignum, and Gauthier Picard, editors, *Engineering Societies in the Agents World X, 10th International Workshop, Proceedings ESAW 2009*, volume 5881 of *Lecture Notes in Computer Science*, pages 116–131. Springer, 2009.

[Aldewereld *et al.*, 2013] H. Aldewereld, V. Dignum, and W. Vasconcelos. We ought to; they do; blame the management! – a conceptualisation of group norms. In *Proc. 15th Int. Workshop on Coordination, Organisations, Institutions and Norms (COIN 2013)*, 2013.

[Alechina *et al.*, 2012] Natasha Alechina, Mehdi Dastani, and Brian Logan. Programming norm-aware agents. In *Proceedings of the 11th International Conference on Autonomous Agents and Multiagent Systems (AAMAS 2012)*, pages 1057–1064. IFAAMAS, 2012.

[Alechina *et al.*, 2013] Natasha Alechina, Mehdi Dastani, and Brian Logan. Reasoning about normative update. In *Proceedings of the Twenty Third International Joint Conference on Artificial Intelligence (IJCAI 2013)*, pages 20–26. AAAI Press, 2013.

[Alechina *et al.*, 2014] Natasha Alechina, Wiebe van der Hoek, and Brian Logan. Fair allocation of group tasks according to social norms. In Nils Bulling, Leendert van der Torre, Serena Villata, Wojtek Jamroga, and Wamberto Vasconcelos, editors, *Computational Logic in Multi-Agent Systems, 15th International Workshop, CLIMA XV, Prague, Czech Republic, August 18-19, 2014*, volume 8624 of *Lecture Notes in Computer Science*, pages 19–34, Prague, Czech Republic, August 2014. Springer.

[Alechina *et al.*, 2015] Natasha Alechina, Nils Bulling, Mehdi Dastani, and Brian Logan. Practical run-time norm enforcement with bounded lookahead. In *Proceedings of the 2015 International Conference on Autonomous Agents and Multiagent Systems, AAMAS 2015, Istanbul, Turkey, May 4-8, 2015*, pages 443–451, 2015.

[Allan et al., 2005] Chris Allan, Pavel Avgustinov, Aske Simon Christensen, Laurie Hendren, Sascha Kuzins, Ondřej Lhoták, Oege de Moor, Damien Sereni, Ganesh Sittampalam, and Julian Tibble. Adding trace matching with free variables to aspectj. In *Proceedings of the 20th Annual ACM SIGPLAN Conference on Object-oriented Programming, Systems, Languages, and Applications*, OOPSLA '05, pages 345–364, New York, NY, USA, 2005. ACM.

[Alur et al., 2002] Rajeev Alur, Thomas Henzinger, and Orna Kupferman. Alternating-time temporal logic. *Journal of the ACM*, 49(5):672–713, 2002.

[Anderson, 1957] A.R. Anderson. The formal analysis of normative concepts. *American Sociological Review*, 22:9–17, 1957.

[Anderson, 1958a] A.R. Anderson. The logic of norms. *Logique et Analyse*, 2:84–91, 1958.

[Anderson, 1958b] A.R. Anderson. A reduction of deontic logic to alethic modal logic. *Mind*, 22:100–103, 1958.

[Astefanoaei et al., 2009] L. Astefanoaei, M. Dastani, J.J. Meyer, and F. de Boer. On the semantics and verification of normative multi-agent systems. *International Journal of Universal Computer Science*, 15(13):2629–2652, 2009.

[Baier and Katoen, 2007] Christel Baier and Joost-Pieter Katoen. *Principles of Model Checking*. The MIT Press, 2007.

[Barringer et al., 2007] Howard Barringer, David Rydeheard, and Klaus Havelund. Rule systems for runtime monitoring: From Eagle to RuleR. In Oleg Sokolsky and Serdar Taşıran, editors, *Runtime Verification: 7th International Workshop, RV 2007, Vancover, Canada, March 13, 2007, Revised Selected Papers*, pages 111–125, Berlin, Heidelberg, 2007. Springer Berlin Heidelberg.

[Bauer et al., 2011] Andreas Bauer, Martin Leucker, and Christian Schallhart. Runtime verification for LTL and TLTL. *ACM Transactions on Software Engineering and Methodology*, 20(4):14:1–14:68, 2011.

[Bicchieri and Mercier, 2014] Cristina Bicchieri and Hugo Mercier. Norms and beliefs: How change occurs. In Maria Xenitidou and Bruce Edmonds, editors, *The Complexity of Social Norms*, pages 37–54. Springer International Publishing, 2014.

[Bicchieri, 2006] Cristina Bicchieri. *The Grammar of Society: The Nature and Dynamics of Social Norms*. Cambridge University Press, March 2006.

[Boella and van der Torre, 2004] Guido Boella and Leendert van der Torre. Regulative and constitutive norms in normative multiagent systems. In *Proceedings of the Ninth International Conference on Principles of Knowledge Representation and Reasoning (KR'04)*, pages 255–266, 2004.

[Broersen et al., 2001] Jan Broersen, Mehdi Dastani, Joris Hulstijn, Zisheng Huang, and Leendert van der Torre. The boid architecture: Conflicts between beliefs, obligations, intentions and desires. In *Proceedings of the Fifth International Conference on Autonomous Agents*, AGENTS '01, pages 9–16, New York, NY, USA, 2001. ACM.

[Broersen et al., 2002] J. Broersen, M. Dastani, J. Hulstijn, and L. van der Torre. Goal generation in the BOID architecture. *Cognitive Science Quarterly*, 2(3-4):428–447, 2002.

[Bulling and Dastani, 2011] Nils Bulling and Mehdi Dastani. Verification and implementation of normative behaviours in multi-agent systems. In *Proc. of the 22nd Int. Joint Conf. on Artificial Intelligence (IJCAI)*, pages 103–108, Barcelona, Spain, July 2011.

[Bulling et al., 2013] Nils Bulling, Mehdi Dastani, and Max Knobbout. Monitoring norm violations in multi-agent systems. In *Twelfth International conference on Autonomous Agents and Multi-Agent Systems (AAMAS'13)*, pages 491–498, 2013.

[Castelfranchi, 2004] C. Castelfranchi. Formalizing the informal?: Dynamic social order, bottom-up social control, and spontaneous normative relations. *JAL*, 1(1-2):47–92, 2004.

[Chang et al., 1993] Edward Chang, Zohar Manna, and Amir Pnueli. The safety-progress classification. In Friedrich L. Bauer, Brauer Wilfried, and Helmut Schwichtenberg, editors, *Logic and Algebra of Specification*, volume 94 of *NATO ASI Series*, pages 143–202. Springer, 1993.

[Clarke et al., 1986] E. M. Clarke, E. A. Emerson, and A. P. Sistla. Automatic verification of finite-state concurrent systems using temporal logic specifications. *ACM Transactions on Programming Languages and Systems*, 8(2):244–263, 1986.

[Cliffe et al., 2007] O. Cliffe, M. de Vos, and J. Padget. Specifying and reasoning about multiple institutions. In *Coordination, Organizations, Institutions, and Norms in Agent Systems II*, volume 4386, pages 67–85. Springer LNCS, 2007.

[Criado et al., 2010] Natalia Criado, Vicente Julián, Vicente Botti, and Estefania Argente. A norm-based organization management system. In Julian Padget, Alexander Artikis, Wamberto Vasconcelos, Kostas Stathis, VivianeTorres Silva, Eric Matson, and Axel Polleres, editors, *Coordination, Organizations, Institutions and Norms in Agent Systems V*, volume 6069 of *Lecture Notes in Computer Science*, pages 19–35. Springer Berlin Heidelberg, 2010.

[Dastani *et al.*, 2009] Mehdi Dastani, Nick A.M. Tinnemeier, and John-Jules Ch. Meyer. A programming language for normative multi-agent systems. *Multi-Agent Systems: Semantics and Dynamics of Organizational Models*, pages 397–417, 2009.

[Dastani *et al.*, 2013] Mehdi Dastani, Davide Grossi, and John-Jules Meyer. A logic for normative multi-agent programs. *Journal of Logic and Computation, special issue on Normative Multiagent Systems*, 23(2):335–354, 2013.

[Elster, 2009] J. Elster. Social norms and the explanation of behavior. *The Oxford handbook of analytical sociology*, pages 195–217, 2009.

[Esteva *et al.*, 2002] M. Esteva, D. de la Cruz, and C. Sierra. ISLANDER: an electronic institutions editor. In *Proceedings of the First International Joint Conference on Autonomous Agents and MultiAgent Systems (AAMAS 2002)*, pages 1045–1052, Bologna, Italy, 2002.

[Esteva *et al.*, 2004a] M. Esteva, J.A. Rodríguez-Aguilar, B. Rosell, and J.L. Arcos. AMELI: An agent-based middleware for electronic institutions. In *Proceedings of AAMAS 2004*, pages 236–243, New York, US, July 2004.

[Esteva *et al.*, 2004b] Marc Esteva, Juan Rodriguez-Aguilar, Carles Sierra, and Wamberto Vasconcelos. Verifying norm consistency in electronic instiutions. In Virginia Dignum, Daniel Corkill, Catholijn Jonker, and Frank Dignum, editors, *Proceedings of the AAAI-04 Workshop on Agent Organizations: Theory And Practice*, volume Technical Report WS-04-02, San Jose, July 2004. AAAI, AAAI Press.

[Falcone *et al.*, 2011] Yliès Falcone, Laurent Mounier, Jean-Claude Fernandez, and Jean-Luc Richier. Runtime enforcement monitors: composition, synthesis, and enforcement abilities. *Formal Methods in System Design*, 38(3):223–262, 2011.

[Gabbay, 1987] Dov Gabbay. The declarative past and imperative future: Executable temporal logic for interactive systems. In *Proceedings of the 1st Conference of Temporal Logic in Specification (TLS'1987)*, volume 398 of *Lecture Notes in Computer Science*, pages 409–448. Springer-Verlag, 1987.

[Garcia-Camino *et al.*, 2005] A. Garcia-Camino, P. Noriega, and J. A. Rodriguez-Aguilar. Implementing norms in electronic institutions. In *Proceedings of the Fourth International Joint Conference on Autonomous Agents and MultiAgent Systems (AAMAS'05)*, pages 667–673, New York, NY, USA, 2005.

[García-Camino *et al.*, 2009] A. García-Camino, J. Rodríguez-Aguilar, C. Sierra, and W. Vasconcelos. Constraint rule-based programming of norms for electronic institutions. *Autonomous Agents and Multi-Agent Systems*, 18:186–217, 2009.

[Giannakopoulou and Havelund, 2001] Dimitra Giannakopoulou and Klaus Havelund. Automata-Based Verification of Temporal Properties on Running Programs. In *Proceedings of the 16th IEEE International Conference on Automated Software Engineering*, pages 412–416. IEEE Computer Society, 2001.

[Governatori *et al.*, 2013] Guido Governatori, Francesco Olivieri, Antonino Rotolo, and Simone Scannapieco. Computing strong and weak permissions in defeasible logic. *Journal of Philosophical Logic*, 42(6):799–829, 2013.

[Governatori, 2015] Guido Governatori. Thou shalt is not you will. In Ted Sichelman and Katie Atkinson, editors, *Proceedings of the 15th International Conference on Artificial Intelligence and Law, ICAIL 2015*, pages 63–68. ACM, 2015.

[Grossi *et al.*, 2004] D. Grossi, F. Dignum, L. M. M. Royakkers, and J-J. Ch. Meyer. Collective obligations and agents: Who gets the blame? In *Deontic Logic in Computer Science*, volume 3065, pages 129–145. Springer LNCS, 2004.

[Grossi, 2007] D. Grossi. *Designing Invisible Handcuffs*. PhD thesis, Utrecht University, SIKS, 2007.

[Harel, 1984] David Harel. Dynamic logic. In D. Gabbay and F. Guenthner, editors, *Handbook of Philosophical Logic, Volume II: Extensions of Classical Logic*, volume 165 of *Synthese Library*, chapter II.10, pages 497–604. D. Reidel Publishing Co., Dordrecht, 1984.

[Havelund and Rosu, 2004] Klaus Havelund and Grigore Rosu. Efficient monitoring of safety properties. *International Journal on Software Tools for Technology Transfer*, 6(2):158–173, August 2004.

[Hübner *et al.*, 2010a] Jomi F. Hübner, Olivier Boissier, Rosine Kitio, and Alessandro Ricci. Instrumenting multi-agent organisations with organisational artifacts and agents. *Autonomous Agents and Multi-Agent Systems*, 20:369–400, 2010.

[Hübner *et al.*, 2010b] Jomi Fred Hübner, Olivier Boissier, and Rafael H. Bordini. From organisation specification to normative programming in multi-agent organisations. In Jürgen Dix, João Leite, Guido Governatori, and Wojtek Jamroga, editors, *Computational Logic in Multi-Agent Systems, 11th International Workshop, CLIMA XI, Lisbon, Portugal, August 16-17, 2010. Proceedings*, volume 6245 of *Lecture Notes in Computer Science*, pages 117–134. Springer, 2010.

[Hübner *et al.*, 2011] Jomi Hübner, Olivier Boissier, and Rafael Bordini. A normative programming language for multi-agent organisations. *Annals of Mathematics and Artificial Intelligence*, 62:27–53, 2011.

[Jones and Sergot, 1993] A. J. I. Jones and M. Sergot. On the characterization of law and computer systems. In J.-J. Ch. Meyer and R.J. Wieringa, editors, *Deontic Logic in Computer Science: Normative System Specification*, pages 275–307. John Wiley & Sons, 1993.

[Kanger, 1971] S. Kanger. New foundations for ethical theory. In R. Hilpinen, editor, *Deontic Logic: Introductory and Systematic Readings*, pages 36–58. Reidel Publishing Company, 1971.

[Knobbout and Dastani, 2012] Max Knobbout and Mehdi Dastani. Reasoning under compliance assumptions in normative multiagent systems. In Wiebe van der Hoek, Lin Padgham, Vincent Conitzer, and Michael Winikoff, editors, *Proceedings of the 11th International Conference on Autonomous Agents and Multiagent Systems (AAMAS 2012)*, pages 331–340. IFAAMAS, 2012.

[Knobbout et al., 2016] Max Knobbout, Mehdi Dastani, and John-Jules Meyer. A dynamic logic of norm change. In *22nd European Conference on Artificial Intelligence, The Hague, The Netherlands, August 29 - September 2, 2016*. To be published, 2016.

[Ligatti et al., 2005] Jay Ligatti, Lujo Bauer, and David Walker. Edit automata: Enforcement mechanisms for run-time security policies. *International Journal of Information Security*, 4(1-2):2–16, 2005.

[Ligatti et al., 2009] Jay Ligatti, Lujo Bauer, and David Walker. Run-time enforcement of nonsafety policies. *ACM Trans. Inf. Syst. Secur.*, 12(3):19:1–19:41, January 2009.

[Martin et al., 2005] Michael Martin, Benjamin Livshits, and Monica S. Lam. Finding application errors and security flaws using PQL: A program query language. In *Proceedings of the 20th Annual ACM SIGPLAN Conference on Object-oriented Programming, Systems, Languages, and Applications*, OOPSLA '05, pages 365–383, New York, NY, USA, 2005. ACM.

[Meneguzzi and Luck, 2009] Felipe Rech Meneguzzi and Michael Luck. Norm-based behaviour modification in BDI agents. In Carles Sierra, Cristiano Castelfranchi, Keith S. Decker, and Jaime Simão Sichman, editors, *Proceedings of the 8th International Joint Conference on Autonomous Agents and Multiagent Systems (AAMAS 2009)*, pages 177–184. IFAAMAS, 2009.

[Pnueli, 1977] Amir Pnueli. The temporal logic of programs. In *Proceedings of the 18th Annual Symposium on Foundations of Computer Science (FOCS)*, pages 46–57, 1977.

[Schneider, 2000] Fred B. Schneider. Enforceable security policies. *ACM Trans. Inf. Syst. Secur.*, 3(1):30–50, February 2000.

[Schnoebelen, 2003] Ph. Schnoebelen. The complexity of temporal logic model checking. In Philippe Balbiani, Nobu-Yuki Suzuki, Frank Wolter, and Michael Zakharyaschev, editors, *Advances in Modal Logic 4*, pages 393–436. King's College Publications, 2003.

[Searle, 1995] J. Searle. *The Construction of Social Reality*. Free, 1995.

[Sen and Roşu, 2003] Koushik Sen and Grigore Roşu. Generating optimal monitors for extended regular expressions. *Electronic Notes in Theoretical Computer Science*, 89(2):226 – 245, 2003. RV'2003, Run-time Verification (Satellite Workshop of CAV '03).

[Silva, 2008] V. Torres Silva. From the specification to the implementation of norms: an automatic approach to generate rules from norms to govern the behavior of agents. *International Journal of Autonomous Agents and Multiagent Systems (JAAMAS)*, 17(1):113–155, 2008.

[Stolz and Bodden, 2006] Volker Stolz and Eric Bodden. Temporal assertions using aspectj. *Electronic Notes in Theoretical Computer Science*, 144(4):109 – 124, 2006. Proceedings of the Fifth Workshop on Runtime Verification (RV 2005).

[Tinnemeier et al., 2009] Nick Tinnemeier, Mehdi Dastani, John-Jules Meyer, and Leon van der Torre. Programming normative artifacts with declarative obligations and prohibitions. In *Proceedings of the IEEE/WIC/ACM International Conference on Intelligent Agent Technology (IAT'09)*, pages 69–78, 2009.

3
Modeling Normative Conflicts in Multiagent Systems

VIVIANE T. SILVA, WAMBERTO W. VASCONCELOS,
JÉSSICA S. SANTOS, JEAN O. ZAHN AND MAIRON BELCHIOR

1 Introduction

Norms have been used in open multiagent systems (MASs) as a mechanism to regulate such systems by influencing or restricting the behavior of their agents without directly interfering with their autonomy. Norms define the actions that must be performed (obligations), actions that can be performed (permissions), and actions that should not be performed (prohibitions); similarly, norms may address states (situations) instead of actions. Norms provide a means to specify and obtain desirable system behaviors [Grossi *et al.*, 2010].

Nevertheless, there might exist the possibility of normative conflicts. A normative conflict arises when the fulfilment of one norm causes the violation of another, and vice-versa. The most common cases of conflict occur (i) between a norm that obliges and another one that prohibits the same behavior, or (ii) between a norm that prohibits and another one that permits the same behavior. When there is a normative conflict, whatever the agents do or refrain from doing may lead to a state that is not norm-compliant [Kollingbaum *et al.*, 2007c; Vasconcelos and Norman, 2009].

In this chapter, we present the elements used by several approaches to represent a norm and the techniques found in the literature for the detection and resolution of conflicts between norms in multiagent systems.

Although, there is no consensus in the literature about the elements used to represent a norm, a norm is commonly associated with the following elements: *deontic concept*, *entity* and *action*. The *deontic concept* describes behavior restrictions for the agents in the form of obligations, permissions and prohibitions. The *entity* is the norm addressee whose behavior is being regulated and the *action* is indeed the behavior being regulated by the norm.

The techniques used to detect normative conflicts were classified in this chapter in two main groups (i) approaches that deal with normative conflicts at *design time* and (ii) approaches that deal with normative conflicts at *runtime*. In addition, we have also classified the approaches according to the kind of conflicts they can detect: (iii) approaches that can identify *direct conflicts*, and (iv) approaches that can also identify *indirect conflicts*. In *direct conflicts* the elements of the norms are the same but in *indirect conflicts* the elements are related by domain-dependent or other kinds of relationships.

The approaches used to resolve normative conflicts were divided in two kinds: (i) *norm prioritization* and (ii) *norm update*. The *norm prioritization* strategy prioritizes one of the norms in conflict by overriding the other under particular circumstances. In the *norm update* strategy, one of the norms in conflict is updated in order to eliminate the conflict.

2 Normative Multiagent Systems

Norms of a normative MAS can regulate access to resources of the system, interactions between agents (norms that regulate dialogs), the performance of an action, the achievement of a state of affairs, and so on. Since software agents are endowed with autonomy, they can reason about norms and they may choose to comply or not with these norms. In this context, normative MASs can adopt mechanisms to enforce that the agents act according to their norms. The strategy used for norm enforcement can associate sanctions with norm violation and rewards with norm fulfillment in order to influence the agents' behavior [Boella *et al.*, 2006; Meneguzzi and Luck, 2009].

Normative concepts have been studied in several areas, such as philosophy, legal theory, and sociology, among others. So far, there is not a consensus in the literature about which elements a norm should represent. Each work presents a norm definition with different components. However, some components are common to several approaches. A norm is usually associated with a deontic concept (obligation, prohibition, permission), is addressed to an entity or a group of entities and regulates a given behavior.

Deontic logic [Wright and Henrik, 1951] is a modal logic for reasoning about ideal behavior. The concepts of deontic logic have traditionally been used for the analysis and reasoning about normative law and are widely used in normative MASs to formally describe norms and their modalities [Wieringa and Meyer, 1993; Meyer *et al.*, 1993]. In deontic logic, the modality of a norm is called a deontic concept. It represents the nature of the regulation defined by the norm, namely, a prohibition, a permission or an obligation.

A norm can regulate the performance of an atomic action, a complex action (parameterized action or composed one) [Cholvy and Cuppens, 1995; Torres da Silva *et al.*, 2015; Fenech *et al.*, 2009; Gaertner *et al.*, 2007; Garcia *et al.*, 2013; Giannikis and Daskalopulu, 2011; Kagal and Finin, 2007; Li, 2014; Vasconcelos *et al.*, 2012] or the achievement of a state of affairs [dos Santos Neto *et al.*, 2012; Günay and Yolum, 2013b]. Norms can be classified as individual norms, when they are applied to a given entity [Broersen *et al.*, 2001b; Broersen *et al.*, 2001a; dos Santos Neto *et al.*, 2012; García-Camino *et al.*, 2006; Kagal and Finin, 2007; Oren *et al.*, 2008], or as collective norms, when they regulate the behavior of an organization [Torres da Silva *et al.*, 2015; Garcia *et al.*, 2013; Zahn, 2015; Zahn and da Silva, 2014] or a role [Aphale *et al.*, 2012; Cholvy and Cuppens, 1995; Torres da Silva *et al.*, 2015; dos Santos Neto *et al.*, 2012; Gaertner *et al.*, 2007; Giannikis and Daskalopulu, 2011; Günay and Yolum, 2013b; Kollingbaum *et al.*, 2007a; Li, 2014; Sensoy *et al.*, 2012]. A role is an abstract representation of a social position within a given group. When a norm is addressed to a role, all agents playing the role must comply with the norm. When a norm is addressed to an organization (or group of agents), all agents that belong to the given organization must comply with the norm.

The norm representation of some approaches also considers that a norm is only valid within a context (an organization or an environment, for instance) [Zahn and da Silva, 2014; Torres da Silva *et al.*, 2015]. In this case, the agent only must comply with the norm inside the given context. Outside this context, the norm is not applicable. If the context is not defined, the norm is applicable in all contexts of the MAS. In such case, the context may be an organization, an environment [Zahn and da Silva, 2014; Torres da Silva *et al.*, 2015], for example.

Other approaches consider that a norm is only valid during a given period and associate activation and deactivation conditions with each norm [Criado *et al.*, 2010; Torres da Silva *et al.*, 2015; Gaertner *et al.*, 2007; Kagal and Finin, 2007; Vasconcelos *et al.*, 2012]. In this case, the norm becomes active and must be fulfilled when the activation condition is satisfied and it becomes inactive when the deactivation condition arises. The activation and deactivation conditions can be states or events, such as a date, the execution of an action, the fulfilment of a norm, and so on. If activation and deactivation conditions are not defined, the norm is always applicable.

Some approaches allow variables in some of its components, conferring generality and compactness on the representation, as variables parameterize a norm specification. Some approaches associate variables with constraints [Aphale *et al.*, 2012; Sensoy *et al.*, 2012; Vasconcelos and Norman, 2009; Vasconcelos *et al.*, 2009], restricting their value. Tables 1 and 2 illustrate the approaches using the same components to define a norm. Note that we only list those components that are considered by more than one paper by different authors, leaving out norm components that are particular to one proposal. Examples of those components are: (i) the state of the norm (described in [Da Silva and Zahn, 2013; Zahn and da Silva, 2014] that determines whether the norm has been violated or fulfilled), (ii) social context (defined in [Oren *et al.*, 2008] as the social entity that imposes the norm), (iii) declaration time (presented in [Vasconcelos *et al.*, 2009] that establishes the time when the norm was introduced in the system), and (iv) regulation (described in [Cholvy and Cuppens, 1995] to establish that norms are addressed to regulations). Similarly, Table 2 lists the kinds of activation condition that are common to more than one approach (time, state, action, and context). Some approaches are mentioned in Table 1 but are not mentioned in Table 2 because they do not consider any of the norm elements listed in Table 2 (namely, activation and deactivation conditions, rewards, sanctions, context, constraints).

2.1 Conflict Detection

Conflict detection may be done either at *design time* or at *runtime*. At *design time*, the detection mechanism is performed before the execution of the MAS in order to identify potential conflicts and avoid their occurrence during the execution of the system. At *runtime*, the MAS or the agents must be able to solve conflicts dynamically, as the system executes/runs. The efficacy of tools used to detect normative conflicts depends on some factors, such as the norm expressiveness supported by the detection method and the ability of the proposed mechanism to reason about the relationships of the application domain.

In the literature addressing normative conflict detection in MAS, it is common to

Table 1. Norm components of different approaches 1-2

Reference	Deontic Concept	Regulated Entity			Regulated behaviour		Norm language		
		Agent	Role	Organization	Action	State	Parameters 1st order	Propositions	Logical Operators
[Aphale et al., 2012]	•	•			•				
[Boella et al., 2012]	•	•				•		•	•
[Broersen et al., 2001a]	•				•	•		•	
[Broersen et al., 2001b]	•				•	•		•	•
[Cholvy and Cuppens, 1995]	•	•			•			•	
[Cholvy and Garion, 2004]	•	•			•			•	•
[Criado et al., 2010]	•		•		•			•	
[Torres da Silva et al., 2015]	•		•	•	•			•	
[Zahn and da Silva, 2014]	•		•	•	•			•	
[dos Santos Neto et al., 2011]	•		•					•	
[dos Santos Neto et al., 2012]	•		•			•		•	
[Fenech et al., 2009]	•		•		•	•		•	
[Fenech et al., 2008]	•				•			•	•
[Gaertner et al., 2007]	•		•		•			•	
[Garcia-Camino et al., 2006]	•				•			•	
[Garcia et al., 2013]	•			•			•		
[Giannikis and Daskalopulu, 2011]	•	•			•		•		
[Giannikis and Daskalopulu, 2009]	•	•					•		•
[Günay and Yolum, 2013a]	•		•					•	
[Kagal and Finin, 2004]	•				•	•	•		
[Kagal and Finin, 2007]	•		•		•	•	•		
[Kollingbaum and Norman, 2004]	•					•	•		
[Kollingbaum et al., 2006]	•	•			•	•	•		
[Kollingbaum et al., 2007a]	•	•			•		•		
[Kollingbaum et al., 2007c]	•				•		•		
[Li, 2014]	•	•						•	•
[Oren et al., 2008]	•	•			•			•	•
[Sensoy et al., 2012]	•				•				•
[Vasconcelos et al., 2012]	•				•				
[Vasconcelos et al., 2009]	•		•				•		
[Vasconcelos and Norman, 2009]	•		•						
[Zahn, 2015]	•	•				•		•	
[Da Silva and Zahn, 2013]	•	•	•					•	•

Table 2. Norm components of different approaches 2-2

	Activation Condition					Regulated behaviour				Rewards		Sanctions		Context		Constrs.
	Time	State	Event	Action	Context	Time	State	Event	Context	Action	state	Action	State	Org.	Env.	
[Aphale et al., 2012]																•
[Criado et al., 2010]										•		•				
[Torres da Silva et al., 2015]	•					•								•	•	
[Zahn and da Silva, 2014]	•				•	•								•	•	
[dos Santos Neto et al., 2012]									•	•	•		•			
[dos Santos Neto et al., 2011]							•									
[Fenech et al., 2009]			•	•												
[Fenech et al., 2008]			•	•												
[Gaertner et al., 2007]						•										
[Kagal and Finin, 2004]						•										
[Kagal and Finin, 2007]						•										
[Kollingbaum and Norman, 2004]		•														
[Li, 2014]		•						•								
[Sensoy et al., 2012]						•										•
[Vasconcelos et al., 2012]	•					•										•
[Vasconcelos et al., 2009]																•
[Vasconcelos and Norman, 2009]														•	•	
[Zahn, 2015]	•					•										
[Da Silva and Zahn, 2013]	•					•								•	•	

61

find a distinction between *deontic conflicts* and *deontic inconsistences* [Elhag et al., 2000]. When a norm prohibits a certain behavior that is obliged by another norm at the same time, a *deontic conflict* arises. In such case, norm-compliant agents are unable to comply with the two norms in a consistent way and a violation will always occur. When there is a norm permitting a certain behavior that is prohibited – simultaneously – by another norm, a *deontic inconsistency* occurs. When that happen, the agent may choose to comply with the prohibition and not perform the behavior defined in the permission, avoiding violations. A permissive norm does not force its addressee to perform an action, i.e., a permission may be not acted upon. This point of view is based on the impossibility-of-joint-compliance test for conflict (IJC-test, for short) [Hill, 1987], which determines that compliance statements represent the fulfilment of the content of a norm. For instance, the norm "*obligatory p*" corresponds to the compliance statement "*p is done*". Permissions are not susceptible for constructing compliance statements and for that reason, a conflict analysis involving permissive norms may only indicate a possibility of conflict.

Normative conflicts may also be classified as *direct* or *indirect*. The *direct conflicts* are those that arise when the norms regulate the behavior of the same agent, control the execution of the same behavior in the same context, and have contradictory deontic modalities, i.e., obligation versus prohibition, or permission versus prohibition. For instance, suppose that a norm is of the form: Db, where D is a deontic concept that represents an obligation (O) or a prohibition (F); and b represents the behavior (action or state) being regulated. If an agent is associated with the norms Op and Fp, a direct comparison between the norm components can detect a conflict. On the other hand, *indirect conflicts* are those that arise between norms addressing different norm elements, but related. To detect an indirect conflict, it is necessary to considerer the characteristics of the application domain [Zahn and da Silva, 2014]. For instance, if an agent is associated with the norms Oq and Fp and $q \rightarrow p$, then there is an *indirect conflict* between them. The actions q and p are not the same, but they are related somehow by the implication operator.

Normative conflicts may also arise between norms that have the same deontic concept. The authors in [Oren *et al.*, 2008; Fenech *et al.*, 2009; Giannikis and Daskalopulu, 2011; Torres da Silva *et al.*, 2015] propose a mechanism that is able to detect conflicts that occur when norms are simultaneously obligating the same agent to execute actions that cannot be performed at the same time (commonly called *orthogonal actions*). For instance, considering that a norm is of the form: Db, where D is a *deontic concept* that represents an obligation (O) or a prohibition (F); and b represents the behavior (action or state) being regulated. If an agent is associated with the norms Op and Os, a conflict arises if p is a orthogonal behavior to s, i.e., the agent cannot adopt both behaviors at the same time.

Table 1 shows a classification of the papers that check normative conflicts at *design time* and those that check conflicts at *runtime*; the table also classifies the papers by their ability to check *direct* and *indirect* normative conflicts. Each row shows a set of approaches (with their bibliographic references), and it describes when the approach is used – *design time* and *runtime* – and the kind of conflict detected – *direct* and *indirect conflicts*. All approaches mentioned in Table 1 are able to detect direct normative conflicts.

There are many strategies to detect normative conflicts. Each approach has a set of characteristics (e.g., detect *indirect* conflicts at *design time*) and has a set of strategies (e.g., the use of trace analysis method). For instance, in [Da Silva and Zahn, 2013; Zahn and da Silva, 2014] it was adopted the *normalization* method to detect normative conflict. The method consists of locally rewriting norms in order to transform these into an alternative format and finding out if the norms overlap with one other. This approach is similar to the *unification* method presented elsewhere, e.g., in [Gaertner *et al.*, 2007; Vasconcelos *et al.*, 2007; Vasconcelos *et al.*, 2009; Vasconcelos and Norman, 2009; Vasconcelos *et al.*, 2012].

Indirect normative conflicts can be detected by taking into account an ontology that describes the characteristics of the application domain. Such characteristics involves the set of norms and the relationships defined between the norms components. In [Da Silva and Zahn, 2013; Zahn and da Silva, 2014; Torres da Silva *et al.*, 2015; Zahn, 2015], a conflict checker algorithm is able to detect direct and *indirect* normative conflicts through the analysis of the *system ontology*. The approaches [Aphale *et al.*, 2012; Sensoy *et al.*, 2012] propose a detection algorithm that identifies conflicting norms by context anticipating in which conflicts may arise. The problem of anticipating conflicts between two norms is transformed into an ontology consistency-checking problem. In [Giannikis and Daskalopulu, 2009; Giannikis and Daskalopulu, 2011] the authors also use the *context anticipating* as a complementary technique to conflict detection.

The use of the *first-order unification* to check overlapping variables is a technique used by some researchers, such as [Gaertner *et al.*, 2007; Vasconcelos *et al.*, 2009; Vasconcelos and Norman, 2009; Vasconcelos *et al.*, 2012]. The *first-order unification* consists of checking if variables of a prohibition overlap with the variables of an obligation (or permission). In [Vasconcelos *et al.*, 2009], the authors stated that the detection algorithm always terminates (possibly failing, if a unifier cannot be found), is correct and has a linear computational complexity. Similarly, the computational model presented in [Vasconcelos and Norman, 2009] uses unification and constraint satisfaction to detect potential conflicts among norms. Besides, such approach stores all the actions performed and all the norms that currently hold in order to maintain a global history of the enactment of the society of agents.

We also report on another technique to detect normative conflict, namely the trace analysis. In the work presented in [Fenech *et al.*, 2008], in order to detect conflicts, the authors extend the trace semantic presented in [Kyas *et al.*, 2008], which enables checking whether a trace satisfies a contract or not. The detection mode is complete and always terminates and it is more detailed in [Fenech *et al.*, 2009]. The trace analysis used in the detection mechanism is explained in [Li, 2014], where the possible event traces are generated in order to check which traces lead to conflicts. In this work, a conflict arises when an institutional fact is true in one institution and false in another one at the same time. Institutional facts are results of the events and the norms of the institutions. Such approach is able to check direct normative conflicts among institutions at design time.

In [Kollingbaum et al., 2006], the authors state their proposed architecture NoA is capable of informing agents about conflicts and inconsistencies. Normative conflicts, in general, can be detected by investigating the intersection of sets of behaviors (action or states) that are addressed by partially of fully instantiated activity statements within norm specification. The detection of *indirect* normative conflicts is performed by investigating all possible plans that can be chosen to achieve a state of affairs and by analyzing the side-effects of plans. This conflict detection strategy is called as *plan investigation* and it is similarly adopted in [Kollingbaum and Norman, 2004; Kollingbaum and Norman, 2005; Kollingbaum et al., 2007a].

The Regulated Open Multiagent Systems (ROMAS) reported in [Garcia et al., 2013], is a CASE tool for designing systems and can detect *direct* conflicts among organizational norms and agent norms at *design time*. The ROMAS CASE tool presents a graphical interface to model ROMAS systems and provides a means to verify the model created using model-checking techniques.

The strategies to detect conflicts are summarized in Table 2. Each row presents a set of approaches that make use of similar strategies. The columns of the table determine the classification of the approaches according to the strategy adopted.

3 Conflict Resolution

Once the normative conflict is identified, resolution mechanisms can be applied in order to solve the conflict. Several approaches propose means to resolve conflicts among norms. The strategies used by those proposals can be divided into two kinds: *norm prioritization* and *norm update*. The *norm prioritization* strategy prioritizes one of the norms in conflict by overriding the other under particular circumstances. In the *norm update* strategy, one of the norms in conflict is updated in order to eliminate the conflict.

Most of the proposed approaches to resolve conflicts among norms define a *prioritization* order between norms to specify which norm is more relevant. In the literature, there are three classical principles that have been used to solve deontic conflicts: *lex posterior*, *lex superior* and *lex specialis*. The *lex posterior* principle assumes that newer norms are preferred over the older ones. According to [Sensoy et al., 2012], this principle is useful if the conflicting norms are issued by the same authorities, because temporal relationships between different authorities may be misleading. This conflict resolution principle has been adopted by many approaches, such as [García-Camino et al., 2006; Kagal and Finin, 2007; Kollingbaum and Norman, 2004; Vasconcelos et al., 2009; Giannikis and Daskalopulu, 2011]. The *lex superior* principle considers that norms issued by a more important authority have priority over those norms issued by less important authorities. It is necessary to have a hierarchical relationship between the authorities who issue the norm in order to adopt this principle. The *lex superior* principle has been adopted, for instance, in [Sensoy et al., 2012; Kagal and Finin, 2007; Vasconcelos et al., 2009; Giannikis and Daskalopulu, 2011]. The third principle, *lex specialis*, states that specific norms should override the more general ones. This principle requires non-trivial reasoning process about subsumption relationships between norms, and it may be applied if the norms belong to the same organization [Sensoy et al., 2012]. This principle was adopted, for instance, in [Sensoy et al., 2012; García-Camino et al., 2006; Vasconcelos et al., 2009; Giannikis and Daskalopulu, 2011].

There are other conflict resolution strategies in the literature based on *norm prioritization*. For instance, in [Cholvy and Cuppens, 1995], the authors presented a conflict resolution approach that establishes an order of *prioritization* between the roles mentioned in the conflicting norms. For instance, if there is a conflict between roles $r1$ and $r2$ and the judgment of priorities states that $r1 > r2$, then all norms associated with $r1$ takes precedence on the norms associated with $r2$. The judgment of priorities may depend on the individual (especially in moral dilemmas), or may be derived from the hierarchy of roles. [Oren et al., 2008] represent conflicts between norms as a normative conflict graph in which norms are nodes and edges are conflicts between norms. If this graph contains no edges, then no normative conflict exists. The conflict resolution approach removes nodes (and their associated edges) until no normative conflicts exist. The authors considered three separate heuristics for norm removal: (i) *random drop heuristic*, (ii) *maximal conflict-free set heuristic* and (iii) *preferred extension based norm conflict resolution heuristic*.

Modality precedence can also be used to resolve normative conflicts. In [Kagal and Finin, 2007], negative authorizations can have precedence over positive authorizations, or vice versa. [Kollingbaum and Norman, 2004] suggested to take into account the social position and power of the norm issuer in order to solve the normative conflict. According to the work presented in [dos Santos Neto et al., 2012], when there is a conflict between two norms, the algorithm to solve the conflict prioritizes the norm with the highest contribution to the achievement of the agent's desires (and intentions). This norm's contribution is calculated by verifying if the contribution coming from the fulfilment of the first norm plus the contribution coming from the violation of the second norm is greater to or equal than the contribution coming from the fulfilment of the second norm plus the contribution coming from the violation of the first norm. If that happens, the first norm is selected to be fulfilled and the second one to be violated. Otherwise, the second norm is selected to be fulfilled and the first to be violated. If the norm contributions have equal values, the algorithm can choose either one.

On the other hand, for conflict resolution strategies based on the kind norm update, there are also many proposals found in literature and briefly pointed out as follows. [Vasconcelos et al., 2009] presented a conflict resolution mechanism that consists of a curtailment of the conflicting norms by adding constraints to their scope of influence. In order to decide which norm to curtail, the classic principle of lex superior can be adopted. [Kollingbaum et al., 2006] also proposed to modify the norm's scope of influence to resolve conflicts between norms. Three approaches were presented: (i) extending the scope of influence of an obligation, (ii) reducing the scope of influence of a prohibition and (iii) introducing new permissions that override temporarily the prohibitions in order to make the action permitted and allow the fulfilment of the obligation. In the work described in [Gaertner et al., 2007], the authors propose a fine-grained way of resolving normative conflicts via unification. It curtails the influence of the normative position by using the annotations when checking if the norm applies to illocutions.

The approach in [Sensoy et al., 2012] propose an approach that resolve conflicts between two norms, without making one norm override the other. The approach uses automated planning in order to find a plan whose actions will cause the state of the word to change in a way that the expiration condition of one of the conflicting norms holds. In [Günay and Yolum, 2013b; Günay and Yolum, 2013a], the authors suggested to use delegations to resolve conflicts between commitments. According to the authors, commitment conflicts and norm conflicts are closely related. The delegation will resolve the conflict only if the agent has the power to delegate some of their responsibilities and if the deputy agent has enough resources and are able to fulfill the delegated commitment.

There is no particular resolution strategy that is the most adequate to resolve the normative conflict in multiagent system. The decision of what approach is more adequate will depend on many factors, such as the norm expressiveness, computational complexity of the strategy, application domain, and others. Table 3 shows a complete list of conflict resolution approaches found in the literature according to which strategy were used by them.

4 Discussion and Conclusions

Multiagent systems are composed of software agents, which are autonomous, self-interested and possible heterogeneous entities. Due to the autonomy, self-interest, and heterogeneity of the software agents, predicting or controlling the overall behavior of a MAS is a challenging issue. Therefore, norms have been used to guide and model the behavior of software agents, without restricting their autonomy. These norms are commonly associated with deontic concepts, such as obligations, permissions and prohibitions. However, in a system governed by multiple norms, normative conflicts may arise, i.e., sometimes the fulfilment of a norm automatically violates another one. For this reason, the agents or the MAS must have mechanisms to detect and resolve conflicts among norms.

The detection of all potential normative conflicts that may occur in MAS is a task that depends on many factors, such as the analysis of the characteristics of the application domain and the investigation involving the relationships between behaviors (actions and states), agents and contexts in which the norms are applied. Furthermore, the majority of approaches that can detect and resolve indirect conflicts between norms are applied only within a specific architecture. The mechanisms that identify indirect conflicts at design time require a pre-determination of all relationships between the elements of the system, i.e., the designer must stablish all relationships of the system in the moment that was designed.

There are many limitations and challenges in approaches involving normative MAS. Some approaches can only deal with norms regulating atomic actions; others support the regulation of parameterized actions, a set of actions as well as states of affairs (resulting from the action executions). The common thread is that all approaches can check normative conflicts involving only two norms (BOID identifies conflicts among

Table 3. Classification of approaches to detect normative conflicts

	When the approach is used		Kind of conflict detected	
	Design Time	Runtime	Direct conflicts	Indirect conflicts
[Aphale et al., 2012] [Cholvy and Cuppens, 1995] [Cholvy and Garion, 2004] [Da Silva and Zahn, 2013] [Torres da Silva et al., 2015] [Fenech et al., 2008] [Fenech et al., 2009] [Sensoy et al., 2012] [Zahn, 2015] [Zahn and da Silva, 2014]	•		•	•
[Garcia et al., 2013] [Li, 2014] [Vasconcelos and Norman, 2009]	•		•	
[Giannikis and Daskalopulu, 2009] [Giannikis and Daskalopulu, 2011] [Kollingbaum and Norman, 2004] [Kollingbaum et al., 2006] [Kollingbaum and Norman, 2005] [Kollingbaum et al., 2007c] [Kollingbaum et al., 2007b] [Oren et al., 2008] [Vasconcelos et al., 2009]		•	•	•
[dos Santos Neto et al., 2012] [dos Santos Neto et al., 2011] [Gaertner et al., 2007] [Günay and Yolum, 2013b] [Günay and Yolum, 2013a] [Vasconcelos et al., 2012]		•	•	

Table 4. Different strategies for normative conflict detection

Reference	Norm. Ontology	Analysis	Cons. Check	Unific.	Cons. Satisf.	Subst.	Cont. Anticip.	Trace Analysis	Comp. Analysis	Plan Investig.	Model.
[Zahn and da Silva, 2014]	•										
[Torres da Silva et al., 2015]		•									
[Zahn, 2015]		•									
[Da Silva and Zahn, 2013]											
[Vasconcelos et al., 2009]				•	•	•					
[Aphale et al., 2012]									•		
[Sensoy et al., 2012]			•	•	•	•					
[Cholvy and Cuppens, 1995]							•		•		
[Cholvy and Garion, 2004]						•			•		
[Fenech et al., 2008]									•		
[Fenech et al., 2009]								•			
[Li, 2014]								•			
[Gaertner et al., 2007]											
[Vasconcelos et al., 2012]				•		•					
[Kollingbaum and Norman, 2004]											
[Kollingbaum et al., 2006]											
[Kollingbaum et al., 2007a]											
[Kollingbaum et al., 2007b]											
[Kollingbaum et al., 2007c]											
[Giannikis and Daskalopulu, 2009]									•		
[Giannikis and Daskalopulu, 2011]							•				
[dos Santos Neto et al., 2012]									•		
[dos Santos Neto et al., 2011]									•		
[Günay and Yolum, 2013b]									•		
[Günay and Yolum, 2013a]					•				•		
[Oren et al., 2008]										•	
[Garcia et al., 2013]											•

attitudes, though – [Broersen *et al.*, 2001b; Broersen *et al.*, 2001a]). However, sometimes a normative conflict can only be detected when the analysis occur with more than two norms simultaneously, giving rise to the necessity of checking/solving normative conflict among multiple norms.

BIBLIOGRAPHY

[Aphale *et al.*, 2012] Mukta S Aphale, Timothy J Norman, and Murat Şensoy. Goal-directed policy conflict detection and prioritisation. In *International Workshop on Coordination, Organizations, Institutions, and Norms in Agent Systems*, pages 87–104. Springer, 2012.

[Boella *et al.*, 2006] Guido Boella, Leendert Van Der Torre, and Harko Verhagen. Introduction to normative multiagent systems. *Computational & Mathematical Organization Theory*, 12(2-3):71–79, 2006.

[Boella *et al.*, 2012] Guido Boella, Silvano Colombo Tosatto, Artur D'Avila Garcez, Valerio Genovese, Alan Perotti, and Leendert Van Den Torre. Learning and reasoning about norms using neural-symbolic systems. In *Proceedings of the 11th International Conference on Autonomous Agents and Multiagent Systems-Volume 2*, pages 1023–1030. International Foundation for Autonomous Agents and Multiagent Systems, 2012.

[Broersen *et al.*, 2001a] Jan Broersen, Mehdi Dastani, Joris Hulstijn, Zisheng Huang, and Leendert van der Torre. The boid architecture: conflicts between beliefs, obligations, intentions and desires. In *Proceedings of the fifth international conference on Autonomous agents*, pages 9–16. ACM, 2001.

[Broersen *et al.*, 2001b] Jan Broersen, Mehdi Dastani, and Leendert Van Der Torre. Resolving conflicts between beliefs, obligations, intentions, and desires. In *European Conference on Symbolic and Quantitative Approaches to Reasoning and Uncertainty*, pages 568–579. Springer, 2001.

[Cholvy and Cuppens, 1995] Laurence Cholvy and Frédéric Cuppens. Solving normative conflicts by merging roles. In *Proceedings of the 5th international conference on Artificial intelligence and law*, pages 201–209. ACM, 1995.

[Cholvy and Garion, 2004] Laurence Cholvy and Christophe Garion. Answering queries addressed to several databases according to a majority merging approach. *Journal of Intelligent Information Systems*, 22(2):175–201, 2004.

[Criado *et al.*, 2010] Natalia Criado, Estefania Argente, and V Botti. A bdi architecture for normative decision making. In *Proceedings of the 9th International Conference on Autonomous Agents and Multiagent Systems: volume 1-Volume 1*, pages 1383–1384. International Foundation for Autonomous Agents and Multiagent Systems, 2010.

[Da Silva and Zahn, 2013] Viviane Torres Da Silva and Jean Zahn. Normative conflicts that depend on the domain. In *International Workshop on Coordination, Organizations, Institutions, and Norms in Agent Systems*, pages 311–326. Springer, 2013.

[dos Santos Neto *et al.*, 2011] Baldoino F dos Santos Neto, Viviane Torres da Silva, and Carlos JP de Lucena. Developing goal-oriented normative agents: The nbdi architecture. In *International Conference on Agents and Artificial Intelligence*, pages 176–191. Springer, 2011.

[dos Santos Neto *et al.*, 2012] Baldoino F dos Santos Neto, Viviane Torres da Silva, and Carlos JP de Lucena. An architectural model for autonomous normative agents. In *Advances in Artificial Intelligence-SBIA 2012*, pages 152–161. Springer, 2012.

[Elhag *et al.*, 2000] Abdullatif AO Elhag, Joost APJ Breuker, and PW Brouwer. On the formal analysis of normative conflicts. *Information & Communications Technology Law*, 9(3):207–217, 2000.

[Fenech *et al.*, 2008] Stephen Fenech, Gordon J Pace, and Gerardo Schneider. Detection of conflicts in electronic contracts. *NWPT 2008*, 34, 2008.

[Fenech *et al.*, 2009] Stephen Fenech, Gordon J Pace, and Gerardo Schneider. Automatic conflict detection on contracts. In *International Colloquium on Theoretical Aspects of Computing*, pages 200–214. Springer, 2009.

[Gaertner *et al.*, 2007] Dorian Gaertner, Andres Garcia-Camino, Pablo Noriega, J-A Rodriguez-Aguilar, and Wamberto Vasconcelos. Distributed norm management in regulated multiagent systems. In *Proceedings of the 6th international joint conference on Autonomous agents and multiagent systems*, page 90. ACM, 2007.

[Garcia *et al.*, 2013] Emilia Garcia, Adriana Giret, and Vicente Botti. A model-driven case tool for developing and verifying regulated open mas. *Science of Computer Programming*, 78(6):695–704, 2013.

[García-Camino *et al.*, 2006] Andrés García-Camino, Pablo Noriega, and Juan-Antonio Rodríguez-Aguilar. An algorithm for conflict resolution in regulated compound activities. In *International Workshop on Engineering Societies in the Agents World*, pages 193–208. Springer, 2006.

[Giannikis and Daskalopulu, 2009] Georgios K Giannikis and Aspassia Daskalopulu. Normative conflicts: Patterns, detection and resolution. *WEBIST*, pages 527—532, 2009.

[Giannikis and Daskalopulu, 2011] Georgios K Giannikis and Aspassia Daskalopulu. Normative conflicts in electronic contracts. *Electronic Commerce Research and Applications*, 10(2):247–267, 2011.

[Grossi et al., 2010] Davide Grossi, Dov Gabbay, and Leendert Van Der Torre. The norm implementation problem in normative multi-agent systems. In *Specification and verification of multi-agent systems*, pages 195–224. Springer, 2010.

[Günay and Yolum, 2013a] Akın Günay and Pınar Yolum. Constraint satisfaction as a tool for modeling and checking feasibility of multiagent commitments. *Applied intelligence*, 39(3):489–509, 2013.

[Günay and Yolum, 2013b] Akın Günay and Pınar Yolum. Engineering conflict-free multiagent systems. In *First international workshop on engineering multiagent systems (EMAS)*, 2013.

[Hill, 1987] H Hamner Hill. A functional taxonomy of normative conflict. *Law and Philosophy*, 6(2):227–247, 1987.

[Kagal and Finin, 2004] Lalana Kagal and Tim Finin. Modeling communicative behavior using permissions and obligations. In *International Workshop on Agent Communication*, pages 120–133. Springer, 2004.

[Kagal and Finin, 2007] Lalana Kagal and Tim Finin. Modeling conversation policies using permissions and obligations. *Autonomous Agents and Multi-Agent Systems*, 14(2):187, 2007.

[Kollingbaum and Norman, 2004] M Kollingbaum and T Norman. Strategies for resolving norm conflict in practical reasoning. In *ECAI Workshop Coordination in Emergent Agent Societies*, volume 2004, 2004.

[Kollingbaum and Norman, 2005] Martin J Kollingbaum and Timothy J Norman. Informed deliberation during norm-governed practical reasoning. In *International Conference on Autonomous Agents and Multiagent Systems*, pages 183–197. Springer, 2005.

[Kollingbaum et al., 2006] M Kollingbaum, T Norman, Alun Preece, and Derek Sleeman. Norm refinement: Informing the re-negotiation of contracts. In *ECAI 2006 Workshop on Coordination, Organization, Institutions and Norms in Agent Systems, COIN@ ECAI*, volume 2006, pages 46–51, 2006.

[Kollingbaum et al., 2007a] Martin J Kollingbaum, Timothy J Norman, Alun Preece, and Derek Sleeman. Norm conflicts and inconsistencies in virtual organisations. In *Coordination, organizations, institutions, and norms in agent systems II*, pages 245–258. Springer, 2007.

[Kollingbaum et al., 2007b] Martin J Kollingbaum, Wamberto Vasconcelos, Andres García-Camino, and Timothy J Norman. Conflict resolution in norm-regulated environments via unification and constraints. In *International Workshop on Declarative Agent Languages and Technologies*, pages 158–174. Springer, 2007.

[Kollingbaum et al., 2007c] Martin J Kollingbaum, Wamberto W Vasconcelos, Andres García-Camino, and Tim J Norman. Managing conflict resolution in norm-regulated environments. In *International Workshop on Engineering Societies in the Agents World*, pages 55–71. Springer, 2007.

[Kyas et al., 2008] Marcel Kyas, Cristian Prisacariu, and Gerardo Schneider. Run-time monitoring of electronic contracts. In *International Symposium on Automated Technology for Verification and Analysis*, pages 397–407. Springer, 2008.

[Li, 2014] Tingting Li. *Normative Conflict Detection and Resolution in Cooperating Institutions*. PhD thesis, University of Bath, Somerset, UK, 2014.

[Meneguzzi and Luck, 2009] Felipe Meneguzzi and Michael Luck. Norm-based behaviour modification in bdi agents. In *Proceedings of The 8th International Conference on Autonomous Agents and Multiagent Systems-Volume 1*, pages 177–184. International Foundation for Autonomous Agents and Multiagent Systems, 2009.

[Meyer et al., 1993] John-Jules Ch Meyer, Roel J Wieringa, et al. Deontic logic in computer science normative system specification. 1993.

[Oren et al., 2008] Nir Oren, Michael Luck, Simon Miles, and Timothy J Norman. An argumentation inspired heuristic for resolving normative conflict. In *Proceedings of the Fifth Workshop on Coordination, Organizations, Institutions and Norms in Agent Systems*, 2008.

[Sensoy et al., 2012] Murat Sensoy, Timothy J Norman, Wamberto W Vasconcelos, and Katia Sycara. Owl-polar: A framework for semantic policy representation and reasoning. *Web Semantics: Science, Services and Agents on the World Wide Web*, 12:148–160, 2012.

[Torres da Silva et al., 2015] Viviane Torres da Silva, Christiano Braga, and Jean de Oliveira Zahn. Indirect normative conflict. In *Proceedings of the 17th International Conference on Enterprise Information Systems-Volume 1*, pages 452–461. SCITEPRESS-Science and Technology Publications, Lda, 2015.

[Vasconcelos and Norman, 2009] Wamberto Vasconcelos and Timothy J Norman. Contract formation through preemptive normative conflict resolution. In *Dagstuhl Seminar Proceedings*. Schloss Dagstuhl-Leibniz-Zentrum für Informatik, 2009.

[Vasconcelos et al., 2007] Wamberto Vasconcelos, Martin J Kollingbaum, and Timothy J Norman. Resolving conflict and inconsistency in norm-regulated virtual organizations. In *Proceedings of the 6th international joint conference on Autonomous agents and multiagent systems*, page 91. ACM, 2007.

[Vasconcelos et al., 2009] Wamberto W Vasconcelos, Martin J Kollingbaum, and Timothy J Norman. Normative conflict resolution in multi-agent systems. *Autonomous agents and multi-agent systems*, 19(2):124–152, 2009.

[Vasconcelos et al., 2012] Wamberto W Vasconcelos, Andrés García-Camino, Dorian Gaertner, Juan A Rodríguez-Aguilar, and Pablo Noriega. Distributed norm management for multi-agent systems. *Expert Systems with Applications*, 39(5):5990–5999, 2012.

[Wieringa and Meyer, 1993] Roel J Wieringa and John-Jules Ch Meyer. Applications of deontic logic in computer science: A concise overview. *Deontic logic in computer science*, pages 17–40, 1993.

[Wright and Henrik, 1951] Von Wright and Georg Henrik. Deontic logic. *Mind*, 60(237):1–15, 1951.

[Zahn and da Silva, 2014] Jean O. Zahn and Viviane Torres da Silva. On the checking of indirect normative conflicts. In Viviane Torres da Silva, Rafael H. Bordini, and Felipe Meneguzzi, editors, *Procs. of Workshop Escola de Sistemas de Agentes, seus Ambientes e Aplicações, Porto Alegre, Brazil*, 2014.

[Zahn, 2015] Jean O. Zahn. Um mecanismo de verificação de conflitos normativos indiretos. Master's thesis, Instituto de Computação, Universidade Federal Fluminense, Rio de Janeiro, Brazil, March 2015.

4
Modeling Norm Dynamics in Multiagent Systems

CHRISTOPHER K. FRANTZ AND GABRIELLA PIGOZZI

1 Introduction and Motivation

Since multiagent systems are inspired by human societies, they do not only borrow their coordination mechanisms such as conventions and norms, but also need to consider the processes that describe *how norms come about*, *how they propagate in the society*, and *how they change over time*.

In the NorMAS community, this is best reflected in various norm life cycle conceptions that look at normative processes from a holistic perspective. While the earliest life cycle model emerged in the research field of international relations, the first life cycle model in the AI community has been proposed at the 2009 NorMAS Dagstuhl workshop by Savarimuthu and Cranefield [2009] and is based on a comprehensive survey of then existing contributions to the research field. Subsequently, two further models have been proposed that offer more refined accounts of the fundamental underlying processes.

In this article, we review all existing norm life cycle models (Section 2), including the introduction of the individual life cycle models and their contextualization with specific contributions that exemplify life cycle processes. In addition, we provide a *comprehensive contemporary overview of individual contributions to the area of NorMAS* and a *systematic comparison of the discussed life cycle models* (Section 2.6). Based on this analysis, we propose a refined *general norm life cycle model* that resolves terminological ambiguities and ontological inconsistencies of the existing models while reflecting the contemporary view on norm formation and emergence.

This comprehensive review of life cycle models represents the birds-eye perspective on dynamics in normative multiagent systems, which is complemented by research areas that operate at the intersection of normative processes captured by life cycle models. In addition to this holistic perspective, we thus discuss two active research fields that deal with norm dynamics: norm change and norm synthesis.

In human societies, norms change over time: new norms can be created to face changes in the society, old norms can be retracted either because they became obsolete or were superseded by others, and also norms can be modified. Thus, multiagent systems too need mechanisms to model and reason about norm change. The field of *norm change* (Section 3) puts a specific focus on the definition of mechanisms that describe and regulate the change of norms over time. Essential aspects include the *translation of legal to logical specifications*, the *definition of a normative approach to norm change*, and the *tuning of computational mechanisms for norm change*. This research area is rather recent and to date there is still no consensus on a common account for norm change. This section retraces the historical development and debates within this field and provides an outlook on future directions.

The second subfield, *norm synthesis* (Section 4), has a longer history that has its roots in the systems engineering domain and is concerned with the use of norms and social laws as scalable coordination mechanisms in open systems. The associated challenges are twofold and have led to the development of distinct branches, with one concentrating on the analysis of factors that mitigate the *emergence of norms or conventions*, and the second one focusing on the *identification and classification of norms* in existing normative environments. This section identifies a taxonomy of norm synthesis approaches based on a comprehensive literature overview of the field, and illustrates contemporary developments using selected contributions.

We conclude this article by contextualizing the discussed subfields with the proposed general norm life cycle model, reflecting on the progression of research on norm dynamics, and finally, by providing an outlook on contemporary and future challenges of modeling of norm dynamics.

2 Norm Life Cycle Models

In the following sections, we introduce four norm life cycle models discussed in the literature to date. The models are organized chronologically, and, with exception of the last model by Mahmoud *et al.* (Section 2.4), are of increasing complexity. The first model by Finnemore and Sikkink (Section 2.1) describes normative processes to capture the dynamics of international relations, whereas the models by Savarimuthu and Cranefield (Section 2.2), Hollander and Wu (Section 2.3), and Mahmoud *et al.* (Section 2.4) have been proposed in the research field of normative multiagent systems. Since the identified individual processes that constitute all models are supported by relevant literature contributions, we provide an updated review of associated literature. The later three models represent incremental extensions of earlier models, and, in consequence, feature redundant elementary processes. In such cases, we refer the reader to the corresponding processes in earlier life cycle models.

2.1 Model 1: Finnemore & Sikkink
2.1.1 Overview

Norms have been traditionally studied in the social sciences [Crawford and Ostrom, 1995] (see also Finnemore and Sikkink [1998], Elster [1989], Bicchieri [2006]), but no consensus yet exists on how norms emerge and are subsequently adopted in a society. In order to understand the role that norms play in international politics, Finnemore and Sikkink [1998] introduced the concept of "life cycle" to model the origin and the dynamics of norms. They claimed that norms follow a specific pattern and that each portion of the life cycle is characterized by different actors and mechanisms. The term of life cycle was later imported and became particularly relevant for the study and modeling of normative multiagent systems.

Finnemore and Sikkink's norm life cycle is a three-stage process, as shown in Figure 1: the first step is norm *emergence*, followed by norm *acceptance* (following Sunstein [1996], also called norm *cascade*), and the last stage is norm *internalization*. The move from norm emergence to norm cascade happens once the norm has been accepted by a certain amount of actors (the threshold point).

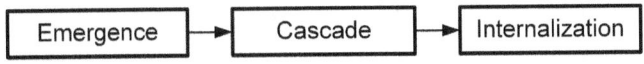

Figure 1. Finnemore and Sikkink's Norm Life Cycle Model

It is important to mention that a norm does not necessarily complete a life cycle. If, for instance, a norm does not reach the threshold point, it will not move from the emergence stage to the cascade stage. The different stages of Finnemore and Sikkink's model are supported by examples coming from women's movement of suffragettes and laws of war.

2.1.2 Stage 1: Norm Emergence

At the origin of norms we find norm *entrepreneurs*, agents committed to persuade a critical mass to support new norms or to alter existing ones in order to achieve desirable behavior in a state or community. As Hoffmann [2003] notes, leaders and entrepreneurs are not novel concepts in political science [Nadelman, 1990; Young, 1990; Schneider and Teske, 1992; Bianco and Bates, 1990]: "Entrepreneurship is a popular factor for explaining solutions to collective action problems, equilibrium choice, the emergence of cooperation as well as norms" (Hoffmann [2003], p. 8). As an example of a norm entrepreneur, Finnemore and Sikkink mention Henry Dunant, who played a crucial role in forming the norm that, in time of war, doctors and wounded soldiers should be treated as noncombatants and, by consequence, granted immunity.

The task of norm promoters is rarely easy. More often proposing a new norm implies competing with existing social contexts and established states of affairs.

This means that one has to be ready to battle with competing norms or conflicting interests. The mechanisms by which individuals manage to convince other individuals is debated [Checkel, 1998; Risse et al., 1999]. Finnemore and Sikkink argue that the difficulty of the task explains why norm entrepreneurs frequently resumed to controversial or even illegal acts (such as the protests engaged by suffragettes, who refused to pay taxes and went on hunger strikes, among other things). Altruism, empathy and commitment to an ideal are the motives that Finnemore and Sikkink attribute to norm entrepreneurs to explain their dedication.

Observing norm emergence in international relations, Finnemore and Sikkink stress that norm entrepreneurs act within organizational platforms, like nongovernmental organizations. This facilitates the reaching of the threshold point and thus the emergence of the norm. In the context of international politics, empirical studies fix such threshold around one-third of the total states, even though some states are more critical to the adoption of a norm than others. The second stage (norm cascade) is reached when the threshold is passed.

Subsequent models, like Hollander and Wu [2011a], will refine Finnemore and Sikkink's norm life cycle and will replace entrepreneurs by machine learning and cognitive approaches (Section 2.3).

2.1.3 Stage 2: Norm Cascade

We have seen that once the threshold of the critical mass is passed, according to the Finnemore and Sikkink's model, we move to the stage of norm cascade. This is called so because the acceptance rate of the new norm among the individuals increases rapidly. The mechanism that seems to govern the acceptance of a norm is *socialization*, a kind of persuasion by some agents to others to embrace a certain norm. In the case of states, such persuasion appears to lean against the need of a state to be recognized as a member of an international organization. In other words, exactly as it happens to people, countries would be exposed to peer pressure. In particular, the desire to acquire or increment internal and international legitimation, the pressure of conformity and the need for norm leaders to increase their esteem seem to be the reason to respond to such a pressure.

2.1.4 Stage 3: Internalization

If a norm reaches the third and last stage, it becomes internalized. This means that such norm is acquired and not object of debate anymore. As Epstein stated, once a norm is accepted, people "conform without really thinking about it" (Epstein [2001], p.1). Examples of nowadays internalized norms are the abolition of slavery or the right to vote for women. But internalized norms can also be specific to certain professions. Finnemore and Sikkink mention the examples of doctors and soldiers, who become acquainted with different "normative biases": "Doctors are trained to value life above all else. Soldiers are trained to sacrifice life for certain strategic goals" (Finnemore and Sikkink [1998], p.905).

2.1.5 Discussion

Constructivists (to which Finnemore and Sikkink's approach belongs) have been criticized for failing to account how entrepreneurs hammer new norms or come to propose the alteration of existing ones, as well as how they manage to convince other critical agents in their vision. Hoffmann [2003] partially addresses such criticisms by building an agent-based model to explore the role of norm entrepreneurs. His model does not tackle the question of how entrepreneurs convince other agents, but focuses "on the unexamined assumption that a persuasive entrepreneur can influence the outcomes that arise from the interactions of heterogeneous, interdependent agents" (Hoffmann [2003], p. 13). His model shows that the constructivist's hypothesis of the role of norm entrepreneurs is indeed plausible. In particular, his aim is to understand under what conditions a norm entrepreneur can function as a norm catalyzer for the emergence of new norms and the alteration of existing ones. Norm entrepreneurs turn out to be able to influence norm emergence even when they can reach only a small portion of the population (around 30%), and their influence increases with their reach. Hoffmann's model suffers (as the author himself acknowledges) from some limitations, like the assumption of a unique norm entrepreneur, the lack of communication among agents, agents' power is not modeled, and only non-complex norms are considered.

2.2 Model 2: Savarimuthu & Cranefield

2.2.1 Overview

The first life cycle model for norms we have encountered was proposed in the context of international relations. As we have seen, Finnemore and Sikkink [1998] directed their attention to human societies and to processes that can explain how norms emerge and spread within and among states. Ten years separate Finnemore and Sikkink's work from the second model we consider here, the life cycle model proposed by Savarimuthu and Cranefield [2009; 2011].

Savarimuthu and Cranefield's model comes from the study of simulation-based works on norms in the context of multiagent systems. By looking at the various mechanisms employed by the researchers working on simulation on norms, they extend the three-stage model of Finnemore and Sikkink.

Savarimuthu and Cranefield's contribution came in two papers: the first one [Savarimuthu and Cranefield, 2009] presented a four phases norm life cycle (*norm creation, spreading, enforcement and emergence*), whereas the subsequent one [Savarimuthu and Cranefield, 2011] included one additional stage (*identification*). For this reason, in the present section we will focus on the latter, more recent, contribution. For each step Savarimuthu and Cranefield provide a categorization of the mechanisms that have been employed in the simulation-based works on norms, as shown in Figure 2.

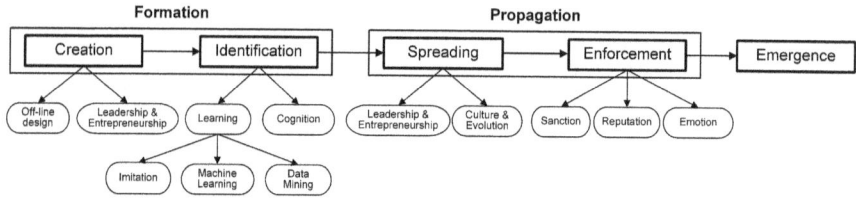

Figure 2. Savarimuthu and Cranefield's Norm Life Cycle Model

2.2.2 Norm Creation

Unlike Finnemore and Sikkink [1998], who acknowledged only the role of norm entrepreneurs for the creation of norms, Savarimuthu and Cranefield [2011] realize that in the context of multiagent systems, norms can be created by three different approaches: off-line design, norm leaders and norm entrepreneurs. In off-line design the norm is introduced by an external designer and is hard-wired into the agents. Norm leaders, on the other hand, are powerful agents of the system that (following a democratic or an authoritarian process) create norms for the other agents to follow. Finally, norm entrepreneurs are not necessarily norm leaders. Similarly, as seen in Finnemore and Sikkink's model, an entrepreneur can propose a new norm that he thinks is beneficial to the society. But until the entrepreneur does not succeed to persuade the other agents to accept such norm, the norm is not a social norm.

Off-line Design One of the most well-known works in the area of off-line design is Shoham and Tennenholtz [1995]'s work on synthesising social laws, specifically in the traffic domain. In this specific context, off-line design implies that mobile robots (as traffic participants) are initialized with a set of traffic laws ('rules of the road') that have been computed at design time in order to prevent collisions at runtime. Such rules allow to minimize the need of a central coordinator on the one hand and that of a negotiation mechanism among agents on the other hand. Traffic laws provide the agents with a set of social laws that help them avoiding collisions. A multi-robot grid system is considered, where m robots can move on an $n \times n$ grid. Shoham and Tennenholtz suggest one can imagine rows and columns of that grid as lanes in a supermarket. In order to avoid the collision between robots (which happens when more than one robot occupies the same coordinate), some traffic laws are given. For instance, one may impose that in odd rows agents can move only right, in even rows they can move only left, in odd columns they can move only down and in even columns the only movement possible is up. Rules define also the priority when two or more robots approach a junction and how robots can change their movement direction Shoham and Tennenholtz [1995]'s work has subsequently been extended (e.g. to consider the minimality of social laws [Fitoussi and Tennenholtz, 2000]) and has found various adaptations in works on norm emergence (e.g. [Sen and Airiau, 2007; Mukherjee *et al.*, 2007]).

A similarly influential model from the sociological domain is Conte and Castelfranchi [1995b]'s evaluation of norms for the purpose of aggression control to facilitate cooperation in a stylized food-gathering society. In their model societies are selectively initialized as either *strategic* or *normative*, where strategic agents systematically attack fellow food-carrying agents, whereas normative ones accept a notion of possession, thus promoting a higher level of survival at the macro level. Results have shown that normative populations do better than strategic ones. However, in mixed populations strategic agents do much better than the normatives. The reason is that non-normative agents benefit from the behavior of normatives.

Castelfranchi *et al.* [1998] have further extended the model to consider the role of reputation (see Section 2.2.5). The role of reputation is considered also in Hales [2002], which extended Castelfranchi *et al.*'s food-consumption problem by assigning agents to the group of normative agents or to the group of cheaters.

Walker and Wooldridge [1995] observe that the simplicity of off-line design models comes at a price. To be truly beneficial, such approach requires that all characteristics of a system should be known a priori (which is not the case for open systems, for example). Another difficulty is that it is extremely costly and time-consuming to constantly reprogram agents, which is required in case agents' goals change, as it happens in complex systems. Moreover, Savarimuthu and Cranefield [2011] note that it is not realistic to assume that all agents follow a given norm.

Leadership and Entrepreneurship Mechanisms Leaders are agents who have the social power and abilities to persuade other agents to accept a norm. Leadership mechanisms have been employed for norm emergence and norm spreading (see Section 2.2.4).

Verhagen [2001] proposes a model in which individuals have varying degrees of decision-making autonomy in normative settings. Agents interact with a normative advisor that influences agents' decisions to follow or abandon norms. Once an agent decides to follow a specific norm, it announces this decision to the whole society. The normative advisor, as well as other agents, can send feedback to each other. To reflect the hierarchical social structure and promote norm convergence, greater weight can be placed on the information received from the leader, which is a member of the agent society itself. Earlier work by Boman [1999] operationalizes norms as constraints for individuals' decision-making, and likewise has the role of a normative advisor. In contrast to Verhagen, Boman's normative advisor is external to the agent society.

In Savarimuthu *et al.* [2008a] a society can have several normative advisors (or role models) who give advice to agents who are their followers. Agents are connected to each other through one social network topology among fully connected networks, random networks and scale-free networks. The interesting twist is that an agent can be at the same time a role model for some agents and a follower of some other agent. Since several norm leaders can exist, different norms can emerge in the society.

Norm entrepreneurs were notably introduced in Finnemore and Sikkink's norm life cycle model, presented in Section 2.1. Hoffmann [2003] has experimented on the notion of norm entrepreneurs, as seen in the Discussion subsection 2.1.5.

2.2.3 Norm Identification

The first norm life cycle model proposed by Savarimuthu and Cranefield [2009] consisted of four stages (norm creation, spreading, enforcement and emergence). The idea being that, as in [Finnemore and Sikkink, 1998], once a norm is created, it may spread in a society if certain conditions are satisfied. However, in [Savarimuthu and Cranefield, 2011], they added the *identification* step between norm creation and spreading. Such step is needed in all those situations in which a norm has not been explicitly created, for example when a norm results from the interaction process among agents. In those cases, agents have first to be able to identify the created norms. Simulation-based works on norms have explored two approaches for norm identification: agents can learn new norms by imitation, machine learning or data mining mechanisms; alternatively, agents can use their cognitive abilities to infer and recognize the norms of a system.

Learning Mechanisms – Imitation Among the simulation models that experimented on learning mechanisms based on imitation is that of Epstein [2001]. Using a driving setting in which agents can observe whether other agents (within a certain radius) drive on the right or on the left, Epstein showed that agents conform to the driving preferences of the majority of the observed agents. Imitation mechanisms can explain the identification and the spreading of a norm.

Yet, some authors, like López y López and Márquez [2004] as well as Campenní *et al.* [2009], cast some doubts on the claim that such mechanisms can explain the co-existence of different norms in a group of agents. Instead of seeing norms are hard-wired in the agents, Campenní *et al.* [2009] imagine the interaction between agents coming from different societies. Their goal is to investigate the role of cognition in norm recognition: How do agents tell that something is a norm? In their model, there are four scenarios, some actions that are context-specific and one action that is common to all scenarios. In one set of simulations, agents can change contexts, whereas in another set of simulations, at a certain moment, agents must stay in the context they have reached and can interact only with agents that are in the same context (imagine a situation in which a population is split into two groups and each group is constrained to not have contacts with the other group). The purpose of this second set of simulations is to show that frequency may be a sufficient (but not necessary) condition for agents to converge to the same action. Results show that new norms can emerge, eventually giving rise to the competition between two rival norms.

Learning Mechanisms – Machine Learning Shoham and Tennenholtz [1992a] employed co-learning, a reinforcement learning mechanism that makes an agent choose the strategy that revealed to be the most successful in the past. They showed that norm emergence decreases with the decrease of the frequency of the updates of an agent's strategy. The efficiency of norm emergence turned out to decrease also with the increase of an agent's memory flush.

Building on the scenario introduced in [Conte and Castelfranchi, 1995b], Walker and Wooldridge [1995] ran 16 experiments with different parameters for the size of the majority and the update function (the latter could depend on the majority rule, on the memory flush or on communication mechanisms). Results showed that the network topology and communication may play an important role and, hence, more simulations are needed to better understand mechanisms for norm emergence.

More recently, norm emergence has been investigated using social learning in a model in which agents repeatedly interact with other agents by Sen and Airiau [2007]. Experiments took into account different population sizes, various learning strategies, and number of available actions. The specific situation is that of learning of which side of the road to drive on but also the problem of who has the priority if two agents gain a junction at the same time. The outcomes confirm that such a mode of learning is a robust mechanism for the emergence of social norms.

Learning Mechanisms – Data Mining An approach to norm identification that uses association rule mining to identify obligation norms is Savarimuthu et al. [2010b]'s *Obligation Norm Inference* (ONI) algorithm. Such model enables agents to sense their environment, memorize experiences and observations as well as normative signals, which build the basis for the identification of personal norms (p-norms) and group norms (g-norms). The memorized event episodes are then mined for obligation norms using association rules algorithms. The agent-based simulation experiment considers a virtual restaurant in which agents may not know whether the restaurant expects the customers to order and pay for the food at the counter before eating or if they are expected to order, consume the food and pay only before leaving. Another protocol agents may need to identify is the tipping norm: in some countries, for example, tipping is expected (in the USA, for instance), whereas in others (like most countries in Europe) it is not expected. The difficulty in identifying an obligation norm is that a sanction is triggered by the absence of an action (a customer in a restaurant may be sanctioned if he is not tipping the waiter). Savarimuthu et al. [2013a] propose a corresponding approach for the identification of prohibition norms.

Savarimuthu and Cranefield [2011] observe that data mining is a promising approach. However, explicit signals for sanctions or reward have to be present in order for norms to be easily identified.

Cognition The EMIL-A architecture [Andrighetto *et al.*, 2007; Campenní *et al.*, 2009; Andrighetto *et al.*, 2010][1] is a cognitive architecture to explore how agents' mental abilities may explain the acquisition of new norms. Reinforced candidate norms are identified from observed normative information (represented as normative frame) that traverses different memory layers, representing the transition from short-term experiences to long-term memory. Once established, normative beliefs are held in a *Normative Board*, along with associated normative action plans. These internalized normative beliefs inform the agent's goal generation, decision-making and action planning. The previously discussed work by Savarimuthu *et al.* [2010b] also proposed an architecture for agents to identify norms using agents' cognition abilities.

2.2.4 Norm Spreading

Once a norm has been explicitly created or agents have identified it, the norm can start being spread in the society. Among the different mechanisms that can serve this purpose, there are leadership and entrepreneurship that we already encountered in the norm creation stage, but also cultural and evolutionary mechanisms.

Culture and Evolution Cultural and evolutionary mechanisms have been considered in [Boyd and Richerson, 1985; Chalub *et al.*, 2006]. According to Boyd and Richerson [1985] social norms can be propagated along three types of transmissions: vertical, horizontal and oblique. *Vertical relationships* describe the intergenerational transmission of norms by parents to offspring, whereas *horizontal transmission* occurs among peers of a given generation. *Oblique relationships* combine the former two and describe the unidirectional dissemination of norms by authority figures towards their contemporary subalterns. Vertical relationships are constrained to the intergenerational sharing of norms which makes them particularly applicable to evolutionary models such as Axelrod's norm game [Axelrod, 1986]. Horizontal approaches assume a uniform social structure, which limits this approach to abstract group or society representations, as is the case for large parts of the norm emergence work (e.g. [Sen and Airiau, 2007; Villatoro *et al.*, 2011a; Mihaylov *et al.*, 2014; Airiau *et al.*, 2014]; Section 4). The last relationship type lends itself well to model inter- and intra-generational norm transmission for comprehensive society representations that consider power and authority structures. Examples for this include Franks *et al.* [2014]'s use of *Influencer Agents* to drive the norm convergence, or Yu *et al.* [2015]'s hierarchical approach to information sharing.

Savarimuthu and Cranefield [2011] note that if cultural and evolutionary mechanisms can explain how a norm is spread, they cannot answer the question of how a norm is internalized in the first place.

---

[1] The contribution in Campenní *et al.* [2009]is a notable extension of Andrighetto *et al.* [2010]'s work.

2.2.5 Norm Enforcement

The existence of a norm presupposes that such norm can be violated. Norm enforcement mechanisms serve to deter agents from violating a norm. This can be done through punishment, via some mechanisms that negatively affect the reputation of a norm violator, or again by affecting the agent's emotions (for example, by instilling a sense of guilt in the norm violator). Savarimuthu and Cranefield [2011] stress that norm enforcement can also play a role in the spreading process of a norm. Observing the punishment of a norm violator can either discourage other agents from violating that norm or identifying that norm, in case it was not explicitly created.

Sanctions The most well-known work on external sanctions is Axelrod [1986]'s norm game that specifically explores the notion of metanorms, i.e. the sanctioning of non-sanctioning observers of violations, to sustain a society's norm.[2] An essential challenge of normative regulation (in artificial systems as in real life) is the balance of cost and effect of sanctions, both to minimize the cost of enforcement, while maximizing the effect in order to regulate behavior effectively [Axelrod, 1986; Horne, 2001; Savarimuthu et al., 2008a]. Mahmoud et al. [2012; 2015] refine Axelrod's model by investigating the effect of dynamic punishment, and ultimately propose an alternative to Axelrod's evolutionary approach based on individual learning to produce a model in which norms can stabilize within a given generation.

In López y López [2002; 2003] a model where agents have goals and different personalities is developed. Punishments and rewards are considered only when they affect an agent's goals.

Reputation A positive or negative opinion about one agent from the interacting agents in a society can play a substantial role in the norm compliance in a group of agents.

In Castelfranchi et al. [1998]'s and Younger [2004]'s models, ostracism is an implicit result of reputation sharing, which leads to the exclusion of individuals from future interaction. In particular, Castelfranchi et al. [1998]'s game reconsiders Conte and Castelfranchi [1995b]'s stylized food-gathering society seen in Section 2.2.2, with the addition of normative reputation: agents learn the reputation of other agents, that is, they learn whether an agent is normative or strategic (i.e. a cheater). However, in order to be profitable, the information about cheaters must be communicated to other agents. In the context of multiagent systems Perreau de Pinninck et al. [2010] propose a distributed mechanism that affords the

---

[2] Axelrod's contribution was impressive and extremely influential. However, it should be noted that Galan and Izquierdo [2005] have shown that his results are not stable. When running many more simulations of Axelrod's model and for longer, opposite results can be obtained. As the authors also stress, one should not forget that their analysis required computational power which was not available when Axelrod proposed his model.

isolation of violating nodes in the context of peer-to-peer applications. They evaluate its properties for various network topologies.

Emotion Staller and Petta [2001] introduce an extension of the cognitive agent architecture JAM [Huber, 1999] with components to augment the rational agent model with emotion appraisal processes, an aspect considered essential to mediate any form of norm enforcement [Scheve et al., 2006]. Fix et al. [2006] propose a model of normative agents that include the display of emotional responses to normative actions. In this work the agents' internal states are represented using reference nets [Valk, 1998], a variant of Petri nets.

2.2.6 Norm Emergence

Once a norm has spread across a certain proportion of the society (according to different simulation results, the minimum required is a third of the population), it is said that the norm has emerged. This implies that a significant proportion of the population recognizes and follows that norm. It is worth noticing, however, that such process can be reverted. A norm may lose its appeal in a group and is hence either abandoned, replaced or modified by a competing one.

No specific category of empirical work on norms is associated with norm emergence. However, there is one category whose impact is notable across all stages of norm development. This is the consideration of network topology, as described in the Transmission part in Section 2.3.2.

2.2.7 Discussion

Savarimuthu and Cranefield [2011]'s life cycle model is an extension of the life cycle introduced by Finnemore and Sikkink in [Finnemore and Sikkink, 1998]. There are, however, two main differences.

The first one is that, whereas Finnemore and Sikkink's model was thought for human societies, Savarimuthu and Cranefield direct their attention to normative multiagent systems and to simulation studies of norms using software agents. The second difference is that Savarimuthu and Cranefield not only capture two additional steps in their model, but also that for each phase, they consider more mechanisms.

2.3 Model 3: Hollander & Wu

2.3.1 Overview

To date, the most complex norm life cycle model has been proposed by Hollander and Wu [2011a]. Their model refines the ones initially introduced by Finnemore and Sikkink [1998] (Section 2.1) and Savarimuthu and Cranefield [2011] (Section 2.2), resulting in a total of ten *normative processes*, namely *creation, transmission, recognition, enforcement, acceptance, modification, internalization, emergence, forgetting,* and *evolution*. In contrast to the earlier models, Hollander and Wu identify three superprocesses (*enforcement, internalization,* and *emergence*) that combine elementary processes and characterize their high-level function. Note that the superprocess labels are borrowed from the most essential elementary process out of all processes they combine. A further novelty is the interpretation of *emergence* as an iterative process, and *evolution* as a metaprocess the authors refer to as "end-to-end process" [Hollander and Wu, 2011a]. The schema in Figure 3 provides a systematic overview of the complete life cycle. Where existing, the superprocesses are represented as boxes comprising their elementary processes, with the corresponding superprocess label highlighted in bold font. We will briefly outline the entire life cycle before introducing the individual processes in detail.

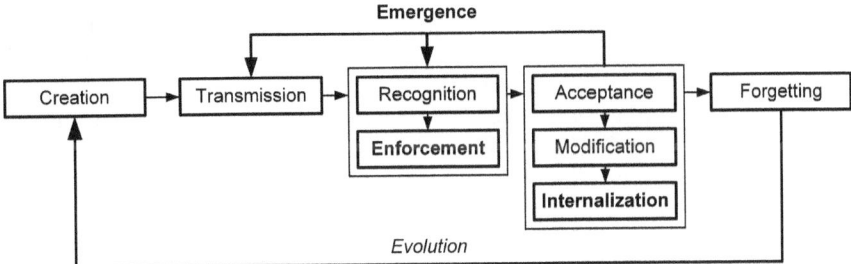

Figure 3. Hollander and Wu's Norm Life Cycle Model

Initially, potential norms are explicitly created, before being transmitted to the wider society, and rely on recognition and enforcement processes (captured in the superprocess *enforcement*) to promote their adoption. The superprocess *internalization* involves the decision whether to accept a norm, potentially modifying it, and finally, internalizing it, and thus becoming an enforcer of the norm itself. The subsequent cyclic reinforcement of the norm, including transmission, enforcement and internalization (tagged *emergence*), determines whether the initial *potential norm* becomes a norm. If attaining normative status, norms undergo a continuous refinement that requires reiteration through the elementary processes to gain salience. Any norm modification, such as the adaptation to new circumstances, implies that some normative content is forgotten. Swipe-card payments for bus services, for example, make it increasingly permissible for individuals to enter buses through arbitrary doors, instead of requiring the traditional entry through specific doors for payment. Contrasting the gradual forgetting of normative content, norms can be superseded by alternative norms, in which case the original norm is forgotten in its entirety. For example, over the past decades in many Western countries the general tolerance towards smoking in public places has been progressively replaced with general rejection.

In the following, we will discuss selected processes in greater detail and contextualize those with the earlier life cycle models as well as recent developments.

2.3.2 Life Cycle Processes

Creation Similar to Savarimuthu and Cranefield [2011], Hollander and Wu acknowledge that norm creation involves a wide range of different processes, including methods found in the natural world [Boella *et al.*, 2008; Finnemore and Sikkink, 1998; López y López *et al.*, 2007; Savarimuthu and Cranefield, 2009] - such as spontaneous emergence from social interaction, decree by an agent in power, or negotiation within a group of agents. However, in the context of work on NorMAS, Hollander and Wu identify two primary methods of norm creation, namely off-line design [Conte and Castelfranchi, 1995a; Shoham and Tennenholtz, 1995] and autonomous innovation [Hollander and Wu, 2011a]. While off-line design assumes that experimenters create the norms a priori and inject those into instantiated agents, autonomous innovation (akin to 'on-line design') assigns the role of norm creation to agents themselves.

Notable works in the area of off-line design include Shoham and Tennenholtz [1995] and Conte and Castelfranchi [1995b], as already discussed in Section 2.2.2.

Autonomous innovation covers a broader range of approaches, ranging from the adoption of specific strategies to the challenging problem of ideation, namely giving agents the ability to produce novel ideas without external input.

In contrast to previous models' norm creation mechanisms in the form of norm leadership/entrepreneurship (see Sections 2.1 and 2.2), Hollander and Wu [2011a] identify two types of mechanisms used for autonomous innovation, namely:

- Game-theoretical and machine learning approaches (e.g. Sen and Airiau [2007], Mukherjee *et al.* [2007], Perreau de Pinninck *et al.* [2008], Urbano *et al.* [2009], Sen and Sen [2010], Savarimuthu *et al.* [2010b]), and

- Cognitive approaches (e.g. Savarimuthu *et al.* [2010b], Andrighetto *et al.* [2007]).

Even though many models use a combination of those mechanisms,[3] their application tends to serve distinctive purposes. Game-theoretical approaches emphasize the identification of optimal strategies from a set of given strategies, thus representing an incremental step from off-line design towards autonomous norm innovation. Machine learning is generally used in conjunction with game-theoretical approaches, mostly to represent a notion of memory (e.g. Sen and Airiau [2007], Mukherjee *et al.* [2007]).

Essential work that combines game-theoretical and machine learning approaches is the research field of *norm emergence* or *convention emergence*. This field concentrates on the identification of factors that promote high convergence levels for

---

[3]Examples for combining game-theoretical and machine learning approaches are provided by Sen and Airiau [2007] and Mukherjee *et al.* [2007]; an example for the combined use of machine learning and cognitive approaches is Savarimuthu *et al.* [2010b]'s work.

norms within the observed society. While decision-making itself is modeled as some form of game (with 'rules of the road' [Shoham and Tennenholtz, 1995] as the preferred coordination game), agent components such as memory are represented using machine learning (commonly reinforcement learning in the form of Q-learning [Watkins and Dayan, 1992]). Depending on the aspect of interest, the model is augmented with additional mechanisms to investigate the influence of memory (e.g. Villatoro *et al.* [2009]), characteristics of network topologies and structural dynamics (e.g. Savarimuthu *et al.* [2007], Villatoro *et al.* [2009], Sen and Sen [2010], Villatoro *et al.* [2013]), norm transmission mediated by social learning (e.g. Sen and Airiau [2007], Mukherjee *et al.* [2007; 2008], Airiau *et al.* [2014]), as well as adaptive sanctioning (e.g. Mahmoud *et al.* [2012; 2015]).

Sen and Airiau [2007], for example, let agents engage in social interaction in the context of the 'rules of the road' scenario (described in Section 2.2), in which cars approach an unregulated intersection and have to identify an optimal coordination mechanism, such as 'yield to the right', and prevent deadlocks (both cars yield) or collision.[4] Agents memorize past encounters and adjust their behavior based on the success of their action. As part of their evaluation, Sen and Airiau explore different population sizes, action spaces and learning algorithms to show how agent societies can autonomously arrive at stable norms.

Further approaches investigate the influence of hierarchical structures on the distribution of norms (e.g. Franks *et al.* [2013; 2014], Yu *et al.* [2013; 2015]).[5]

While work in the area of norm emergence concentrates on the interactions and corresponding macro-level outcomes, cognitive approaches concentrate on the mechanics of normative agent architectures. Cognitive norm architectures contextualize perceived behavior with existing beliefs to infer normative content and/or consider normative beliefs in their deliberation process. Approaches of this kind generally consider more complex forms of learning. They further invoke semantically rich norm representations and processes that come closest to what we can describe as *ideation* [Ehrlich and Levin, 2005], i.e. proposing behaviors that potentially qualify as normative, and selectively filtering those.

Representative works that apply cognitive approaches include the *Beliefs-Obligations-Intentions-Desires* (BOID) architecture [Broersen *et al.*, 2001; Broersen *et al.*, 2002] which extends the widely adopted Belief-Desire-Intention (BDI) architecture [Bratman, 1987; Rao and Georgeff, 1995] with an obligation component that preempts the goal generation and prioritizes the individuals' obligations. In this approach, obligations are statically embedded in an agent's belief base.

While BOID emphasizes normative reasoning, alternative approaches propose mechanisms to facilitate norm identification and decision-making, along with the

---

[4] We will come back to this scenario in greater detail in Section 4, given of its relevance in the area of norm synthesis.
[5] We will discuss the field of norm emergence in more detail in Section 4.

involved micro-/macro-level interaction, as in the cognitive architecture EMIL [Andrighetto *et al.*, 2007; Campenní *et al.*, 2009; Andrighetto *et al.*, 2010], that extends the BDI concept with the ability to acquire new norms, which we discussed in Section 2.2.3.

Cognitive approaches, such as Savarimuthu *et al.*'s norm identification frameworks for the detection of obligation [Savarimuthu *et al.*, 2010b] and prohibition norms [Savarimuthu *et al.*, 2013a], rely on notions of machine learning to afford realistic agent representations [Savarimuthu *et al.*, 2011; Ossowski, 2013]. Further examples for the combined use of cognitive and machine learning components include the identification of normative content from action and/or event sequences (e.g. Savarimuthu *et al.* [2010a]), the implementation of alternative learning mechanisms beyond experiential learning or 'learning by doing', such as social/observational learning [Bandura, 1977] as applied by Epstein [2001], Hoffmann [2003], as well as Sen and Airiau [2007]. Another combined use of cognitive and machine learning is to facilitate the use of direct communication (e.g. used by Verhagen [2001] as well as Walker and Wooldridge [1995]).

Transmission The norm transmission process in Hollander and Wu's model (equivalent to the spreading process in Savarimuthu and Cranefield [2011]'s model), considers three components that characterize how information is spread. Those include:

- the nature of *Agent Relationships*,

- the applied *Transmission Techniques*, and

- the underlying *Network Structure*.

Agent Relationships Similar to Savarimuthu and Cranefield [2011], Hollander and Wu share Boyd and Richerson [1985]'s observation of relationship types as either being vertical, horizontal or oblique, an aspect we discussed in the context of Savarimuthu and Cranefield's model (Section 2.2.4).

Transmission Techniques Beyond the identification of relationships, Hollander and Wu [2011a] identify two transmission techniques for norms, the first being *active transmission* in which norms are actively broadcast throughout the relationship networks. Alternatively, agents can use *passive transmission* and absorb perceived normative information. Examples of mechanisms to facilitate active transmission include direct communication, whereas observation of the social environment (on the part of a norm recipient) is an example of passive transmission.

In most simulation works, active transmission is used to convey normative content by direct communication or in the form of sanctions. Examples include Hoffmann [2005], who uses proactively communicating norm entrepreneurs to promote convergence, as well as the previously mentioned work by Franks *et al.* [2013], or

Yu et al. [2010; 2015]'s use of *supervisors* to model hierarchical communication between networked multiagent systems. Further examples from the sociological domain include Castelfranchi et al. [1998]'s and Younger [2004]'s society models that rely on reputation sharing for the purpose of promoting cooperation.

Examples of passive communication are used to represent notions of imitation or social learning. Examples include Verhagen [2001]'s work on norms learning, as well as the work on the impact of social learning on norm convergence (e.g. Nakamaru and Levin [2004], Sen and Airiau [2007], Airiau et al. [2014]) and synthesis (e.g. Frantz et al. [2015]). An example of the use of passive transmission in social scenarios is Flentge et al. [2001]'s representation of imitation by copying memes from successful neighbors.

Network Structure The third aspect of norm transmission is the nature of the underlying connectivity structure that acts as an information transport medium. Depending on the objective, the connectivity structure is conceived as a multi-dimensional grid environment or as network topology of varying complexity.

In grid environments, agents are stationary or mobile, and observe agents within their specified neighborhoods, and can, depending on their neighborhood configuration, perceive adjacent cells. Agents' grid environments are generally modeled as von Neumann neighborhoods – in which agents can sense orthogonally adjacent cells – or Moore neighborhoods – in which agents can sense all adjacent cells.

The modeling of norm transmission via network structures permits the configuration of more complex relationship networks, with network topologies of equal degrees of connectedness (e.g. as fully connected networks), as well as random connectivity (random networks [Erdős and Rényi, 1959]). Alternatively, networks can display varying degrees of connectedness, such as small world networks [Watts and Strogatz, 1998] that simulate sparse links between communities characterized by dense internal connectedness. Scale-free network topologies [Barabási and Albert, 1999] work on the far end of the spectrum and produce a structure characterized by power law distributions, with individuals being centered around densely-connected hubs.

In analogy to the stationary or mobile configuration in a grid environment, a further important aspect is whether network topologies are static or dynamic at runtime. Effects of complex network topologies on norm emergence have been explored by Zhang and Leezer [2009], Franks et al. [2014], and Sen and Sen [2010]. Villatoro et al. [2009] put specific emphasis on the interaction between memory size and the chosen topology, whereas Airiau et al. [2014] concentrate on the effect of social learning across different topologies. Savarimuthu et al. [2007] and Villatoro et al. [2011a; 2013] explore the effect of dynamic topologies on norm emergence.

Recognition In Hollander and Wu's model, the processes *creation* and *transmission* are followed by the superprocess *enforcement* that consists of the subprocesses *recognition* and *enforcement* (see Figure 3). Norm recognition is similar to Savarimuthu and Cranefield's account of *norm identification* and describes the agent's ability to recognize the norms enacted in the observed society or group. Means to do so include communication with norm participants (as is the case with human societies [Henderson, 2005]) as well as observational learning. Similar to technological approaches in the context of norm creation, earlier models relied on off-line identification of agents as norm followers and deviants (e.g. Castelfranchi *et al.* [1998], Hales [2002]), whereas recent models apply more sophisticated mechanisms to identify norms, which include machine learning [Sen and Airiau, 2007; Mukherjee *et al.*, 2007; Savarimuthu *et al.*, 2013b; Frantz *et al.*, 2015] and/or cognitive approaches [Savarimuthu *et al.*, 2010b; Andrighetto *et al.*, 2007]. Since the recognition of norms may involve the observation of sanctions, it is closely related to enforcement.

Enforcement Norm enforcement describes the application of sanctions to stimulate adherence to the normative content. Sanctions can be positive (in the form of rewards) or negative in nature and can further be differentiated by their source, that is whether they originate from internal or external sources.

For this purpose Hollander and Wu differentiate three types of enforcements:

- Externally Directed Enforcement
- Internally Directed Enforcement
- Motivational Enforcement

Externally Directed Enforcement Externally directed enforcement describes the sanctioning by an outside observer that witnesses and reacts to a norm violation or an agent's refusal to accept a transmitted norm (e.g. a follower rejecting a leader's imposed norm) [Flentge *et al.*, 2001; Galan and Izquierdo, 2005; Savarimuthu *et al.*, 2008b].

Applied sanctions can be of economic nature (e.g. reducing or limiting access to resources), affect the violator's reputation (e.g. shunning, ostracism) [Axelrod, 1986; Castelfranchi *et al.*, 1998; Hales, 2002; Younger, 2004] (as seen in Section 2.2.5), or prevent it from propagating deviance to others (e.g. by preventing procreation in the case of vertical norm transmission [Flentge *et al.*, 2001]) [Caldas and Coelho, 1999].

The prototypical example for external sanctions is Axelrod's norm game [Axelrod, 1986], as discussed in Section 2.2.5 in the context of Savarimuthu and Cranefield's life cycle model.

Internally Directed Enforcement Sanctions of internal origin rely on an individual's self-enforcement triggered by the violation of internalized norms. The prototypical mechanism for internally motivated norm enforcement is the activation of emotions (discussed in greater detail in Section 2.2.5).

Motivational Enforcement Hollander and Wu further identify the notion of motivational enforcement, which is essentially a special case of internally directed enforcement. It describes the implicit commitment of all individuals to follow system-wide norms if they are aligned with an individual best interest, an aspect understood as conventions [Lewis, 1969]. A classical example is the convention of uniform road side use: the precise strategy (i.e. whether to drive on the left or right side) is secondary to the complete acceptance and internalization by the society since unilateral deviation produces suboptimal outcomes (i.e. accidents caused by ghost drivers).

Internalization Essential processes for norm emergence in Hollander and Wu's model are associated with the superprocess norm internalization. Hollander and Wu differentiate between *Acceptance*, *Modification*, and *Internalization* (as the terminating subprocess of the superprocess *Internalization*).

The acceptance of enforced norms is the starting point for the internalization of norms by individuals and decisive for the emergence of norms, since individuals either decide to accept or reject socially imposed norms based on the compatibility with their personal beliefs, desires and intentions. Possible outcomes are the acceptance of a new norm, the substitution of an existing conflicting norm, or its rejection. Acceptance is operationalized as some form of cost-benefit analysis [Meneguzzi and Luck, 2009].

If agents decide to accept norms, their integration into the internal cognitive structures requires the transformation of norms from an objectified outside perspective to a subjective representation that involves an individual's biases, inaccuracies of perception, etc. This potentially leads to a modified understanding of that norm, an aspect that affects the norm during its further progression in the life cycle.

Finally, the accepted and potentially modified norm is internalized by the receiving agent. Compared to the other stages of the norm life cycle, this process has found limited explicit attention. In most applications, individuals simply adopt the accepted norms without further refinement or adaptation. From a motivational perspective, this is compatible with measures that suggest that the absence of external pressures is indicative of complete norm internalization [Epstein, 2001]. However, this view only accounts for subsequent norm adherence, but cannot explain violations further down the track. Refined approaches evaluate the effect of the internalized norm and on the decision-making process. An important example is Verhagen [2001]'s work, in which agents seek increasing alignment with their associated group by comparing and internalizing corresponding action probabilities. Alternatively, as done in the BOID architecture [Broersen *et al.*, 2001], internalized norms can be maintained separately from personal strategies and activated selectively depending on situation-specific autonomy values [Broersen *et al.*, 2002].

In their original survey, Hollander and Wu [2011a] highlighted the limited explicit focus on internalization, especially in comparison to life cycle processes such as enforcement. However, recent works in the area of NorMAS reveal more explicit treatments of internalization, generally in the form of continuous probabilistic adaptation of strategy choices based on reinforcement learning (e.g. Salazar *et al.* [2010], Villatoro *et al.* [2013], Franks *et al.* [2014], Airiau *et al.* [2014], Frantz *et al.* [2014b; 2015], Yu *et al.* [2015]), or by using thresholds for the adoption of new strategies (e.g. Hollander and Wu [2011b], Mihaylov *et al.* [2014]). In Section 2.6 we provide a comprehensive overview of internalization mechanisms used in works on normative multiagent systems.

Emergence In contrast to all earlier models, Hollander and Wu conceive emergence as a dynamic macro-level process that describes a cyclic iteration involving the transmission of the internalized norm to new participants. This is followed by enforcement (based on the subprocesses *Recognition* and *Enforcement*) to drive the internalization (composed of subprocesses *Acceptance*, potential *Modification*, and *Internalization*) of the norm by new subjects, who themselves participate in the spreading of the norm – ultimately leading to the norm's emergence as a macro-level phenomenon. This emergence understanding is aligned with Savarimuthu and Cranefield's, who interpret emergence as the final stage of the norm life cycle, but do not explicitly reflect the cyclic reinforcement of norms by reiterating through the formation stage. Finnemore and Sikkink's life cycle model maintains a different emergence interpretation and associates emergence with the micro-level creation of a norm, e.g. via entrepreneurship, before sharing and penetrating the wider society.

The exploration of emergence characteristics is strongly tied to the applied modeling technique. Game-theoretical approaches evaluate emergence by identifying stabilizing strategy choices (equilibria) chosen from a set of given alternative strategies. The dominant strategy choice is then interpreted as the emergent norm (see e.g. Axelrod [1986], Mukherjee *et al.* [2007], Zhang and Leezer [2009]). Since agents are represented as structurally uniform selfish rationalizers with a minimal action repertoire, the exploration is focused on macro-level outcomes. Cognitive approaches, on the other hand, do permit a macro-level observation of specific norms, but furthermore, allow a more realistic reconstruction of micro-level processes. This includes detail and diversity of individuals' cognitive structures, the precise level and nature of enforcement (see e.g. Caldas and Coelho [1999], Savarimuthu *et al.* [2008b]), the use of richer norm representations, diverse action sets, and a variety of norm learning mechanisms (e.g. based on experiential learning, social learning and direct communication) [Savarimuthu *et al.*, 2011].

Models can further address infrastructural aspects, such as the impact of different connectivity structures on normative outcomes. Related findings suggest that scenarios in which normative behavior is transmitted from neighbors (e.g. in grid environments) tend to result in the dominance of a single norm, whereas individualized learning promotes the emergence of diverse normative configurations [Boyd and Richerson, 1985; Boyd and Richerson, 2005; Nakamaru and Levin, 2004]. While the application of network structures can lead to stronger normative diversity, experimental results suggest that the impact of the actual network topology is secondary to its dynamic nature (as opposed to static networks) [Bravo *et al.*, 2012]. However, the convergence of conventions (and emergence of local subconventions) can be controlled by maintaining links to distant nodes [Villatoro *et al.*, 2009].

Forgetting & Evolution In contrast to the earlier models by Finnemore and Sikkink as well as Savarimuthu and Cranefield, Hollander and Wu are the first to complete the norm life cycle by explicitly considering the process of *Forgetting*. In this conception forgetting is essential to sponsor the evolutionary refinement of norms, since continuously changing norm contexts may render existing norms irrelevant. An example is the normalized use of smart devices in education, with proactive integration of social media platforms such as Facebook into the learning environment. This is in opposition, or at least in competition, to traditional norms that ban the use of mobile devices in classroom environments. Once forgotten, norms make space for new norms that are better adapted to environmental needs, which constitutes the end-to-end process that closes the evolutionary loop of the norm life cycle.

2.3.3 Discussion

As mentioned at the outset of this section, this model proposed by Hollander and Wu introduces the to date most comprehensive life cycle model. The model not only considers abstract high-level processes (superprocesses), but decomposes those into elementary processes that capture large parts of contemporary research and, beyond this, identify gaps in normative agent architectures (such as the explicit consideration of *Norm Acceptance*) to produce more comprehensive representations of human reasoning processes. In addition to the fine-grained nature, this model further deviates from the linear operation of previous models by identifying emergence as a metaprocess that links individual processes and results in a continuous iteration through elementary processes. Beyond the 'completion' of the life cycle by considering the abandoning of norms, a further essential novelty is the consideration of norm evolution as a continuous process that affords both the modification and the substitution of norms over time.

2.4 Model 4: Mahmoud et al.

Overview The latest life cycle model has been proposed by Mahmoud *et al.* [2014b]. Similar to the earlier life cycle models developed in the context of NorMAS, their work is based on a comprehensive literature review, both considering individual works as well as previous life cycle models. In contrast to Hollander and Wu's detailed model, their approach identifies five core processes (*Creation, Emergence, Assimilation, Internalization, Removal*) with a further decomposition of selected processes as shown in Figure 4. Since this model has only been briefly described by the original authors themselves and strongly builds on concepts introduced in the context of Hollander and Wu's earlier, more detailed model, we provide a concise overview at this stage, before discussing the novel contributions in more detail.

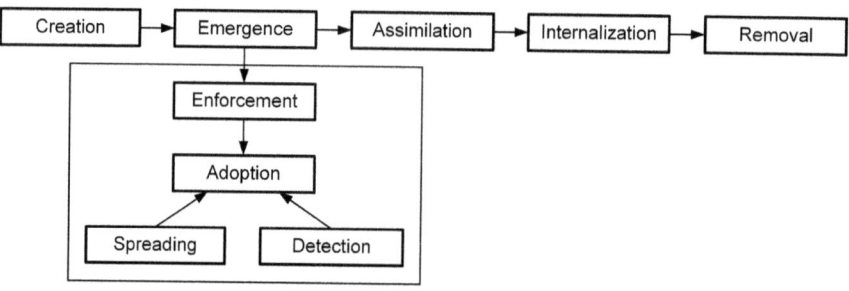

Figure 4. Mahmoud et al.'s Norm Life Cycle Model

Processes The initial process, as with most other life cycle models is *Creation*, which operates based on mechanisms described by Savarimuthu and Cranefield [2011], namely off-line design, autonomous innovation and social power (see Section 2.2).

A central deviation from previous models is the process of *Emergence*, which Mahmoud *et al.* decompose into two individual processes, *Norm Enforcement* and *Norm Adoption*. The latter of those is further decomposed into the processes *Norm Detection* and *Norm Spreading*. Unlike Hollander and Wu's model, emergence is considered a sequential process.

In Mahmoud *et al.*'s model, *Enforcement* consists of direct and indirect sanctioning, where direct sanctioning is the conventional application of reward or punishment, whereas indirect sanctioning is reflected in an individual's reputation and emotions (e.g. guilt).

The *Adoption* process is a composite process that consists of the spreading of new norms and the detection of norms. The *Spreading* process captures the transmission directions outlined by Savarimuthu and Cranefield [2011] (vertical, horizontal and oblique). The *Detection* of new norms captures all forms of norm learning to identify new norms, including imitation, social learning, case-based reasoning and data mining. The model further emphasizes the essential nature of network topologies to facilitate the spreading of norms, including the differentiation of static and dynamic networks, but does not consider alternative mechanisms such as sensing in grid-based environments.

Following the *Emergence* process, the model introduces a novel *Assimilation* process. The authors follow Eguia [2011]'s definition of assimilation "as the process in which agents embrace new social norms, habits, and customs, which is costly but offers greater opportunities" ([Mahmoud *et al.*, 2014b], p.15). In their conception, assimilation involves deciding whether to adopt new social norms by trading off associated costs and benefits.

This next process is the *Internalization* that, similar to Hollander and Wu's conception, includes the *Acceptance*, *Transcription* and *Reinforcement* of the newly acquired norm, with the purpose of embedding it in the agent's behavior.

The final *Removal* process is equivalent to Hollander and Wu [2011a]'s process of forgetting norms. The purpose is the removal of obsolete norms, as well as being an implicit consequence of norm modification. Mahmoud *et al.* further adopt an unspecified end-to-end process that links *Removal* and *Creation*, possibly implying the evolutionary process introduced by Hollander and Wu.

Discussion The model by Mahmoud *et al.* breaks the trend of proposing progressively more detailed models and attempts to identify the essential processes instead. This condensed conception produces an incoherent understanding of the norm life cycle and semantic ambiguities, the causes of which we will explore in the following section.

Despite the authors' awareness of previous models, in this model emergence only considers the enforcement and adoption of norms (which captures aspects such as spreading and detection), but does not consider the internalization of norms essential for their emergence. How norms can emerge without being internalized is left unexplained. This leaves unclear whether internalization is implied as part of the *Adoption* process that concentrates on spreading and detection of norms. If this were the case, this would produce an ambiguous understanding of the subsequent internalization process.

A similar problem relates to the novel *Assimilation* process, which represents the authors' own substantive contribution [Mahmoud *et al.*, 2014a] to the field of NorMAS. Since assimilation describes the process of deciding whether to adopt given norms, it is unclear in how far this is different from the *Acceptance* process that is part of norm internalization [Mahmoud *et al.*, 2014b], or if it is meant to replace the acceptance component of internalization. The authors' related contribution [Mahmoud *et al.*, 2014a] discusses the assimilation of norms in heterogeneous communities and suggests that the norm internalization itself is a *subprocess* of norm assimilation, an aspect that is not reflected in the sequential organization of both processes in the life cycle model (see Figure 4). The inspection of the authors' related work suggests that assimilation not so much describes a norm-centered life cycle process. Instead, it characterizes an agent's capability since it describes the *ability and willingness of agents to integrate into their social envi-*

ronment [Mahmoud *et al.*, 2014a], which entails the adoption of norms, customs, habits, etc.

Overall, the model attempts to rationalize the existing norm life cycle models, leading to a refined but insufficiently specified and contextualized life cycle model, specifically with respect to the emergence process as well as the assimilation component – aspects that challenge its coherence and, in consequence, applicability.

2.5 Comprehensive Literature Overview

In the previous sections, we introduced the most relevant life cycle models known in the literature and discussed associated significant contributions. Table 1 integrates the mentioned literature into a comprehensive chronological overview that spans across selected life cycle processes.[6] Whereas the process characteristics of creation, identification, spreading, and enforcement are based on the criteria and approaches discussed in the context of the individual life cycle models (specifically in Sections 2.2 and 2.3), this overview puts particular focus on capturing internalization mechanisms and emergence characteristics, both of which have found limited recognition in previous surveys.

Earlier works on norm internalization apply the specification of norms at design time, which occurs in conjunction with off-line norm creation (which we labeled 'embedded'). However, in the majority of contributions, the adoption and internalization of norms generally occur unreflected (labeled 'immediate'). In more recent approaches, we can observe a shift towards more continuous internalization of norms based on observation ('social learning') as well as probabilistic or threshold-based adoption based on sustained reinforcement ('threshold-based learning', 'Q-learning').

Another category that is characterized by a range of varying, often scenario-dependent measures is the notion of emergence. Examples include convergence thresholds on shared equilibrium strategies in the case of coordination games. In alternative approaches emergence refers to the alignment of sets of norms, both including crisp (e.g. Campenní *et al.* [2009], Andrighetto *et al.* [2010], Griffiths and Luck [2010]) and fuzzy set conceptions (e.g. Frantz *et al.* [2014b; 2016]), or the identification of a shared normative understanding, e.g. by election (Riveret *et al.* [2014]) or by generalization (Frantz *et al.* [2015]). Another group of approaches interpret emergence as the convergence on shared conceptualizations of lexica (e.g. Salazar *et al.* [2010], Franks *et al.* [2013]).

---

[6]This overview refines and extends an earlier survey produced by Savarimuthu and Cranefield [2011].

Table 1. Chronological Overview of Literature and Associated Life Cycle Characteristics

Publication	Creation	Identification	Spreading	Enforcement	Internalization	Emergence
Axelrod [1986]	-	-	vertical	Sanctions	immediate	Converging strategy choice
Kittock [1995]	Off-line design	Machine learning	-	Sanctions	memorizing strategy	Converging strategy choice
Conte and Castelfranchi [1995b]	Off-line design	Machine learning	-	-	embedded	Survival under different strategies
Walker and Wooldridge [1995]	Off-line design	Machine learning	-	-	-	-
Shoham and Tennenholtz [1992b; 1995]	Off-line design	Machine learning	-	-	-	Converging strategy choice
Shoham and Tennenholtz [1997]	Off-line design	-	-	-	-	-
Castelfranchi et al. [1998]	Off-line design	-	-	Sanctions	immediate	-
Saam and Harrer [1999]	Off-line design	-	oblique	Reputation	embedded	Social alignment of action probabilities
Verhagen [2001]	Leadership	-	oblique	Sanctions	embedded	Converging on action choice
Epstein [2001]	-	Machine learning	horizontal	Leader/group feedback	alignment with group	-
Flentge et al. [2001]	-	-	vertical	Sanctions	imitation	-
Hales [2002]	-	-	-	Reputation	inherited	Converging on chosen value
Hoffmann [2003]	Off-line design	-	oblique	Reward	immediate	Converging on state
Delgado [2002; 2003]	Entrepreneurship	-	horizontal	Payoff	immediate	-
López y López and Luck [2004]	Off-line design	Machine learning	-	Sanction/Reward	-	Stabilising norms
Nakamaru and Levin [2004]	-	Machine learning	horizontal	-	immediate	Converging strategy choice
Pujol et al. [2005]	-	Machine learning	-	Reputation	immediate	Converging norms
Chalub et al. [2006]	-	Machine learning	vertical	Emotion	immediate	-
Fix et al. [2006]	-	-	-	Sanction/Reward	immediate	-
López y López et al [2006; 2007]	Off-line design	Machine learning	-	-	immediate	Converging strategy choice
Sen and Airiau [2007]	-	Machine learning	-	Payoff	immediate	Converging strategy choice
Mukherjee et al [2007; 2008]	-	Machine learning	-	Payoff	immediate	Converging on value
Savarimuthu et al [2007; 2008a]	-	Cognition, social learning	-	Payoff	Learning	Shared event-action trees
Campenni et al [2009; 2010]	-	-	oblique	Payoff	immediate	Converging on shared norm
Savarimuthu et al. [2009]	-	-	oblique	Payoff	immediate	Converging towards joint strategy
Urbano et al. [2009]	-	Machine learning	-	Reward	immediate	Converging action choice
Villatoro et al. [2009]	-	Machine learning	horizontal	-	immediate	Converging towards joint action
Sen and Sen [2010]	-	-	vertical	-	immediate	Converging on multiple norms
Griffiths and Luck [2010]	-	Cognition, data mining	-	Sanction signal	immediate	Identification of event sequences as norms
Savarimuthu et al [2010b; 2010a; 2011]	Off-line design	-	-	Ostracism	-	Minimal norm violations
Perreau de Pinninck et al. [2010]	-	-	horizontal	-	probabilistic	Convergence on shared word/concept lexicon
Salazar et al. [2010]	-	-	oblique	Payoff	social learning	Convergence on shared goal state
Yu et al. [2010]	-	Machine learning	horizontal	Payoff	Q-learning	Convergence on non-conflicting conventions
Sugawana [2011]	-	Machine learning	horizontal	Payoff	social learning	-
Villatoro et al. [2011b]	-	-	horizontal	internal	threshold-based learning	Equilibrium strategies (various scopes)
Hollander and Wu [2011b]	-	Machine learning	oblique	dynamic sanctions	-	Converging action choice
Mahmoud et al [2012; 2015]	-	-	horizontal	Sanction	Learning	-
Riveret et al [2012; 2013]	-	Machine learning	-	-	-	-
Savarimuthu et al. [2013b]	-	Machine learning	horizontal	Payoff	Q-learning	Converging action choice
Villatoro et al. [2013]	-	Evolutionary algorithm	oblique	Leadership	probabilistic	Convergence on shared word/concept lexicon
Franks et al. [2013]	-	Machine learning	oblique	Sanction	immediate	Synthesized set of norms
Morales et al [2013; 2014; 2015b]	-	Machine learning	horizontal	Payoff	threshold-based learning	Converging strategy choice
Mihaylov et al. [2014]	-	Machine learning	horizontal	-	Q-learning	Converging action choice
Airiau et al. [2014]	-	Machine learning	-	Payoff	Q-learning	Alignment of fuzzy normative understanding
Frantz et al [2014b; 2016]	-	Machine learning	oblique	Payoff	Q-learning	Shared normative understanding
Riveret et al. [2014]	-	Machine learning	horizontal	Sanction/Reward	Q-learning	Multi-level norm generalization
Frantz et al [2014c; 2015]	-	Machine learning	oblique	-	Q-learning	Converging strategy choice
Yu et al. [2015]	-	-	-	-	Q-learning	-
Beheshti et al. [2015]	-	Cognition	horizontal	Sanction/Reward	Q-learning	Norm-conforming behavior

2.6 Systematic Comparison of Norm Life Cycle Models

To this stage, we have introduced a diverse set of life cycle models along with associated contributions, but have yet to relate those systematically. Finnemore and Sikkink [1998]'s model (Section 2.1), proposed in the field of international relations, identifies three processes in a norm's life, starting with its explicit creation (*Emergence*), its spreading (*Cascade*), leading to wide-ranging adoption (*Internalization*). In contrast to all other models, their model looks at states as central players and emphasizes the long-term perspective of normative change (e.g. embedding the changing societal normative view in professional ethics).

The remaining three ones are products of systematic reviews of contemporary research in the area of NorMAS, an approach spearheaded by Savarimuthu and Cranefield [2011]. Their model (Section 2.2) provides a refined account of the beginning of a norm's development, with a particular focus on the initial formation and propagation. Their model interprets emergence as an outcome measure and does not include a long-term perspective on norms, such as their decay and substitution over time.[7] However, since their model is grounded in a systematic review of existing works, this does not indicate a principle shortcoming of the model, but rather reflects the contemporary state of the research field.

Hollander and Wu [2011a]'s model (Section 2.3) provides the most comprehensive account of norms' life cycles, and, similar to Savarimuthu and Cranefield's grouping of processes into stages, identifies essential superprocesses that are composed of refined subprocesses. Their model goes beyond previous accounts and proposes processes that are only weakly reflected in literature, thus identifying presumed research gaps. The most important contribution of their model is the recognition of cycles of recurring processes, an example of which is the characterization of norm emergence as a reiteration of transmission, enforcement and internalization. The second essential contribution is the integration of a long-term perspective on normative change, which they reflect as an evolution process.

Finally, Mahmoud *et al.* [2014b] (Section 2.4) describe a model that condenses the number of relevant processes of the normative life cycle to five. Their model puts specific emphasis on norm assimilation, i.e. an individual's decision whether to accept (and subsequently internalize) a given norm. They further decompose the emergence process into enforcement and adoption (which in itself consists of the processes *Norm Spreading* and *Norm Detection*). Similar to Finnemore and Sikkink, as well as Savarimuthu and Cranefield, Mahmoud *et al.* conceive a linear norm life cycle; they do not consider iterative processes.

An aspect that challenges the systematic comparison of all four models is not only the varying level of detail, but the observable terminological ambiguity. In the different life cycle models the sharing or spreading of norms is selectively captured by the terms 'cascade' (Finnemore and Sikkink), 'spreading' (Savarimuthu

---

[7]They consider those as part of a refined set of life stages in later work [Savarimuthu *et al.*, 2013b].

and Cranefield, Mahmoud *et al.*), and 'transmission' (Hollander and Wu). A further notable example is the norm 'identification' (Savarimuthu and Cranefield) that is alternatively characterized either as 'recognition' (Hollander and Wu) or 'detection' (Mahmoud *et al.*).

Beyond those synonyms, the specific processes in different models have semantic overlappings. To facilitate a systematic comparison of content and semantic relationship, in Figure 5 we provide an overview of all life cycle models, with individual processes roughly aligned by semantic relationship. Process labels are formatted and grouped to reflected their nature and importance in the respective life cycle model:

- Savarimuthu and Cranefield differentiate between individual processes and stages. Consequently, the *life cycle stage names* are held in bold font.

- Hollander and Wu's *superprocess labels* are held in bold font. The emergence and evolution processes are further explicitly included in the schematic overview.

- Mahmoud *et al.*'s model composes the emergence process from two elementary processes and is thus held in bold font, along with all further *processes of the same conceptual weight*.

Dotted lines indicate the semantic relationships between individual processes of the corresponding life cycle models. For example, Finnemore and Sikkink's *Cascade* process combines components of Savarimuthu and Cranefield's *Spreading* and *Enforcement* processes.

Despite the diversity of norm life cycles, the systematic review of all models reveals clusters of processes that have similar or identical functions (identified as solid horizontal lines in Figure 5). We can generalize four such clusters, or phases, of norm life cycles, and label those by complementing the labels of the initial two life cycle stages in Savarimuthu and Cranefield [2011]'s model:

- Formation – Processes associated with the creation and inference of norms

- Propagation – Processes associated with the communication of norms

- Manifestation – Processes associated with the general acceptance and entrenchment of norms

- Evolution – Processes associated with the evolutionary refinement of norms

The identified phases correspond to the abstract phases proposed by Andrighetto *et al.* [2013], namely *Generation*, *Spreading*, *Stability* and *Evolution*, an aspect that supports the semantic process clusters proposed above. Our terminological

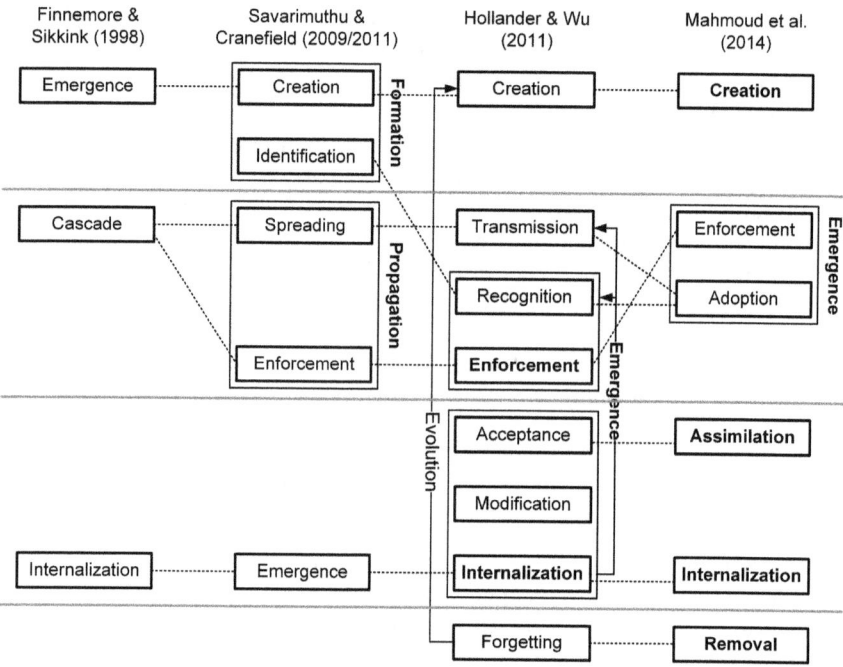

Figure 5. Schematic Comparison of Discussed Norm Life Cycle Models

choice is driven by the goal to comprehensively capture the semantics of associated processes of all discussed norm life cycle models (e.g. operations associated with norm internalization extend beyond the characterization of a norm as stable – see discussion below). In the following, we will use the identified phases to compare and contextualize the norm life cycle models.

Phase 1: Formation All models identify norm creation as the initial life cycle step. In contrast to all other models, Finnemore and Sikkink [1998] employ a different emergence understanding. In their conception emergence entails the initial creation of a norm (which Hollander and Wu [2011a] describe as "norm creation on a micro scale" [Hollander and Wu, 2011a]), whereas life cycle models from the area of NorMAS (henceforth referred to as NorMAS models) understand emergence as "norm establishment on a macro scale" [Hollander and Wu, 2011a]. However, the underlying understanding of this initial phase – the explicit creation of a norm – is identical for all models. Despite this uniform characterization, we label this phase as *Formation* in order to capture a more general understanding of norm creation, widening the scope to approaches that do not rely on explicit norm creation such as the identification of existing/unknown norms by observation, an aspect implicitly captured by Savarimuthu and Cranefield's notion of *Norm Identification* (which we discuss in Section 4).

Phase 2: Propagation Following the creation, all models describe some sort of norm communication, or propagation (*Cascade, Spreading, Transmission,* and *Adoption*). A special case is Mahmoud *et al.* [2014b]'s *Adoption* process, which entails both norm spreading and detection. All NorMAS models recognize a notion of norm identification (*Identification, Recognition,* and *Adoption*), but have a varying sequential organization. While Savarimuthu and Cranefield's early allocation of norm identification is driven by the understanding that agents need to identify norms in their environment, all subsequent models interpret it as a step that follows the transmission of a norm. Similarly, all NorMAS models recognize enforcement as an essential determinant of a norm's success.

Phase 3: Manifestation The propagation of norms is followed by their *Internalization*. In Finnemore and Sikkink's model that refers to the wide-ranging adoption of a norm within society and its embedding in societal institutional structures. In addition to gaining stability, at this stage norms thus manifest themselves in the social fabric which implicitly reinforces their persistence, constrains future action, but also limits the potential of competing norms. Manifested norms can attain quasi-legal status, e.g. by shaping the codes of ethics for specific occupations, which are subsequently absorbed into the discipline's professional training and practices. This understanding is compatible with Savarimuthu and Cranefield's *Emergence* interpretation, which represents the extent to which a norm is able to penetrate the affected society.

While these first two models describe norm manifestation as a macro-level process, the models of Hollander and Wu as well as Mahmoud *et al.* describe refined sets of micro-level processes that lead to the internalization of norms. Hollander and Wu differentiate between *Acceptance, Modification,* and *Internalization*, including the decision whether to adopt a norm in the first place, and further take into account individual biases introduced during internalization. Mahmoud *et al.* reduce those to two processes, namely *Assimilation* and *Internalization*. As discussed in Section 2.4, the authors borrow the notion of *Acceptance* (which is identical to Hollander and Wu's *Acceptance*[8]), and consider it part of the *Internalization* process. However, they introduce a preceding *Assimilation* process[9] (whose function is not clear, since it is insufficiently contrasted to *Acceptance*) and *Internalization*. At the end of this manifestation phase, all models assume that

---

[8]"Norm acceptance is a conflict resolution process in which external social enforcements compete against the internal desires and motivations of the agent. If the new norm is in conflict with existing norms and may lead to inconsistent behaviors, or if the cost of accepting the new norm is too high, it will be rejected ..." [Hollander and Wu, 2011a], paragraph 3.24.

[9]"[Norm assimilation is] ... the process in which agents embrace new social norms, habits and customs, which is costly but offers greater opportunities." [Mahmoud *et al.*, 2014b], p.15 with reference to Eguia [2011].

individuals have embraced the promoted norms.

Phase 4: Evolution The fourth phase which we tag *Evolutionary Phase* is only reflected in the later life cycle models which introduce the processes *Forgetting* and *Removal* that reflect the end of the normative life cycle. However, more important than their function to 'complete' the norm life cycle is their role as starting point for an evolutionary process (as introduced by Hollander and Wu [2011a]; Section 2.3) in which norms are refined or substituted by more relevant or efficient norms; forgetting old norms is a by-product of this evolutionary refinement and technical necessity to maintain efficient but also realistic architecture implementations.

The 'Special Case' Emergence Only exception to the uniform organization of processes into general phases is the notion of emergence, which reflects the terminological ambiguity surrounding this concept. Whereas Finnemore and Sikkink's micro-level interpretation of emergence is associated with the *Formation Phase*, Mahmoud *et al.* see the *Propagation Phase* with the processes of *Enforcement* and *Adoption* as decisive for emergence. Hollander and Wu see emergence as an iterative process that spans across *Formation* and *Manifestation Phase*. Savarimuthu and Cranefield associate emergence with the third phase of norm manifestation and interpret it as a result of *Formation* and *Propagation*.

We believe that Hollander and Wu's cyclic representation represents the most accurate characterization of the emergence process, since it links the macro-level emergence process with the underlying propagation and internalization processes, an aspect we will revisit in the context of proposed refinements (see Section 2.7). Savarimuthu and Cranefield's interpretation as outcome measure only reflects a quantifiable macro-level phenomenon, but does not maintain its relationship to the underlying processes that produce it. Mahmoud *et al.* inherently rely on propagation processes to determine a norm's emergence. Their model neither considers the cyclic nature of emergence nor does it consider the internalization of norms as a precursor for their further spread (see discussion in Section 2.4).

Norm Life Cycle Models and Levels of Analysis Comparing the individual models leaves the general impression that later models (with exception of Mahmoud *et al.*) are increasingly detailed and comprehensive. However, while this observation is warranted, it rather reflects the operational levels the life cycle models represent. Finnemore and Sikkink's, as well as Savarimuthu and Cranefield's models, describe the adoption and implementation of norms on the macro level, i.e. group or society level. This is well captured in Finnemore and Sikkink's understanding of internalization as the process of embedding the norm in a society's social structures and institutions. Similarly, Savarimuthu and Cranefield describe emergence as a macro-level outcome that describes the adoption of a norm across the wider society. Hollander and Wu's model introduces a shift from the macro-level norm perspective to an individual-centered micro-perspective, an aspect that is particularly apparent in the elementary processes they describe in the context of the establishment phase. Micro-level processes include *Acceptance* (the decision whether or not to accept norms), *Modification* (the modification of norms during internalization based on individual biases), and finally *Internalization*, which describes an individual's integration of norms into its existing belief structure. Only the subsequent *Emergence* and *Evolution* processes operate on the macro level, since they shift the perspective from individual to society level. Mahmoud *et al.*'s model similarly emphasizes individual-level processes such as *Assimilation* and *Internalization*, which they decompose into operational steps that are similar to Hollander and Wu's processes (Mahmoud *et al.* : *Acceptance, Transcription, Reinforcement*; Hollander and Wu: *Acceptance, Modification, Internalization*). In both models forgetting and removal of norms emphasizes a micro-level operation and is considered a technological necessity (in the light of limited computational resources), but obscures the macro-level function of facilitating an evolutionary refinement [Hollander and Wu, 2011a] of the normative landscape.

Understanding the different operation levels of the introduced models is helpful, since it allows their selective consultation. For the modeling and analysis of macro-level phenomena, the use of Savarimuthu and Cranefield's model may provide sufficient conceptual backdrop, whereas detailed cognitive agent models will find the most comprehensive structural blueprint in Hollander and Wu's model, with other models providing even higher levels of abstraction (Finnemore and Sikkink) or varying emphasis of individual-level processes (Mahmoud *et al.*).

2.7 General Norm Life Cycle Model

As a result of reviewing the existing life cycle models and their respective biases, we propose a general life cycle model that harmonizes various inconsistencies of the introduced approaches (e.g. micro- vs. macro-level operation, emergence understanding), but also addresses explicit conceptual omissions that are of increasing importance in recent developments (see Sections 3 and 4).

As such, the proposed general norm life cycle model introduces five essential revisions, which we discuss in the following:

- Distinction between micro-level processes and macro-level phenomena

- Norm Identification as an alternative life cycle entry point (in addition to explicit norm creation)

- Enforcement as a dynamic process with norm emergence as a resulting phenomenon

- Norm Forgetting as by-product of norm evolution

- Potential norm modification throughout all life cycle processes

Distinction between Micro-Level Processes and Macro-Level Phenomena As discussed in great detail in the previous Section 2.6, the existing norm life cycle models operate on varying levels of abstraction, with the initial models identifying coarsely-structured processes, whereas the latter two models describe processes of varying granularity (e.g. Hollander and Wu's end-to-end processes, superprocesses in addition to regular processes). We propose a systematic distinction by separating the micro-level processes (e.g. Transmission, Identification and Internalization) that find explicit representation in normative architectures, from macro-level phenomena that arise from the cyclic operation of the underlying processes. We believe that differentiating between a processual and phenomenological perspective on norms is useful to inform modeling considerations in different problem domains, such as the engineering of a process-centric normative agent architecture, in contrast to macro-level processes such as the emergence of norms within agent societies or their evolution over time. However, at the same time, these perspectives should not be dissociated in order to retain the links between the phenomena and the underlying processes. *Norm Emergence* is thus a result of iterative Transmission, Identification, Internalization and Enforcement processes. *Norm Evolution* extends across the entire norm life cycle, additionally involving the inception of new norms (Norm Creation) as well as the forgetting of decaying norms (Norm Forgetting).

Norm Identification as a Life Cycle Entry Point To date, the existing approaches assume the explicit creation of a norm. Proposed mechanisms include norm leadership, entrepreneurship, autonomous innovation and social power. However, in reality, norms may not necessarily be explicitly created, of unknown origin, but be rooted from behavioral regularities based on individuals' necessity to act in the first place (described as "urgency of practice" [Bourdieu, 1977]). In principle, a situational strategy choice to coordinate behavior (e.g. chosen means of greeting, roadside choice) can emerge as self-enforcing convention (without intentional explicit conceptualization), before finding recognition as a fully fledged norm.[10] Previous works acknowledge the existence of natural emergence processes[11] (Boella *et al.* [2008], Finnemore and Sikkink [1998], López y López *et al.* [2007], Savarimuthu and Cranefield [2009]), but assume an explicit creation as the starting point of the normative life cycle. We propose that a comprehensive norm life cycle should reflect the unplanned inception of norms based on social interaction as a possible alternative starting point of a norm's life – in addition to the explicit creation.

Enforcement as a Dynamic Process A further aspect relates to the role of enforcement. All NorMAS life cycle models represent enforcement as an explicit process that appears independent of notions such as spreading. However, enforce-

---

[10] Examples for works that showcase this characteristic (e.g. Morales *et al.* [2015a], Riveret *et al.* [2014], Frantz *et al.* [2015]) are discussed in the context of the upcoming Section 4.

[11] Here emergence should be understood as the micro-level process of norm inception.

ment itself can be interpreted as a dynamic process that promotes the cyclic reinforcement of norms, leading to their spread and thus their increasing adoption, producing emergence as an associated phenomenon (as discussed in the previous paragraph). Some form of enforcement – whether implicitly (e.g. serving as a guiding role model or influence based on shared values) or explicitly (e.g. by engaging in overt sanctioning) – is a prerequisite for the transmission of norms. In this context, it is further important to note that enforcement does not carry a specific valence, but can bear positive associations, such as providing a reward for a norm-compliant employee, or represent an explicit punishment, such as humiliating an individual in front of her reference group (e.g. an employee amongst fellow co-workers). Apart from such forms of overt *external enforcement*, enforcement can further be directed at oneself (internal enforcement), reflected in emotions such as the "warm glow" [Andreoni, 1989] of compliance (i.e. 'doing the right thing') or the guilt of violation (e.g. engaging in jaywalking despite conventional compliance).

Whether implicit or explicit, positive or negative, internal or external, enforcement relies on the prior internalization by the potential enforcer. This does not necessarily imply that the enforcer applies this norm to her- or himself or even 'believes' in it. As such, individuals can be tasked with the enforcement or feel pressured to defend norms they object to (such as not engaging in jaywalking in the presence of bystanders). Similarly, not violating a norm when facing the opportunity (without actively promoting it) can act as norm reinforcement. An example for this is the rejection of a bribe, especially if the actor holds a role model function (e.g. as a manager) [Hogg, 2001]. Conversely, the observation of violation by an authority figure (e.g. taking a bribe) can accelerate norm erosion. Whether compliant or not, essential for any positive or negative enforcement is some internalized conceptualization of the enforced norm in order to make its compliance and violation detectable. Consequently, we do not see emergence as a process in itself, but as a phenomenon that results from a sustained cyclic reinforcement based on the transmission, identification, internalization, and subsequent enforcement of norms, leading to their manifestation.

Forgetting as a By-Product of Norm Evolution A final aspect relates to the notion of forgetting. Hollander and Wu introduce forgetting as an end point of an evolutionary cycle that affords a norms refinement. However, the conceptualization as an 'end-to-end process' presents it as a sequential step in a series of processes. Similar to the conception of emergence laid out before, we see evolution as a phenomenon that arises from the continuous reinforcement of norms, their change during identification and internalization, as well as their potential to become obsolete and ultimately forgotten. This process cannot be conceived as sequential but operates concurrently, with newly identified norms gaining more salience and potentially leading to existing norms' adaptation or decay. Though forgetting is an essential endpoint in the normative life cycle, it does not represent the starting point for a continuously operating evolution process; 'forgetting' is a by-product of evolving norms.

Schematic Overview In Figure 6 we show a schematic overview of the proposed refined norm life cycle that condenses elements of the previously introduced models, but incorporates essential revisions. We will briefly explore the processes in the following.

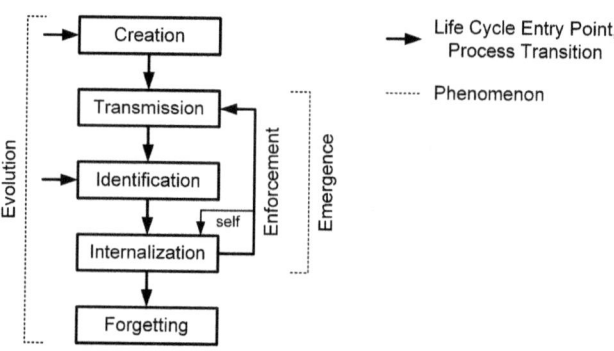

Figure 6. General Norm Life Cycle

As stated before, norms can either be explicitly created or identified at runtime (the corresponding right-facing arrows in Figure 6 mark these life cycles starting points). If created, norms are transmitted and identified.[12] As mentioned above, identification is not only initiated by transmission, but may involve the identification of an existing norm (e.g. by observation). Once internalized (by a complex internalization process that may contain elementary processes as laid out by Hollander and Wu [2011a]), norms can be reinforced, which may operate internally (e.g. based on motivational enforcement or elicited emotions), or be directed towards external targets. External enforcement requires the transmission of normative content, the subsequent identification and internalization by enforcement targets, and so on. This constitutes the norm's emergence. At any time, new norms can be created or identified, potentially causing change in the normative system by emerging and becoming salient. If cyclic reinforcements of a given norm cease, the norm loses its relevance and is incrementally forgotten. This second phenomenon can be understood as norm evolution. Both, emergence and evolution, are similar in that they represent phenomena (and could be construed as meta-processes in the epistemological sense[13]), but they vary in scope regarding the involved processes.

---

[12] Note that we use terms synonymously for the ambiguous terminology in existing life cycle models as discussed before. In this case, the notion of 'identification' is identical to 'recognition'.

[13] Our interpretation is in contrast to Hollander and Wu (Section 2.3) who use the term to describe

Norm Modification throughout Norm Life Cycle Hollander and Wu [2011a] discuss the modification of norms as part of the internalization process. However, we believe that the potential for norm modification, whether intentionally and systematic or not, arises during *any* form of transmission, internalization, or subsequent externalization (e.g. enforcement) of normative content.

This can involve the loss of information during transmission or simply transmission errors, leading to partial or simply wrong information. For example, ambient traffic noise may prevent bystanders from perceiving the scolding of jaywalkers or lead them to misconstrue the normative content (e.g. as a heated discussion).

Complementing potential modification sources during transmission, the identification of norms can be challenged by sensory biases that lead to a modified reproduction of normative content. Visual impairment, for example, may challenge or prevent an individual from capturing normative signals of relevance, such as the inability to observe a norm violation in the form of jaywalking.

During the internalization of norms, individuals can intentionally modify their interpretation of norms based on individual experience, background and aspirations. Hand-shaking, for example, can be interpreted as an acknowledgement of social status or objected to on the grounds of potential disease transmission. While the perceived action may be unambiguous (i.e. not manipulated during transmission and sensing), the individual may introduce an intentional bias, such as building a negative connotation with an internalized norm with an intent to change or abandon it.

This subjective perception of social reality extends to the unconscious realm, with an abundance of further mechanisms at work that drive individuals' biases in decision-making, belief formation and behavior, as well as memory and social biases. Decision-making biases can be introduced by the oftentimes disproportionate perception of rewards and sanctions as well as an asymmetric risk tolerance (see e.g. Prospect Theory [Kahneman and Tversky, 1972]). An illustrative fact in line with this observation is that individuals are by magnitudes of thousands more likely to succumb to diseases from behavioral causes (e.g. lack of exercise, smoking) than terrorist attacks, yet fear the latter disproportionally more.

Further behavioral biases, for example, include paying selective attention to favorable information, as well as seeking for confirmation of conceptions and beliefs that we already hold (confirmation bias), such as the focus on information that 'validates' an existing norm. Memory biases are fundamentally concerned with humans' limited information processing capabilities (bounded rationality [Simon, 1955]), including limited information recall, the fading of memory over time, as well as our brain's ability to fill in of memory from imagination (false memories),

---

end-to-end connections between elementary processes. They essentially consider regular processes and end-to-end processes as same natural kinds, and consequently do not allocate the operation of meta processes on a higher level of abstraction.

all of which can lead to the distortion of internalized norms. Similarly, social effects can lead to biases with respect to the normative content, such as biases towards conformity with authority figures or ingroup members. Many of these systemic biases interact with human mechanisms for operating under uncertainty. Examples for such mechanisms include the use of stereotypes to ascribe characteristics to unknown individuals (implicit social cognition [Greenwald et al., 2002]), or the application of irrational decision-making heuristics when acting under pressure ('gut feeling').

The presented selection of the cognitive biases is non-exhaustive, of course, but it offers a starting point for the exploration of cognitive influences that distort the interpretation of normative content during the norm internalization process.

Finally, norms can be modified based on the *characteristics of enforcement and enforcer*, generally affecting the salience and predictability of norms.

One fundamental determinant is the *valence* of enforcement, i.e. whether a norm is reinforced by rewards (such as a 'pat on the back') or punishments (such as scolding). As indicated earlier in the context of discussing cognitive biases, the nature of enforcement can modify norms. This includes the asymmetric impact of positive and negative sanctions (see e.g. Kahneman and Tversky [1972] and Baldwin [1971]), but also frequency, intensity and variation in enforcement. Infrequently reinforced norms are unlikely to gain high salience and may thus be easily foregone. Highly variable or inconsistent enforcement, however, interacts with individuals' risk affinity (e.g. promoting probabilistic norm compliance) but also involves the perceived level of fairness (e.g. inconsistent leadership behavior in organizational environments [Sims and Brinkmann, 2003]), which can lead to the loss of norm commitment by norm subjects, or even active opposition.

Other influence factors on enforcement that can lead to norm modification include the *social relationship between enforcer and subjects*, but also the *nature of the enforcer*. As shown by Goette et al. [2006] and Horne [2007], increased social relationship (e.g. shared group membership) between enforcer and subjects correlates with the enforcement practice. However, the central or distributed nature of the enforcer can be decisive for the enforcement. Enforcers can be quasi-centralized and self-appointed (e.g. such as rules regarding dish washing procedures imposed by administrative secretary) and show predictable enforcement strategies ('conventional sanctions'), whereas decentralized enforcement can be unpredictable with respect to the number of enforcers (e.g. unknown number of enforcers objecting to jaywalking), the applied strategies (e.g. gestures vs. scolding) and emerging dynamics (e.g. eruption into collective participation in humiliation), and thus lead to nuanced reinforcement and conceptualization of the norm as more or less serious.

Complementing the misinterpretation of normative content based on sensory bias, enforcers can likewise cause a modification of normative content by sending ambiguous signals. Examples include the insufficient command of language to express a sanction appropriately or the confusion of terminology for reward and punishment (e.g. 'awesome' vs. 'awful').

Table 2 highlights the discussed potential causes for norm modification and associates those with individual processes. While this selection identifies potential modification sources, specific factors depend on the scenario, the capabilities of the transmission medium, as well as sensory and cognitive agent models and corresponding action capabilities. In addition to intentional modification, norms can thus essentially be modified whenever an individual interacts with its social environment, the effects of which can accumulate and drive the continuous evolution norms are subjected to, providing a starting point for exploring the emergence of divergent norms within separated social clusters.

Table 2. Potential Sources of Norm Modification

Process	Causes for Modification
Transmission	Information Loss; Transmission Errors
Identification	Sensory Biases/Constraints
Internalization	Cognitive Biases; Intentional Modification
Enforcement	Choice of Enforcement; Characteristics of Enforcer(s); Relationship to Enforcement Target(s)

Summary In this section, we have proposed a general norm life cycle model that builds on the systematic comparison of existing life cycle models, harmonizes identified terminological and conceptual inconsistencies (see Section 2.6 for details), and introduces additional characteristics we deem relevant for a *general* norm life cycle model (e.g. norm identification as an alternative life cycle entry point).

While this proposed model highlights the essential processes of a general norm life cycle that we believe are necessary for its operationalization, it leaves the potential for the domain- or model-dependent refinement of individual processes, similar to Hollander and Wu's model. However, this model integrates the commonalities of existing models, while offering a comprehensive and consistent reflection of norm dynamics found in the contemporary literature. It further provides a clear differentiation between processes and associated phenomena.

2.8 Discussion

Based on the condensed, yet comprehensive overview of selected existing normative life cycle models[14], we provided a systematic comparison and synthesized

---

[14] Further life cycle models have been proposed by Andrighetto *et al.* [2013] and Singh [2014], but have been excluded from this comparison because of their highly abstract perspective (similar to the

the identified essential components into a refined interpretation of the normative life cycle. However, the focus on individual processes of the life cycle models obscures two areas of development that *combine* individual processes to model norm dynamics comprehensively – the areas of *norm change* and *norm synthesis*. We will explore those specific areas in the following, before contextualizing those with the proposed life cycle concept at the end of this chapter.

3 Norm Change
3.1 Overview

In the previous sections, we have seen different models that have been introduced in the literature to capture the life cycle of norms. These models consider the creation of norms, the processes that can facilitate their spreading, and the recognition (or learning) of norms by agents. Yet, we also know that in human societies norms can *change* over time. For example, on the occasion of the G8 summit in 2009 in Italy the Schengen treaty was suspended to guarantee the security of the local population and of the delegations, and then reinstated. In a similar way, normative systems in multi-agent systems must be able to evolve over time, for example due to actions of creating or removing norms in the system. However, the dynamic nature of norms in artificial systems is often not addressed in the simulation work on norms.

Norms are crucial in modeling agents' interactions. The definition of a normative multi-agent system that the community put forward at the first NorMAS workshop in 2005 is that "Normative MultiAgent Systems are multi-agent systems with normative systems in which agents can decide whether to follow the explicitly represented norms, and the normative systems specify how and in which extent the agents can modify the norms" [Boella *et al.*, 2006]. In order to ensure systems with autonomous agents, it is essential that norms can be violated (even though non-compliant agents are sanctioned). Because of the accent on the ability of the agents to modify norms, this definition was then known as "the normchange definition" of normative multi-agent systems.

The central problem of changing norms lead to two workshops on the dynamics of norms, the first one in 2007 in Luxembourg and the second one in 2010 in Amsterdam[15]. These two international workshops brought together researchers working on norm change from different perspectives. The revision of norms was also one of the ten open philosophical problems in deontic logic highlighted in Hansen *et al.* [2007] and further extended in Pigozzi and van der Torre [2017]. As we will see in the pages that follow, a consensus on a common framework to model norm change is still lacking.

'phases' discussed in Section 2.6) and state-centric technical focus, respectively.
[15] http://www.cs.uu.nl/events/normchange2/

3.2 From Law to Logic

Historically, the first approaches to norm change were driven by lawyers. For instance, at the 1981 international conference 'Logica, Informatica, Diritto' held in Florence (Italy), one of the conference sessions was explicitly dedicated to the problem of the abrogation of rules[16]:

> The abrogation of rules creates special problems in determining which is the 'legal system in force', as in the case of abrogation of the consequences of explicit rules and not of the rules themselves.

In the same years, a logic study of the changes of a legal code brought together three researchers coming from different backgrounds: Alchourrón, Gärdenfors and Makinson, respectively a legal theorist, a philosopher and a logician.

At the beginning, it was Alchourrón and Makinson who started investigating three types of change (Alchourrón and Makinson [1981; 1982]). The first type consists of the addition of a new norm (consistent with the other norms in the code) to an existing code. Such enlargement leads to the addition of the new norm to the code along with all the consequences that can be derived from it. The second type of change occurs again when a new norm is added, but now the new item is inconsistent with the ones already in the code. In this case we have an *amendment* of the code: in order to coherently add the new regulation, we need to reject those norms that conflict with the new one. Finally, the third change occurs when a norm is eliminated (technically, a *derogation*). In order for the elimination to be successful, however, also all other norms of the existing code that imply that norm have to be eliminated.

The approach of Alchourrón and Makinson was general: in the definition of change operators for a set of norms of some legal systems, the only assumption was that a norm is a formula in propositional logic. Thus, they suggested that "the same concepts and techniques may be taken up in other areas, wherever problems akin to inconsistency and derogation arise" ([Alchourrón and Makinson, 1981], p.147).

When Gärdenfors joined (at that time he was mainly working on counterfactuals), the trio became the founders of the well-known AGM theory, and started the fruitful research area of belief revision [Alchourrón *et al.*, 1985]. Belief revision is the formal study of how a theory (a deductively closed set of propositional formulas) may change in view of new information, which may cause an inconsistency with the existing beliefs.

Expansion, revision and contraction are the three belief change operations that Alchourrón, Gärdenfors and Makinson identified. *Expansion* is the addition of

---

[16]When a norm is abrogated, its effects in the past still hold. This is different from the annulment of a norm, which also eliminates its effects in the past.

a new proposition that is not in conflict with the existing formulas in the theory. *Revision* is the addition of information that is inconsistent with the existing beliefs. In order to consistently add such information, all conflicting formulas have to be removed. Finally, *contraction* is the elimination of a formula from the theory.

The AGM theory provides a set of postulates for each type of theory change. There is an obvious correspondence between the three types of belief change and the three changes in a system of norms mentioned above. The link between theory change and change of a legal code was explicitly acknowledged by Alchourrón, Gärdenfors and Makinson:

> [...] theory *contraction*, where a proposition x which was earlier in a theory A, is rejected. When A is a code of norms, this process is known among legal theorists as the *derogation* of x from A. [...] Another kind of change is *revision*. [...] In normative contexts this kind of change is also known as *amendment*. ([Alchourrón *et al.*, 1985], p. 510)

It should be noted, however, that the AGM theory was mainly used for belief change. This is because beliefs and norms were both represented as formulas in propositional logic.

One of the first attempts to specify the AGM framework to tackle norm change was a paper by Maranhão [2001], presented at the 2001 ICAIL conference. The approach was inspired by Fermé and Hansson [1999]'s selective revision, where only part of the input information is accepted. Maranhão introduced a *refinement* operator, which refines an agent's belief set by accepting the new input under certain conditions. Refinement provides a tool to represent the introduction of exceptions to rules in order to avoid conflicts in normative systems (for instance in those cases where judges face new conditions which were not mentioned in the legal statute but turn out to be relevant in practical situations).

As we will see in the following pages, the belief revision approach has been recently reconsidered to represent and reason about norm change (see Section 3.4).

3.3 Semantic Approaches

Two main approaches to model norm change have been developed in the literature: semantic approaches inspired by the dynamic logic approach [van Ditmarsch *et al.*, 2007], and syntactic approaches where norm change is performed directly on the set of norms.

Among semantic approaches we find the dynamic context logic proposed by Aucher *et al.* [2009], which represents norm change (in particular the dynamics of constitutive norms[17]) as a form of model update. Starting from a modal logic

---

[17] Constitutive norms are rules that define an activity. For example, the institutions of marriage, money, and promising are systems of constitutive rules or conventions. As another example, a signature may count as a legal contract, and a legal contract may define a permission to use a resource and an obligation to pay.

of context [Grossi *et al.*, 2008], context expansion and context contraction operators are introduced. The intuition is that contexts can be seen as set of models of theories. Context expansion is thus linked to the promulgation of counts-as conditionals while context contraction is used for the abrogation of constitutive norms. Norms are statements of the kind "the fact α implies a violation". One of the advantages of this approach is that it can be used for the formal specification and verification of computational models of interactions based on norms.

A similar proposal is by Pucella and Weissman [2004], where operations for granting or revoking extensions are defined in a dynamic logic of permission. Aucher *et al.* [2009]'s framework is more general. Changes in the granting and revoking of permissions and obligations are more specific than the normative system change captured in Pucella and Weissman [2004]'s article.

3.4 Syntactic Approaches
3.4.1 Defeasible Logic
When new norms are created or old norms are retracted from a normative system, the changes have repercussions on obligations and permissions that such norms established. Obligations can change without removing or adding norms. For example, change in the world can lead to new obligations without changing the legal norms. For this reason, Governatori and Rotolo [2010] insist on the need to distinguish norms from obligations and permissions (as done in deontic logic).

Inspired by the legal practice, Governatori and Rotolo aim at a formal account of legal modifications. They use a syntactic approach, where norm change is an operation performed on the rules contained in the code. Such modifications can be implicit or explicit. Implicit modifications are the most common. They arise when new norms are introduced in the legal system and such norms conflict with existing ones. The new norms enforce a retraction of the old ones because, for example, have a higher ranking status, like a national law can derogate a regional law. Explicit modifications are obtained when norms that define how other existing norms have to be modified are added to the legal code.

In particular, the mechanisms of annulments and abrogations are studied. Annulment removes a norm from the code. It operates *ex tunc*: all effects (past and future) are canceled. Abrogation too is a kind of norm removal but, unlike annulments, it applies *ex nunc*: it cannot operate retroactively, leaving their effects in the past hold.

The notion of abrogation is complex and there is no agreement among jurists on whether abrogations actually remove norms or not. In order to illustrate the difficulties, Governatori and Rotolo give the following example:

> If a norm n_1 is abrogated in 2007, its effects are no longer obtained after then. But, if a case should be decided in 2008 but the facts of the case are dated 2006, n_1, if applicable, will anyway produce its

effects because the facts held in 2006, when n_1 was still in force (and abrogations are not retroactive). Accordingly, n_1 is still in the legal system, even though is no longer in force after 2007. ([Governatori and Rotolo, 2010], p. 159)

As seen in this example, the difficulty of abrogations comes from the fact that, in most cases, direct effects should be removed, but this is not necessarily the case for indirect effects. Clearly the temporal dimension is crucial in their formal representation, but it also makes the formalization more cumbersome.

So Governatori and Rotolo first try to capture annulments and abrogations with theory revision in defeasible logic without temporal reasoning. Unfortunately, the result is not fully satisfactory as retroactivity cannot be captured. This is a crucial aspect as retroactivity allows to distinguish abrogation from annulment.

In the second part of the paper then, they use a temporal extension of defeasible logic to keep track of the changes in a normative system and to deal with retroactivity.

Norms have two temporal dimensions: the time of validity of a norm (when the norm enters in the normative system) and the time of effectiveness (when the norm can produce legal effects). As a consequence, multiple versions of a normative system are needed. In order to illustrate the problem, we recall this example from a hypothetical taxation law discussed in [Governatori and Rotolo, 2010]:

> If the taxable income of a person at January 31, for the previous year is in excess on 100,000\$, then the top marginal rate computed at February 28 is 50% of the total taxable income. And this provision is in force from January 1. This rule can be written as follows:

$$(Threshold^{31Jan} \to HighMarginalRate^{28Feb})^{1Jan}$$

> Let us suppose that the last installment for the salary was paid to an employee on January 4, and that it makes the total taxable income greater than the threshold stated above. We use $Threshold^{4Jan}$ to signal that the threshold of 100,000\$ has been certified on January 4. [...] So let us ask what the top marginal rate for the employee is if she lodges a tax return on January 20. [...] [From] the point of view of January 20, the top marginal rate is 50%. Suppose now that there is a change in the legislation and that the above norm is changed on February 15, and the change is that the top marginal rate is 30%.

$$(Threshold^{31Jan} \to MediumMarginalRate^{28Feb})^{15Feb}$$

In this case if the employee lodges her tax return after February 15, the top marginal rate is 30% instead of 50%. ([Governatori and Rotolo, 2010], p. 173-174)

This example shows that what can be derived depends on which rules are valid at the time when we do the derivation, especially if rules can be changed. Thus, in order to keep track of the norm changes, Governatori and Rotolo represent different versions of a legal system.

3.4.2 Back to AGM

On May 19th, 1988 a three kilometers long bridge connecting the de Ré island in the Atlantic Ocean to France was inaugurated. Among the effects of such a convenient connection was that the price per square meters on the island flared up. Suddenly, farmers whose families had been living on the island sometimes since the XVth century, found they had to pay the wealth and large fortune tax, a tax directed to individuals who own assets of high net worth. Most of those farmers are retired people with low pension, living on the products on their fields of potatoes, asparagus and vines. In order to pay the wealth and large fortune tax, some had to sell part of their fields and endangered their retirements plans. This raised serious concerns on the unexpected implications of such tax and some people advocated a change of such law.

As we have seen, one of the motivations of the AGM theory of belief revision was the study of norm change. One may also argue that some of the AGM axioms (that have been criticized in the belief revision context) appear reasonable when applied to the legal discourse. The *success* postulate for revision, for example, imposes to always accept the new input. This postulate has been heavily criticized in the belief revision literature as irrational behaviors may result from it (consider, for example, an agent who receives a stream of contradicting inputs like $\phi, \neg\phi, \phi, \neg\phi, \ldots$). The success makes however sense in the legal context, when we wish to enforce a new norm.

As we have seen in the previous subsection, the explicit temporal representation and the use of meta-rules of Governatori and Rotolo [2010]'s approach resulted in complex logics. In order to reduce such complexity, Governatori *et al.* [2013] explored three AGM-like contraction operators to remove rules, add exceptions and revise rule priorities. Similarly to Governatori and Rotolo, this approach is rooted in the legal practice. The operators and the principles are illustrated with examples taken from the Italian Constitution and real decisions taken by the Italian Constitutional Court.

Boella *et al.* [2009] (subsequently extended in [Boella *et al.*, 2016b]) also reconsidered the original inspiration of the AGM theory of belief revision as framework to evaluate the dynamics of rule-based systems. Boella *et al.* [2016b] observe that if we wish to weaken a rule-based system from which we derive too much, we can use the theory of belief base dynamics [Hansson, 1993] to select a subset of the rules as the contraction of the rule-based system.

EXAMPLE 1.1 ([Boella *et al.*, 2016b], p.274) *Consider a rule-based system consisting of the following two rules:*

1. *If a then b*

2. *If b then c*

Assume we do not want to have c in context {a}, whereas c can be derived by iteratively applying the first and the second rule. We can define rule base contraction operators that drop either the first or the second rule, or both.

However, the next example illustrates that such rule contraction operators may not be sufficient.

EXAMPLE 1.2 ([Boella *et al.*, 2016b], p.274) *Assume d is an exception to c in context a. In that case, we may want to end up with a rule base consisting of the following two rules:*

1. *If $a \wedge \neg d$ then b, and*

2. *If b then c*

or a rule base consisting of the following two rules:

1. *If a then b, and*

2. *If $b \wedge \neg d$ then c.*

In other words, in some applications, we may need to change *some of the rules. In particular, rule contraction may assume a* rule logic *which informs us that the rule 'if a then b' implies the rule 'if $a \wedge \neg d$ then b', or that 'if b then c' implies the rule 'if $b \wedge \neg d$ then c'.*

Thus, even if base contraction is the most straightforward and safe way to perform a contraction, it always results in a subset of the original base, which sometimes means removing too much. Take, for example $\{(a,x)\} \div (a,x) = \{\}$, where $\div$ denotes the contraction operator. Thus, under base contraction, the only result is to throw away the rule. But under AGM one can put a weaker rule. For instance, if (a,x) is the rule "If an individual owns land for more than 1.3 million Euros (a), then he must pay the wealth and large fortune tax (x)". To avoid problems as those on de Ré island, we may wish to change the law by introducing an exception, like $\{(a,x)\} \div (a,x) = \{(a \wedge b, x)\}$, where b stays for people with high income.

This was one of the motivations of Boella *et al.* [2016b]. In their abstract approach, rules are pairs (a,x) of propositional formulas and a normative system R is a set of pairs. Several logics for rules are considered by resorting to the input/output logic framework developed by Makinson and van der Torre [2000; 2003].[18]

Rules allow to derive formulas, that is, obligations and prohibitions in a normative system. The factual situation (called *context* or *input*) determines which obligations and prohibitions can be derived in a normative system. Formally, in the input/output notation: if $(a,x) \in R$ then $x \in out(R,a)$. This means that, according to the normative system R, in context a, the formula x is obligatory. The idea is that a is the input (or context) and x is the output. Of the operations defined semantically and characterized by derivation rules in Makinson and van der Torre [2000], three operations are considered in Boella *et al.* [2009; 2016b]: simple-minded, basic, and simple-minded reusable.

In order to generalize the AGM postulates for normative change, a rule set is taken to be a set of rules closed under an input/output logic. Rule expansion, rule contraction and rule revision in the input/output framework are then defined. Similarly as for the belief change case, the definition of rule expansion is unproblematic. Here, the legislator wishes to add a new rule that does not conflict with the existing ones. Rule contraction and rule revision, on the other hand, are more interesting.

AGM postulates for expansion, contraction and revision are reformulated for rule expansion, rule contraction and rule revision. It turns out that (surprisingly) the postulates for rule contraction are consistent only for some input/output logics, but not for others. On the positive side, the proof theory of rule change was shown to be closely related to the proof theory of permissions from an input/output perspective [Boella *et al.*, 2016b].

The translation from the AGM contraction postulates to the postulates for rule revision turned out to be more difficult. One of the difficulties was the definition of the negated input (roughly corresponding to $\neg(a,x)$) and the inconsistent set of rules in input/output logic (which would correspond to an 'incoherent' system of rules in the normative systems paradigm).

Postulates for (belief and rule) revision and (belief and rule) contraction are independent. No contraction operator appears in the revision postulates, and no revision operator appears in the postulates for contraction. Yet, the Levi identity and the Harper identity defined respectively the belief revision operator as a sequence of contraction and expansion, and the belief contraction is defined in terms of belief revision.

---

[18] Maranhão [2017] employs input/output logics and belief revision principles to model legal interpretation. Judicial doctrine is seen as theory change, where rules and values need to be revised to obtain a coherent system.

Using the Levi identity, rule revision was defined in terms of rule contraction. The operators so defined were shown to satisfy the AGM postulates. For the Harper identity, however, the question is still open [Boella *et al.*, 2016b].

A similar approach to Boella *et al.* [2009; 2016b]'s has been proposed by Stolpe [2010]. There, AGM contractions and revision are used to define derogation and amendment of norms. In particular, the derogation operation is an AGM partial meet contraction obtained by defining a selection function for a set of norms in input/output logic. Norm revision defined via the Levi identity characterize the amendment of norms. Stolpe can thus show that derogation and amendment operators are in one-to-one correspondence with the Harper and Levi identities as inverse bijective maps.

3.5 Computational Mechanisms of Norm Change

Beside the theoretical investigations to norm change presented in the previous sections, few work exist on the computational mechanisms of norm change.

The drawback of determining norms at design time is that unforeseen situations may occur and the system cannot adapt to the new circumstances. The approach proposed by Tinnemeier *et al.* [2010] tackles this problem by allowing the modification of norms at runtime, so that a programmer can stipulate when and how norms can be modified. In Tinnemeier *et al.* [2010]'s framework norms can be modified by external agents as well as the normative framework.

The proposed norm change mechanism is system-dependent and enforcement-independent. The first principle states that who can change norms, how and when norms may be changed depends on the system. The authors justify this first principle by recalling the clause that a normative system must "specify how and in which extent the agents can modify the norms", as in the definition proposed at the first NorMAS workshop in 2005. The second principle ensures that the norm change and the norm enforcement mechanisms should be defined independently. This is to increase the readability and manageability of the program.

Two types of norm change rules are defined. The first type is used to change instances of norms without modifying the norm scheme. These rules define the circumstances under which some norm instances have to be removed to be replaced by other norm instances. The second type of rules is used to alter norm schemes. As for the first type, these rules define under which circumstances norm schemes are to be changed by retracting some norm schemes and adding others.

What happens to the instances already instantiated, when the underlying norm scheme is changed? Tinnemeier *et al.* [2010] observe that there are situations in which we want to leave the instantiated instances unchanged, and others in which it makes sense to apply the change retroactively. Thus, two types of norm scheme change rules are given. Finally, building on [Tinnemeier *et al.*, 2009], the syntax and operational semantics of the programming language are given.

Previous work on norm change at runtime includes [Bou *et al.*, 2007; Campos *et al.*, 2009]. Bou *et al.* [2007] also consider the problem of adapting a system to novel and unpredictable circumstances. To this end, they present an approach to enable normative frameworks (called "electronic institutions" in [Bou *et al.*, 2007; Campos *et al.*, 2009]) to adapt norms to agents' behaviour changes as well as to comply with institutional goals. The norm change mechanisms of Bou *et al.* [2007] allow to modify existing norms. Unlike [Tinnemeier *et al.*, 2009], new norms cannot be introduced nor can existing norms be removed. Another difference is that Bou *et al.* [2007] use a quantitative approach to represent the environment and the agents.

Campos *et al.* [2009] approached the difficulty of how to adapt a normative system to the changes of its agents' behavior by adding situatedness and adaptation (two properties usually characterizing agents) to the system. The result is a system that can make changes and that can also adapt to changes. As in Bou *et al.* [2007]'s approach, the aim is to modify agent coordination to enhance the system's performance in attaining institutional goals.

Even though Boella and van der Torre [2004]'s approach is theoretical, it shares some similarities to the works presented here. Starting from the distinction between regulative norms (that indicate what is obligatory or permitted) and constitutive (or count-as) rules (that define an activity), they use constitutive rules to create new norms as well as to define what changes the agents can introduce. As in the norm instance change rules and norm scheme change rules of Tinnemeier *et al.* [2010], constitutive and regulative rules in Boella and van der Torre [2004] are modeled as conditional rules specifying when a norm can be changed and what the consequences are.

3.6 Discussion

In this short excursus we have seen that the first formal investigations of changes in a legal code had roots in logic, namely in the AGM framework. This line of research has been reconsidered, notably in the works of Governatori and Rotolo [2010; 2013], Stolpe [2010], and Boella *et al.* [2009; 2016b], often coupled with non-classical logics such as defeasible logic or input/output. Another direction has been to follow a semantic approach inspired by dynamic logic, as done in Pucella and Weissman [2004] and Aucher *et al.* [2009]. Finally, besides the theoretical investigations, work has been done on the computational mechanisms of norm change, like Tinnemeier *et al.* [2010], Bou *et al.* [2007] and Campos *et al.* [2009].

Norm change is a fairly recent research theme in the NorMAS community. The first international workshop explicitly dedicated to the dynamics of norms was held in 2007. This observation can in part explain the lack of consensus around a common theoretical framework. But it probably does not explain it completely. Other reasons may reside in the limits of abstract frameworks like AGM, even when combined with with richer rule-based logical systems, in the difficulty to capture and distinguish norm change from changes in obligations, and again in the elusive character of legal changes in the real world. Recent developments in legal informatics may help casting light on norm dynamics. Works on legal document and knowledge management systems (like the EUNOMOS project [Boella et al., 2016a]) allow, for example, to keep track of (implicit and explicit) changes in the legislation. Although these works provide some first steps in the understanding of the dynamics of normative systems, much still remains unexplored.

4 Norm Synthesis

The second theme of norm synthesis has a long-standing history but has experienced a recent revival of attention. While norm change primarily focuses on the logical implications of the modification of existing (legal) norms over time, norm synthesis puts a stronger emphasis on how (social) norms emerge and converge in the first place, and how they can be identified.

4.1 Foundations

Norm synthesis is inspired by the area of program synthesis (i.e. generating a program from a given specification [Manna and Waldinger, 1980]), but, in contrast to the former, shifts the focus to the coordination of autonomously operating agents. The specific purpose of norm synthesis is thus to identify an optimal set of norms (a normative system) to coordinate individuals' behaviors in a multiagent system. The 'optimality' of a solution depends on the specified objectives, such as the minimal set of norms to facilitate coordination [Fitoussi and Tennenholtz, 2000].

Shoham and Tennenholtz [1992b; 1995]'s work on synthesis of social laws is considered the initial work in the area of norm synthesis. They propose a general formal model to identify a set of social laws at design time (offline) to assure the coordinated operation of structurally uniform agents. They showcase this approach by 'handcrafting' a set of social laws that guarantee collision-free coordination in a grid-based traffic scenario ('rules of the road'[19]), instead of determining action prescriptions for each possible system state. However, they also show that the automated synthesis for offline approaches is NP-hard [Shoham and Tennenholtz, 1995], challenging the generalizable application. Onn and Tennenholtz [1997] propose a general solution for the synthesis problem for scenarios that can be represented as biconnected graphs by reducing synthesis to a graph routing problem. Fitoussi and Tennenholtz [2000] further introduce qualitative characteristics for synthesized social laws, such as their *Minimality* and *Simplicity*. As alluded to before, minimal social laws seek to specify fewest possible restrictions on agents' behaviours, thus giving individuals the greatest possible autonomy, while main-

---

[19]This *de facto* reference scenario has been adopted and refined in large parts of subsequent work on norm synthesis.

taining coordination in the overall system. An extremely restrictive social law would effectively prescribe any action an agent could take in any given situation (e.g. to walk on the right side of a footpath in a given direction, or even more restrictive, prescribing specific navigation routes between different locations), thus removing any form of autonomy on the part of the agent. A minimal social law (e.g. not to step on the road), in contrast, would retain the agent's ability to pursue its own goals, as long as it is compatible with the system objectives (e.g. avoiding collisions between cars and pedestrians). In a more recent approach, Christelis and Rovatsos [2009]'s work on automated offline norm synthesis addresses the complexity problem by identifying prohibitive states in incomplete state specifications that are generalized across the entire state space. It is important to note that these early approaches to norm synthesis do not consider or tolerate any form of violation; unlike most subsequent work, their conceptions of social laws describe hard constraints agents cannot forgo.

The shift towards refined norm interpretations that emphasizes the interactionist over legal perspective (and thus regulation over regimentation) [Boella et al., 2008] has stimulated a differentiated treatment of rewards and sanctions as mechanisms of social enforcement. This sociologically-inspired norm perspective drove the exploration of associated influence factors (such as memory and connectivity), along with a movement from *offline* to *online* norm synthesis, resulting in two subfields. *Convention/Norm Emergence* (which we will differentiate later) emphasize mechanisms that influence the convergence on norms or conventions, whereas work we cluster under the label *Identification* concentrates on the mechanics of detecting and synthesizing norms in the first place. The latter can further be subdivided into approaches that rely on a centralized or decentralized operation, that is, approaches that use a central entity to synthesize norms, or delegate the generalization and integration of identified norms to the agents themselves. Figure 7 provides a schematic overview of the outlined structure of the research field. Overall, the subfields of norm synthesis cover the notion of norms in the broad sense (i.e. as institutions), ranging from self-enforcing conventions via socially enforced norms to centrally enforced social laws or rules. In the following, we will discuss selected contributions to the area of norm synthesis, with a particular focus on approaches that emphasize the detection and identification of norms.

4.2 Synthesis as Norm/Convention Emergence

Research efforts in the area of *norm emergence* put particular concentration on an understanding of the contextual conditions and mechanisms that bring norms about, including their distributed nature. Instead of relying on a centralized entity to determine norms a priori or embedding hard-coded (offline designed) norms into individuals, norm emergence affords the decentralized collaboration of agents to converge on commonly accepted social norms.

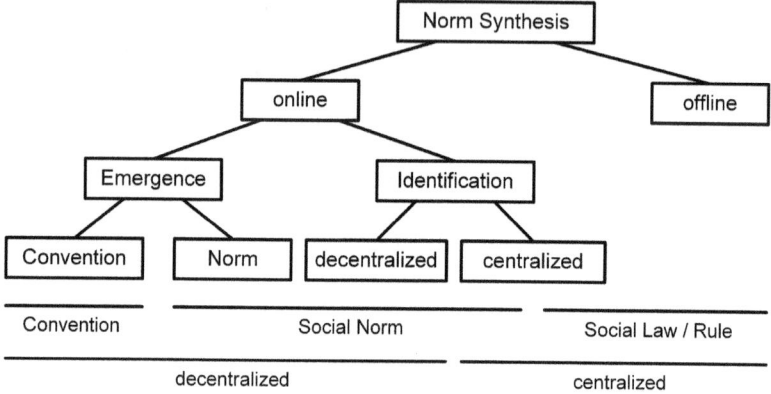

Figure 7. Taxonomy of Norm Synthesis Approaches

Explored mechanisms include:

- Memory size (e.g. Villatoro et al. [2009])

- Network topologies and dynamics of relationships (e.g. Savarimuthu et al. [2009], Villatoro et al. [2009], Sen and Sen [2010], Sugawara [2011], Villatoro et al. [2013])

- Clusters (e.g. Pujol et al. [2005])

- Interaction-based social learning (e.g. Sen and Airiau [2007], Mukherjee et al. [2007; 2008], Airiau et al. [2014])

- Lying (e.g. Savarimuthu et al. [2011])

- Dynamic sanctions (e.g. Mahmoud et al. [2012; 2015])

- Hierarchical structures with varying levels of influence (e.g. Franks et al. [2013; 2014], Yu et al. [2013; 2015])

Further contributions in the area of norm emergence include algorithms for distributed decision-making to arrive at a shared lexicon [Salazar et al., 2010] or shared sets of tags [Griffiths and Luck, 2010].

The decentralized operation of norm emergence places an emphasis on larger number of agents and their direct interaction in favour of cognitive capability and central coordination. Consequently, the computational complexity of individual agents is limited and the applied norm representations are mostly abstract in the form of converging strategy choices in coordination games or string-based representations; the normative content is symbolic and can only be inferred from the motivating scenario. In addition to the abstract normative content, in most cases, agents converge on a single norm (with exception of Savarimuthu *et al.* [2009] and Sen and Sen [2010]). In addition, most scenarios sustain the emerging norm without explicit enforcement, thus representing *self-enforcing conventions* as opposed to *externally enforced social norms*, affording the differentiation into *Convention Emergence* and *Norm Emergence*.

Following the exploration of the emergence strand of norm synthesis, we will turn to the identification strand that captures norm synthesis processes in a narrow sense, primarily focusing on the detection, identification, and integration of norms into consistent normative systems.

4.3 Synthesis as Identification

Work that identifies and synthesizes norms at runtime can be differentiated into centralized approaches, which interpret norm synthesis in the original spirit of identifying centrally managed system-wide norms, and decentralized ones that analyze the inception of norms from a bottom-up perspective.

A series of centralized online norm synthesis approaches that follow the tradition of Shoham and Tennenholtz has been spearheaded by Morales *et al.*. In their work, Morales *et al.* [2013] propose the *Intelligent Robust Norm Synthesis* mechanism dubbed IRON in an adapted version of the grid-based 'rules of the road' scenario originally introduced by Shoham and Tennenholtz that focuses on coordination in traffic junctions. Agents have a limited observational range and move in travel direction, unless constrained by imposed norms. IRON continually monitors traffic participants' behavior. When detecting collisions, IRON identifies the underlying conditions (e.g. car approaching from the right) and introduces a norm that prevents a similar event from reoccurring (e.g. by introducing an obligation to stop whenever facing a car to one's right). These centrally generated and managed norms (which make those effectively rules or social laws) are imposed upon all traffic participants, thus progressively moving towards a stable collision-free normative system.

To prevent over-regulation from introducing too many specific norms based on individual observations, IRON attempts to *generalize* norms based on their shared preconditions by selectively ignoring a specific norm's partial precondition. The generalized norm is evaluated at runtime by detecting eventual recurring collisions, in which case the original specific norms are deemed relevant and are reinstated. To determine the *effectiveness* of given norms, IRON further monitors their activation, and ascribes frequently applied norms higher effectiveness. To identify *necessary* norms, Morales *et al.* [2013] (unlike Shoham and Tennenholtz [1995]'s social law approach) make use of the agents' ability to violate norms, which enables

IRON to identify imposed norms that are actually necessary to maintain coordination and remove unnecessary ones (i.e. norms whose violation does not produce collisions).

Morales *et al.* [2014] successively introduce further iterations of their approach (dubbed SIMON) that consider structural diversity of norm participants (e.g. by introducing emergency vehicles) and refined mechanisms for norm generalization with specific focus on minimizing the necessary simulation runtime to produce a collision-free normative system Morales *et al.* [2014; 2015c]. Their following system iteration, LION [Morales *et al.*, 2015b], includes the focus on the identification of semantic relationships between norms, so as to produce fewer, more general norms (liberal norms) that maximize the norm participants' autonomy.

This series of works on norm synthesis highlights the advantages of centralized approaches not only to identify norms, but to integrate those. In this interpretation, synthesis involves an explicit analytical effort to integrate individual norms into a coherent normative system, producing semantically meaningful complex coordination outcomes, beyond a coordinated strategy choice as observed in most norm emergence approaches. Consequently, a comprehensive approach to norm synthesis captures life cycle processes that include identification, as well as internalization and forgetting of norms, thus covering processes that are associated with the evolution of norms over time (see Section 2.6). Processes such as spreading and enforcement, characteristically associated with the work on norm emergence, are secondary.

Riveret *et al.* [2014]'s transfiguration approach takes an incremental step towards decentralized systems by endowing individual agents with learning capabilities enabling them to infer behavioral prescriptions from stochastic games. Being grounded in the field of computational justice, their approach marries bottom-up dynamics (transfiguration of experience into prescriptions) with notions of self-governance by means of collective action (voting). The voting process is initiated once all agents have submitted their inferred (and preferred) prescriptions, the most common of which is put forth as a motion. Agents are then invited to vote based on the perceived purposefulness of the prescription content, which is abstractly represented using a notion of global and individually perceived *potential*. Since the purpose of the voting process (in the spirit of self-governance) is to promote globally useful prescriptions, the agents decide probabilistically based on the alignment of the candidate prescription's individual and global potential. Once adopted, the prescription becomes a self-imposed rule of that society.

This work emphasizes the computational representation of social processes that enable self-governance by retaining high levels of decisional autonomy on the part of the society members, while abstractly providing centralized decision-making and enforcement inspired by real societies. Beyond the conceptual integration of bottom-up and top-down governance processes, this contribution emphasizes the efficiency benefits associated with centrally coordinated collective decision-making.

Contributions that shift the perspective away from approaches that emphasize effective coordination towards individual-centric operations can be captured under the umbrella of *decentralized online norm synthesis*. In addition to the focus on the individual as an entity of concern, in principle these approaches lend themselves well for explorative scenarios with a broader (if not open) range of actions than used in the centralized coordination scenarios. Research efforts related to this cluster include Andrighetto et al. [2007; 2010] as well as Savarimuthu et al. [2010b; 2013a]. We will not discuss these works in great detail at this stage as we covered those in the context of norm creation in Hollander and Wu [2011a]'s life cycle model (see Section 2.3). Instead, we will concentrate on contributions that treat norm synthesis as a holistic process involving multiple life cycle processes.

An important work in this area is Savarimuthu et al. [2013b]'s work on norm recommendation. Their approach is motivated by the identification and recommendation of an existing system's norms to newcomers, which can operate in a centralized or decentralized fashion. Their system combines norm identification, classification and life cycle stage detection in order to recommend the existence and relevance of observed norms. The initial step of norm detection operates on a continuous stream of events by identifying recurring event episodes that are terminated with a sanction signal. The algorithm collects event episodes of varying window sizes in order to establish the subset of actions that provoke a sanction signal and identifies those as candidate norms. In the second step, norms are classified with respect to their salience. For this purpose, the mechanism tracks both the invocation of actions contained in the candidate norms as well as the frequency of punishments as a response to action activation. By ranking these measures, the mechanism classifies norms by salience, where the existence of punishment is indicative of higher levels of salience, as opposed to mere action activation. A further step emphasizes the long-term perspective and attempts to identify a norm's life cycle stage (*life stage*), with possible stages being emerging, growing, maturing, declining, and decaying. The system monitors norms' punishment probabilities over time and evaluates those with respect to given successive thresholds associated with emergence (frequency of activation) and growth, based on which it infers the life stage. For example, norms that have experienced an increase in punishment between two time intervals but remain between the emergence and growth thresholds, are considered growing. The system uses heuristics that use the established measures for salience and life stage as an input to *recommend* the existence of a given norm.

Similar to Morales et al. [2015a]'s works, Savarimuthu et al. [2013b]'s synthesis approach allows the identification of multiple norms, along with a quantitative measure of salience that is comparable with Morales et al. [2013]'s notion of effectiveness and necessity. Savarimuthu *et al.* [2013b]'s approach further includes a systematic classification of norms with respect to their life cycle stage, thus emphasizing the long-term perspective. However, unlike Morales *et al.* [2015a], this work relies on an abstract string-based norm representation and does not consider semantic relationships between norms, thus preventing operations such as generalization or substitution of norms.

The final approach we present under the umbrella of norm synthesis takes an intermediate stance by operating decentralized while maintaining meaningful norm representations. Frantz *et al.* [2014c; 2015] propose a norm generalization approach that operates on individual observations. At its core, this approach is motivated by individuals' tendency to subconsciously develop stereotypes as decision-making shortcuts they can use when encountering unknown interaction partners. To facilitate this generalization, the mechanism relies on uniform structural representations of actors, actions and norms based on Nested ADICO (nADICO) [Frantz *et al.*, 2013; Frantz *et al.*, 2015], a rule-based norm representation that builds on the *Grammar of Institutions* [Crawford and Ostrom, 1995] and affords the explicit representation of structural institutional regress [Frantz, 2015], i.e. the nested interdependency of sanctions and corresponding metanorms. As a first step, observations are aggregated based on shared observable attributes as well as subsets thereof (higher generalization levels), forming the basis to synthesize descriptive norms (or conventions) the observer attributes to observed groups of individuals. To infer injunctive norms from observations, individuals further track corresponding reactions to ascribe the generalized action sequences normative character and interpret the generalized reactions as social consequences (i.e. rewards or sanctions). The frequency and intensity of observations indicate a norm's salience by mapping it onto a continuous deontics conception (*Dynamic Deontics* [Frantz *et al.*, 2014a]) that spans from prohibition via permission to obligation, the *deontic range* of which is unique for each agent and determined by its previous experience. In addition to the extremal cases, this concept introduces intermediate stages along this continuum (e.g. obligations that are omissible and can be exceptionally foregone), a principle that is used to reflect the subjectively perceived priority of a given norm, and implicitly solves potential norm conflicts.

In contrast to the approach by Morales *et al.* [2015a], this work does not solve a specific coordination problem, but introduces a fully decentralized approach to understand agents' behaviors by inspecting their *subjective understanding* of a scenario's normative content, thus shifting it into closer proximity to emergence-based approaches. Similar to Morales *et al.* [2015a] (but unlike Savarimuthu *et al.* [2013b]), this approach uses a comprehensive human-readable norm representation (as *institutional statements*) and allows the identification of norm relationships by generalizing individual observations. The uniform norm representation further permits the analysis on arbitrary social aggregation levels (e.g. group, society).

Table 3 provides a chronological overview of all identified norm synthesis approaches based on the characteristics introduced at the beginning of this subsection (see Figure 7), including the nature of norm (convention, norm, rule, social law), central coordination and the ability to produce or identify multiple norms.

Table 3. Overview of Norm Synthesis Approaches

Contribution	Institution Type	Centralized	Offline	Single Norm
Shoham and Tennenholtz [1995]	Social Law	yes	yes	no
Pujol et al. [2005]	Convention	no	no	yes
Sen and Airiau [2007]	Convention	no	no	yes
Savarimuthu et al. [2007; 2008a]	Norm	no	no	no
Mukherjee et al. [2007; 2008]	Convention	no	no	yes
Christelis and Rovatsos [2009]	Social Law	yes	yes	no
Villatoro et al. [2009]	Convention	no	no	yes
Urbano et al. [2009]	Convention	no	no	yes
Sen and Sen [2010]	Convention	no	no	yes
Griffiths and Luck [2010]	Norm	no	no	no
Sugawara [2011]	Convention	no	no	no
Mahmoud et al. [2012]	Norm	no	no	yes
Morales et al. [2013]	Social Law	yes	no	no
Franks et al. [2013]	Convention	no	no	yes
Villatoro et al. [2013]	Convention	no	no	yes
Savarimuthu et al. [2013b]	Norm	both	no	no
Mihaylov et al. [2014]	Convention	no	no	yes
Airiau et al. [2014]	Convention	no	no	yes
Morales et al. [2014]	Social Law	yes	no	no
Riveret et al. [2014]	Norm / Rule	yes	no	no
Frantz et al. [2014c; 2015]	Norm	no	no	no
Morales et al. [2015b]	Social Law	yes	no	no
Mahmoud et al. [2015]	Norm	no	no	yes

4.4 Contextualization with the General Norm Life Cycle Model

At the current stage, norm synthesis presents itself as a diverse field that is driven by varying objectives. Apart from the historical separation into offline and online approaches, we can identify a cluster of existing approaches that either concentrate on the:

- Investigation of factors and circumstances that promote norm adoption (emphasizing macro-level outcomes), or

- Mechanisms for the runtime identification, generalization, implementation, and integration with established norms (emphasizing micro-level mechanisms).

Relating these approaches to individual life cycle processes of the general norm life cycle model (see Section 2.7) as shown in Figure 8, we can observe that emergence-based approaches emphasize spreading/transmission mechanisms (e.g. type and dynamic nature of network topologies, hierarchical structures, social learning, memory size) along enforcement characteristics (e.g. sanctioning, lying).

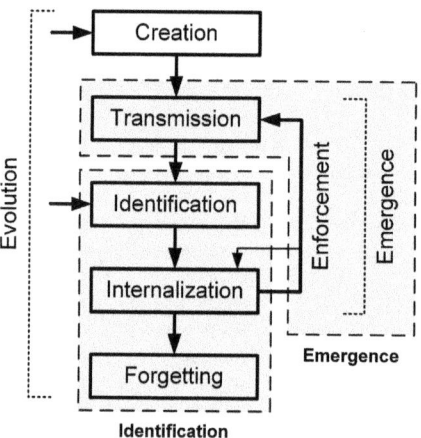

Figure 8. Norm Synthesis Approaches and Related Life Cycle Processes

The second group of mechanisms emphasize the detection and identification of existing norms. Deductive tasks for the generalization of comprehensive normative systems are related to a complex norm internalization process (such as the one conceptualized by Hollander and Wu [2011a]), since it represents a composite process that merges newly discovered norms and existing sets of norms, which requires the ability to modify, generalize and integrate norms. The synthesis of normative systems further relies on the ability to discard or forget norms.

Despite the comprehensive coverage of different life cycle stages, the review of existing approaches indicates gaps. An important central topic that has found limited explicit attention in current approaches is the detection of norm conflicts, an aspect with a strong relation to the norm internalization process. Riveret et al. [2014], as well as Savarimuthu et al. [2013b], treat norms independently without considering their relationship to existing norms. Frantz et al. [2015] and Morales et al. [2015a] include generalization processes and mechanisms to accommodate conflicting or competing norms, but only Morales et al. [2015b] perform explicit detection of norm relationships such as complementarity and substitutability. An area that has found recent attention is the focus on *dynamic normative systems* [Huang et al., 2016] in which the normative environment itself is not considered static, but changes over time, and thus requires agents to revise their normative understanding in order to accommodate those changes. Initial work by Huang et al. [2016] analyzes the associated complexity of norm recognition and synthesis.

4.5 Discussion of Challenges and Future Directions

In this section, we provided a comprehensive discussion of the historical roots of norm synthesis, the shifts from offline to online synthesis, and the subsequent differentiation into more implicit emergence-focused and more explicit identification-centric approaches. We further discussed a set of relevant contributions to the latter identification strand of norm synthesis. However, apart from surveying the field, this comprehensive overview of the area of norm synthesis allows us to identify areas which we believe deserve further attention.

Reviewing the strands of (online) norm synthesis, an outstanding development is the systematic integration of both strands by enriching emergence-based approaches with richer micro-level architectures that incorporate components of identification-based mechanisms. For identification-based approaches, this implies a stronger focus on generalizable representations of norms and social structures beyond specific scenarios. The marriage of both strands provides a basis for more realistic representations of social scenarios, with *emergence* sponsoring the insight on how to structure interaction in social environments, and *identification* providing mechanisms to develop complex, yet consistent normative systems as we encounter them in the real world.

We further believe that the exploration of dynamic normative systems represents an important research direction if we aim towards the use of norm synthesis in real-world applications (e.g. robotics). It further has the potential to link the theoretical contributions developed in the area of norm change, e.g. modeling changes in legal systems (as discussed in Section 3), with the mechanisms that facilitate the identification, generalization and integration of corresponding operational norms developed in the area of norm synthesis.

Looking beyond the scope of contemporary work, an important challenge for the successful adoption of norm synthesis is the identification and development of application domains that enable the use of these techniques in realistic scenarios, both involving the extent and complexity of available data. In this context, a challenge that all contemporary approaches to norm synthesis share is their operation on structured data. Making unstructured, noisy or semi-structured data (such as found in big data) accessible under consideration of the complexity limitations of current norm synthesis approaches will increase its relevance for real-world applications. Specific examples include the automated the extraction of norms from large and diverse real-world data corpi, as well as performing online norm synthesis, e.g. for the ad hoc inference of normative understanding in the context of robotics or digital assistants.

5 Summary, Conclusions and Outlook

In this chapter, we have provided an overview of the contemporary perspective on norm dynamics, with a particular focus on norm change and norm synthesis as important active research fields in multiagent systems.

The research around norm change (Section 3) has resulted in a comprehensive exploration of logical challenges associated with the representation of changing social and legal norms, such as temporal implications of changing laws and an adequate formal translation of the notion of an incoherent normative system. At this stage, the relatively young but promising field has yet to find a shared consensus on the theoretical foundation to provide the platform for the systematic application of its contributions in the context of normative multiagent systems as well as other disciplines.

Research in the area of norm synthesis (Section 4) concentrates on the analysis of factors that contribute to emerging norms (norm emergence) as well mechanisms to detect existing norms (norm identification). This field has experienced a revival with the recent focus on the synthesis of normative systems at runtime (online) – as opposed to the traditional offline approach. In addition, the field features an increasing number of approaches that favor decentralized over centralized approaches or combine both approaches and use social choice mechanisms for the integration of bottom-up and top-down perspectives on norm synthesis.

To understand the developments in both fields, we initially presented an overview of approaches that define the norm life cycles (Section 2), while providing an overview of the contemporary state of current contributions associated with individual life cycle processes. We further systematically compared the surveyed life cycles based on involved processes and norm characteristics, while identifying abstract phases of the norm life cycle. From this analysis, we extracted the essential processes and integrated those in a *general norm life cycle model* that reflects the contemporary view on norm emergence. The refined model resolves terminological and conceptual inconsistencies/omissions identified in the existing life cycle models. It further suggests that external influence factors can lead to norm modification throughout all stages of the norm life cycle, and, unlike earlier models, distinguishes between normative processes and associated phenomena.

Since this chapter specifically concentrates on the *modeling of norm dynamics*, we do not capture the wider technical and philosophical implications of norm dynamics, such as the dealing with normative conflicts and violations (see chapter 'Modeling Normative Conflicts in Multiagent Systems' in this volume), aspects of norm autonomy (see Verhagen [2000]), and the role of trust for the functioning of norms (see Andrighetto *et al.* [2013]).

Surveying individual contributions to the field of NorMAS in general – and to the areas of norm change and synthesis in particular – we can observe a tendency to apply richer norm conceptions that span across multiple norm life cycle processes. As a result, developed systems produce increasingly dynamic behavior. This includes a) the identification of norms at runtime, b) the change of norms over time, and c) their potential decay and substitution.

These observations highlight an important progression for the wider discipline, since it positions the current development on the roadmap laid out in the 2007 Dagstuhl NorMAS workshop that identified five levels in the development of normative multiagent systems (see Boella *et al.* [2008]):

- Level 1 – Off-line designed norms
- Level 2 – Explicit norm representations that can be used for communication and negotiation
- Level 3 – Runtime addition, removal and modification of norms
- Level 4 – Embeddedness in social reality
- Level 5 – Development of moral reality

The first three levels are undisputed – the shift towards dynamic creation (Level 3) is reflected in numerous contributions to the field. However, the ability of agents to identify and synthesize norms in their social environment at runtime, the ability to engage in social choice processes, as well as agents' compliance in dynamic normative systems provide the basis to make agents active participants in shaping social reality, and thus moves them closer to the fourth development level (without discussing the associated challenges here – for details see Boella *et al.* [2008]).

Fundamentally, this integration of normative concepts in social reality cannot be dissociated from the consideration of ethical and moral concerns as suggested for the last level – the development of moral reality by assuming moral agency. This resonates with contemporary developments, such as the productive use of autonomous cars, increasing automation of the workforce via robotics, decentralization of autonomy (e.g. in distributed ledger technology), along with the revived societal debates around the impact of artificial intelligence (e.g. recall the debates around universal base income). This necessity to address the embeddedness in social reality and moral reality at the same time is reflected in calls for future research directions in artificial intelligence (e.g. Russell *et al.* [2015]) and visible in initial contributions towards that end (e.g. Conitzer *et al.* [2017]).

These general AI challenges provide a unique opportunity for the interdisciplinary field of normative multiagent systems. This field studies the very dynamics that allow systems to address fuzzy and complex problems conventional rule-based systems are not prepared to deal with. It does so by exploiting two central features of norms, a) their adaptiveness towards changing social and technological environments, and b) their innate scalability based on their decentralized operation.

Independent of the application domain, this leaves us researchers with the task to foster and establish an interdisciplinary operationalization of norms as dynamic decentralized coordination mechanisms. This, in consequence, makes norm dynamics an integral component for the modeling of realistic social behavior within and beyond normative multiagent systems.

Acknowledgement

The authors would like to thank Bastin Tony Roy Savarimuthu for bringing them together to collaborate on this chapter.

BIBLIOGRAPHY

[Airiau et al., 2014] S. Airiau, S. Sen, and D. Villatoro. Emergence of conventions through social learning. *Autonomous Agents and Multi-Agent Systems*, 28(5):779–804, 2014.

[Alchourrón and Makinson, 1981] C. E. Alchourrón and D. Makinson. *Hierarchies of Regulations and their Logic*, pages 125–148. Springer Netherlands, Dordrecht, 1981.

[Alchourrón and Makinson, 1982] C. E. Alchourrón and D. Makinson. On the logic of theory change: Contraction functions and their associated revision functions. *Theoria*, 48:14–37, 1982.

[Alchourrón et al., 1985] C. E. Alchourrón, P. Gärdenfors, and D. Makinson. On the logic of theory change: Partial meet contraction and revision functions. *J. Symb. Log.*, 50(2):510–530, 1985.

[Andreoni, 1989] J. Andreoni. Giving with impure altruism: Applications to charity and ricardian equivalence. *Journal of Political Economy*, 97(6):1447–1458, 1989.

[Andrighetto et al., 2007] G. Andrighetto, M. Campenní, R. Conte, and M. Paolucci. On the immergence of norms: A normative agent architecture. In *Proceedings of the AAAI Symposium, Social and Organizational Aspects of Intelligence*, Washington DC, 2007.

[Andrighetto et al., 2010] G. Andrighetto, M. Campenní, F. Cecconi, and R. Conte. The complex loop of norm emergence: A simulation model. In Keiki Takadama, Claudio Cioffi-Revilla, and Guillaume Deffuant, editors, *Simulating Interacting Agents and Social Phenomena*, pages 19–35. Springer, Berlin, 2010.

[Andrighetto et al., 2013] G. Andrighetto, C. Castelfranchi, E. Mayor, J. McBreen, M. Lopez-Sanchez, and S. Parsons. (Social) norm dynamics. In Giulia Andrighetto, Guido Governatori, Pablo Noriega, and Leendert van der Torre, editors, *Normative Multi-Agent Systems. Vol. 4 of Dagstuhl Follow-Ups*, pages 135–170, 2013.

[Aucher et al., 2009] G. Aucher, D. Grossi, A. Herzig, and E. Lorini. Dynamic context logic. In Xiangdong He, John Horty, and Eric Pacuit, editors, *Logic, Rationality, and Interaction: Second International Workshop, LORI 2009, Chongqing, China, October 8-11, 2009. Proceedings*, pages 15–26, Berlin, Heidelberg, 2009. Springer Berlin Heidelberg.

[Axelrod, 1986] R. Axelrod. An evolutionary approach to norms. *The American Political Science Review*, 80(4):1095–1111, 1986.

[Baldwin, 1971] D. A. Baldwin. The power of positive sanctions. *World Politics*, 24(1):19–38, 1971.

[Bandura, 1977] A. Bandura. *Social Learning Theory*. General Learning Press, New York (NY), 1977.

[Barabási and Albert, 1999] A.-L. Barabási and R. Albert. Emergence of scaling in random networks. *Science*, 286(5439):509–512, October 1999.

[Beheshti et al., 2015] R. Beheshti, A. M. Ali, and G. Sukthankar. Cognitive social learners: An architecture for modeling normative behavior. In *Proceedings of the Twenty-Ninth AAAI Conference on Artificial Intelligence*, AAAI'15, pages 2017–2023. AAAI Press, 2015.

[Bianco and Bates, 1990] W. T. Bianco and R. Bates. Cooperation by design: Leadership, structure, and collective dilemmas. *American Political Science Review*, 84(1):133–147, 1990.

[Bicchieri, 2006] C. Bicchieri. *The Grammar of Society: The Nature and Dynamics of Social Norms*. Cambridge University Press, New York, 2006.

[Boella and van der Torre, 2004] G. Boella and L. van der Torre. Regulative and constitutive norms in normative multiagent systems. In *Proceedings of the Ninth International Conference (KR2004)*, pages 255–265. AAAI Press, 2004.

[Boella et al., 2006] G. Boella, L. van der Torre, and H. Verhagen. Introduction to normative multiagent systems. *Computation and Mathematical Organizational Theory*, 12(2-3):71–79, 2006.

[Boella et al., 2008] G. Boella, L. van der Torre, and H. Verhagen. Introduction to special issue on normative multiagent systems. *Autonomous Agents and Multi-Agent Systems*, 17(1):1–10, 2008.

[Boella et al., 2009] G. Boella, G. Pigozzi, and L. van der Torre. Normative framework for normative system change. In *8th International Joint Conference on Autonomous Agents and Multiagent Systems (AAMAS 2009), Budapest, Hungary, May 10-15, 2009, Volume 1*, pages 169–176, 2009.

[Boella et al., 2016a] G. Boella, L. Di Caro, L. Humphreys, L. Robaldo, P. Rossi, and L. van der Torre. Eunomos, a legal document and knowledge management system for the web to provide relevant, reliable and up-to-date information on the law. *Artificial Intelligence and Law*, 24(3):245–283, 2016.

[Boella et al., 2016b] G. Boella, G. Pigozzi, and L. van der Torre. AGM contraction and revision of rules. *Journal of Logic, Language and Information*, 25(3-4):273–297, 2016.

[Boman, 1999] M. Boman. Norms in artificial decision making. *Artificial Intelligence and Law*, 7(1):17–35, 1999.

[Bou et al., 2007] E. Bou, M. López-Sánchez, and J. A. Rodríguez-Aguilar. Adaptation of autonomic electronic institutions through norms and institutional agents. In Gregory M. P. O'Hare, Alessandro Ricci, Michael J. O'Grady, and Oğuz Dikenelli, editors, *Engineering Societies in the Agents World VII: 7th International Workshop, ESAW 2006 Dublin, Ireland, September 6-8, 2006 Revised Selected and Invited Papers*, pages 300–319, Berlin, Heidelberg, 2007. Springer Berlin Heidelberg.

[Bourdieu, 1977] P. Bourdieu. *An Outline of a Theory of Practice*. Cambridge University Press, London, 1977.

[Boyd and Richerson, 1985] R. Boyd and P. Richerson. *Culture and the Evolutionary Process*. University of Chicago Press, Chicago (IL), 1985.

[Boyd and Richerson, 2005] R. Boyd and P. Richerson. *The Origin and Evolution of Cultures*. Oxford University Press, New York (NY), 2005.

[Bratman, 1987] M. Bratman. *Intentions, Plans, and Practical Reason*. Harvard University Press, Cambridge (MA), 1987.

[Bravo et al., 2012] G. Bravo, F. Squazzoni, and R. Boero. Trust and partner selection in social networks: An experimentally grounded model. *Social Networks*, 34(4):481–492, 2012.

[Broersen et al., 2001] J. Broersen, M. Dastani, J. Hulstijn, Z. Huang, and L. van der Torre. The BOID architecture: Conflicts between beliefs, obligations, intentions and desires. In *Proceedings of the Fifth International Conference on Autonomous Agents*, AGENTS '01, pages 9–16, New York, NY, USA, 2001. ACM.

[Broersen et al., 2002] J. Broersen, M. Dastani, J. Hulstijn, and L. van der Torre. Goal generation in the BOID architecture. *Cognitive Science Quarterly*, 2(3-4):428–447, 2002.

[Caldas and Coelho, 1999] J. C. Caldas and H. Coelho. The origin of institutions: socio-economic processes, choice, norms and conventions. *Journal of Artificial Societies and Social Simulation*, 2(2):1, 1999.

[Campenní et al., 2009] M. Campenní, G. Andrighetto, F. Cecconi, and R. Conte. Normal = Normative? The role of intelligent agents in norm innovation. *Mind & Society*, 8(2):153–172, 2009.

[Campos et al., 2009] J. Campos, M. López-Sánchez, J. A. Rodríguez-Aguilar, and M. Esteva. Formalising situatedness and adaptation in electronic institutions. In Jomi Fred Hübner, Eric Matson, Olivier Boissier, and Virginia Dignum, editors, *Coordination, Organizations, Institutions and Norms in Agent Systems IV : COIN 2008 International Workshops, COIN@AAMAS 2008, Estoril, Portugal, May 12, 2008. COIN@AAAI 2008, Chicago, USA, July 14, 2008. Revised Selected Papers*, pages 126–139, Berlin, Heidelberg, 2009. Springer Berlin Heidelberg.

[Castelfranchi et al., 1998] C. Castelfranchi, R. Conte, and M. Paolucci. Normative reputation and the costs of compliance. *Journal of Artificial Societies and Social Simulation*, 1(3):3, 1998.

[Chalub et al., 2006] F. A. C. C. Chalub, F. C. Santos, and J. M. Pacheco. The evolution of norms. *Journal of Theoretical Biology*, 241(2):233–240, 2006.

[Checkel, 1998] J. Checkel. The constructivist turn in international relations theory. *World Politics*, 50(2):324–348, 1998.

[Christelis and Rovatsos, 2009] G. Christelis and M. Rovatsos. Automated norm synthesis in an agent-based planning environment. In *Proceedings of The 8th International Conference on Autonomous Agents and Multiagent Systems - Volume 1*, AAMAS '09, pages 161–168, Richland, SC, 2009. International Foundation for Autonomous Agents and Multiagent Systems.

[Conitzer et al., 2017] V. Conitzer, W. Sinnott-Armstrong, J. S. Borg, Y. Deng, and M. Kramer. Moral decision making frameworks for artificial intelligence. In *AAAI Conference on Artificial Intelligence*, 2017.

[Conte and Castelfranchi, 1995a] R. Conte and C. Castelfranchi. *Cognitive and Social Action*. UCL Press, London, 1995.

[Conte and Castelfranchi, 1995b] R. Conte and C. Castelfranchi. Understanding the effects of norms in social groups through simulation. In Nigel Gilbert and Rosaria Conte, editors, *Artificial Societies: The Computer Simulation of Social Life*, pages 252–267. UCL Press, London, 1995.

[Crawford and Ostrom, 1995] S. E. S. Crawford and E. Ostrom. A Grammar of Institutions. *The American Political Science Review*, 89(3):582–600, September 1995.

[Delgado et al., 2003] J. Delgado, J. M. Pujol, and R. Sangüesa. Emergence of coordination in scale-free networks. *Web Intelligence and Agent Systems*, 1(2):131–138, April 2003.

[Delgado, 2002] J. Delgado. Emergence of social conventions in complex networks. *Artificial Intelligence*, 141(1–2):171–185, 2002.

[Eguia, 2011] J. X. Eguia. *A Theory of Discrimination and Assimilation*. New York University Press, New York (NY), 2011.

[Ehrlich and Levin, 2005] P. R. Ehrlich and S. A. Levin. The evolution of norms. *PLoS Biology*, 3(6), 06 2005.

[Elster, 1989] J. Elster. Social norms and economic theory. *Journal of Economic Perspectives*, 3(4):99–117, 1989.

[Epstein, 2001] J. M. Epstein. Learning to be thoughtless: Social norms and individual computation. *Computational Economics*, 18(1):9–24, 2001.

[Erdős and Rényi, 1959] P. Erdős and A. Rényi. On random graphs i. *Publicationes Mathematicae*, 6:290–297, 1959.

[Fermé and Hansson, 1999] E. L. Fermé and S. O. Hansson. Selective revision. *Studia Logica: An International Journal for Symbolic Logic*, 63(3):331–342, 1999.

[Finnemore and Sikkink, 1998] M. Finnemore and K. Sikkink. International norm dynamics and political change. *International Organization*, 52(4):887–917, 1998.

[Fitoussi and Tennenholtz, 2000] D. Fitoussi and M. Tennenholtz. Choosing social laws for multi-agent systems: Minimality and simplicity. *Artificial Intelligence*, 119(1–2):61–101, 2000.

[Fix et al., 2006] J. Fix, C. von Scheve, and D. Moldt. Emotion-based norm enforcement and maintenance in multi-agent systems: Foundations and Petri net modeling. In Hideyuki Nakashima, Michael P. Wellman, Gerhard Weiss, and Peter Stone, editors, *Proceedings of the 5th International Joint Conference on Autonomous Agents and Multi-Agent Systems*, AAMAS '06, pages 105–107, New York (NY), 2006. ACM Press.

[Flentge et al., 2001] F. Flentge, D. Polani, and T. Uthmann. Modelling the emergence of possession norms using memes. *Journal of Artificial Societies and Social Simulation*, 4(4):3, 2001.

[Franks et al., 2013] H. Franks, N. Griffiths, and S. S. Anand. Learning influence in complex social networks. In Ito, Jonker, Gini, and Shehory, editors, *Proceedings of the 12th International Conference on Autonomous Agents and Multi-Agent Systems*, AAMAS '13, pages 447–454, 2013.

[Franks et al., 2014] H. Franks, N. Griffiths, and S. S. Anand. Learning agent influence in MAS with complex social networks. *Autonomous Agents and Multi-Agent Systems*, 28:836–866, 2014.

[Frantz et al., 2013] C. Frantz, M. K. Purvis, M. Nowostawski, and B. T. R. Savarimuthu. nADICO: A nested grammar of institutions. In G. Boella, E. Elkind, B. T. R. Savarimuthu, F. Dignum, and M. K. Purvis, editors, *PRIMA 2013: Principles and Practice of Multi-Agent Systems*, volume 8291 of *Lecture Notes in Artificial Intelligence*, pages 429–436, Berlin, 2013. Springer.

[Frantz et al., 2014a] C. Frantz, M. K. Purvis, M. Nowostawski, and B. T. R. Savarimuthu. Modelling institutions using dynamic deontics. In Tina Balke, Frank Dignum, M. Birna van Riemsdijk, and Amit K. Chopra, editors, *Coordination, Organizations, Institutions and Norms in Agent Systems IX*, volume 8386 of *Lecture Notes in Artificial Intelligence*, pages 211–233, Berlin, 2014. Springer.

[Frantz et al., 2014b] C. Frantz, M. K. Purvis, B. T. R. Savarimuthu, and M. Nowostawski. Analysing the dynamics of norm evolution using interval type-2 fuzzy sets. In *WI-IAT '14 Proceedings of the 2014 IEEE/WIC/ACM International Joint Conferences on Web Intelligence (WI) and Intelligent Agent Technologies (IAT)*, volume 3, pages 230–237, 2014.

[Frantz et al., 2014c] C. Frantz, M. K. Purvis, B. T. R. Savarimuthu, and M. Nowostawski. Modelling dynamic normative understanding in agent societies. In Hoa Khanh Dam, Jeremy Pitt, Yang Xu, Guido Governatori, and Takayuki Ito, editors, *Principles and Practice of Multi-Agent Systems - 17th International Conference, PRIMA 2014*, volume 8861 of *Lecture Notes in Artificial Intelligence*, pages 294–310, Berlin, 2014. Springer.

[Frantz et al., 2015] C. K. Frantz, M. K. Purvis, B. T. R. Savarimuthu, and M. Nowostawski. Modelling dynamic normative understanding in agent societies. *Scalable Computing: Practice and Experience*, 16(4):355–378, 2015.

[Frantz et al., 2016] C. K. Frantz, B. T. R. Savarimuthu, M. K. Purvis, and M. Nowostawski. Generalising social structure using interval type-2 fuzzy sets. In Matteo Baldoni, Amit K. Chopra, Tran Cao Son, Katsutoshi Hirayama, and Paolo Torroni, editors, *PRIMA 2016: Principles and Practice of Multi-Agent Systems: 19th International Conference, Phuket, Thailand, August 22-26, 2016, Proceedings*, pages 344–354. Springer International Publishing, Cham, 2016.

[Frantz, 2015] C. K. Frantz. *Agent-Based Institutional Modelling: Novel Techniques for Deriving Structure from Behaviour*. PhD thesis, University of Otago, Dunedin, New Zealand, 2015. Available under: http://hdl.handle.net/10523/5906.

[Galan and Izquierdo, 2005] J. M. Galan and L. R. Izquierdo. Appearances can be deceiving: lessons learned re-implementing Axelrod's 'evolutionary approach to norms'. *Journal of Artificial Societies and Social Simulation*, 8(3):2, 2005.

[Goette et al., 2006] L. Goette, D. Huffman, and S. Meier. The impact of group membership on cooperation and norm enforcement: Evidence using random assignment to real social groups. *American Economic Review*, 96(2):212–216, May 2006.

[Governatori and Rotolo, 2010] G. Governatori and A. Rotolo. Changing legal systems: legal abrogations and annulments in defeasible logic. *Logic Journal of IGPL*, 18(1):157–194, 2010.

[Governatori et al., 2013] G. Governatori, A. Rotolo, F. Olivieri, and S. Scannapieco. Legal contractions: A logical analysis. In Enrico Francesconi and Bart Verheij, editors, *ICAIL*, pages 63–72. ACM, 2013.

[Greenwald et al., 2002] A. G. Greenwald, M. R. Banaji, L. A. Rudman, S. D. Farnham, B. A. Nosek, and D. S. Mellott. A unified theory of implicit attitudes, stereotypes, self-esteem, and self-concept. *Psychological Review*, 109(1):3–25, 2002.

[Griffiths and Luck, 2010] N. Griffiths and M. Luck. Norm diversity and emergence in tag-based cooperation. In M. De Vos, N. Fornara, J. V. Pitt, and G. A. Vouros, editors, *Coordination, Organizations, Institutions, and Norms in Agent Systems VI - COIN 2010 International Workshops, COIN@AAMAS 2010, Toronto, Canada, May 2010, COIN@MALLOW 2010, Lyon, France, August 2010, Revised Selected Papers*, volume 6541 of *Lecture Notes in Computer Science*, pages 230–249. Springer, 2010.

[Grossi et al., 2008] D. Grossi, J.-J. C. Meyer, and F. Dignum. The many faces of counts-as: A formal analysis of constitutive-rules. *J. of Applied Logic*, 6(2):192–217, 2008.

[Hales, 2002] D. Hales. Group reputation supports beneficent norms. *Journal of Artificial Societies and Social Simulation*, 5(4):4, 2002.

[Hansen et al., 2007] J. Hansen, G. Pigozzi, and L. van der Torre. Ten philosophical problems in deontic logic. In G. Boella, L. van der Torre, and H. Verhagen, editors, *Normative Multi-Agent Systems. Dagstuhl Seminar Proc. 07122*, 2007.

[Hansson, 1993] S. O. Hansson. Reversing the Levi identity. *Journal of Philosophical Logic*, 22:637–669, 1993.

[Henderson, 2005] D. Henderson. Norms, invariance, and explanatory relevance. *Philosophy of the Social Sciences*, 35(3):324–338, 2005.

[Hoffmann, 2003] M. Hoffmann. Entrepreneurs and norm dynamics: An agent-based model of the norm life cycle. Technical report, Department of Political Science and International Relations, University of Delaware, Newark (DE), 2003.

[Hoffmann, 2005] M. Hoffmann. Self-organized criticality and norm avalanches. In *Proceedings of the Symposium on Normative Multi-Agent Systems*, Hatfield (UK), 2005. AISB.

[Hogg, 2001] M. A. Hogg. A social identity theory of leadership. *Personality and Social Psychology Review*, 5(3):184–200, 2001.

[Hollander and Wu, 2011a] C. D. Hollander and A. S. Wu. The current state of normative agent-based systems. *Journal of Artificial Societies and Social Simulation*, 14(2):6, 2011.

[Hollander and Wu, 2011b] C. D. Hollander and A. S. Wu. Using the process of norm emergence to model consensus formation. In *Fifth IEEE International Conference on Self-Adaptive and Self-Organizing Systems (SASO)*, pages 148–157, Oct 2011.

[Horne, 2001] C. Horne. Sociological perspectives on the emergence of norms. In M. Hechter and K. Opp, editors, *Social Norms*, pages 3–34. Russell Sage Foundation, New York (NY), 2001.

[Horne, 2007] C. Horne. Explaining norm enforcement. *Rationality and Society*, 19(2):139–170, 2007.

[Huang et al., 2016] X. Huang, J. Ruan, Q. Chen, and K. Su. Normative multiagent systems: A dynamic generalization. In *Proceedings of the Twenty-Fifth International Joint Conference on Artificial Intelligence*, IJCAI'16, pages 1123–1129. AAAI Press, 2016.

[Huber, 1999] M. J. Huber. JAM: A BDI-theoretic mobile agent architecture. In Oren Etzioni, Jörg P. Müller, and Jeffrey M. Bradshaw, editors, *Proceedings of the Third International Conference on Autonomous Agents (Agents '99)*, pages 236–243, Seattle (WA), 1999.

[Kahneman and Tversky, 1972] D. Kahneman and A. Tversky. Subjective probability: A judgment of representativeness. *Cognitive Psychology*, 3(3):430 – 454, 1972.

[Kittock, 1995] J. Kittock. Emergent conventions and the structure of multi-agent systems. In *Lectures in Complex systems: the proceedings of the 1993 Complex systems summer school, Santa Fe Institute Studies in the Sciences of Complexity Lecture Volume VI, Santa Fe Institute*, pages 507–521. Addison-Wesley, 1995.

[Lewis, 1969] D. K. Lewis. *Convention: A Philosophical Study*. Harvard University Press, Cambridge (MA), 1969.

[López y López and Luck, 2004] F. López y López and M. Luck. Towards a model of the dynamics of normative multi-agent systems. In G. Lindemann, D. Moldt, and M. Paolucci, editors, *RASTA 2002*, volume 2934 of *Lecture Notes in Artificial Intelligence*, pages 175–194, Heidelberg, 2004. Springer.

[López y López and Márquez, 2004] F. López y López and A. A. Márquez. An architecture for autonomous normative agents. In *Proceedings of the Fifth Mexican International Conference in Computer Science - ENC*, pages 96–103, Los Alamitos, CA, USA, 2004. IEEE Computer Society.

[López y López et al., 2002] F. López y López, M. Luck, and M. d'Inverno. Constraining autonomy through norms. In *Proceedings of the First International Joint Conference on Autonomous Agents and Multiagent Systems AAMAS*, pages 674–681, New York, NY, USA, 2002. ACM.

[López y López et al., 2006] F. López y López, M. Luck, and M. d'Inverno. A normative framework for agent-based systems. *Computational & Mathematical Organization Theory*, 12(2):227–250, 2006.

[López y López et al., 2007] F. López y López, M. Luck, and M. d'Inverno. A normative framework for agent-based systems. In Guido Boella, Leon van der Torre, and Harko Verhagen, editors, *Normative Multi-agent Systems, Dagstuhl Seminar Proceedings 07122*, Dagstuhl Seminar Proceedings. Internationales Begegnungs- und Forschungszentrum für Informatik (IBFI), Schloss Dagstuhl, Germany, 2007.

[López y López, 2003] F. López y López. *Social Powers and Norms: Impact on Agent Behaviour*. PhD thesis, Department of Electronics and Computer Science, University of Southampton, United Kingdom, 2003.

[Mahmoud et al., 2012] S. Mahmoud, N. Griffiths, J. Keppens, and M. Luck. Efficient norm emergence through experiential dynamic punishment. In *ECAI'12*, pages 576–581, 2012.

[Mahmoud et al., 2014a] M. A. Mahmoud, M. S. Ahmad, M. Z. M. Yusoff, and A. Mustapha. Norms assimilation in heterogeneous agent community. In Hoa Khanh Dam, Jeremy Pitt, Yang Xu, Guido Governatori, and Takayuki Ito, editors, *Principles and Practice of Multi-Agent Systems - 17th International Conference, PRIMA 2014*, volume 8861 of *Lecture Notes in Artificial Intelligence*, pages 311–318, Berlin, 2014. Springer.

[Mahmoud et al., 2014b] M. A. Mahmoud, M. S. Ahmad, M. Z. M. Yusoff, and A. Mustapha. A review of norms and normative multiagent systems. *The Scientific World Journal*, 2014:23 pages, 2014. Article ID 684587.

[Mahmoud et al., 2015] S. Mahmoud, N. Griffiths, J. Keppens, A. Taweel, T. J. Bench-Capon, and M. Luck. Establishing norms with metanorms in distributed computational systems. *Artif. Intell. Law*, 23(4):367–407, December 2015.

[Makinson and van der Torre, 2000] D. Makinson and L. van der Torre. Input/output logics. *Journal of Philosophical Logic*, 29:383–408, 2000.

[Makinson and van der Torre, 2003] D. Makinson and L. van der Torre. What is input/output logic. In Benedikt Löwe, Wolfgang Malzkom, and Thoralf Räsch, editors, *Foundations of the Formal Sciences II : Applications of Mathematical Logic in Philosophy and Linguistics (Papers of a conference held in Bonn, November 10-13, 2000)*, Trends in Logic, vol. 17, pages 163–174, Dordrecht, 2003. Kluwer. Reprinted in this volume.

[Manna and Waldinger, 1980] Z. Manna and R. Waldinger. A deductive approach to program synthesis. *ACM Transactions on Programming Languages and Systems*, 2(1):90–121, 1980.

[Maranhão, 2001] J. Maranhão. Refinement. A tool to deal with inconsistencies. In *Proceedings of the 8th ICAIL*, pages 52–59, 2001.

[Maranhão, 2017] J. Maranhão. A logical architecture for dynamic legal interpretation. In *Proceedings of 16th ICAIL*, pages 129–139, 2017.

[Meneguzzi and Luck, 2009] F. Meneguzzi and M. Luck. Norm-based behaviour modification in bdi agents. In *Proceedings of The 8th International Conference on Autonomous Agents and Multiagent Systems - Volume 1*, AAMAS '09, pages 177–184, Richland, SC, 2009. International Foundation for Autonomous Agents and Multiagent Systems.

[Mihaylov et al., 2014] M. Mihaylov, K. Tuyls, and A. Nowé. A decentralized approach for convention emergence in multi-agent systems. *Autonomous Agents and Multi-Agent Systems*, 28(5):749–778, 2014.

[Morales et al., 2013] J. Morales, M. López-Sánchez, J. A. Rodríguez-Aguilar, M. Wooldridge, and W. Vasconcelos. Automated synthesis of normative systems. In *Proceedings of the 2013 International Conference on Autonomous Agents and Multi-Agent Systems*, AAMAS '13, pages 483–490, 2013.

[Morales et al., 2014] J. Morales, M. López-Sánchez, J. A. Rodríguez-Aguilar, M. Wooldridge, and W. Vasconcelos. Minimality and simplicity in the on-line automated synthesis of normative systems. In *Proceedings of the 2014 International Conference on Autonomous Agents and Multi-Agent Systems*, AAMAS '14, pages 109–116, 2014.

[Morales et al., 2015a] J. Morales, M. López-Sánchez, J. A. Rodríguez-Aguilar, W. Vasconcelos, and M. Wooldridge. Online automated synthesis of compact normative systems. *ACM Trans. Auton. Adapt. Syst.*, 10(1):2:1–2:33, March 2015.

[Morales et al., 2015b] J. Morales, M. López-Sánchez, J. A. Rodríguez-Aguilar, M. Wooldridge, and W. Vasconcelos. Synthesising liberal normative systems. In *Proceedings of the 2015 International Conference on Autonomous Agents and Multiagent Systems*, AAMAS '15, pages 433–441, Richland, SC, 2015. International Foundation for Autonomous Agents and Multiagent Systems.

[Morales et al., 2015c] J. Morales, I. Mendizabal, D. Sanchez-Pinsach, M. López-Sánchez, and J. A. Rodríguez-Aguilar. Using IRON to build frictionless on-line communities. *AI Commun.*, 28(1):55–71, 2015.

[Mukherjee et al., 2007] P. Mukherjee, S. Sen, and S. Airiau. Emergence of norms with biased interaction in heterogeneous agent societies. In *Proceedings of the 2007 IEEE/WIC/ACM International Conferences on Web Intelligence and Intelligent Agent Technology*, pages 512–515, 2007.

[Mukherjee et al., 2008] P. Mukherjee, S. Sen, and S. Airiau. Norm emergence under constrained interactions in diverse societies. In *Proceedings of the 7th International Joint Conference on Autonomous Agents and Multiagent Systems - Volume 2*, AAMAS '08, pages 779–786, Richland, SC, 2008. International Foundation for Autonomous Agents and Multiagent Systems.

[Nadelman, 1990] E. Nadelman. Global prohibition regimes: The evolution of norms in international society. *International Organization*, 44(4):479–526, 1990.

[Nakamaru and Levin, 2004] M. Nakamaru and S. A. Levin. Spread of two linked social norms on complex interaction networks. *Journal of Theoretical Biology*, 230(1):57–64, 2004.

[Onn and Tennenholtz, 1997] S. Onn and M. Tennenholtz. Determination of social laws for multi-agent mobilization. *Artificial Intelligence*, 95(1):155–167, 1997.

[Ossowski, 2013] S. Ossowski. *Agreement Technologies*. Springer, Dordrecht (NL), 2013.

[Perreau de Pinninck et al., 2008] A. Perreau de Pinninck, C. Sierra, and M. Schorlemmer. Distributed norm enforcement via ostracism. In JaimeSimpo Sichman, Julian Padget, Sascha Ossowski, and Pablo Noriega, editors, *Coordination, Organizations, Institutions, and Norms in Agent Systems III*, volume 4870 of *Lecture Notes in Computer Science*, pages 301–315. Springer, Berlin, 2008.

[Perreau de Pinninck et al., 2010] A. Perreau de Pinninck, C. Sierra, and M. Schorlemmer. A multiagent network for peer norm enforcement. *Autonomous Agents and Multi-Agent Systems*, 21(3):397–424, 2010.

[Pigozzi and van der Torre, 2017] G. Pigozzi and L. van der Torre. Multiagent deontic logic and its challenges from a normative systems perspective. *The IfCoLog Journal of Logics and their Applications*, 4(9), 2017.

[Pucella and Weissman, 2004] R. Pucella and V. Weissman. Reasoning about dynamic policies. In Igor Walukiewicz, editor, *Foundations of Software Science and Computation Structures*, pages 453–467, Berlin, Heidelberg, 2004. Springer Berlin Heidelberg.

[Pujol et al., 2005] J. M. Pujol, J. Delgado, R. Sangüesa, and A. Flache. The role of clustering on the emergence of efficient social conventions. In *Proceedings of the 19th International Joint Conference on Artificial Intelligence*, IJCAI'05, pages 965–970, San Francisco, CA, USA, 2005. Morgan Kaufmann Publishers Inc.

[Rao and Georgeff, 1995] A. S. Rao and M. P. Georgeff. BDI agents: From theory to practice. *Proceedings of the First International Conference on Multi-Agent Systems (ICMAS-95)*, pages 312–319, 1995.

[Risse et al., 1999] T. Risse, S. Ropp, and K. Sikkink. *The Power of Human Rights: International Norms and Domestic Change*. Cambridge University Press, Cambridge, 1999.

[Riveret et al., 2012] R. Riveret, A. Rotolo, and G. Sartor. Probabilistic rule-based argumentation for norm-governed behaviour. *Artificial Intelligence*, 20(4):383–420, 2012.

[Riveret et al., 2013] R. Riveret, G. Contissa, D. Busquets, A. Rotolo, J. Pitt, and G. Sartor. Vicarious reinforcement and ex ante law enforcement: A study in norm-governed learning agents. In *Proceedings of the Fourteenth International Conference on Artificial Intelligence and Law*, ICAIL '13, pages 222–226, New York, NY, USA, 2013. ACM.

[Riveret et al., 2014] R. Riveret, A. Artikis, D. Busquets, and J. Pitt. Self-governance by transfiguration: From learning to prescriptions. In Fabrizio Cariani, Davide Grossi, Joke Meheus, and Xavier Parent, editors, *Deontic Logic and Normative Systems*, volume 8554 of *Lecture Notes in Computer Science*, pages 177–191. Springer, Springer, 2014.

[Russell et al., 2015] S. Russell, D. Dewey, and M. Tegmark. Research priorities for robust and beneficial artificial intelligence. *AI Magazine*, 36(4):105–114, 2015.

[Saam and Harrer, 1999] N. J. Saam and A. Harrer. Simulating norms, social inequality, and functional change in artificial societies. *Journal of Artificial Societies and Social Simulation*, 2(1):2, 1999.

[Salazar et al., 2010] N. Salazar, J. A. Rodríguez-Aguilar, and J. L. Arcos. Robust coordination in large convention spaces. *AI Communications*, 23(4):357–372, 2010.

[Savarimuthu et al., 2011] B. T. R. Savarimuthu, R. Arulanandam, and M. Purvis. Aspects of active norm learning and the effect of lying on norm emergence in agent societies. In David Kinny, JaneYung-jen Hsu, Guido Governatori, and AdityaK. Ghose, editors, *Agents in Principle, Agents in Practice*, volume 7047 of *Lecture Notes in Computer Science*, pages 36–50. Springer, Berlin, 2011.

[Savarimuthu and Cranefield, 2009] B. T. R. Savarimuthu and S. Cranefield. A categorization of simulation works on norms. In Guido Boella, Gabriella Pigozzi, and Leendert van der Torre, editors, *Normative Multi-agent Systems, Dagstuhl Seminar Proceedings 09121*, pages 39–58, Internationales Begegnungs- und Forschungszentrum für Informatik (IBFI), Schloss Dagstuhl, Germany, 2009.

[Savarimuthu and Cranefield, 2011] B. T. R. Savarimuthu and S. Cranefield. Norm creation, spreading and emergence: A survey of simulation models of norms in multi-agent systems. *Multiagent and Grid Systems*, 7(1):21–54, January 2011.

[Savarimuthu et al., 2007] B. T. R. Savarimuthu, S. Cranefield, M. Purvis, and M. Purvis. Norm emergence in agent societies formed by dynamically changing networks. In *2007 IEEE/WIC/ACM International Conference on Intelligent Agent Technology*, pages 464–470, 2007.

[Savarimuthu et al., 2008a] B. T. R. Savarimuthu, S. Cranefield, M. Purvis, and M. Purvis. Role model based mechanism for norm emergence in artificial agent societies. In JaimeSimpo Sichman, Julian Padget, Sascha Ossowski, and Pablo Noriega, editors, *Coordination, Organizations, Institutions, and Norms in Agent Systems III*, volume 4870 of *Lecture Notes in Computer Science*, pages 203–217. Springer, Berlin, 2008.

[Savarimuthu et al., 2008b] B. T. R. Savarimuthu, S. Cranefield, M. A. Purvis, and M. K. Purvis. Social norm emergence in virtual agent societies. In *Proceedings of the 7th International Conference on Autonomous Agents and Multi-Agent Systems*, 2008.

[Savarimuthu et al., 2009] B. T. R. Savarimuthu, S. Cranefield, M. A. Purvis, and M. K. Purvis. Norm emergence in agent societies formed by dynamically changing networks. *Web Intelligence and Agent Systems*, 7(3):223–232, 2009.

[Savarimuthu et al., 2010a] B. T. R. Savarimuthu, S. Cranefield, M. A. Purvis, and M. K. Purvis. A data mining approach to identify obligation norms in agent societies. In *Proceedings of the International Workshop on Agents and Data Mining Interaction (ADMI@AAMAS 2010), Toronto, Canada*, pages 54–69, May 2010.

[Savarimuthu et al., 2010b] B. T. R. Savarimuthu, S. Cranefield, M. A. Purvis, and M. K. Purvis. Obligation norm identification in agent societies. *Journal of Artificial Societies and Social Simulation*, 13(4):3, 2010.

[Savarimuthu et al., 2013a] B. T. R. Savarimuthu, S. Cranefield, M. A. Purvis, and M. K. Purvis. Identifying prohibition norms in agent societies. *Artificial Intelligence and Law*, 21:1–46, 2013.

[Savarimuthu et al., 2013b] B. T. R. Savarimuthu, J. Padget, and M. A. Purvis. Social norm recommendation for virtual agent societies. In Guido Boella, Edith Elkind, Bastin Tony Roy Savarimuthu, Frank Dignum, and Martin K. Purvis, editors, *PRIMA 2013: Principles and Practice of Multi-Agent Systems*, volume 8291 of *Lecture Notes in Computer Science*, pages 308–323. Springer, Berlin, 2013.

[Savarimuthu, 2011] B. T. R. Savarimuthu. *Mechanisms for Norm Emergence and Norm Identification in Multi-Agent Societies*. PhD thesis, University of Otago, Dunedin, New Zealand, 2011.

[Scheve et al., 2006] C. von Scheve, D. Moldt, J. Fix, and R. von Luede. My agents love to conform: Norms and emotion in the micro-macro link. *Computational & Mathematical Organization Theory*, 12(2):81–100, 2006.

[Schneider and Teske, 1992] M. Schneider and P. Teske. Toward a theory of the political entrepreneur: Evidence from local government. *American Political Science Review*, 86(3):737–747, 1992.

[Sen and Airiau, 2007] S. Sen and S. Airiau. Emergence of norms through social learning. In Manuela Veloso, editor, *Proceedings of the 20th International Joint Conference on Artifical Intelligence*, IJCAI'07, pages 1507–1512, San Francisco (CA), 2007. Morgan Kaufmann Publishers Inc.

[Sen and Sen, 2010] O. Sen and S. Sen. Effects of social network topology and options on norm emergence. In Julian Padget, Alexander Artikis, Wamberto Vasconcelos, Kostas Stathis, VivianeTorres da Silva, Eric Matson, and Axel Polleres, editors, *Coordination, Organizations, Institutions and Norms in Agent Systems V*, volume 6069 of *Lecture Notes in Computer Science*, pages 211–222. Springer, Berlin, 2010.

[Shoham and Tennenholtz, 1992a] J. Shoham and M. Tennenholtz. Emergent conventions in multiagent systems: Initial experimental results and observations. In *Proceedings of the Third International Conference on the Principles of Knowledge Representation and Reasoning KR*, pages 225–231, San Mateo, CA, USA, 1992. Morgan Kaufmann.

[Shoham and Tennenholtz, 1992b] Y. Shoham and M. Tennenholtz. On the synthesis of useful social laws for artificial agent societies. In *Proceedings of the 10th National Conference on Artificial Intelligence (AAAI '92)*, pages 276–281, San Jose (CA), July 1992.

[Shoham and Tennenholtz, 1995] Y. Shoham and M. Tennenholtz. On social laws for artificial agent societies: off-line design. *Artificial Intelligence*, 73(1-2):231 – 252, 1995. Computational Research on Interaction and Agency, Part 2.

[Shoham and Tennenholtz, 1997] Y. Shoham and M. Tennenholtz. On the emergence of social conventions: modeling, analysis, and simulations. *Artificial Intelligence*, 94(1–2):139–166, July 1997.

[Simon, 1955] H. A. Simon. A Behavioral Model of Rational Choice. *The Quarterly Journal of Economics*, 69(1):99–118, 1955.

[Sims and Brinkmann, 2003] R. R. Sims and J. Brinkmann. Enron ethics (or: Culture matters more than codes). *Journal of Business Ethics*, 45(3):243–256, Jul 2003.

[Singh, 2014] M. P. Singh. Norms as a basis for governing sociotechnical systems. *ACM Trans. Intell. Syst. Technol.*, 5(1):21:1–21:23, January 2014.

[Staller and Petta, 2001] A. Staller and P. Petta. Introducing emotions into the computational study of social norms: A first evaluation. *Journal of Artificial Societies and Social Simulation*, 4(1):2, 2001.

[Stolpe, 2010] A. Stolpe. Norm-system revision: Theory and application. *Artificial Intelligence and Law*, 18:247–283, 2010.

[Sugawara, 2011] T. Sugawara. Emergence and stability of social conventions in conflict situations. In *Proceedings of the Twenty-Second International Joint Conference on Artificial Intelligence - Volume Volume One*, IJCAI'11, pages 371–378. AAAI Press, 2011.

[Sunstein, 1996] C. R. Sunstein. Social norms and social roles. *Columbia Law Review*, 96(4):903–968, 1996.

[Tinnemeier *et al.*, 2009] N. Tinnemeier, M. Dastani, J.-J. C. Meyer, and L. van der Torre. Programming normative artifacts with declarative obligations and prohibitions. In *Proceedings of the 2009 IEEE/WIC/ACM International Conference on Intelligent Agent Technology (IAT 2009)*, pages 145–152. IEEE Computer Society, 2009.

[Tinnemeier *et al.*, 2010] N. Tinnemeier, M. Dastani, and J.-J. C. Meyer. Programming norm change. In *Proceedings of the 9th International Conference on Autonomous Agents and Multiagent Systems: Volume 1 - Volume 1*, AAMAS '10, pages 957–964, Richland, SC, 2010. International Foundation for Autonomous Agents and Multiagent Systems.

[Urbano *et al.*, 2009] P. Urbano, J. Balsa, L. Antunes, and L. Moniz. Force versus majority: A comparison in convention emergence efficiency. In JomiFred Hübner, Eric Matson, Olivier Boissier, and Virginia Dignum, editors, *Coordination, Organizations, Institutions and Norms in Agent Systems IV*, volume 5428 of *Lecture Notes in Computer Science*, pages 48–63. Springer, Berlin, 2009.

[Valk, 1998] R. Valk. Petri nets as token objects. an introduction to elementary. In *Proceedings of Application and Theory of Petri Nets*, pages 1–25, Berlin, 1998. Springer.

[van Ditmarsch *et al.*, 2007] H. van Ditmarsch, W. van der Hoek, and B. Kooi. *Dynamic Epistemic Logic*. Synthese Library Series, vol. 337, Springer, Heidelberg, 2007.

[Verhagen, 2000] H. J. E. Verhagen. *Norm Autonomous Agents*. PhD thesis, The Royal Institute of Technology and Stockholm University, Stockholm, Sweden, 2000.

[Verhagen, 2001] H. Verhagen. Simulation of the learning of norms. *Social Science Computer Review*, 19(3):296–306, 2001.

[Villatoro *et al.*, 2009] D. Villatoro, S. Sen, and J. Sabater-Mir. Topology and memory effect on convention emergence. In *Proceedings of the 2009 IEEE/WIC/ACM International Joint Conference on Web Intelligence and Intelligent Agent Technology - Volume 02*, WI-IAT '09, pages 233–240, 2009.

[Villatoro *et al.*, 2011a] D. Villatoro, G. Andrighetto, J. Sabater-Mir, and R. Conte. Dynamic sanctioning for robust and cost-efficient norm compliance. In Toby Walsh, editor, *Proceedings of the 22nd International Joint Conference on Artificial Intelligence*, volume 1 of *IJCAI'11*, pages 414–419. AAAI Press, 2011.

[Villatoro *et al.*, 2011b] D. Villatoro, J. Sabater-Mir, and S. Sen. Social instruments for robust convention emergence. In Toby Walsh, editor, *Proceedings of the 22nd International Joint Conference on Artificial Intelligence*, volume 1 of *IJCAI'11*, pages 420–425. AAAI Press, 2011.

[Villatoro *et al.*, 2013] D. Villatoro, J. Sabater-Mir, and S. Sen. Robust convention emergence in social networks through self-reinforcing structures dissolution. *ACM Trans. Auton. Adapt. Syst.*, 8(1):2:1–2:21, April 2013.

[Walker and Wooldridge, 1995] A. Walker and M. Wooldridge. Understanding the emergence of conventions in multi-agent systems. In *Proceedings of the first international conference on multi-agent systems (ICMAS)*, pages 384–389, Menlo Park (CA), 1995. AAAI Press.

[Watkins and Dayan, 1992] C. J. C. H. Watkins and P. Dayan. Technical note: Q-learning. *Machine Learning*, 8(3-4):279–292, 1992.

[Watts and Strogatz, 1998] D. J. Watts and S. H. Strogatz. Collective dynamics of 'small-world' networks. *Nature*, 393(6684):440–442, June 1998.

[Young, 1990] O. Young. Political leadership and regime formation: On the development of institutions in international society. *International Organization*, 45(3):281–308, 1990.

[Younger, 2004] S. Younger. Reciprocity, normative reputation, and the development of mutual obligation in gift-giving societies. *Journal of Artificial Societies and Social Simulation*, 7(1):5, 2004.

[Yu *et al.*, 2010] C.-H. Yu, J. Werfel, and R. Nagpal. Collective decision-making in multi-agent systems by implicit leadership. In *Proceedings of the 9th International Conference on Autonomous Agents and Multiagent Systems: Volume 3 - Volume 3*, AAMAS '10, pages 1189–1196, Richland, SC, 2010. International Foundation for Autonomous Agents and Multiagent Systems.

[Yu *et al.*, 2013] C. Yu, M. Zhang, F. Ren, and X. Luo. Emergence of social norms through collective learning in networked agent societies. In *Proceedings of the 2013 International Conference on Autonomous Agents and Multi-agent Systems*, AAMAS '13, pages 475–482, 2013.

[Yu *et al.*, 2015] C. Yu, H. Lv, F. Ren, H. Bao, and J. Hao. Hierarchical learning for emergence of social norms in networked multiagent systems. In *AI 2015: Advances in Artificial Intelligence: 28th Australasian Joint Conference, Canberra, ACT, Australia, November 30 – December 4, 2015, Proceedings*, pages 630–643, Cham, 2015. Springer International Publishing.

[Zhang and Leezer, 2009] Y. Zhang and J. Leezer. Emergence of social norms in complex networks. In *International Conference on Computational Science and Engineering*, pages 549–555, Vancouver, 2009.

5
Modeling Organizations and Institutions in MAS

NICOLETTA FORNARA, TINA BALKE-VISSER

ABSTRACT. *Institutions* and *Organizations* are two concepts within the MAS community that are commonly referred to when the question arises on how to ensure that an (open) MAS exhibits some desired properties, while the agents interacting in that MAS have some degree of autonomy at the same time. This chapter gives a brief introduction to the two concepts and its related ideas. It outlines research done in the area of normative MAS and gives pointers on current challenges for modeling institutions and organizations.

1 Introduction

In Multiagents systems (MAS), software agents that enjoy some degree of autonomy interact [Wooldridge and Jennings, 1995]. As a consequence, similar to human societies, the problem arises on how to ensure that the MAS exhibits some desired global property, without compromising the agent's autonomy at the same time [Grossi, 2007, p. 2]. Leaning on existing works such as for example in sociology, psychology and organizational theory, in recent years MAS researchers have been starting to incorporate and model concepts such organizations and institutions in computational systems, as demonstrated by several publications on the topic in the AAMAS conference series[1], the COIN workshop series[2], and the Normative Multiagent Systems seminars[3].

In this Chapter, we will provide an introduction to the concepts of institutions and organizations and their modeling. This chapter is not aiming to be an in-depth literature review and it will not give details on all aspects of modeling institutions and organizations in MAS, but it rather aims to point the interest reader to topics and areas of interest and give him or her starting points for further studies.

Wanting to model "institutions" and "organizations" a first step is to understand what the two words means and what concept they refer to. As

---

[1] http://www.ifaamas.org/proceedings.html
[2] http://www.pcs.usp.br/~coin
[3] http://icr.uni.lu/normas/history.html

simple as this sounds, this task is not an easy one as (i) not only are the two concepts interlinked - they are both broadly speaking, coordinate means [Grossi, 2007] - but (ii) in the agents community, the words are often used as synonymous. One of the reasons for the latter is that different research communities started to use the terms differently, sometimes borrowing concepts from other disciplines. Researcher wanting to publish/work in the respective communities – in order to pass review processes for their papers – had to use the communities jargon, i.e. use the terms the community was using. This resulted in situations where researchers used different words for one and the same idea they described, in order to publish in the different communities they were working in.

Taking a step back, a popular source, often cited by the agent community when it comes to the definition of the terms "institutions" and "organizations", is North [North, 1996, p. 4f]:

> "A crucial distinction in this study is made between institutions and organizations. [...] Conceptually what must be clearly differentiated are the rules from the players. The purpose of the rules is to define the way the game is played. But the objective of the team within that set of rules is to win the game – by a combination of skills, strategy and coordination; by fair means and sometimes by foul means. Modeling the strategies and the skills of the team as it develops is a separate process from modeling the creation, evolution, and consequences of the rules."

The distinction indicated by North is the idea that organizations are agents like households, firms and states that have preferences and objectives, whereas institution are formal and informal societal constraints such as laws, conventions, constitutions, habits and rules, which reduce the total scarce resources available [Khalil, 1995]. Broadly speaking, both institutions as well as organizations are means of coordination and provide some form of structure, but whereas institutions focus on the structure of the rules and norms, organizational structure of a MAS concerns the agents, their roles and their relationships by which the overall behavior of the MAS is defined [Grossi, 2007].

Based on this abstract distinction, this chapter tries to give an overview of both, modeling organizations as well as institutions (and the differences between them). For this purpose, we start by looking at the modeling of institutions first, by discussing regulative and constitutive norms, as well as agents communication languages as means to communicate and thereby share norms. Afterwards the focus of the chapter will turn to organizations with the topics of modeling agents (and their roles) as well as their relations to one another in terms of organizational structures are being addressed.

2 Survey on Modeling Institutions in MAS

The formalization, realization, and management of open distributed interaction systems where autonomous software agents (operating on behalf of one or more human users) may interact for exchanging resources or providing services is widely recognized to be an important research problem that is becoming more and more relevant with the massive development of distributed social network systems on the Internet.

One approach, which may be followed for the formalization of those systems, consists in modeling them as a set of *Artificial Institutions* (AI). Human social institutions [Searle, 1995], like for example the institution of *marriage* or *family*, the institution of *money*, or the institution of *education*, have been used, in Multiagent Systems (MAS) research, as a source of inspiration for the definition of the various abstract concepts and software components required for the concrete realization of artificial/electronic institutions. Artificial institutions are fundamental because:

> "Their main purpose is to **enable** and **regulate** the interaction among autonomous agents in order to achieve some collective endeavour" [Fornara et al., 2013, p. 278].

In *open distributed systems* or *socio-technical systems* [Chopra et al., 2014], which support the interaction of various components (autonomous software called agents or humans) with different, often competitive, goals, there is the need to *enable* and to *regulate* such interactions. This with the objective to keep the evolution of those systems within certain boundaries, and to make it possible for the system itself to reach certain social goals. There is also the need to create in the interacting agents an *expectation* on the reasonable future evolution of the interaction, this in order to enable the agents to coherently plan their actions for reaching their own goals. This can be done by formalizing an open distributed system using multiple, sometime interconnected, *artificial institutions*, and by modelling and realizing certain software components for their management.

In order to create *open spaces* where interactions among autonomous agents may happen and where those interactions may be constrained without being always regimented, it is necessary to analyze and formally specify the various interconnected *static and dynamic components* that enable and regulate those interactions in real life. It is therefore necessary to:

1. Formally define the application-independent abstract concepts that are relevant for the specification of agents' institutions, for example the notion of *norm* or *regulation*, *institutional power*, and *constitutive rule*;

2. Specify the software components required for the management of the concrete objects created from the abstract concepts, like for example

a *norm monitoring* component, or a component for computing the state of the interaction among agents on the basis of the concrete institutional powers assigned to the agents in a system;

3. Formally define the conceptual and logical model of the application dependent knowledge/data used by those software components.

The choice of the formalism to adopt for the specification and implementation of the various software components and for the specification of the abstract concepts and the concrete definition of instances of those concepts is an important aspect in the definition of a model for artificial institutions.

In MAS literature, as discussed below, there are various proposals for the formalization of artificial institutions, which are also called electronic institutions or agent institutions. The conceptual model of the fundamental concepts required for the formal specification of those institutions is usually placed side by side with the specification of an institutional framework required for the actual implementation of institutions and of the software components for their management.

A recent and very interesting discussion on the analogies and differences between artificial/electronic institutions (AI) and virtual organizations plus an extensive comparison of three models for the formal specification of institutions are presented in [Fornara et al., 2013][4]. The three discussed and compared models are: (i) the ANTE Framework [Lopes Cardoso, 2010]; (ii) the OCeAN metamodel for the specification of Artificial Institutions (AI) [Fornara and Colombetti, 2009a; Fornara, 2011] that has been extended into the MANET model for the specification of AI situated in environments [Tampitsikas et al., 2012]; (iii) and the conceptual model and computational architecture for Electronic Institutions (EIs) extended with the EIDE development environment [Arcos et al., 2005; Esteva et al., 2008].

Other interesting frameworks, which are briefly discussed in [Fornara et al., 2013], are: (i) the *OMNI* model [Dignum, 2004] which allows the description of MAS-based organizations followed by the *OperettA* framework [Aldewereld and Dignum, 2010] which supports the implementation of real systems; (ii) the *instAL* normative framework [Cliffe et al., 2007c; Corapi et al., 2011] that may be used to specify, verify and reason about norms used to regulate MAS; (iii) and the definition of Norm-Governed Computational Societies [Artikis et al., 2009] followed by the specification of Sustainable

---

[4]This chapter is part of the book "Agreement Technologies" [Ossowski, 2013], published as result of the COST Action on Agreement Technologies[5] (2008-2012) which involved many researchers in the field of institutional and normative multiagent systems.

Institutions [Pitt *et al.*, 2011] which is influenced by Olstrom definition of institution [Ostrom, 1990].

Another relevant book chapter on regulated MAS, which can be used to create an institutional reality where autonomous agents may interact, is the chapter "Regulated MAS: Social Perspective" [Noriega *et al.*, 2013]. It is part of the book on Normative Multiagent Systems that was published as result of NorMAS 2012 Seminar, which took place in March 2012 in Dagstuhl, Germany.

Crucial abstract concepts which are required for the specification of every artificial institution and of the software components for their management are discussed in the following sections.

2.1 Regulative rules - Norms

A very important characteristic of the autonomous agents developed by different users that interact on an open network, like for example a peer to peer network in Internet, is that no assumption can be made on their internal design and, like for human beings, it is impossible to assume that they will always fulfill their norms. In particular as discussed in [Fornara and Colombetti, 2009b] it is not always possible and advantageous to regiment all obligations. Think for example to the obligation to pay for an ordered product when it is received, at least, it is reasonable to sanction irregular behaviors, but it is difficult to regiment it. In MAS research there are numerous proposals to formally and declaratively specify norms or policies (these two terms are very often used as synonymous). They are used for expressing *obligations*, *permissions*, or *prohibitions* to perform certain actions when certain specific conditions, related to state of affairs or to specific events, are satisfied. Norms/policies are usually characterized by the following attributes: the *type*, for distinguishing between obligations, permissions and prohibitions, the *debtor* or the *addressee* which may be expressed using *roles*, the *activation* and the *expiration* or *deactivation conditions* which describe the events or the state of affairs that activate or deactivate the norm, the *content* that is the prohibited, permitted or obliged action or state of affair, and the sanctions and reward for norms fulfillment or violation.

An attribute that may be useful in the specification of abstract norms at design time, is the *roles* that an agent may play in an artificial institution, for example in a auction the role of participant or auctioneer. When one or more institutions are instantiated for the realization of a concrete open interaction system, the various roles defined in those institutions have to be replaced by their concrete counterpart that an agent may play in a concrete realization of an artificial institution, like the participant of the auction number run01. When a norm, related to various agents by using a concrete role attribute, becomes active, various instances of that norm (one for each agent that plays the concrete role specified in the norm) need to be created and managed by the interaction system.

An important software component of every normative system is the *monitor of norms*. That is, a component able to compute the state of the norms on the basis of their activation condition and content and on the basis of the actions and events that happen in the system. This component is necessary in order to be able to compute or deduce if a given norm is fulfilled or violated and therefore apply sanctions or reward with the goal of enforcing norms fulfillment for all those norms that are not regimented. In order that the monitoring component can use the knowledge about the state of the interaction for computing the fulfillment or violation of norms it necessary to realize a *synchronization component* able to dynamically update the *knowledge base* used for representing the state of the interaction with the observable changes due to events or agents actions. The realization of this components may be challenging, in particular for those events that are nor easily observable by means of sensors or that are not directly connected to the actions of the agents. It is also crucial that the *knowledge base* used for representing the state of the interaction has a conceptual model of the concepts and properties (or relations) that are relevant for the description and regulation of the interaction. Starting from general concepts like time, action, event to the specification of application dependent part like for instance in an auction the notion of offer, owner, and the action of paying and delivering.

In an open normative multiagent system, it is crucial to specify the norms using formal declarative languages (like logics or logic programming languages). This choice has many important advantages, because it makes possible to:

- Represent the norms as data, instead of coding them into the software, with the advantage of making possible to add, remove, or change the norms both when the system is off line, and at run-time, without the need to reprogram some components of the interaction system or the software agents that use the system;

- Develop agents able to reason and plan their actions by taking into considerations the correlations between their goals and external social constraints expressed also in terms of norms, this by reasoning on what norms apply in a given situation, what activities are obligatory, permitted or prohibited, using for example some form of what-if reasoning [Uszok et al., 2008] for deciding whether or not to comply to norms by taking into account norm rewards and sanctions;

- Develop agents able to interact within different systems without the need of being reprogrammed [Fornara, 2011];

- Realize an application-independent monitoring component able to keep trace of the state of norms on the basis of the events that happen in the system, on the basis of the agents' actions, and on the basis of the state of the interaction (this mechanism can also be able to react to norm fulfillments or violations);

- Realize mechanisms for checking norm conflicts, understanding when conflicts may arise, and solving or avoiding them, like for example by introducing priority ordering between norms [Elagh, 2000; Sensoy *et al.*, 2012].

The choice of the formal language used for the declarative specification of norms is difficult because many aspects have to be taken into account. The most important are: the expressive power of the language, its computational complexity, the fact that the underlying logic is decidable, the diffusion of the language among software practitioners and research communities, its feasibility for fast prototyping, and its adoption as an international standard an crucial aspect for having good interoperability between separately engineered software agents.

Semantic Web Technologies [Hitzler *et al.*, 2009] may be the successfully adopted for an efficient and effective representation of norms/policies for open interaction systems running on the Internet. A relevant advantage of this choice is that Semantic Web technologies are increasingly becoming a standard for Internet applications and therefore are supported by many tools: many efficient reasoners (like Fact++, Pellet, Racer Pro, HermiT), tools for ontology editing (like Protg), and libraries for automatic ontology management (like OWL-API and JENA).

Currently there are some works, in the multiagent community, that adopt Semantic Web Technologies for the formalization and management of norms. One is represented by the works of Fornaras group who has investigated the possibility to use OWL 2 and SWRL rules for the specification of agents commitments and obligations [Fornara and Colombetti, 2010; Fornara, 2011; Marfia *et al.*, 2016]. In those papers an OWL ontology of obligations. The *activation condition* of an obligation is a class of events that when happen trigger the activation of the obligation. The *content* of an obligation is a class of possible actions that have to be performed within a given deadline. The proposed model of obligations allow also to specify the relation between obligations and time, it is therefore possible to specify deadlines and interval of time. The *monitoring* of those obligations, that is checking if they are fulfilled of violated on the basis of the actions of the agents, can be realized thanks to a specific framework required for managing the elapsing of time and to perform closed-world reasoning on certain classes. A similar ontological formalization of obligations has been also extended for being used in a complete OWL 2 model of artificial institutions (initially called OCeAN and subsequently MANET) instantiated at run-time by dynamically creating in the environment *spaces of interaction* [Fornara and Tampitsikas, 2013]. An updated version of such a model of obligations has been also applied in the field of access control where policies has to be specified for regulating the access and the use of data [Nguyen et al., 2015; Marfia et al., 2016].

Another interesting approach that uses Semantic Web Technologies for norms formalization and management is the OWL-POLAR framework for semantic policy representation and reasoning [Sensoy et al., 2012]. This framework investigates the possibility of using OWL ontologies for representing the state of the interaction among agents and SPARQL queries for reasoning on policies activation, for anticipating possible conflicts among policies, and for conflicts avoidance and resolution.

Another relevant proposal where Semantic Web technologies are used for policy specification and management is the KAoS policy management framework [Uszok et al., 2008], which is composed by three layers: (i) the human interface layer where policies, expressing authorizations and obligations, are specified in the form of constrained English sentences; (ii) the policy management layer used for encoding in OWL the policy-related information; (iii) the policy monitoring and enforcing layer used to compile OWL policies to an efficient format usable for monitoring and enforcement.

2.2 Constitutive rules

As clearly presented by John Searle [Searle, 1995] in his book "the construction of social reality", in human interactions are involved brute facts and facts that exist only thanks to an institutional setting. This second type of facts are called *institutional facts*. They exist thanks to the existence of a system of *constitutive rules* collectively created that define and create them. *Constitutive rules* have the form X counts as Y in context C, where X can be brute fact, Y is an institutional fact and C is the context where the rule holds. Constitutive rules can be used for mapping brute facts into institutional facts. Like for example mapping the raise of one hand into a bid in an auction. Once an institutional fact has been defined, a constitutive rule can be used for specifying that an institutional fact counts as another institutional fact. Like for example mapping the highest bid in an auction into a commitment to pay a given amount of money to the owner of the product sold in the auction.

In artificial institutions that are the digital counterpart or extension of a human institution, the software designers are in charge of deciding: (i) which facts they want to formalize as brute facts, by defining causal effect rules able to change them;, (ii) which facts they want to formalize as institutional facts, which exist only thanks to common agreement between the interacting agents or between their designers, agreement that may be specified in a system of constitutive rules.

In AI *institutional actions* are a special type of actions that change institutional attributes [Fornara and Colombetti, 2009a]. Given that institutional attributes exist thanks to the common agreement of the interacting parties, institutional actions can be performed, if certain conditions hold, by means of suitable public communicative acts: declarations. A very important application independent contextual condition that an agent must satisfy in order to successfully perform an institutional action is to hold the *institutional power* to perform it. This connection between a public *declaration* (X) to perform an institutional action by an agent, the holding of various contextual conditions (C) including also the correct institutional power of the agent who is attempting to perform the institutional action, and the correct happening of the institutional action (Y), can be formalized with a *special constitutive rule*. For example an agent can perform a bid in an auction by performing a declaration of the amount of money that it wants to offer, the declaration counts as a bid if the agent is a regular registered participant in the auction.

Like for norms, it is very common to specify the institutional powers of agents at design time by using a set of *roles*, this allows to abstract from the specific agents that will interact within an institutional framework and requires to develop a module for the correct instantiation of the institutional powers at run time. Moreover, similarly to what discussed for norms, it is crucial to implement a *synchronization component* able to dynamically update the *knowledge base* used for representing the state of the interaction with the observable changes due to events or actions that will trigger the various constitutive rules.

The notion of institutional power has been analysed in [Jones and Sergot, 1996] and it has been discussed and formalized in Event Calculus in [Fornara and Colombetti, 2009a]. Another logic formalization of constitutive rules can be found in [Boella and van der Torre, 2004] and in [Grossi et al., 2006]. In [Brito et al., 2014] a recent interesting analysis has been performed on the similarities and differences of using events and states as brute facts for modelling institutional facts.

2.3 Agent Communication

In MAS literature it is possible to identify two different and complementary approaches for supporting and enabling the direct and indirect interactions among agents. A direct interaction may be realized by means of the definition of an Agent Communication Language (ACL) and realized with the direct message passing between agents. An indirect interaction may be realized by using blackboard systems, in which every agent can put information on a common information space, the blackboard, and any agent can read the information from the blackboard at any moment. The main negative aspect of blackboard systems is that they have a centralized structure that is not well suited for the realization of open interaction systems. In the following subsection we will present in more details ACLs.

2.3.1 Agent Communication Languages

In order to interact in an open environment autonomous agents need to adopt a common language, therefore they need to define a standard Agent Communication Language. The most relevant proposals of standard ACL in MAS literature are based on speech act theory [Austin, 1962; Searle, 1969], an approach that views language use as a form of action, making it possible to treat communicative acts and other types of action in a uniform way.

The first studies on ACLs follow what we can call a *mentalistic* approach, that is, they defines the meaning of a set of communicative acts having different performative by using agents mental states, like beliefs, desires (or goals) and intentions. Two well-known ACLs that follow this approach are KQML[Finin *et al.*, 1994] and FIPA ACL[6], proposed by the Foundation for Intelligent Physical Agent (FIPA). Using agents' mental states may be adequate in cooperative multiagent system, but it is not appropriate for open interaction systems composed by heterogeneous and often competitive agents made by different vendors [Singh, 1998]. In this kind of context it is impossible to trust other agents completely or make assumption about their internal design.

Therefore at the beginning of 2000s, a new approach to the definition of ACLs based on the social, objective consequences, and new obligations of performing a speech act were proposed [Colombetti, 2000; Fornara and Colombetti, 2002]. In this approach, the semantics of different type of speech acts is expressed using *commitments* directed from one agent to another, and this type of ACL is called commitment-based ACL. In particular in this approach, following the taxonomy of speech acts defined by John Searle [Searle, 1976] (which classifies illocutionary acts into five categories: declarations, assertives, commissives, directives and expressives) the seman-

---

[6]http://www.fipa.org/repository/aclspecs.html

tics of every type of communicative acts is defined using the notion of social commitment, conditional commitment, and pre-commitment.

Agents' commitments to one another can also be used for expressing the semantics of the messages exchanged in the specification of protocols (where protocols represent the allowed interactions among communicating agents). The use of commitments in the specification of the meaning of the messages exchanged in a given protocol makes the specification more flexible with respect to the traditional approaches, which model protocols as fixed action sequences. Two interesting papers on this topic are: [Yolum and Singh, 2002] where a slightly different model of commitments is used; and [Fornara and Colombetti, 2003] where the model of commitment used for the definition of the semantics of commitment-based ACLs is used for the flexible specifications of interactions protocols like the English Auction protocol. In this work it is also showed how it is possible to combine basic acts (belonging to the taxonomy of speech acts) for defining new type of acts that are frequently used in certain applications. For example it is showed that a proposal, used in a lot of e-commerce applications, can be formalized combining a request and a conditional promise to do something on condition that the request will be accepted, this means that it is not simply a combination of two types of communicative acts, but also their content is strongly related.

The approach of using a fixed set of primitives for expressing the semantics of many different types of communicative acts is criticized in a recently published chapter on Agent Communication Languages [Chopra and Singh, 2013, p.8]. The reason of the critique is based on the fact that in specific application contexts (like business applications) it is not always enough to use basic communicative acts type but it is necessary to define new primitive communicative acts like for example "quote price" or "stock quote". Therefore, in this chapter the semantics of domain-specific primitives, needed by a MAS, is given using a set of abstractions, among them the most important is the notion of commitment. Given that also in the approach based on a fixed set of primitives it is possible to define new communicative acts combining the existing one whose semantics is based on commitments, the two approaches results to be quite similar.

The initial approaches to the definitions of commitment-based ACLs did not define the semantics of a very important type of speech acts: declarations. Declarations are the particular category of communicative acts whose point is to bring about a change in the institutional reality in virtue of their successful performance. Declarations are fundamental in artificial institutions because, as previously discussed, they are the means for performing institutional actions. Their formalization was initially sketched in [Fornara and Colombetti, 2004] and improved in [Fornara et al., 2007] where the definition of an ACL is strictly connected with the definition of various artificial institutions.

Finally it is worth to mention that in 2013 a collection of six manifestos, each of which identifies important concerns and directions in agent communication has been published [Chopra et al., 2013].

2.4 Artificial Institutions situated in Environments

Taking into account the literature on modelling agent environment as a first-class abstraction [Weyns et al., 2007], it is possible to extend the model of artificial institutions and the model of the distributed system where AI are used. A crucial task of the environment, in a MAS model, is to register the events or actions that happen in the system and notify them to the agents registered for the template of such events/actions. The realization of this task can be combined with the idea a designing a MAS using different AIs.

In fact, the specification of an artificial institution (AI) consists in the abstract specification at design time of the concepts introduced so far, for example the formalization of norms and constitutive rules in terms of roles. The advantage of such an abstract specification is that it can be re-used in the specification of different MAS in different applications. Once one or more AIs are designed, it is necessary to describe how they can be concretely instantiated at run-time for the realization of a real open interaction system. One possible approach consists in proposing, coherently with the theory on agent environments, to instantiate AIs by introducing in the model of open distributed system the notion of *institutional space* of interaction [Tampitsikas et al., 2012; Fornara and Tampitsikas, 2013]. Institutional spaces are crucial because they allow to represent the boundaries of the effects of institutional and physical events or actions performed by agents, secondly they are the component in charge of enforcing the norms in response to the happened events/actions. Institutional spaces can be created and destroyed run-time on the basis of the agents' interactions.

An interesting aspect of the research on AI and environment is due to the fact that the same AI or different AIs may be instantiated in different institutional spaces. Those spaces may exists in *parallel* (inside the same container, like for example different auctions inside a marketplace), and they may also contain *sub-spaces* (like for example the space of a marketplace that contains different spaces each one corresponding to a running auction). Therefore, it is relevant to study the *inter-dependencies* between AIs (at design time) and between different institutional spaces at run time.

One possible approach for managing those interdependencies is described in the Multiagent Normative EnvironmenTs (MANET) meta-model [Tampitsikas *et al.*, 2012]. It consists in introducing the notion of *observability* of an event outside the boundaries of the institutional space where it happen. For example a norm defined in a marketplace may regulate the actions that an agent is allowed to perform in an auction (which is a sub-space of the marketplace). Another useful functionality is the *notification* of events among parallel institutional spaces. In fact, for example a norm inside one auction may regulate the action of an agent on the basis of the role that the same agent play in another auction.

Another interesting study on the specification and reasoning on multiple institutions is [Cliffe *et al.*, 2007b]. In this paper the formal specification of single institution [Cliffe *et al.*, 2007a] is extended to multiple institutions. This by extending the notion of institutional power to perform an action inside a single institution to the possibility for another institution to be empowered to change directly the state of another institution.

2.5 Comparison of Artificial Institutions Models

In Table 1 and 2, it is schematically summarized the support given by three relevant models of Artificial Institutions to the specification of the various components and concepts described in the previous sections; a more detailed comparison can be found in [Fornara *et al.*, 2013].

Those models of artificial institutions have been used for the realization of different prototype systems for solving different type of problems. For example the OCeAN + MANET model has been used for modelling an e-Energy Marketplace [Tampitsikas *et al.*, 2012] and the norms that regulate the Dutch-auction [Fornara and Tampitsikas, 2013]. The EIs model + EIDE framework has been used for modelling and develop a regulated open MAS able to manage water demand [Botti *et al.*, 2009], a MAS decision support tool for water-right markets [Giret *et al.*, 2011], an open MAS for realizing an hotel information system [Robles *et al.*, 2006], and EIs for running the Spanish Fish Market [Cuní *et al.*, 2004].

3 Survey on Modeling Organizations in MAS

Besides institutions, organizations have also obtained increasing attention from the MAS community in the last years. An organization can be seen as a set of entities and their interactions, which are regulated by mechanisms of social order and created by more or less autonomous actors to achieve common goals. These agents and their goals are interlinked by some form of organizational structure and in most MAS research this structure is seen as a means to manage the complex dynamics in open MAS.

Looking at MAS literature especially the OPERA [Dignum, 2004], TROPOS [Bresciani *et al.*, 2004], GAIA [Zambonelli *et al.*, 2003] and MOISE+

	Roles	**Norms**	**Constitutive Rules**
ANTE	Two type of roles: institutional roles and generic roles subject to norms	One normative environment with a set of regulations, which checks whether agents follow the norms, applies correction measures, and enables the run-time establishment of new normative relationships	Institutional facts are connected to brute facts (mainly agent illocutions) through appropriate constitutive rules
OCeAN + MANET	Roles are labels defined by one AI, at design time they are associated to norms and powers	Specification of norms at design time associated to roles and dynamic creation of instances of norms at run-time associated to specific agents	One special constitutive rule for performing institutional actions by means of declarations
EIs + EIDE	Specification of role subsumption, and two forms of compatibility among roles	There is the possibility of explicitly expressing norms as production rules that are triggered whenever an illocution is uttered, thus allowing the specification of regimented and not-regimented conventions	The are not basic institutional facts, there is a domain language used in illocutionary formulas and whose terms correspond with physical facts and actions

Table 1. Comparison of Artificial Institution Models - Part I

	Organization of the interactions	**Communication Language**	**Implementation**
ANTE	Different *normative contexts* are established at run-time	Agents are free to interact with any other agents with any language, illocutionary actions may be performed by agents towards the normative environment as attempts to create institutional facts	Jade FIPA-compliant platform and Jess rule-based inference engine
OCeAN + MANET	The activities are realized into different *institutional spaces* or in *physical spaces* of interaction	Commitment-based semantics for assertives, commissives, and directives communicative acts	OCeAN: (i) Event Calculus; (ii) Java + Semantic Web Technologies; MANET: PROLOG + GOLEM environment framework
EIs + EIDE	The activities are realized into *scenes*, which are connected by *transitions* creating a *performative structure*	All communications between an agent and the institution are mediated by a *governor*, utterance is admitted if and only if it complies with the institutional conventions	Z specification language and an ad-hoc peer-to-peer architecture

Table 2. Comparison of Artificial Institution Models - Part II

and ORA4MAS methodologies [Hübner et al., 2010] are often cited[7].

All of the just mentioned methodologies have in comment that they are based on organizational structures as their cornerstones. As such they recognize that when modeling the interaction of agents within a (open) MAS, it is not sufficient to simply focus on the architecture of the agents as well as their (communicative) abilities, and that one cannot assume autonomous agents to act according to the needs and expectations of the system design [Grossi, 2007].

Organizations "represent rationally ordered instruments for the achievement of stated goals" [Selznick, 1948]. As such they are being used to achieve specific objectives, which are defined by the specification of a number of sub-goals that are related to the overall goal of the organization. Looking at real world organizations, in business environments, an organization must furthermore consider the environment it is located in and exhibit characteristics such as a certain degree of predictability, stability over time as well as a focus on the organization's goals and strategies. Traditionally (as early as August Comte (1798-1854)), organizations therefore are considered to have two dimensions (that one needs to think of when wanting to model them): a factual dimension and a procedural one [Sichman and Conte, 1998]. Whereas the factual dimension focuses on the observable behavior of the organization - and thus takes a more high level view on its goals as well as output - the procedural dimension has its focus on the question on how that behavior of the organization is achieved. In the procedural dimension therefore the view shifts to the division of labor to roles, the determination of authority and power as well as the establishment of communication links [Dignum, 2004]. [Argente et al., 2006, Fig. 1] provides an overview of different MAS organizational architectures and functional and procedural / static dynamic features they exhibit.

3.1 Modeling Organizations: Between Top-Down and Bottom-Up

Wanting to model an organization at the very end of the spectrum, there are two opposite ways of designing it: (1) establish the organizational design off-line beforehand (design-time) or (2) let the organization be grown on-line from the bottom up by its participating agents (run-time). In the first case, the agents have no say in the global aims of the society. In the second case, which is more favoured by researchers working on open MAS, the agents are the key and their goals as well as negotiations between them result in organizations being dynamically formed. A simple example of such

---

[7]For a general survey on organization works in MAS see [Horling and Lesser, 2005] or [Luciano et al., 2005] for example.

an organization is twitter for example. Whereas the overall role of twitter is somewhat set (communication platform in the wider sense), the content of the communication is completely up to the users (within certain legal bounds). Thus, twitter hash-tags are not pre-define, but they are emerging based as a result of people using (and other people copying) them.

One highly discussed question within the MAS community is what is required to enable agents to do the latter. Main discussion point thereby is how rigid structures need to be that agents can use to use or establish organizations. Whereas in some works (e.g. [Esteva, 2003]) advocate very rigid structures at instruction level that do not allow agents to deviate from expected behavior, other approaches attempt to aim for more flexible systems where agents can reason about deviating from expected behavior. The aim of this more flexible approach thereby is to enable agents (and indirectly thereby the organizations they are are situated in) to adapt to changes and extensions to the environment or to allow for 'foreign' agents to join [Giorgini et al., 2002].

3.2 Organizational Structure

The just outlined differences in organizational design that MAS designers face, have also been studied in the traditional organizational theory research area, where the two types of organizations are typically distinguished for human organizations: mechanical (sometimes also called mechanistic) and organic organizations [Robbins and Judge, 2017].

In mechanical organizations - that closely relate to the design-time idea of MAS organizational design - tasks are precisely defined in advance, and they are broken down into separately specialized parts. Real world example of these kind of organizations are typically manufacturing companies, but also there are other groups that benefit from mechanistic organization; like universities.

In mechanical organizations, there is a strictly hierarchical structure both between the parts, but also between the knowledge and reasoning processes within the organization. In mechanical organizations communication tends to be mainly vertical, i.e. from the top (typical a centralized role) to the bottom of the hierarchy. Typical examples human organizations the have this kind of setup are bureaucracies and matrix structures [Argente et al., 2006].

Organic organizations in contrast are following concept of growing the organization from the bottom up. As such the members of/agents within such an organization can collaboratively (e.g. in groups) redefine and adjust the tasks, sub-goals and roles related to the organization, thereby possibly changing the whole organization and its goals. In organic organizations, less levels of authority and control are being present and communication is mainly horizontal rather then vertical. Knowledge and task control also tend to be distributed and reaction time to changes in such organizations is said to be shorter. Typical examples of organic organizations are team structures and virtual organizations [Argente et al., 2006]. Good examples of this type of structure would be Google and the coveted positions that lie within the Facebook Corporation.

From the above, for a designer wanting to decide on what the most suitable structure is for the system that he aims to model, he need to answer several questions first:

- Can the goals and tasks be divided into independent, formalized and standardized sub-tasks? An if so, how to approach this best?

- Which of the tasks and sub-tasks have dependencies that need considering?

- Can tasks be grouped together and what are good means to group tasks (function, geographical location, client, process, etc.)?

- At what level have decisions to be made and controls to be set up?

- What kind of environment is the organization located in (open, closed, static, dynamic)?

- What is the line of reporting in the organization? Who has authority and what is the chain of command?

- What rules and formal processes are being required in the organization?

- What level of predictability is the organization to have?

Goal of answering these questions is to enable the modeler of the organization to determine the main organizational features in order to develop the initial design of the organizational structure of the organization independently from the final use of agent concepts[Argente et al., 2006].

As a note to the reader, though we presented organizational structure as a single term above, in the real world and as a consequence also in organizational models, it can be multi-dimensional and consider several structured aspects at the same time that all need to be represented. Some of the once already having been mentioned are "authority", "communication", "delegation", "responsibility", "control", "decision-making", "power", etc. [Grossi, 2007].

3.3 Roles

The difference between the two concepts of textitInstitutions and *Organizations* can be exemplified when looking at the notion of 'roles'. From an institutional point of view roles are typically studied in terms of the set of norms associated with them, whereas from an organizational point of view the focus tends to be on the roles as a position in an organizational structure.

As organizations are typically established with some form of goal that is executed by agents enacting certain roles in mind, from an organizational perspective focus tends to be on roles that contribute to the achievement of the overall goal, rather then on the specific actors performing the particular roles. Thus, although it is the agent's capabilities that allow him to perform a role in a certain way, most of the time for an organization it is irrelevant who performs the role as long as it is performed. Think of a restaurant for example which will have several employees with the job title "cook". For a customer it is not relevant which (group of) cook(s) actually cooked his dish, as long as the dish tastes good and is delivered on time. This so called "role-oriented" approach is advocated by many works on organizations in MAS, including for example [Argente et al., 2006; Jureta et al., 2007; Ferber et al., 2003; Dignum, 2004].

Grossi [Grossi, 2007] argues that from the structural point of view a role is just a position in a structure, that is to say, a set of links, whereas from the institutional perspective instead, they can be seen as a set of norms. Following the argumentation in [Grossi, 2007] these two set partially overlap w.r.t the properties they express for transition systems. Whereas roles as set of norms specify how the role can be enacted, deacted, and what kind of status the agents acquire by enacting the role; roles as set of links specify the status acquired by agents playing certain roles (while disregarding how that role can be enacted or deacted), specify the the activities (e.g., delegation or information) that can be executed while enacting the role and, possibly, also their mental effects on the interacting agents.

Recapping, both institutions and organizations specify what an agent ought to, is permitted to, or has the right to do as well as have means to specify the status an agent playing a certain tole has acquired. What has to be noted however is that there are differences in this status specification. Whereas institutions connect abstract activities and state of affairs (i.e., transition and state types) to concrete ones, this is not the case for organizations. In contrast these have the activities that can be executed by agents playing a certain role as their main focus of attention. Thus, whereas institutions consider how a certain role can be reached, organizations are taking a look at what can be done while playing a role.

3.4 Comparison of Artificial Organization Models

Table 4 gives an overview of how some MAS organizational models mentioned here relate to the above. It is not a conclusive overview, but rather a short glimpse into the different organizational approaches. A more detailed overview can for example be found in [Argente et al., 2005].

	Top-down/Bottom-Up	**Structure**	**Dynamics**
OperA	In opera the organizational model can be define whereas the social and the interaction model are consequences of the agent interactions. The OMNI approach [Dignum et al., 2005], an extension of OperA introduces normative aspects and translates norms from an abstract level (in which organizational statutes and values are defined to a procedural level (where norms are implemented).	OperA describes the desired behaviour of the society and its general structure by means of an organizational model, where roles, interactions and social norms are described.	Organizational dynamics are detailed using a social model (in which agents are assigned roles using social contracts that describe the agreed behaviour inside the society) and an interaction model (that described the actual behaviour of a society during its interaction).
Tropos	An organizational model is used that details the organizations main actors, goals and dependencies at design time. This is done on a level of agent patterns (for particular roles) that are assigned to organizational topologies. The definition of social rules or global rules that apply to the whole organization are not considered.	Several organizational topologies and roles within these structures are considered. The structures include for example bureaucracy, matrix structures ad virtual organizations.	Social agent patterns are assigned to organizational topologies at design time. At run-time the effects of this assignment are analysed based on the pre-set rules.

Table 3. Comparison of Artificial Institutions Models

	Top-down/Bottom-Up	Structure	Dynamics
GAIA (with Organizational Abstractions)	Main organizational goals of the system and its expected global behaviour are specified at design phase. Based on this organizations are established which can be divided into sub-organizations if needed (each of which can have its own structure). The environment, roles, interactions and social rules are also pre-defined.	GAIA considers that a specific topology for the system will force the use of several roles that depend on the selected topology pattern.	GAIA uses the concept of role-enactment, where the roles are depending on the chosen topology. Thus a change in the topology can alter the roles and their enactment, but not the other way around.

Table 4. Comparison of Artificial Organization Model

4 Conclusions and Forward Looking

Openness, decentralization, and heterogeneity of software components are fundamental characteristics of distributed systems operating on the Internet and in particular in the World Wide Web. At least since the 1990s, models and experimental implementations of open, decentralized and heterogeneous systems have been the main concern of the area of Computer Science research on Multiagent Systems (MASs). More recently, those studies went on with the proposal of numerous conceptual models of institutions and organizations. It is therefore not surprising this long tradition of studies may represent a fundamental source of ideas and methods for developing Web-oriented applications. In particular the sub-area of MAS research known as NorMAS (Normative Multiagent Systems) has been concerned with modeling, monitoring, and enforcing norms and policies in open distributed environments, producing solutions that have already been empirically tested with success, although mainly in the context of academic prototypes.

In this chapter we have provided an introduction to the basic concepts of modeling organizations and institutions in MAS and gave pointers to the work that has been done in the various NorMAS communities already. We started out by looking into institutions and discussed fundamental concepts such as regulative and constitutive norms, as well as ACLs and the challenges arising from institutions being situated in an environment that can impact on the institution. Afterwards our focus shifted to organizations. After briefly detailing the basic differences between organizations and institutions, the focus of the chapter afterwards turned to modeling organizational structure as well as modeling roles in organizations.

Looking forward, we believe that times are ripe for adapting institutions and organizations MAS models and techniques for solving real-world problems that arise on the Internet. Indeed, the development of advanced Web applications is already providing significant examples of actual applications on which the capabilities of MAS solutions can be put to be tested, evaluated, and improved.

For example, such solutions may be relevant for the regulation of access to generic datasets on the Internet or to datasets in the Web of Data, provided that they are implemented coherently with currently available Web technologies [Finin et al., 2008; Nguyen et al., 2015]. Organizational and institutional models may be crucial also for the realization of automatic machine-to-machine exchange of datasets when norms/policies and may be institutional concepts can be used for expressing the licenses [Governatori et al., 2013] or ad hoc contracts/agreements that regulate the access and use of those data.

However, considerable research is still necessary before this approach can be adopted by industry-level products solving realistic problems. Moreover a deep comparison with approaches proposed in other fields of research is required.

BIBLIOGRAPHY

[Aldewereld and Dignum, 2010] Huib Aldewereld and Virginia Dignum. Operetta: Organization-oriented development environment. In Mehdi Dastani, Amal El Fallah-Seghrouchni, Jomi Hübner, and João Leite, editors, *LADS*, volume 6822 of *Lecture Notes in Computer Science*, pages 1–18. Springer, 2010.

[Arcos et al., 2005] J. Lluís Arcos, M. Esteva, P. Noriega, J. A. Rodríguez-Aguilar, and C. Sierra. Engineering open environments with electronic institutions. *Eng. Appl. of AI*, 18(2):191–204, 2005.

[Argente et al., 2005] Estefania Argente, Vicente Julian, Soledad Valero, and V. J. Botti. Towards an organizational mas methodology. In *Recent Advances in Artificial Intelligence Research and Development*, Frontiers in Artificial Intelligence and Applications, pages 397–404. IOS Press, 2005.

[Argente et al., 2006] Estefania Argente, Vicente Julian, and Vicente Botti. Multi-agent system development based on organizations. *Electronic Notes in Theoretical Computer Science*, 150(3):55–71, 2006.

[Artikis et al., 2009] Alexander Artikis, Marek Sergot, and Jeremy Pitt. Specifying norm-governed computational societies. *ACM Trans. Comput. Logic*, 10(1):1:1–1:42, January 2009.

[Austin, 1962] John Langshaw Austin. *How to do things with words*. William James Lectures. Oxford University Press, 1962.

[Boella and van der Torre, 2004] Guido Boella and Leendert W. N. van der Torre. Regulative and constitutive norms in normative multiagent systems. In Didier Dubois, Christopher A. Welty, and Mary-Anne Williams, editors, *Principles of Knowledge Representation and Reasoning: Proceedings of the Ninth International Conference (KR2004), Whistler, Canada, June 2-5, 2004*, pages 255–266. AAAI Press, 2004.

[Botti et al., 2009] Vicente J. Botti, Antonio Garrido, Adriana Giret, and Pablo Noriega. Managing water demand as a regulated open MAS. In Matteo Baldoni, Cristina Baroglio, Jamal Bentahar, Guido Boella, Massimo Cossentino, Mehdi Dastani, Barbara Dunin-Keplicz, Giancarlo Fortino, Marie Pierre Gleizes, João Leite, Viviana Mascardi, Julian A. Padget, Juan Pavón, Axel Polleres, Amal El Fallah-Seghrouchni, Paolo Torroni, and Rineke Verbrugge, editors, *Proceedings of the Second Multi-Agent Logics, Languages, and Organisations Federated Workshops, Turin, Italy, September 7-10, 2009*, volume 494 of *CEUR Workshop Proceedings*. CEUR-WS.org, 2009.

[Bresciani et al., 2004] Paolo Bresciani, Anna Perini, Paolo Giorgini, Fausto Giunchiglia, and John Mylopoulos. Tropos: An agent-oriented software development methodology. *Autonomous Agents and Multi-Agent Systems*, 8(3):203–236, 2004.

[Brito et al., 2014] Maiquel Brito, Jomi Fred Hübner, and Rafael H. Bordini. Analysis of the use of events and states as brute facts in modelling of institutional facts. In *Revised Selected Papers of the COIN 2013 International Workshop on Coordination, Organizations, Institutions, and Norms in Agent Systems IX - Volume 8386*, pages 177–192, New York, NY, USA, 2014. Springer-Verlag New York, Inc.

[Chopra and Singh, 2013] Amit K. Chopra and Munindar P. Singh. *Agent Communication*, volume Multiagent Systems, chapter 3. The MIT Press, 2013.

[Chopra et al., 2013] Amit K. Chopra, Alexander Artikis, Jamal Bentahar, Marco Colombetti, Frank Dignum, Nicoletta Fornara, Andrew J. I. Jones, Munindar P. Singh, and Pinar Yolum. Research directions in agent communication. *ACM Trans. Intell. Syst. Technol.*, 4(2):20:1–20:23, April 2013.

[Chopra et al., 2014] Amit K. Chopra, Fabiano Dalpiaz, Fatma Başak Aydemir, Paolo Giorgini, John Mylopoulos, and Munindar P. Singh. Protos: Foundations for Engineering Innovative Sociotechnical Systems. In *Proceedings of the 22nd IEEE International Requirements Engineering Conference (RE'14)*. IEEE, 2014.

[Cliffe et al., 2007a] Owen Cliffe, Marina De Vos, and Julian Padget. *Computational Logic in Multi-Agent Systems: 7th International Workshop, CLIMA VII, Hakodate, Japan, May 8-9, 2006, Revised Selected and Invited Papers*, chapter Answer Set Programming for Representing and Reasoning About Virtual Institutions, pages 60–79. Springer Berlin Heidelberg, Berlin, Heidelberg, 2007.

[Cliffe et al., 2007b] Owen Cliffe, Marina De Vos, and Julian Padget. *Coordination, Organizations, Institutions, and Norms in Agent Systems II: AAMAS 2006 and ECAI 2006 International Workshops, COIN 2006 Hakodate, Japan, May 9, 2006 Riva del Garda, Italy, August 28, 2006. Revised Selected Papers*, chapter Specifying and Reasoning About Multiple Institutions, pages 67–85. Springer Berlin Heidelberg, Berlin, Heidelberg, 2007.

[Cliffe et al., 2007c] Owen Cliffe, Marina De Vos, and Julian Padget. Specifying and reasoning about multiple institutions. In Pablo Noriega, Javier Vazquez-Salceda, Guido Boella, Olivier Boissier, Virginia Dignum, Nicoletta Fornara, and Eric Matson, editors, *Coordination, Organization, Institutions and Norms in Agent Systems II - AAMAS 2006 and ECAI 2006 International Workshops, COIN 2006 Hakodate, Japan, May 9, 2006 Riva del Garda, Italy, August 28, 2006*, volume 4386 of *Lecture Notes in Computer Science*, pages 67–85. Springer Berlin / Heidelberg, 2007.

[Colombetti, 2000] M. Colombetti. A commitment-based approach to agent speech acts and conversations. In *Proceedings of the Workshop on Agent Languages and Conversational Policies*, pages 21–29, 2000.

[Corapi et al., 2011] Domenico Corapi, Marina De Vos, Julian Padget, Alessandra Russo, and Ken Satoh. Normative design using inductive learning. *Theory and Practice of Logic Programming*, 11:783–799, 2011.

[Cuní et al., 2004] Guifré Cuní, Marc Esteva, Pere García, Eloi Puertas, Carles Sierra, and Teresa Solchaga. Masfit: Multiagent system for fish trading. In Lorenza Saitta Ramon López de Mántaras, editor, *Proceedings of the 16th Eureopean Conference on Artificial Intelligence, ECAI'2004, including Prestigious Applicants of Intelligent Systems, PAIS 2004, Valencia, Spain, August 22-27, 2004*, pages 710–714. IOS Press, IOS Press, 2004.

[Dignum et al., 2005] V. Dignum, J. Vazquez-Salceda, and F. Dignum. Omni: Introducing social strcuture, norms and ontologies into agent organizations. In *ProMAS*, volume 3346 of *Lecture Notes on Artificial Intelligence*, 2005.

[Dignum, 2004] Virginia Dignum. *A model for organizational interaction: based on agents, founded in logic*. PhD thesis, University Utrecht, 2004.

[Elagh, 2000] A. A. O. Elagh. On the Formal Analysis of Normative Conflicts. *Information & Communications Technology Law*, 9(3):207–217, 2000.

[Esteva et al., 2008] Marc Esteva, Juan A. Rodríguez-Aguilar, Josep Lluís Arcos, Carles Sierra, Pablo Noriega, and Bruno Rosell. Electronic Institutions Development Environment. In *Proceedings of the 7th International Joint Conference on Autonomous Agents and Multiagent Systems (AAMAS '08)*, pages 1657–1658, Estoril, Portugal, 12/05/2008 2008. International Foundation for Autonomous Agents and Multiagent Systems, ACM Press.

[Esteva, 2003] Marc Esteva. *Electronic Institutions: from specification to development*. PhD thesis, Insitut d'Investigaci en Intel-ligncia Artificial, 2003.

[Ferber et al., 2003] Jacques Ferber, Olivier Gutknecht, and Fabien Michel. From agents to organizations: An organizational view of multi-agent systems. In *Agent-Oriented Software Engineering IV*, volume 2935 of *Lecture Notes in Computer Science*, pages 214–230, 2003.

[Finin et al., 1994] Tim Finin, Richard Fritzson, Don McKay, and Robin McEntire. Kqml as an agent communication language. In *Proceedings of the Third International Conference on Information and Knowledge Management*, CIKM '94, pages 456–463, New York, NY, USA, 1994. ACM.

[Finin et al., 2008] T. Finin, A. Joshi, L. Kagal, J. Niu, R. Sandhu, W. Winsborough, and B. Thuraisingham. ROWLBAC: Representing role based access control in OWL. In *Proceedings of the SACMAT*, pages 73–82, New York, NY, USA, 2008. ACM.

[Fornara and Colombetti, 2002] Nicoletta Fornara and Marco Colombetti. Operational specification of a commitment-based agent communication language. In *Proceedings of the First International Joint Conference on Autonomous Agents and Multiagent Systems: Part 2*, AAMAS '02, pages 536–542, New York, NY, USA, 2002. ACM.

[Fornara and Colombetti, 2003] Nicoletta Fornara and Marco Colombetti. Defining interaction protocols using a commitment-based agent communication language. In *Proceedings of the Second International Joint Conference on Autonomous Agents and Multiagent Systems*, AAMAS '03, pages 520–527, New York, NY, USA, 2003. ACM.

[Fornara and Colombetti, 2004] Nicoletta Fornara and Macro Colombetti. A commitment-based approach to agent communication. *Applied Artificial Intelligence*, 18:853–866, 2004.

[Fornara and Colombetti, 2009a] N. Fornara and M. Colombetti. Specifying Artificial Institutions in the Event Calculus. In V. Dignum, editor, *Handbook of Research on Multi-Agent Systems: Semantics and Dynamics of Organizational Models*, Information science reference, chapter XIV, pages 335–366. IGI Global, 2009.

[Fornara and Colombetti, 2009b] Nicoletta Fornara and Marco Colombetti. Specifying and Enforcing Norms in Artificial Institutions. In Matteo Baldoni, Tran Son, M. van Riemsdijk, and Michael Winikoff, editors, *Declarative Agent Languages and Technologies VI*, volume 5397 of *Lecture Notes in Computer Science*, pages 1–17. Springer Berlin / Heidelberg, 2009.

[Fornara and Colombetti, 2010] Nicoletta Fornara and Marco Colombetti. Representation and monitoring of commitments and norms using OWL. *AI Commun.*, 23(4):341–356, 2010.

[Fornara and Tampitsikas, 2013] Nicoletta Fornara and Charalampos Tampitsikas. Semantic technologies for open interaction systems. *Artificial Intelligence Review*, 39:63–79, 2013.

[Fornara et al., 2007] Nicoletta Fornara, Francesco Viganò, and Marco Colombetti. Agent communication and artificial institutions. *Autonomous Agents and Multi-Agent Systems*, 14(2):121–142, 2007.

[Fornara et al., 2013] Nicoletta Fornara, Henrique Lopes Cardoso, Pablo Noriega, Eugénio Oliveira, Charalampos Tampitsikas, and Michael I. Schumacher. *Agreement Technologies*, chapter Modelling Agent Institutions, pages 277–307. Springer Netherlands, Dordrecht, 2013.

[Fornara, 2011] Nicoletta Fornara. *Semantic Agent Systems: Foundations and Applications*, chapter Specifying and Monitoring Obligations in Open Multiagent Systems Using Semantic Web Technology, pages 25–45. Springer Berlin Heidelberg, Berlin, Heidelberg, 2011.

[Giorgini et al., 2002] Paolo Giorgini, Manuel Kolp, and John Mylopoulos. Multi-agent architectures as organizational structures. *Autonomous Agents and Multi-Agent Systems*, 13, 2002.

[Giret et al., 2011] Adriana Giret, Antonio Garrido, Juan A. Gimeno, Vicente J. Botti, and Pablo Noriega. A MAS decision support tool for water-right markets. In Liz Sonenberg, Peter Stone, Kagan Tumer, and Pinar Yolum, editors, *10th International Conference on Autonomous Agents and Multiagent Systems (AAMAS 2011), Taipei, Taiwan, May 2-6, 2011, Volume 1-3*, pages 1305–1306. IFAAMAS, 2011.

[Governatori et al., 2013] Guido Governatori, Antonino Rotolo, Serena Villata, and Fabien Gandon. One license to compose them all - A deontic logic approach to data licensing on the web of data. In Harith Alani, Lalana Kagal, Achille Fokoue, Paul T. Groth, Chris Biemann, Josiane Xavier Parreira, Lora Aroyo, Natasha F. Noy, Chris Welty, and Krzysztof Janowicz, editors, *The Semantic Web - ISWC 2013 - 12th International Semantic Web Conference, Sydney, NSW, Australia, October 21-25, 2013, Proceedings, Part I*, volume 8218 of *Lecture Notes in Computer Science*, pages 151–166. Springer, 2013.

[Grossi et al., 2006] Davide Grossi, John-Jules Ch. Meyer, and Frank Dignum. Classificatory aspects of counts-as: An analysis in modal logic. *J. Log. Comput.*, 16(5):613–643, 2006.

[Grossi, 2007] Davide Grossi. *Designing invisible handcuffs. Formal investigations in institutions and organizations for multi-agent systems*. PhD thesis, Utrecht University, 2007.

[Hitzler et al., 2009] Pascal Hitzler, Markus Krötzsch, and Sebastian Rudolph. *Foundations of Semantic Web Technologies*. Chapman & Hall/CRC, 2009.

[Horling and Lesser, 2005] Bryan Horling and Victor Lesser. Using odml to model and design organizations for multi-agent systems. In Olivier Boissier, Virginia Dignum, Eric Matson, and Jaime Sichman, editors, *Proceedings of the Workshop on From Organizations to Organization Oriented Programming (OOOP 05)*, 2005.

[Hübner et al., 2010] Jomi F. Hübner, Olivier Boissier, Rosine Kitio, and Alessandro Ricci. Instrumenting multi-agent organisations with organisational artifacts and agents. *Autonomous Agents and Multi-Agent Systems*, 20(3):369–400, 2010.

[Jones and Sergot, 1996] Andrew J. I. Jones and Marek J. Sergot. A formal characterisation of institutionalised power. *Logic Journal of the IGPL*, 4(3):427–443, 1996.

[Jureta et al., 2007] Ivan J. Jureta, Stéphane Faulkner, and Pierre-Yves Schobbens. Allocating goals to agent roles during mas requirements engineering. In *Proceedings of the 7th International Conference on Agent-oriented Software Engineering VII*, pages 19–34, 2007.

[Khalil, 1995] Elias L. Khalil. Organizations versus institutions. *Journal of Institutional and Theoretical Economics (JITE)*, 151(3):445–466, September 1995.

[Lopes Cardoso, 2010] H. Lopes Cardoso. *Electronic Institutions with Normative Environments for Agent-based E-contracting*. PhD thesis, Universidade do Porto, 2010.

[Luciano et al., 2005] Luciano, Jaime S. Sichman, and Olivier Boissier. Modeling organization in MAS: a comparison of models. *1st. Workshop on Software Engineering for Agent-Oriented Systems (SEAS'05)*, October 2005.

[Marfia et al., 2016] Fabio Marfia, Nicoletta Fornara, and Truc-Vien T. Nguyen. *Multi-Agent Systems and Agreement Technologies: 13th European Conference, EUMAS 2015, and Third International Conference, AT 2015, Athens, Greece, December 17-18, 2015, Revised Selected Papers*, chapter Modeling and Enforcing Semantic Obligations for Access Control, pages 303–317. Springer International Publishing, Cham, 2016.

[Nguyen et al., 2015] Truc-Vien T. Nguyen, Nicoletta Fornara, and Fabio Marfia. Automatic policy enforcement on semantic social data. *Multiagent and Grid Systems*, 11(3):121–146, 2015.

[Noriega et al., 2013] Pablo Noriega, Amit K. Chopra, Nicoletta Fornara, Henrique Lopes Cardoso, and Munindar P. Singh. Regulated MAS: social perspective. In Giulia Andrighetto, Guido Governatori, Pablo Noriega, and Leendert W. N. van der Torre, editors, *Normative Multi-Agent Systems*, volume 4 of *Dagstuhl Follow-Ups*, pages 93–133. Schloss Dagstuhl - Leibniz-Zentrum fuer Informatik, 2013.

[North, 1996] Douglass C. North. Institutions, organizations and market competition. Economic history, EconWPA, December 1996.

[Ossowski, 2013] Sascha Ossowski, editor. *Agreement Technologies*, volume 8 of *Law, Governance and Technology Series*. Springer Netherlands, Dordrecht, 2013.

[Ostrom, 1990] E. Ostrom. *Governing the commons-The evolution of institutions for collective actions*. Political economy of institutions and decisions, 1990.

[Pitt et al., 2011] Jeremy Pitt, Julia Schaumeier, and Alexander Artikis. *Coordination, Conventions and the Self-organisation of Sustainable Institutions*, pages 202–217. Springer Berlin Heidelberg, Berlin, Heidelberg, 2011.

[Robbins and Judge, 2017] Stephen P. Robbins and Timothy A. Judge. Organizational behavior, 2017.

[Robles et al., 2006] Armando Robles, Pablo Noriega, Marco Julio Robles P., Héctor Hernández T., Victor Soto Ramírez, and Edgar Gutiérrez S. A hotel information system implementation using MAS technology. In Hideyuki Nakashima, Michael P. Wellman, Gerhard Weiss, and Peter Stone, editors, *5th International Joint Conference on Autonomous Agents and Multiagent Systems (AAMAS 2006), Hakodate, Japan, May 8-12, 2006*, pages 1542–1548. ACM, 2006.

[Searle, 1969] John R. Searle. *Speech Acts: An Essay in the Philosophy of Language*. Cambridge University Press, Cambridge, London, 1969.

[Searle, 1976] J.R. Searle. A classification of illocutionary acts. *Language in Society*, 5(1):1–23, 1976.

[Searle, 1995] J. R. Searle. *The construction of social reality*. Free Press, New York, 1995.

[Selznick, 1948] Philip Selznick. Foundations of the theory of organization. *American Sociological Review*, 13:25–35, 1948.

[Sensoy et al., 2012] Murat Sensoy, Timothy J. Norman, Wamberto W. Vasconcelos, and Katia Sycara. OWL-POLAR: A framework for semantic policy representation and reasoning. *Web Semantics: Science, Services and Agents on the World Wide Web*, 12-13:148–160, April 2012.

[Sichman and Conte, 1998] Jaime S. Sichman and Rosaria Conte. On personal and role mental attitudes: a preliminary dependence-based analysis. In *Advances in Artificial Intelligence, 14th Brazilian Symposium on Artificial Intelligence, SBIA '98*, number 1515, 1998.

[Singh, 1998] Munindar P. Singh. Agent communication languages: Rethinking the principles. *Computer*, 31(12):40–47, December 1998.

[Tampitsikas et al., 2012] Charalampos Tampitsikas, Stefano Bromuri, Nicoletta Fornara, and Michael Ignaz Schumacher. Interdependent Artificial Institutions In Agent Environments. *Applied Artificial Intelligence*, 26(4):398–427, 2012.

[Uszok et al., 2008] Andrzej Uszok, Jeffrey M. Bradshaw, James Lott, Maggie R. Breedy, Larry Bunch, Paul J. Feltovich, Matthew Johnson, and Hyuckchul Jung. New developments in ontology-based policy management: Increasing the practicality and comprehensiveness of kaos. In *9th IEEE International Workshop on Policies for Distributed Systems and Networks (POLICY 2008), 2-4 June 2008, Palisades, New York, USA*, pages 145–152. IEEE Computer Society, 2008.

[Weyns et al., 2007] Danny Weyns, Andrea Omicini, and James Odell. Environment as a first class abstraction in multiagent systems. *Autonomous Agents and Multi-Agent Systems*, 14(1):5–30, 2007.

[Wooldridge and Jennings, 1995] M. J. Wooldridge and N. R. Jennings. Intelligent agents: Theory and practice. *The Knowledge Engineering Review*, 10(2):115–152, 1995.

[Yolum and Singh, 2002] Pinar Yolum and Munindar P. Singh. Flexible protocol specification and execution: Applying event calculus planning using commitments. In *Proceedings of the First International Joint Conference on Autonomous Agents and Multiagent Systems: Part 2*, AAMAS '02, pages 527–534, 2002.

[Zambonelli et al., 2003] Franco Zambonelli, Nicholas R. Jennings, and Michael Wooldridge. Developing multiagent systems: The GAIA methodology. *ACM Trans. Softw. Eng. Methodol.*, 12(3):317–370, 2003.

6
Modeling Norms Embedded in Society: Ethics and Sensitive Design

ROB CHRISTIAANSE

1 Introduction

1.1 Acting or not acting

When we start to think about moral norms, morality if you like, we encounter very intriguing problems about situations in which people find themselves facing a choice to make which are of excessive complexity. Take for example "The Trolley problem": "The trolley rounds a bend, and there come into view ahead five track workmen, who have been repairing the track. The track goes through a bit of a valley at that point, and the sides are steep, so you must stop the trolley if you are to avoid running the five men down. You step on the brakes, but alas they don't work. Now you suddenly see a spur of track leading off to the right. You can turn the trolley onto it, and thus save the five men on the straight track ahead. Unfortunately there is one track workman on that spur of track. He can no more get off the track in time than the five can, so you will kill him if you turn the trolley onto him. Is it morally permissible for you to turn the trolley" [Thomson, 1985]. Or the problem addressed by Macintyre in [Macintyre, 1999] where he starts off with the case of *J* (who might be anybody, *jemand*) analyzing whether *J's* defense to the allegation of moral failure holds because *J* failed his responsibility.

1.2 Using (il)legitimate evidence

Let's look into some judicial cases. Regulators use information obtained from different sources to assess whether businesses and citizens comply with applicable laws and regulations. Recently the supreme court of the Netherlands ruled in case number [ECL, b]. The key issue addressed in this case was whether the Tax authority was allowed to use information obtained from Automatic Number Plate Recognition (ANPR) systems checking tax returns from employees driving company leased cars. In defense the defendant plead that article 8 of the ECHR protects ones private live and therefore the tax authority was not allowed to use the aforementioned information to check the tax return as done. The Supreme Court ruled that in this case it was wrongly assumed that the general job description of the tax authority or any (other) provision provides in a sufficient basis to use information from Automatic Number Plate Recognition (ANPR) systems. In 2015 a court procedure was ruled in a similar case [ECL, a]. A vehicle driver was fined for speeding on the A2. The A2 is a motor highway in the Netherlands equipped with a section control system using ANPR for detecting all vehicles driving on the A2. The section control system measures the time and records a time-stamp image of the license plate from the moment the car enters the section until the car leaves the section. During the route (i.e. the section) the process of measuring en recording is repeated at fixed intervals. After the car has left the section the actual speed is calculated and in the case the average speed exceeds some threshold the vehicle driver is fined automatically. In this case the car driver drove 8km/hr. to fast and got fined for EUR45. The man went to court and stated that the section control system infringed his private life so article 8 of the ECHR was violated. In this case the judge ruled that the police law provided in a sufficient basis to use the information obtained from the section control system. As for the claim that the fined man's private life was infringed the judge did not accept this line of reasoning because all vehicle drivers are well informed of the system by means of traffic signs informing car drivers that speed is measured by means of a section control system. Additionally the records of the time stamp images of the license plates are to be disposed of after 72hours counting from the moment the recording of time-stamp images starts. Considering the aforementioned circumstances the judge ruled that the privacy concerns were only violated on a limited level. The fine had to be paid by the car driver.

Both cases share that information obtained from ANPR systems exist and that regulatory bodies use this information as evidence for regulatory oversight purposes. Both share the issue whether usage by regulatory bodies of this evidence infringe fundamental rights that protects ones private live. At first sight the reasoning of the judge in the first case is quite straightforward namely in the first case the "job description" did not provide in a sufficient basis for using the specific information for regulatory oversight purposes so the issue whether the usage of the specific information infringed fundamental rights that protects ones private live need not to be addressed. The nature of the ruling is what we will coin as formal procedural. In the second case the judge ruled that the usage of the specific information did fit the job description and that usage of the specific information was granted and therefore legitimate so the issue whether the usage of the specific information infringed the fundamental rights had to be addressed. Due to the fact that the vehicle driver was well informed at the time he or she drives onto the motor highway A2; he or she should be aware that the car he or she is driving will be under surveillance of the section control system. Due to the fact that the records are only kept for a limited time period the judge reasoned that the fundamental rights preserving ones private live was violated on a limited level, therefore the usage of the specific information for regulatory oversight purposes was legitimate. Hence the judge weighted the consequences for the parties involved. The nature of the ruling is both formal procedural and substantial.

1.3 Ethics and moral

In the case we have to model norms embedded in society with an ethical and sensitive design in mind than the inevitable and pressing question is: "what is an ethical and sensitive design in the first place?" Ethics is a branch of moral philosophy, addressing questions like: "what is a wrong thing to do and what is a right thing to do (in situation x for example) when y happens". In the cases we introduced in the former section, we simply have to ask ourself in simple present future tense "is it a right thing to do?" or "is it a wrong thing to do" or in past perfect tense "was it a right thing to do?" or "was it a wrong thing to do". Wrong is not the opposite from right and right is not the opposite from wrong. In moral theory questions about value play a major role. In a very narrow sense value theory refers to axiology addressing questions whether objects of value are psychological states or objective states of the world. Put in a more broader context the value theory concept addresses questions about the nature of value and its relation to other (moral) categories like naturalistic goods opposed to human made entities i.e. artifacts. With this distinction in mind we get a mechanism enabling us to reason about values and takeoffs, often coined as an evaluation mechanism. In the above illustrated cases we recognize that values are weighted but that the underlying mechanism is often opaque by nature and therefore hard to decipher. How these trade-offs are made is less understood for example when plurality in moral values may exist.

1.4 Judgment

Modeling norms implies that we have to deliberate about the nature of norms to model in the first place. The concept of value seems a intertwining concept among alternatives to choose from. In some cases a trade-off has to be made by some human or technological component if you like. Making trade-offs brings in the question of agency. Agency is the capacity to make choices to act [Macintyre, 1999] [Mische and Ann, 1998]. Moral agency is the ability to make moral judgments based on some notion of right or wrong . Hence there exists some evaluation function to judge. Indeed the evaluation function is a necessary condition, whether the evaluation function is sufficient depends on what is believed to be true and justified. Recognize the epistemic nature of the logic buttressing the evaluation function. Moral responsibility on the other side is about human action and its consequences. The concept of responsibility can be viewed from two distinct though interrelated concepts namely (1) the merit-based view and the (2) consequentialist view [Stanford, 2015; Strawson, 1974]. Broadly speaking the distinction draws the line between responsibility as accountability versus responsibility as attributability. Attributability is a necessary but not a sufficient condition for being accountable. In the case humans have to deal with with moral issues in certain situations, the concept of compartmentalization seems important. "Compartmentalization goes beyond that differentiation of roles and institutional structures that characterizes every social order and it does so by the extent to which each distinct sphere of social activity comes to have its own role structure governed by its own specific norms in relative independence of other such spheres (social space). Within each sphere those norms dictate which kinds of consideration are to be treated as relevant to decision-making and which are to be excluded" [Macintyre, 1999].

1.5 Eliciting requirements in making social and moral values to design

In today's society individuals and institutions act with and in socio-technical systems. Humans and technological components interact with each other and affect each other in contingent ways. Designers of aforementioned socio-technical systems face a tremendous task in how to address moral norms and how to elicit requirements to impose onto the design, built and implementation of a socio-technical system. The type of socio-technical systems we have in mind form the class of normative multi agent systems as defined in [Andrigetto *et al.*, 2013] Making social and moral values central to a design stems from the 1970s at Stanford often referred to as Value Sensitive Design (VSD) first formulated by Friedman [Friedman, 1997; Friedman *et al.*, 2002; Friedman *et al.*, 2008; Friedman, 2003; Hoven, 2015]. Many similar approaches followed coined as "values in design", "values and design" and "design for values " [Hoven, 2015]. VSD is a reaction to the idea and practice that a design of an artifact whatever that may be is a foremost technical and value-neutral task primarily focused on the requirements of users of the artifact. VSD is a theoretically grounded approach to the design of technology that accounts for human values in a principled and comprehensive manner throughout the design process[Friedman *et al.*, 2002]. Artifacts as such are the result of thousands of design decisions. The fact is that these decisions may affect one's health, safety, identity or society at large compare coined as "Diesel Dupe" where the EPA (environmental Protection Agency) detected a "defeat device" or software intentionally designed to cheat with the emission tests in the US. It is the decision to design a "cheat devise" that makes the issue morally questionable. What we would like to point out is that design processes are value sensitive in nature and that it is the choice of the designers whether ethical aspects should be taken into account in the design processes. Therefore we need an explicit interpretation of what is constituted as the tacit understanding, just displayed i.e. showed in practice[Heylighen *et al.*, 2009];"testing a design hypothesis is inextricably bound up with the ethical normative framework of society and with its epistemological principles"[Foque, 2003]. Modeling norms embedded in society in an ethics and sensitive design perspective is not about modeling ethics or moral reasoning but reflects the decisional processes buttressing a design process taking ethical or moral reasoning into account. Indeed a value design perspective is concerned with the mechanisms making a design process ethical and morally sound or unsound. Much of the debates concern the development of information systems technology. Floridi formulated eighteen open problems in the Philosophy of Information [Floridi, 2004] covering fundamental areas like the information definition, information semantics, intelligence/cognition, informational universe/nature and values/ethics. The key question, question P18 is: "does computer ethics have a philosophical foundation?" These types of questions are distinct from the value sensitive design perspectives. The question of philosophical foundation relates to the uniqueness debate [Floridi, 2008].

1.6 Outline

In chapter 2 we start with the introduction of a code of conduct implemented by Nike [NIKE, 2016a; NIKE, 2016b]to frame the design task in a principled approach. In this chapter we address the notion of moral values using categories of type of ethics. First we distinguish descriptive ethics from normative ethics which forms can either be rule-based or virtue-based. Next we address the foundational aspects of any type of ethics, value pluralism and decision procedural aspects to come up with a procedure to classify and analyze characteristics of an ethical system. Ethics is always personal to a human, and in the case a situation is the consequence of human decision making, persons may be under a duty to apply value judgments to the consequences of their decisions, and held responsible for those decisions. Reflexivity shape norms, tastes and wants of an agent and determine the effectiveness of any system. After elaborating on the notions of decision rights, responsibility and accountability we end up in this chapter with rephrasing the original design question into 7 key questions formulated in a principled way using the procedure to classify and analyze an ethical system applied to the code of conduct op Nike. In chapter 3 we address the notion of a model as the start and the result of the design process. Chapter 4 entails some concepts and definitions buttressing normative multi agent systems. Especially we address the problem of mechanism design as formulated by [Hurwicz and Reiter, 2006]. Using a verification scenario which separates the process of finding an equilibrium from recognizing an equilibrium it is possible to design incentive compatible mechanisms which occurs when the incentives that motivate the actions of individual participants are consistent with following the rules established by the group. This notion is paramount in establishing whether the mechanism designed is indeed effective. We end up in chapter 4 with some observations and characterizations of a design in general. In chapter 5 we explore the notion of relationships and values and the commonality of exchange mechanisms when studied from a sociological, anthropological or economical point of view. We analyze an interaction model between two agents addressing the question: How much is either agent willing to "give up" (i.e. to sacrifice) to "get" (i.e. to gain) the wool respectively the cloth? Key problem addressed is how models representing exchange mechanisms can be extended with notions of measurement and valuation. Chapter 6 we elaborate on principles, architectures and state transitions systems. Models are analogous to Janus structures representations with an engineering side facing the real world and an abstract side facing theories[Sowa, 2000]. In design practice it is simpler to formulate theories in first order logics and use explicitly meta reasoning about axioms and postulates; known as the AGM axioms for theory revision [Alchourron et al., 1985]. Finally in chapter 7 we elaborate on ethical sensitive design. First we reflect on the decision right allocation procedure and the verification mechanism. Secondly we introduce the notion of creating a vision from first principles.

2 Moral value(s)

Moral values play a crucial role in our society at large. In the case we were asked to list examples than the list of examples would be endless. In everyday life we all experience issues somehow, direct or indirect, related to norms, morality, and ethical behavior. Can we clarify what is meant by moral values? We start with a short description of a real life example. Nike is required by the The California Transparency in Supply Chains Act of 2010 (SB 657) ("Act") effective from January 1 2012 in the State of California to disclose efforts to eradicate slavery and human trafficking from direct supply chains [NIKE, 2016a; NIKE, 2016b]. Nike raises expectations of their factory partners through standards written down in a Code of Conduct containing a statement of values, intentions and expectations meant to guide decisions in factories. The Code of Conduct of Nike is freely available for the public and expresses on merit grounds what is expected of factory partners [NIKE, 2016b]. We site some parts of the text.

- understanding that our work with contract factories is always evolving, this Code of Conduct clarifies and elevates the expectations we have of our factory suppliers and lays out the minimum standards we expect each factory to meet

- It is our intention to use these standards as an integral component to how we approach NIKE, Inc. sourcing strategies, how we evaluate factory performance, and how we determine with which factories Nike will continue to engage and grow our business

- We believe that partnerships based on transparency, collaboration and mutual respect are integral to making this happen

- Our Code of Conduct binds our contract factories to the following specific minimum standards that we believe are essential to meeting these goals

Next there are eleven "principles" listed such as:

- EMPLOYEES are AGE 16 or OLDER. Contractor's employees are at least age 16 or over the age for completion of compulsory education or country legal working age, whichever is higher. Employees under 18 are not employed in hazardous conditions

- HARASSMENT and ABUSE are NOT TOLERATED. Contractor's employees are treated with respect and dignity. Employees are not subject to physical, sexual, psychological or verbal harassment or abuse

- The CODE is FULLY IMPLEMENTED. As a condition of doing business with Nike, the contractor shall implement and integrate this Code and accompanying Code Leadership Standards and applicable laws into its business and submit to verification and monitoring. The contractor shall post this Code, in the language(s) of its employees, in all major workplaces, train employees on their rights and obligations as defined by this Code and applicable country law; and ensure the compliance of any sub-contractors producing Nike branded or affiliate products

2.1 Design question

Suppose hypothetically that we were asked to design a normative multi agent system based on the Code of Conduct of Nike. We can come up with a design question as follows: "why is a code of conduct needed?" and "does the contents of the Code of Conduct meets its objectives of the public, here the state of California?" Subsequently the next question emerges: "if we implement the current code of conduct will it suffice to ensure Nike and other stakeholders of the company that the suppliers actually realize the objectives as stated in the code of conduct?" A normative multiagent system can be defined as a system by means of mechanisms to represent, communicate, distribute, detect, create, modify, and enforce norms, and mechanisms to deliberate about norms and detect norm violation and fulfillment [Andrigetto et al., 2013]. We adopt the mechanism design definition because is formulates precisely what a normative multiagent system does. Alas our definition does not help us right away how to elicit the norms themselves. We have to look at the design question more in-depth and take a closer look at the content of the code of conduct.

2.2 Descriptive ethics

As we learned Nike is required by the "The California Transparency in Supply Chains Act of 2010 (SB 657) ("Act")" effective from January 1 2012 in the State of California to disclose efforts to eradicate slavery and human trafficking from direct supply chains [NIKE, 2016a]. Apparently this is why we need a code of conduct. Code of conducts come in various forms and are most of the times descriptive in nature: in our example it discloses efforts to do x to achieve y. By means of the code of conduct stakeholders can uncover management's attitude, convictions and conceptions towards values that matter. Indeed it reflects what actions society rewards or punishes; in our case in the first place the law. Descriptive ethics involves empirical investigation often studied in the fields of biology, psychology, sociology, economics and management sciences. Theories and empirical findings find their way in philosophical arguments. As a consequence descriptive ethics is relativist, situational, situated or both. Merely descriptive ethics relate to the discourse of social sciences i.e. cultures and cultural norms: conceived as standards of proper or acceptable behavior [merriam webster, 2016]. Culture can be defined as "Culture is a patterned way of thinking, feeling and reacting, acquired and transmitted mainly by symbols, constituting the distinctive achievements by human groups, including their embodiments in artifacts. the essential core consists of ideas and especially their attached values."[Kluckhohn, 1951].

2.3 Normative ethics

The state of California aims to eradicate slavery and human trafficking from direct supply chains. In this context to eradicate means to destroy, exterminate practices like slavery and human trafficking in supply chains. In the case we interpret the Law normatively than our analysis will not start with the code in conduct in mind, but we would start to ponder about how one ought to act. Normative ethics studies ethical action, more precisely what makes actions wrong or right. Hence the fact that there is a Law stating that slavery and human trafficking has to be exterminated means that there is a practice of slavery and human trafficking; and that the institutions of the state of California has judged that slavery and human trafficking is no longer accepted by society and therefore such practices are no longer accepted. Using the term eradicate does not make a Law normative in nature in the deontological sense. Where the meaning of moral language is concerned lies in the realms of meta-ethics. A meta-ethical question would be "is is possible to eradicate slavery and human trafficking in supply chains in general". Positing this question introduces important notion whether an ethical claims can be judged at all. Deontology is the study of that which is an "obligation or duty", and consequent moral judgment on the actor on whether he or she has complied. Deontology is an approach to ethics that determines whether acts, or the rules and the duties of an agent performing the act is good or right. So goodness or rightness is judged by the act itself and not by the consequences. In deontology it is possible that an act considered as right or good that the act itself produces bad consequences even (!) in the case an agent who performed the act lacks virtues and had a bad intension in performing the act. The same is true in the case an act does not have any consequences at all in terms of the resulting effect pursued by performing the act. Hence it is possible that an agent did perform the act purposely wrong than he or she still did the right thing in deontology approaches to ethics. In the case we want to rule out such behavior than we have to spell out all acts, rules and duties to comply with. Up to this point we stress to point out that there is a distinction between rule-based ethics versus virtue-based ethics. Like in deontological approaches to morality the rule-based ethics focuses on acts and maintains that these rules are moral or not, to the extent of conformity, failures to conform to certain rules or principles. Virtue ethics on the other side holds the position that morality rests upon moral qualities. So it seems correct to impose that rule-based ethics is governed by concepts like acts, moral rules and moral principles and that virtue ethics is governed by moral dispositions, emotions, states of character and the flourishing of human beings. In virtue ethics morality is directly linked i.e. intimately linked to the person who acts, to his or her character and situation. It can be the case that the same act is morally wrong or right subsidiary on who acts and the conditions under which the act is done [Poivet, 2006; Poivet, 2010]. We recognize the concept of compartmentalization stating that differentiation of roles and institutional structures that characterizes every (social) order and it does so by the extent to which each distinct sphere of social activity comes to have its own role structure governed by its own specific norms in relative independence of other such spheres (social space). Within each sphere those norms dictate which kinds of consideration are to be treated as relevant to decision-making and which are to be excluded. [Macintyre, 1999].

2.4 Many values

We could extent our analysis for all categories of ethics like anarchist ethics, pragmatic ethics, role ethics, information ethics, machine ethics, utilitarian ethics, virtue ethics, hedonism, consequentialism, stoicism, evolutionary ethics, applied ethics like business ethics et cetera. An ontological approach seems to fail in getting answers to our design questions in the case we were hypothetically asked to design a normative multi agent system based on the Code of Conduct of Nike. Simply because we have to evaluate whether one of the type(s) of ethics fits our purpose so we can determine design principles which guides our design in setting, implementing and realizing the design objective(s) (i.e. goal(s)). A design principle is a normative principle on the design of the artifact [Greefhorst and Proper, 2011]. In general ethics or moral philosophy is a branch of philosophy that deals with values relating to human conduct, with respect to rightness (goodness) and badness (wrongness) of motives and end(s) of such actions. It needs no elaboration to state that there are many different moral values coined as moral value pluralism. Hence value pluralism is not about different value systems. Our key concern is how to evaluate all these different values in a coherent way, so we can make informed decisions to characterize the ethics adhered by stakeholders. Moral

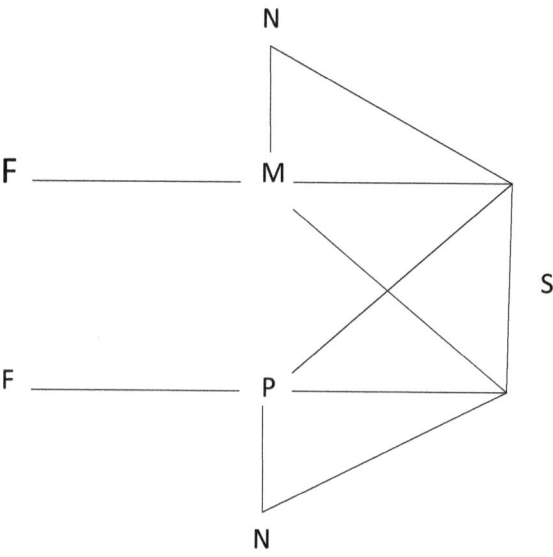

Figure 1. Pluralism

values can be characterized by being monist or pluralist[Mason, 2015] denoted as **M** respectively **P**. Monists claim that there is only one ultimate value. Pluralist defend the position that there are several distinct values. Each value can be classified along three dimensions namely (1) foundational, (2) normative and (3) decision procedural. Foundational entails that there is no one value that subsumes other values denoted as **F**. Normative posits there is a bearer of value denoted as **N**. The third characteristic decision procedural refers to a certain form of consequentialism (i.e. a representation) and has its criterion of goodness or the right action denoted as **S**. The possible relationships are represented in figure 1 value pluralism.

Now we can come up with a procedure to classify a moral value and analyze the characteristics. Reading from left to right we are able to generate a table with all possibilities.

	Found	Mon/Plu	Norm	Scale
	F	M	N	Representation
	F	M	¬N	Representation
	F	P	N	Representation
	F	P	¬N	Representation
Moral Value				
	¬F	M	N	Representation
	¬F	M	¬N	Representation
	¬F	P	N	Representation
	¬F	P	¬N	Representation

Observe that it does not matter how the table is scrambled. Consider the features of **F, M, P** and **scale** as categories in the set theoretical sense. Observe that the set of scales can be empty. Informally we are now able to define a moral value (MV) as a set MV =((F),(N),((P),(M)),(S,∅)).

2.5 Human agency

Human agency is the capacity to make choices and entails the claim that humans do in fact make decisions and enact them on the world [Mische and Ann, 1998]. This particular capacity is always personal to that human, though considerations of the outcomes enacted from private acts of human agency for us and others can then be thought of as an instantiation of moral value of a given situation wherein agents will, or have acted. In this type of situation we speak of moral agency. If a situation is the consequence of human decision making, persons may be under a duty to apply value judgments to the consequences of their decisions, and held to be responsible for those decisions. To understand moral value is to understand how decision rights are dispersed among agents and who decides. Note we are not addressing how agents decide or how they come to decisions. Discussions on the notion of free will and theorizing on the nature of rationality in making choices et cetera are very important to understand the effectiveness of a (moral) value system but it does not affect the question who decides justifying the choices buttressing a (moral) value system. Since the capacity of decision making is always personal to begin with, we have to address an important concept known as reflexivity. Reflexivity has a profound place in social theory and refers to an act of self reference recognizing forces or pressure within the environment and his or her place in the social structure. Agents with a low level of reflexivity are said that the environment shapes the individual norms, tastes, wants et cetera. Agents with a high level of reflexivity shape their own norms and tastes. Reflexivity addresses autonomy and thus autonomous action of an agent. Reflexivity is both a subjective process of self-consciousness inquiry and the study of behavior where relationships are concerned. Reflexivity seen as a subjective process of self-conscious inquiry is phenomenological in nature. Phenomenology studies the structures of consciousness as experienced from the first persons perspective. Here we enter the realms of intentionality, being the central structure of an experience directed towards an object by virtue of its content or meaning which represents the object [Ashmore, 1989]. We will address the meaning of intentionality later on in more detail. Agency and reflexivity play an important role in designing effective and efficient (corporate) organizational structures, information and communication systems [Ouchi, 1978]. It needs no elaboration that with decision rights and the choice to exercise these decision rights (i.e. often coined as (decision) power) there comes a responsibility issue. Decision rights, responsibility and accountability are intertwined concepts hard to decipher. One can feel responsible and act accordingly what the agent sees fit at some moment, although the agent did not have any formal decision rights attributed to him. Did the agent do the proper thing? If the situation turned out to be worsened because of the agent's interference, how would we judge? On the other hand an agent can decide for whatever reasons not to act, although the agent did have formal decision rights attributed to him. Did the agent do the proper thing? If it turns out that it was a bad decision not to act than we might judge the agent to be negligent. How to reason when somebody delegates his or her decision task to another agent and the same situations occur under the condition of delegation? The concept of responsibility can be viewed from two distinct though interrelated concepts namely (1) the merit-based view and the (2) consequentialist view. Following Strawson the distinction draws the line between responsibility as accountability versus responsibility as attributability [Strawson, 1974]. The two categories do not always fit the situation. Contemporary views have introduced what is coined as "The answerability model". In this view attributability and accountability self-disclosure is the target of appraisal and is judgmental sensitive. Indeed reflexive notions can play a role here, like socialization and adaptive behavior et cetera. Here we have the proper considerations and motivation for the need of normative multi agent systems as we defined it above. Moral behavior of agents needs to be monitored and outcomes are judgmental sensitive if we accept a human centered perspective on systems.

2.6 Design question revisited

If there is a moral value than there must be a belief characterized as a propositional attitude, informally defined as the mental states of an agent or a group of agents having some kind of attitude, or opinion about a proposition or about a potential state of affairs in which a proposition is true. In our case it is the belief of the state of California that slavery and human trafficking should be rooted out in the supply chain of companies resident in the state of California. Needless to say that a belief characterized as a propositional attitude is similar to the goal setting processes of enterprises, organizations, forms, soccer clubs et cetera, sharing their vision by means of belief systems [Simons, 1995]. Furthermore there is a decision right allocation procedure buttressing responsibilities and accountability and warrants that the appropriate rules, standards, regulations, rewards and punishment are established. We extent our definition **MV** with beliefs denoted as B and decision right allocation procedure denoted as **DRAP**; we informally define MV = ((B),(F),(N),((P),(M)),(S,∅),(DRAP)).

Now we can return to our design question. We asked ourselves "Why is a code of conduct needed?" and "does the contents of the Code of Conduct meets its objectives of the public, here the state of California?" Subsequently the next question emerged: "if we implement the current code of conduct will it suffice to ensure Nike and other stakeholders of the company that the suppliers actually realize the objectives as stated in the code of conduct?" By rephrasing the original questions we get 7 key questions to address:

	Questions	Sets
1.	What is the believe of the state of California with reference to slavery and human trafficking?	↦B
2.	Are the values expressed by extricating slavery and human trafficking from direct supply chains subsumed in other values?	↦F
3.	Are there several distinct values expressing extricating slavery and human trafficking?	↦P,M
4.	Who are the value bearers in the supply chain?	↦N
5.	How are the decision rights dispersed in the supply chain, who is responsible and accountable ?	↦DRAP,N
6.	What rules, standards, regulations, rewards and punishment are established preserving moral values in the direct supply chains?	↦N
7.	If applicable is there a representation expressing the moral value in communication processes?	↦S

Observe that the object in the case of Nike is the direct supply chain. So from a design perspective the design objective is to come up with **a design** for direct supply chains that warrants that slavery and human trafficking is rooted out from the direct supply chains of Nike that is of the production of sportswear. The seven questions guide the designer to elicit the informational requirements. When we humans design things for the purpose of improving thought or action we actually create an artifact that has a physical presence we can actually manufacture or construct or has a mental presence we can actually learn. Both artifacts are equally artificial since they both would not exist without human invention. Clearly cars, papers, computers, doors, sportswear et cetera are physical artifacts, where reading, listening, logic, language are mental artifacts. Hence mental artifacts produce rules and structures in information structures. It needs no elaboration that mental and physical artifacts are related to each other. Consider in our case the distinction between rule-based ethics versus virtue-based ethics. Clearly rule-based ethics and virtue-based ethics are mental In the case we accept that physical and mental artifacts are equally artificial than we must accept in the real that information structures represented in the mental artifact are equivalent to the physical artifact represented in terms of physical properties[Norman, 1993]. This is why representations express any system; the notion of model plays a central role in any design.

3 The notion of a model

A model is considered to be a representation of some object, behavior, or a system that one wants to understand [Norman, 1993]. In everyday life we are on a day to day basis involved in making decisions about what should a model should look like to become meaningful and therefore useful. For example think of working procedures at your work, deadlines to meet with clients, appointments to make for personal reasons at the doctor, working conditions to respect when we engage in a trade in foreign or domestic countries, appraisal to give in the case good work is delivered by an employee, or punish someone who didn't do a proper job et cetera. In all these cases we have some sort "workflow model", "process model" or "communication model" in mind shared with colleagues, clients, vendors, et cetera in some format, which for example can serve as a plan for organizing ones work, getting things done, or is somehow useful for goal setting purposes or all three together. In the end you want to make sure that people you work and communicate with understand your goals and your wants. A model is always the result and the start of a design process. "A design process is an abductive sensemaking process, a step of adopting a hypothesis as being suggested by the facts ... a form of inference, albeit inference of "best guesses" leaps [..]. *A logic of what might be.* It is not entirely accurate, it is the argument to the best explanation, the hypothesis that makes the most sense given observed phenomena or data and based on prior knowledge" [Kolko, 2010]. In a scientific context testing a hypothesis means confronting statements about an assumed relationship between phenomena with empirical facts. In a design context the terms testing and hypothesis tends to shift in meaning, a design assumption is not a matter of being true or false but given a particular context it is a matter of being the best solution based on vision and believes [Coppens, 2013]. As a consequence very different logics of discovery may be at work in design practices, and the way they are mixed may vary form case to case, from situation to situation, from context to context and so on. Whatever mix or configuration of elements, we will always need (good) theories to account for what happened. Theories are constitutive for every design just because we need to understand why some things do work and other things do not work or will never work. We must explain in advance why. Put in other words, we need a explicit interpretation of what is constituted as the tacit understanding, just displayed i.e.

showed in practice [Heylighen *et al.*, 2009]. Some authors state that "testing a design hypothesis is inextricably bound up with the ethical normative framework of society and with its epistemological principles" [Foque, 2003]. A model represents a justified true believe. A specific issue has to be addressed know as the Gettier problems in epistemology [Gettier, 1963]. There are two generic features that characterize the original Gettier cases, (1) fallibility and (2) luck. Fallibility implies that there are strong indications that the justification favors that the belief is true, without proving conclusively that it is. Luck refers to how the belief manages to combine being true with being justified given the fact that the well but fallibly justified belief in question is true. This notion is very important in the case we have to gather evidence and make inferences whether the behavior of a system, group of systems, human, groups of humans, object or a group of objects can be judged to behave ethical or being ethical.

4 Normative multiagent systems: some concepts and definitions

4.1 Agents

An agent is defined as "a computer system that is situated in some environment, and that is capable of autonomous action in this environment in order to meet its design objectives"[Woolridge, 2000]. An intelligent agent is such a computer system that is capable of flexible autonomous action where flexibility implies (1) reactivity, (2) pro-activeness and (3) social ability. Reactivity means that the computer system is aware of its environment, and is able to respond in a timely fashion to changes. Pro-activeness means that the computer system is able to take initiative. The third implication concerns the ability to interact with other computer systems and humans. A multiagent system is a system composed of multiple, interacting computer systems. What makes a multiagent system intelligible? Imagine that only one computer system has the capability of flexible autonomous action and the other computer systems do not have this capability? What type of agent system do we have? Hence identifying that some system is hybrid is not enough. We need to know precisely. In [Woolridge, 2011] the definition of an agent is slightly altered. Instead of design objectives, agents should meet delegated objectives. This seems trivial from a machine centered point of view but from a human centered point of view this is a big shift in perspective and has major implications. The notion of delegation is directly related to the notion of agency, accountability and responsibility. If it is possible to design, build and implement environmental aware computer systems, and these computer systems become or are empowered than studying multiagent systems from a normative stance raises some deep fundamental theoretical issues with large practical implications. One should bear in mind for example that empowerment in general beholds a management practice of sharing information, rewarding personnel, and share decision power with employees so that they can take initiative and make decisions to solve problems of some kind to improve services and performance [Simons, 1990]. Empowerment is based on the idea that giving employees capabilities, resources, authority, opportunity, motivation, as well as holding them responsible and accountable for output and outcomes of their actions so that they will contribute to their competence and satisfaction. Hence shared vision and shared mental models guide local decision makers [Senge, 1990]. In the case we view a computer system as a local decision maker than we have to make sure that the local decision maker acts according to the shared vision. Needless to say that (normative) control systems need to be in place to guide local decision makers and to make sure that local decision makers act within the boundaries fencing their decision power. Allocation of decision rights buttresses the notion of delegation. This (design) issue will be addressed later in this chapter. First we have to explore the notion of normative multiagent systems.

4.2 Normative multiagent systems: mechanistically viewed

Earlier we defined a normative multiagent system as a system by means of mechanisms to represent, communicate, distribute, detect, create, modify, and enforce norms, and mechanisms to deliberate about norms and detect norm violation and fulfillment [Andrigetto et al., 2013]. This definition takes the mechanism design point of view. In general a mechanism is a mathematical structure that models institutions through which for example economic activity is guided and coordinated. There are many kinds of these institutions for example law makers, administrators, managers of private companies like chief executive officers create institutions in order to achieve their desired goals [Hurwicz and Reiter, 2006]. The problem of mechanism design is: Given a class Θ of environments, an outcome space Z, and a goal function F, find a privacy preserving (i.e. a decentralized) mechanism $\pi = (M, \mu, h)$ that realizes F on Θ, where M is the message space, μ denotes the (group) equilibrium message correspondence $\mu : \Theta \mapsto M$ and h denotes the outcome function $h : M \mapsto Z$. The key insight of [Hurwicz, 1960] was that information about the environment, facts that enable or constrain possibilities are distributed amongst agents. In the case an agent is not able to observe some aspect of the prevailing environment, than the agent does not have the information to guide his or her actions, unless the agent is communicated to by another agent who was able to observe. More specifically an agent is not able to observe the private information of another agent. Hence dispersion of private information amongst agents, known as information asymmetry, gives rise to specific incentive problems. By means of a verification scenario which separates the process of finding an equilibrium from recognizing an equilibrium it is possible to design incentive compatible mechanisms which occurs when the incentives that motivate the actions of individual participants are consistent with following the rules established by the

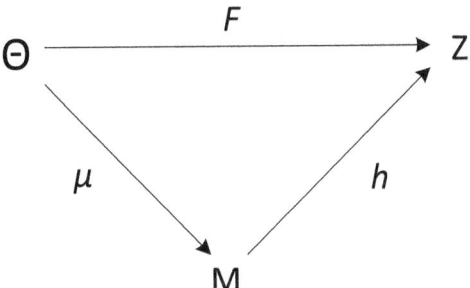

Figure 2. Commuting Diagram

group. Simply put in a verification scenario each agent reads the announced message by saying yes or no. The proposed outcome is judged acceptable if and only if the agents' responses are affirmative. The message exchange process consists of three elements first a message space M, second a group equilibrium message correspondence μ, denoted $\mu : \Theta \mapsto M$ and third outcome function h, denoted as $h : M \mapsto Z$. A message space M consists of the messages available for communication. Messages may include formal written communication like contracts among buyers and sellers, accounting reports, production statistics, emails et cetera. Now it is easy to see that the group equilibrium message correspondence μ associates with each environment, θ, set of messages $\mu(\theta)$, that are equilibrium messages for all the agents. Assume that the messages were proposed actions, than $\mu(\theta)$ consists of all proposals to which each agent would agree in θ. The outcome function h translates messages into outcomes. As we have seen a mechanism π can be defined as an triple $\pi = (M, \mu, h)$. In the case π operates in an environment θ than the result is outcome $h(\mu(\theta))$ in Z. In the given space Θ, for all environments, the mechanism π leads to the desired result by the agent in that particular environment, than we say that π realizes F on Θ. In short π realizes F if for all θ in $\theta, h(\mu(\theta)) = F(\theta)$. Actually the equilibrium message μ represents the behavior of the agents. The concept can be represented in a commuting diagram as shown in figure 2

4.3 Effectiveness

The performance of a mechanism just described depends on elements that constrain the situation, like technological possibilities or those elements that define (i.e. influence) preferences of an agent, that are not subject to control or influence of the designer of the mechanism. The totality of all these elements is coined as the environmental space. In our exposition Θ. Furthermore we know that it is the case that no one, including the designer knows the prevailing environment θ and thus does not know the group equilibrium message μ. Agents know only their own parameters, the designer knows the environmental space Θ and the goal function F, informally defined as the class of environments for which the mechanism is to be designed for and the criterion of desirability. Remember the goal function F, reflects the agents criteria for evaluating the outcomes for example efficiency criteria, fairness criteria and so on. Indeed the mechanism provides in a logic in which ethical and moral criteria can be evaluated. These criteria are widely known to be effectiveness criteria [Paape, 2007]. As we will see later the notion of effectiveness plays a major role in designing all sorts of control and monitoring systems ensuring that goals of the designer are met. There are two crucial aspects we did not pay attention to namely the notion of the game form solutions concept and the revelation mechanism.

4.4 Game form mechanism

More precise the mechanism can be formulated in a game theoretical normal form. The game is defined by the agent's strategies, $S^1...S^N$ and their pay-off functions $\Psi^1...\Psi^N$. The joint strategy space is the Cartesian product of the agents strategies denoted as $S = S^1 \times ... \times S^N$. The pay-off function represents the utility of the agent when the joint strategy $s = (s^1...s^n)$ is used. The value h(s) refers to the outcome when the joint strategy s is used and the value of the pay-off function, when s is used, is the value of the composite $\Psi^i(s) = \psi^i(h(s))$, where $\psi^i(s)$ denotes the utility of h. The game can be written as $G = G(S,h)$, so in the case the environment is θ, the utility function allows the agent to evaluate the pay-off from the joint strategy, when the outcome function is h. The solution concept and the specified message space induces privacy preserving group correspondences from the identified parameter space into the message space M, to be identified with the correspondences μ^i and μ. In the case N-tuple $(\mu^1...\mu^n)$ is an equilibrium, whatever type, the resulting messages in each environment θ, defines the correspondence Θ to M. Now we can easily see that once G implements a goal function F, then there is a mechanism π realizing F only in de case the correspondence μ makes the diagram commute (figure 2).

4.5 Incentive compatible and the revelation principle

Earlier we stated that an agent only knows his own characteristic, that is in the case of environment θ, agent i knows θ^i and his behavior depends only on the private information he has. We do agree on the fact that communication buttresses institutions like markets, organizations et cetera [Hayek, 1945]. Opportunities for mutually hopefully beneficial transactions, social encounters et cetera cannot be found unless individuals share information about their preferences and endowments. The revelation principle states that, for many purposes, it is sufficient to consider only a special class of mechanisms, called incentive-compatible (direct or encoded) revelation mechanisms. As we have seen the key idea is that each individual is asked to report his private information to a mediating mechanism. A direct-revelation mechanism is said to be 'incentive compatible' if, when each individual expects that the other persons (or agents in the sense of computer systems) will be honest and obedient to the mediating mechanism, then no individual (or agent) could ever expect to do better, given the information available to him, by reporting dishonestly to the mediating mechanism. So as a consequence the mechanism is incentive compatible if and only if honesty and obedience is in equilibrium of the resulting communication game. Hence for any equilibrium of any general communication mechanism, there exists an incentive-compatible (direct or encoded) revelation mechanism that is equivalent. This proposition is the revelation principle [Myerson, 1982]. Hence the notion of equivalence (classes) determine under what conditions the commuting diagram commutes. Observe that it is also assumed that the mediating mechanism also needs to be honest and obedient. These values seem to guide actions of all participants in the game and therefore should be accounted for in the design of the mechanism that realizes the game. Participants in the game need to trust one another so each participant must be able to verify the proposed outcome often coined as transparency. In the case agents (human or computer systems) participate in a game than an agent is as well a principal as an agent. Stated otherwise their relationship is by nature reciprocal. Technically from a machine (that is a computational) point of view we recognize the notion of symmetric bi-directionality.

4.6 Incentives

In the case a principal delegates a task to an agent who has different objectives than delegating this task becomes problematic when the information about the agent is imperfect. Hereafter we will explicitly make a distinction between human agents and agents which are computer systems. Following [Laffont and Martimort, 2002] "If the human agent has a different objective function but no private information, the principal could propose a contract that perfectly controls the agent and induces the latters actions to be what he would like to do himself in a world without delegation". As a result incentive problems disappear. Alas conflicting objectives and decentralized information are thus the two basic ingredients of incentive theory. We will argue that even though objectives do not conflict or information is centralized that incentive problems still can occur. Think of fraud, criminal organizations, bribery, market misconduct, CEO compensation, slavery, environmental pollution et cetera. Here we enter the realm of norms and normative behavior. Three types of problems might occur in the case the human agent with private knowledge. First we have moral hazard or hidden action issues. Secondly we have adverse selection of hidden knowledge and thirdly the case of non verifiability. Non verifiability relates to the issue of sharing ex-post the same information but that no third party or no court of law can observe this information.

4.7 Some observations

When μ represents the actual behavior than this mechanism is compatible with the social definition of normative multi agent systems. In [Andrigetto et al., 2013] a normative multi agent system is defined as "a multi agent system governed by restrictions on patterns of behavior of the agents in the system that are actively or passively transmitted and have a social function and impact". Patterns are represented as actions to be preformed, dictating what actions are permitted, empowered, prohibited or obligatory under a set of conditions and the specified effects when compliant and the consequences being not compliant with the set of conditions i.e. the norms. Hence the group equilibrium message correspondence μ associates with each environment, θ, set of messages $\mu(\theta)$, that are equilibrium messages for all the agents. In the case the messages contain proposed actions, than $\mu(\theta)$ consists of all proposals to which each agent would agree in θ. So the equilibrium message μ actually represents the behavior of the agents. The introduction of a verification scenario in the mechanism which separates the process of finding an equilibrium from recognizing an equilibrium makes it possible to design incentive compatible mechanisms which occurs when the incentives that motivate the actions of individual participants are consistent with following the rules established by the group. This is exactly what we are trying i.e. aiming to achieve following the rules of the group. A verification scenario warrants that each agent reads the announced message by saying yes of no. The proposed outcome is judged acceptable if and only if the agents' responses are affirmative. The goal function F reflects the agents criteria for evaluating outcomes, attributes of the outcome concern objectives like efficiency, fairness, effectiveness, compliance, reliability, trust whatever objectives formulated by the designer of the mechanism. It should be noted and emphasized that the equilibrium is founded in the logic of the price mechanism i.e. demand and supply mechanisms. In the case of the model proposed by Hurwicz the application was demonstrated using Walrasian tantonnement mechanism with continue price functions. The price mechanism in general has sound mathematical properties we will explore in the next chapter. The mathematical properties can be revealed using commuting diagrams. Commuting diagrams are a simple means to display some objects, linked together with arrows representing in our case functions. Commutativity means that the two paths depicted in figure 2 amount to the same thing; any two paths of arrows in the diagram that start at the same object and ends at the same object compose to give the same overall function. Instead of arrows the term morphisms is also used. So arrows are all set functions which in each appropriate case satisfy conditions relating to that structure[Goldblatt, 2006]. We are interested in the way the arrows behave and the commonality they all have.

5 Relationships and values

The price mechanism ensures that a market price of a good and or service accurately summarizes the vast array of information held by market participants [Atakan and Ekmekci, 2014]. In the case prices depict i.e. fully reflect all available information than we say that the market is efficient[Fama and Miller, 1972]. Critics argue that markets are imperfect due to a range of all sorts of cognitive biases. But it needs no elaboration that the information aggregation characteristic of the pricing system buttresses many theories about the communicative function of prices and the decisions participants in the marketplace make to exploit business opportunities in the creation of value. It is the notion of value in this respect what is of interest for our purpose designing systems, particularly normative multi agent systems. Prices in economic theory reflect the value of an exchange in the marketplace such as buying and selling transactions. In early social theory the exchange mechanism is also used in analyzing social and anthropological mechanisms. Simmel uses the economic concept of value and argues that we should make a distinction in the exchange of value and the value exchange [Simmel, 1900]. His first observation was that economic value is not just value in general but a definite sum of value, resulting from the commensuration of two intensities of demand to be exact the exchange of sacrifice and gain. An exchange is not a by-product of the mutual valuation of objects but its source [Appadurai, 1986]. For our purposes it suffices to look at value as defined in sociology, economics and anthropology[Graeber, 2001]. Sociological concept of value is merely a conception of what is a good, proper, or desirable way to behave. In economic sense value refers to the degree to which objects are desired i.e. wants as measured how much others are willing "give up" to "get" these objects. Linguistically value might be defined as a meaningful difference. Hence they are all refractions of the same thing. Indeed they have some things in common and they might even share some properties. In the next section we explore the exchange mechanism in more detail.

5.1 Exchange mechanism characteristic

To describe a social encounter in a restaurant or an economic transaction in the marketplace in most cases agent models are used. Assume that we have an agent A who is willing to sell wool in some quantity at some price, given some quality standard. There is another agent B who is willing to sell cloth in some quantity at some price. Indeed there are different standards of quality in cloth, so the cloth for sale has some quality standard. Suppose agent A wants to buy cloth, and agent B wants to buy wool. The key question is then: How much is either agent willing to "give up" (i.e. to sacrifice) to "get" (i.e. to gain) the wool respectively the cloth? Typically this formulation captures precisely what is exchanged. It does not say anything how or when exchanges will actually occur. Exchanges are by definition reciprocal in nature and come in a large variety of what we coin as means like signed contracts, shaking hands et cetera. For example signing a contract by both parties is performative in nature; by the act of signing, we communicate that the exchange is done. Hence a signed contract affords exchanging. An affordance establishes the relationship between an object or an environment and an organism here a (human) agent through a stimulus to perform an action. In our example the stimulus is the signed contract and the detectable change in the external environment. We assume that the agent is sensitive and therefore able to respond to external (or internal) stimuli. This presumption is known as sensitivity. We have to realize that affordances are very special in nature. Following Gibson affordances of an environment are in a sense objective, real, physical unlike values and meanings, which are subjective, phenomenal and mental. But if we look closer than we must assert that affordances are neither an objective property nor a subjective property; hence they are both objective and subjective. It is equally a fact of an environment and a fact of behavior. An affordance points both ways, to the environment and to the agent (observer) [Gibson, 1986]. Agents occupy niches of the environment, where we define a niche as a collection of affordances. Hence often we use terms like habitat or social space of an agent. Affordances can be measured with scales and standard units used for example in physics but they are as we have seen not just physical properties for they have unity relative to the posture and behavior of an agent. In our exposition we are interested in the unity relative to (moral) values. It seems that there are two interrelated notions of cognition at work: (1) experimental cognition and (2) reflective cognition. Experimental cognition leads to a state in which a (human) agent perceives and react to events around us. The reflective mode of cognition is about compulsion, contrast, thought and decision making. The difference lies in the technical details of the information structures of our brain; experimental cognition involves data-driven processing, where reflective cognition involves planning. Observe that via reflective mode of cognition we train i.e. learn to become an expert whose skill is that of experimental cognition[Norman, 1993]. Here we enter the realm of preferences, utility and the notion of bounded rationality. We will explore these notions later on in this chapter. First we return to our example i.e. problem introduced in this chapter.

5.2 The exchange cycle: value exchange - exchange of values

Remember the situation in which an agent A who is willing to sell wool in some quantity at some price, given some quality standard. We have another agent B who is willing to sell cloth in some quantity at some price and that there were different standards of quality in cloth, so the cloth for sale has some quality standard. Now agent A wants to buy cloth, and agent B wants to buy wool. Our key question was: How much is either agent willing to "give up" (i.e. to sacrifice) to "get" (i.e. to gain) the wool respectively the cloth? The situation can be depicted graphically as follows.

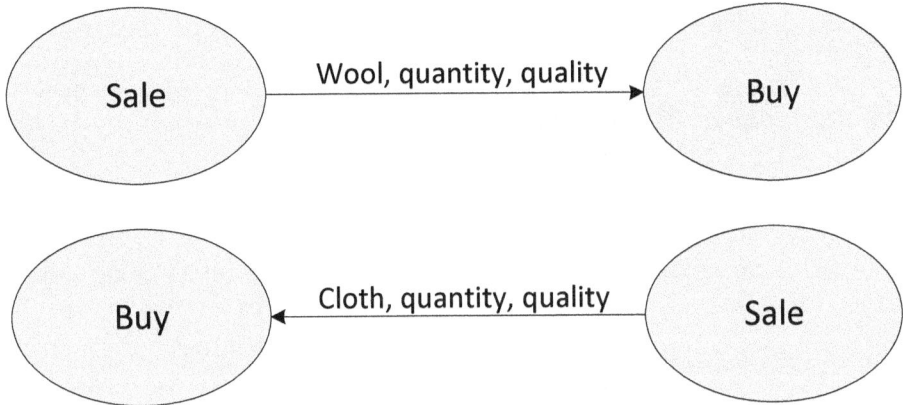

Figure 3. Barter Exchange

We think it needs no elaboration that sacrifices and gains balance when equilibrium is reached. Informally our key question can be rephrased algebraically as:

(1) $$Q^S_{W_{(q)}} : Q^B_{C_{(q)}} = Q^S_{C_{(q)}} : Q^B_{W_{(q)}}$$

Where

- The quantity of some object O, is denoted as Q

- The seller, denoted as superscript S of some object O

- The buyer, denoted as superscript B of some object O

- The object wool, denoted as subscript for some object O here Wool, denoted as subscript W of object Wool

- The object cloth, denoted as subscript for some object O here Cloth, denoted as subscript C of object Cloth

- The quality, denoted as subscript of a object (q)

Now we have to extend the model with the notion of measurement and valuation. In general a quantity is a property of a phenomenon, body or substance. Basically quantities are organized in a system of dimensions - SI. There are so-called base quantities like length and time for example, and quantities that are derived from these base quantities. Hence each base quantity has its own dimension which property has a unique magnitude that can be expressed as a number and a reference. Observe that each derived quantity's dimension follows from the derivation itself [Griffioen, 2013]. Rearranging our equation we get:

(2) $$\frac{Q^S_{W_{(q)}}}{Q^B_{C_{(q)}}} = \frac{Q^S_{C_{(q)}}}{Q^B_{W_{(q)}}}$$

Now we extend our equation with the notion of measurement for the quantity of object O:

(3) $$\frac{Q^S_{W_{(q)}}}{Q^B_{C_{(q)}}} \cdot \frac{Q^{st}_W Q^m_W}{Q^{st}_C Q^m_C} = \frac{Q^S_{C_{(q)}}}{Q^B_{W_{(q)}}} \cdot \frac{Q^{st}_C Q^m_C}{Q^{st}_W Q^m_W}$$

Where the quantity of the object O is measured in some standard unit expressed as a number and a reference denoted as superscript st and superscript m, the dimension quality denoted as (q) of object, the dimension absolute frequency as a number of objects. Standard units expressed as a number and a reference $Q^{st}_O Q^m_O$ in (3) can be denoted as $U_{(O_q)}^S$ for the sell side and $U_{(O_q)}^B$ for the buy side, where U denotes the standard unit expressed as a number and a reference. The quantity of the object O is measured in some standard unit U and the measurement is expressed as a product $Q \cdot U$, the dimension quality denoted as q of object, the dimension absolute frequency as a number of objects. We get:

(4) $$\frac{Q^S_{W_{(q)}}}{Q^B_{C_{(q)}}} \cdot \frac{U^S_{W_q}}{U^B_{C_q}} = \frac{Q^S_{C_{(q)}}}{Q^B_{W_{(q)}}} \cdot \frac{U^S_{C_{(q)}}}{U^B_{W_{(q)}}}$$

Analogue to the definitions of equation (4) we can write (5) with the notion of measurement defined as U for the quality of objects O as:

(5) $$\frac{Q^S_{W_{(q)}}}{Q^B_{C_{(q)}}} \cdot \frac{U^S_{W_q}}{U^B_{C_q}} \cdot \frac{U^S_W}{U^B_C} = \frac{Q^S_{C_{(q)}}}{Q^B_{W_{(q)}}} \cdot \frac{U^S_{C_{(q)}}}{U^B_{W_{(q)}}} \cdot \frac{U^S_C}{U^B_W}$$

It is quite easy where the money part comes in, just multiply the equations with a unit of measurement for money. Let v is the dimension of currency, denoted as v for the money unit. Observe that in our equation each base quantity has its own dimension which property has a unique magnitude that can be expressed as a number and a reference. Hence the unit v is easy extensible to a multitude of currencies. For our purposes we leave this subject to a rest. In the case we define the unit of measurement for money as we get:

(6) $$\frac{Q^S_{W_{(q)}}}{Q^B_{C_{(q)}}} \cdot \frac{U^S_{W_q}}{U^B_{C_q}} \cdot \frac{U^S_W}{U^B_C} \cdot \frac{U_v}{U_v} = \frac{Q^S_{C_{(q)}}}{Q^B_{W_{(q)}}} \cdot \frac{U^S_{C_{(q)}}}{U^B_{W_{(q)}}} \cdot \frac{U^S_C}{U^B_W} \cdot \frac{U_v}{U_v}$$

It is straightforward to see when we multiply all variables i.e. factors that we only have a money measure and that all information is encapsulated in the money measure. When markets are efficient in the way Fama formulated than it is very useful to use market based measures for evaluation procedures[Fama and Miller, 1972; Ouchi, 1980; Ouchi, 1978]). As we mentioned earlier this line of reasoning met some critiques from theorists and experimentalists that the behavioral assumptions underlying the information aggregation characteristic are flawed. In general the critiques concentrate on the rationality assumptions and that one should look for evidence about what humans actually do [Camerer, 2003]. As for economists, sociologist, anthropologists and psychologist equation (5) is quite interesting. Economists are interested in the price behavior and market conditions. For example market regulation issues like monopolistic behavior and transaction cost economics coined as market and organizational failure theories [Williamson, 1975]. Basically Williamson argues that bounded rationality characteristics combined with opportunistic behavior of agents are major concerns classical economic organizational theories overlooked. Economists tend to integrate psychology to economic theory to explain (economic) behavior. Aspects are altruism, happiness, pro-social behavior, the helping hand et cetera [Frey and Stutzer, 2007]. Very interesting is some older work of Mauss [Mauss, 1950]. He studied the actual act of exchange of gifts and rendering of services, and the reciprocating or return of these gifts and services. Although there was no economic system as we know it, Mauss argues that the society he studied can be described by the catalogue of transfers that map all the obligations between its members. The cycling gift system is the society. If we look at equation (5) than we could say that the left part is the weighing function of agent A and the right part is the weighing function of agent B. The weighing function becomes hard to decipher in the case the weighing function is not linearizable. As a consequence we cannot verify whether the calculations are properly conducted. Hence outcomes become unpredictable. Theorists introduce therefore utility functions and conditions that are assumed like preference ordering characteristics and monotonicity. Hence if we are able to measure the goods or services in the correct unit of measurements and the only uncertainty is the outcome of the quality evaluation of agent A and agent B towards the objects sold and bought, than in the case agent A and agent B come to an agreement we know that the equations (7) and (8) must hold:

$$(7) \qquad P_B(Q^B_{W_{(q)}} \cdot U^B_{W_q} \cdot U^B_W) > P_A(Q^S_{W_{(q)}} \cdot U^S_{W_{(q)}} \cdot U^S_W)$$

$$\wedge$$

$$(8) \qquad P_A(Q^B_{C_{(q)}} \cdot U^B_{C_q} \cdot U^B_C) > P_B(Q^S_{C_{(q)}} \cdot U^S_{C_{(q)}} \cdot U^S_C)$$

P_A en P_B denote the preference function outcomes of the objects bought.

5.3 Design characteristics

The structure of the preference function of the agents is what we actually need to elaborate upon in the case norms are modeled for society and the design of the normative multi agent system is value sensitive by design. As we have seen affordances can be measured with scales and standard units used for example in physics but they are not just physical properties for they have unity relative to the posture and behavior of an agent. We are interested in the unity relative to (moral) values. Our equations (7) and (8) formulate precisely the decision rule i.e. the procedure to reason about whether the monitored systems actually behaves in an ethical i.e. moral fashion, actually this is what a normative multi agent system does and foremost the designer can actually formulate which or whenever design choices have to be made, why the choices are inevitable and what consequences these choices actually have for the design and effectiveness of the artifact. Hence the model equation (1) depicts the most elementary mechanism of any exchange relationship. Using the elementary units organized in a system of dimensions we actually enrich the model so we can ensure that no information ever gets lost. Indeed it ensures the minimum informational requirements warranting consistency of the data processing facilities like software, algorithms, communication, hardware, networks, and search and database technologies. Consistency is paramount and buttresses the notion of data quality i.e. data integrity. For example ACID (Atomicity, Consistency, Isolation, Durability) is a set of properties that guarantee that database transactions are processed concurrently. Hence if we design a distributed environment based on web services the key design question is: "How to ensure the ACID principles in transaction processing using web services?"[Gilbert and Lynch, 2002]. The same design question can be formulated for workflow systems such as inter and intra- organizational workflows, contracts nets, value nets et cetera [Christiaanse et al., 2015]. Observe that the first design question provides in a mechanism to ensure the second design question. So we only have to concentrate on the special requirements on the process level ensuring external integrity of the information processing function of the artifact. Observe that the model depicted as equation (4) and the decision rules equation (7) and (8) realizes the mechanism π described in chapter 4, under the conditions that elementary units organized in a system of dimensions are used in the model. We observed that when μ represents the actual behavior that this mechanism is compatible with the social definition of normative multi agent systems. The introduction of a verification scenario in the mechanism which separates the process of finding an equilibrium from recognizing an equilibrium makes it possible to design incentive compatible mechanisms which occurs when the incentives that motivate the actions of individual participants are consistent with following the rules established by the group. There is one (big) difference: we did not use utility functions, but instead we formulated a preference ordering derived from the systems of dimensions. The valuation itself is an empirical question and should be treated as such because we need models that have high predictive value. Otherwise the designed mechanism like normative multi agent systems fails to realize the goal function of the design. We separated the preference and ordering conditions from the object and the subject. Therefore human peculiarities in decision making can be studied in isolation and in combination with the environment. Hence the behavior is influenced by, depends on the environment. We think that the human preference orderings behave on a continuum where complexity and uncertainty plays an important role. Notions of this adaptive toolbox describe mechanisms to model adaptive behavior of agents in the environment they "see" [Gigerenzer, 2001].

6 Principles, architectures and state transition systems

Like we stated earlier a model is always a result and the start of a design process. "A (design) process is an abductive sense making process, a step of adopting a hypothesis as being suggested by the facts ... a form of inference, albeit inference of "best guesses" leaps [..]. A logic of what might be. It is not entirely accurate,... it is the argument to the best explanation, the hypothesis that makes the most sense given observed phenomena or data and based on prior knowledge" [Kolko, 2010]. This is precisely the function of our model depicted as equation (4) and the decision rules equation (7) and (8) realizes the mechanism π described in chapter 4. We return to our Nike example. Suppose that agent A is Nike, and agent B is one of the suppliers in the direct supply chain. Nike wants to be sure that agent B is compliant with applicable laws and regulations ensuring that slavery is rooted out from the supply chain. In the case agent B delivers goods manufactured under conditions of slavery than we would expect that equation (8) fails and thus is not true. What information does agent A (Nike) need to make this assertion? But we have another issue to address simultaneously: how can we be sure that agent A (Nike) will be truthful in their actions and communications. Stated otherwise how to distinguish moral hazard, from adverse selection and non-verifiability problems? We will have to formulate constitutive rules and regulative rules, grounded in our belief, our attitude, et cetera [Andrigetto et al., 2013]. This process is iterative in nature, and the model facilitates current understanding among participants in the direct supply chain and its stakeholders and supervisors. Hence all aspects identified in chapter 2 will be addressed and henceforth all types of ethics will be addressed to reason about the purpose of the normative agent system fueling the question and answering how we can design a mechanism that actually realizes the goal function of Nike and the goal function of in this case the legislator. We have identified seven key questions which have to address in modeling norms. These were:

	Questions	Sets
1.	What is the believe of the state of California with reference to slavery and human trafficking?	↦B
2.	Are the values expressed by extricating slavery and human trafficking from direct supply chains subsumed in other values?	↦F
3.	Are there several distinct values expressing extricating slavery and human trafficking?	↦P,M
4.	Who are the value bearers in the supply chain?	↦N
5.	How are the decision rights dispersed in the supply chain, who is responsible and accountable?	↦DRAP,N
6.	What rules, standards, regulations, rewards and punishment are established preserving moral values in the direct supply chains?	↦N
7.	If applicable is there a representation expressing the moral value in communication processes?	↦S

A design principle is a normative principle on the design of the artifact as such, it is a declarative statement that normatively restricts design freedom [Dietz, 2008; Greefhorst and Proper, 2011]. So by answering the questions summed up above we elicit the normative principles on the design of the artifact. Requirements defined as a required property of an artifact also limits design freedom. Indeed requirements state what properties an artifact should have from the perspective of the goals of stakeholders i.e. institutions, legislators, supervisors, society et cetera. Goals motivate why requirements are imposed on the design. The first three questions address design principles used to express, i.e. buttresses policies to ensure that the design of the artifact meets the aforementioned requirement defined as a property that the artifact should have realizing the goal function of the stakeholder(s). The questions address the What and Why of the design. The questions 4 and 5 address the notion of moral agency. Indeed **who** is responsible as accountable? By answering question 6 we address **how** norms as moral values are enforced where question 7 addresses the informational (infra)structure like information processing, communications and storage of data i.e. how communication processes enables the interaction among agents (human and machines). Observe that the identified questions address governance and management perspectives [ISACA, 2012; OECD, 2015]. The first three questions cover the goal setting processes and objective setting i.e. the governance system, while questions 4 and 5 cover the management system. Question 6 covers the process and control dimension where question 7 covers the information systems and infrastructure. Models in general are to be understood as purposeful abstractions i.e. representations of (some) reality. Usage of models is to represent systems; actually the model can be regarded as a system in itself [Apostel, 1960]. Models are analogous to Janus structures representations with an engineering side facing the real world and an abstract side facing theories[Sowa, 2000]. It is possible even most likely that the model does not fit the empirical data, just because the theory was not appropriate so the theory i.e. our belief has to be revised. The revision process is actually a meta level technique for examining the axioms upon which the theory was "founded". By altering the axioms or postulates new theories are formulated that hopefully forms a better match with the facts. In design practice it is simpler to formulate theories in first order logics and use explicitly meta reasoning about axioms and postulates. Indeed we are interested in mechanisms that realize goal functions. This notion is known as the AGM axioms for theory revision[Alchourron *et al.*, 1985]. In the case our model does not realize the goal function expressed as a belief than we examine whether the pre-conditions i.e. the axioms and postulates buttressing the model are appropriate. Axioms and postulates are directed internally representing the intentional internal point of view. For example when we return to our example of Nike question 1 covers the belief, where questions 2 and 3 cover the intentional point of view, as we have seen being the central structure of an experience directed towards an object by virtue of its content or meaning which represents the object [Ashmore, 1989]. This feature characteristic will be important when we ask ourselves whether a computer system i.e. an artifact can be a moral agent. Next we explore the notion of architecture as a model representing a system.

6.1 Architecture as a model representing a system

The IEEE defines an architecture as "the fundamental organization of a system embodied in its components, their relationships to each other, and to the environment, and the principles guiding its design and evolution" [IEEE, 2011]. Basically every information system is an assembly of 5 basic components known as the von Neuman Architecture [Neuman, 1945]. We have input and output devices, a CPU containing a control unit and ALU and a (internal) memory unit. Computationally a representation defined as a pattern of symbols that stand for values are coined as data and when implemented by a computer system an algorithm controls the representation as input and the representation as output so the algorithm controls the transformation of data representations[Denning and Martell, 2015]. Theoretically there are several models in which the actual behavior of a discrete system can be described. All these models can be described as state transition systems. Formally computation can be studied by means of a state transition systems defined as a pair $(S, \rightarrow)$ where S is a set of states and $\rightarrow$ is a set of transitions; the state p to state q denoted as $(p,q) \in \rightarrow$ we write $p \rightarrow q$. It is easy to label a transition. Labels can mean anything, like expected input conditions, actions to perform during the transition, conditions that must be true before triggering a transition. The state transition system with label's is a tuple $(S, \Lambda, \rightarrow)$ where S is a set of states, $\rightarrow$ is a set of state transitions and Λ is a set of labels;

the state p to q with label α. In the case we are not familiar with the semantics of the label or simply put the semantics of the labels are not known to us, a labeled deductive system (LDS) as a means to be able to reason properly about the representation in the system seems to be a necessary first step to translate the logic of a state transition system with labels via a LDS to classical logics. Hence an LDS is family of logics[Gabbay, 1996]. Observe that the AGM axioms for theory revision can be formulated as an LDS and translated to classical logics. Architecture is the normative restriction of design freedom [Dietz, 2008; Greefhorst and Proper, 2011].

7 Ethical sensitive design

7.1 Decision right allocation procedure (DRAP) and the verification mechanism

We defined human agency as the capacity to make choices and entails the claim that humans do in fact make decisions and enact them on the world. The key design problem in design processes is how to address reflexivity. In chapter 2 we elaborated on the notion of reflexivity as being the mechanism referring to an act of self reference recognizing forces or pressure within the environment and his or her place in the social structure. Agents with a low level of reflexivity are said that the environment shapes the individual norms, tastes, wants et cetera. In the case agents with a high level of reflexivity shape for example their own norms and tastes. Reflexivity addresses autonomy and thus autonomous action of an agent. This is exactly where principles in general and moral principles as moral values come into play, namely principles restrict autonomy of an agent. We like to defend that reflexivity refers to rule-based ethics versus virtue-based ethics. Rule-based ethics is governed by concepts like acts, moral rules and moral principles and virtue ethics is governed by moral dispositions, emotions, states of character and the flourishing of human beings. In virtue ethics morality is directly linked i.e. intimately linked to the person who acts, to his or her character and situation. This notion demarcates social space among agents. Agents occupy niches of the environment seen as the world, where we define a niche as a collection of affordances, often coined as habitat or social space of an agent. An environment is in a sense objective, real, physical unlike values and meanings, which are subjective, phenomenal and mental. Affordances are neither an objective property nor a subjective property; they are both objective and subjective. It is equally a fact of an environment and a fact of behavior. An affordance points both ways, to the environment and to the agent (observer) [Gibson, 1986]. We cannot say in advance that reflexivity behaves on a continuum from rule-based ethics to virtue-based ethics and vise versa. So in the design process we have to make provisions to decide upon how consensus can be reached among agents. Here we must decide what type of rules we adopt to verify whether consensus is reached. Observe consen-

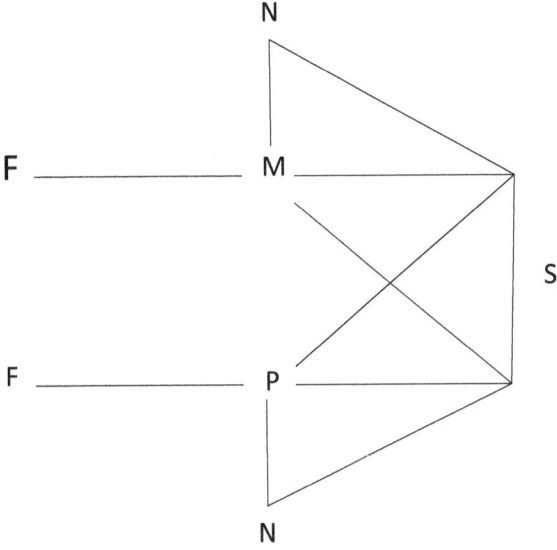

Figure 4. Value pluralism

sus in a design setting addresses the shared values among agents. Each agent will make their private considerations whether to agree or not. Indeed it is possible to reconsider the earlier made choices. Whether this type of rule is accepted is a fundamental design question addressing agency. In some cases it is impossible to decide upfront how a mechanism is to be designed, build and implemented. This means that deciding upon a verification procedure recognizing whether consensus is reached among agents must have provisions for reconsidering earlier made choices. Indeed all seven questions categorized in what, why, who en how are permanently defied by agents and therefore monitoring warrants the soundness of moral choices being made by agents. This is precisely what a normative multi agent system does; monitoring warranting the soundness of moral choices, recognizing whether agents are not compliant, and recognizing whether the designed moral system itself does not have possible negative side effects subverting the moral system actually desired [Budesku and Bruderman, 1995; Hofstede, 1981; Merchant, 1982].

7.2 Creating a vision from first principles

By answering and discussing the seven questions that guide action in designing a normative multi agent system our aim is to create a shared vision and thus shared mental models that guide local decision makers i.e. the agents.

A vision contains i.e. envisions the outcome of the deliberation process discussing and answering the seven questions in a coherent, consistent and sequacious way. Our design question is diagnostic in nature and the reasoning style is abductive. Traditional the diagnostic problem is framed in situations where an observation of the system's behavior is functioning abnormal or even fails to function at all. The issue is than to determine those components, objects et cetera of the system that will explain the difference between observed behavior and the desired correct behavior [Reiter, 1987]. To solve the aforementioned diagnostic problem from first principles only the information of the system description is available together with the observation of the actual behavior. Reiter builds on the work of [Kleer de and Williams, 1987] and provides in a theoretical foundation for diagnosis from first principles. For representation purposes Reiter choose first-order logic for representing systems. As he observed and demonstrated many different logics lead to the same theory of diagnosis. Hence more abstractly Reiter's theory can be formulated as a LDS and then translated into classical logic here first-order logic. In our situation there is one major difference and that is we cannot observe actual system behavior just because the system has to be designed yet. We do have a shared expectation about the expected behavior and we aim that the system after being build and implemented shows in practice the shared expected behavior. Indeed we have to consider that there is a possibility that the actual behavior after having the system built and implemented can actually differ from the expected outcome and we will need safeguards upfront to consider in designing the system. We coined this requirement incentive-compatible (direct or encoded) revelation mechanisms. It needs no elaboration that the design problem and the diagnostic problems both share the same mechanisms and principles. This is easy to see in figure 4 [Kleer de and Williams, 1987].

The main objective is to design a system that minimizes the expected structural discrepancy between the model and the artifact realizing the goal function. For example: if the moral values are not to be debated than a model based on the axioms of such a belief decided to be foundational and strict normative in the deontological sense than the normative multi agent system is to be designed to monitor the behavioral discrepancies between predicted that is normative behavior versus observed normative behavior; strict rules should be enforced upon the agents who are responsible as accountable on merit grounds, whether moral values are in the plural or monist like. Indeed the verification procedure applied by the normative multi agent system communicates the outcome of the verification procedure analogous to the group correspondence message π. In the case the equilibrium is not recognized than the normative multi agent system has to inform the agent whose action is not compliant to the applicable rule so corrective action can be taken or the agent is to be punished by some rule. Punishment can be a blaming and shaming mechanism, dissipation from the group, or group activities, imposition of individual

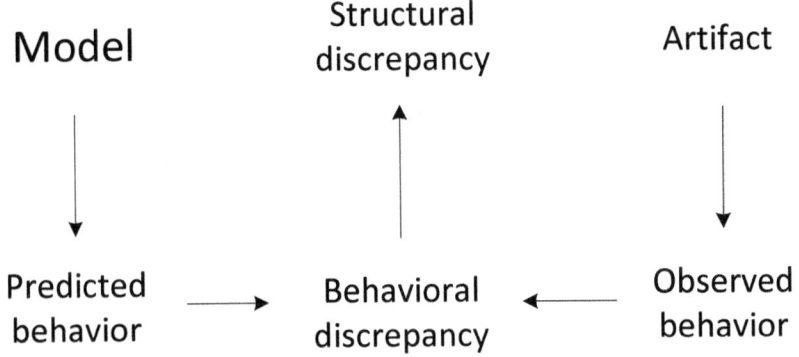

Figure 5. Diagnostic cycle

fines, restrict autonomy et cetera. In the case the equilibrium is recognized than the group correspondence message π actually reflects the behavior of the agents so behavioral discrepancy is not observed and we may infer that the mechanism realizes the goal function of the normative system. Indeed analogous to the punishment ruling we can actually reward the agents for being compliant. Rewarding agents can either be financial, augment autonomy, pat on the back, more privileges et cetera. In this design the value bearer is the agent i.e. the individual. Indeed if we assume that the reflexivity of the agent is low than we might expect that the agent will adapt to his, hers, its environment. If the reflexivity of the agent is high than the key question is whether the agent is willing to comply or subverts the system by lying, cheating or neglects actions to perform and so on. As we have seen the design process provisions in a mechanism under what conditions consensus is reached and maintained. Suppose we have a majority rule, than we cannot rule out that an autonomous agent does not agree to the full extend what has been decided. In this latter case we cannot rule out by design possible behavioral discrepancy of the agent. There are two options for the designer(s). The first option is to take another close(r) look at the actual axioms, presumptions buttressing the model as in our example we started with in the first place. The second option is to introduce more rules and enforce harder. The designers have to make a choice: "which path to follow?" If it is possible to reconsider the earlier made choices than the revision process will be commenced. Whether this type of rule is accepted is a fundamental design question addressing moral agency.

BIBLIOGRAPHY

[Alchourron et al., 1985] C. Alchourron, P. Gardenfors, and D. Makinson. On the logic of theory change: partial meet contraction and revision functions. *Jounal of Symbolic Logic*, 50(2):510–530,

1985.

[Andrigetto et al., 2013] G. Andrigetto, G Governatori, P. Noriega, and L.W.N. van der Torre. *Normative Multi-Agent Systems*, volume 4 of *Dagstuhl follow-Ups*. Schloss Dagstuhl, Saarbrucken / Waadern, first edition, 2013.

[Apostel, 1960] L. Apostel. Towards a formal study of models in the non-formal sciences. *Synthese*, (12):125–161, 1960.

[Appadurai, 1986] A. Appadurai. *Introduction: commodities and the politics of value*, book section 1. Cambridge University Press, 1986.

[Ashmore, 1989] M. Ashmore. *The reflexive thesis*. University of Chigago Press, 1989.

[Atakan and Ekmekci, 2014] A.E. Atakan and M. Ekmekci. Auctions, actions, and the failure of information aggregation. *American Economic Review*, 104(7):2014–2048, 2014.

[Budescu and Bruderman, 1995] D.V. Budescu and M. Bruderman. The relationship between illusion of control and desirability bias. *Journal of Behavioral Decision Making*, 8:109–125, 1995.

[Camerer, 2003] C. Camerer. *Behavioral game theory: experiments in strategic interaction*. Princeton University Press, 2003.

[Christiaanse et al., 2015] R.M.J. Christiaanse, P. Griffioen, and J. Hulstijn. Reliability of electronic evidence: an application for model-based auditing. In *Artificial Intelligence and Law*. AAAI - ACM SIGART, 2015.

[Coppens, 2013] H. Coppens, T. Geel van. Urban architectural design and scientific research: how to save an arranged marriage, 2013.

[Denning and Martell, 2015] P.J. Denning and C.H. Martell. *Great Principles of Computing*. Massachusetts Institute of technology, 1st edition, 2015.

[Dietz, 2008] J.L.G. Dietz. *Architecture - buiding strategy into design*. Academic service / SDU, The Haque, 2008.

[ECL, a] ECL. 2015:ecli:nl:rmmne:2015:3262.

[ECL, b] ECL. Ecli:nl:phr:2016:883.

[Fama and Miller, 1972] E.F. Fama and M.H. Miller. *The theory of finance*. New York, 1972.

[Floridi, 2004] Luciano Floridi. Open problems in the philosophy of information. *Metaphilosophy*, 35(4):29, 2004.

[Floridi, 2008] Luciano Floridi. *Foundations of Information Ethics*. The Handbook of Information and Computer Ethics. John Wiley & SOns, New Jersey, 2008.

[Foque, 2003] R. Foque. Research in design science. *ADSC*, 10-11, 2003.

[Frey and Stutzer, 2007] B.S. Frey and A. Stutzer. *Economics ans Psychology*. MIT Press, London, 2007.

[Friedman et al., 2002] B. Friedman, P.H. Kahn, and A. Borning. Value sensitive design: Theory and methods. Report, Universtity of Washington, 2002.

[Friedman et al., 2008] B. Friedman, P.H. Kahn, and A. Borning. *Value Sensitive Design and Information Systems*. John Wiley & Sons Inc., New Yersey, 2008.

[Friedman, 1997] Friedman. *Human values and the design of computer technology*. Cambridge University Press, 1997.

[Friedman, 2003] Kahn Friedman. *Human values, ethics and design.*, page 1771201. Lawrence Erlbaum, Mahwah, 2003.

[Gabbay, 1996] D.M. Gabbay. *Labelled Deductive Systems*, volume 1. Oxford University Press, Oxford, 1996.

[Gettier, 1963] E.L. Gettier. Is justified true belief knowledge? *Analysis*, 23:(6):121–123, 1963.

[Gibson, 1986] J.J. Gibson. *The ecological approach to visual perception*. Taylor Francis Group, New york, 1986.

[Gigerenzer, 2001] G. Gigerenzer. *Bounded rationality: the adapive toolbox*. MIT Press, Berlin, 2001.

[Gilbert and Lynch, 2002] S. Gilbert and N. Lynch. Brewer's conjecture and the feasibility of consistent, available, partition-tolerant web services. *SIGACT News*, 33(2):51–59, 2002.

[Goldblatt, 2006] R. Goldblatt. *Topoi, the categorial analysis of logic*. Dover, first edition, 2006.

[Graeber, 2001] D. Graeber. *Toward an Anthropological Theory of Value*. Palgrave, New York, 2001.

[Greefhorst and Proper, 2011] D. Greefhorst and E. Proper. *Architecture Principles*. Springer, 2011.

[Griffioen, 2013] P.R. Griffioen. Type inference for lineair algebra with units of measurements. Report, CWI Amsterdam, 2013.

[Hayek, 1945] F. Hayek. The use of knowledge in society. *American Economic Review*, 35:519–530, 1945.

[Heylighen et al., 2009] A. Heylighen, H. Cavellin, and M Bianchin. Design in mind. *Design Issues*, 25(1), 2009.

[Hofstede, 1981] G. Hofstede. Management control of public and not-for-profit activities. *Accounting, Organizations and Society*, 6(3):193–211, 1981.

[Hoven, 2015] Jeroen van den Hoven. *Handbook of Ethics, Values and Technological Design*. Springer, 2015.

[Hurwicz and Reiter, 2006] L. Hurwicz and S. Reiter. *Designing economic mechanisms*. Cambridge University Press, New York, 2006.

[Hurwicz, 1960] L. Hurwicz. *Optimality and informational efficiency in resource allocation processes*. Mathematical methods in social sciences. Stanford Universiy Press, 1960.

[IEEE, 2011] IEEE. Iso architecture, 2011.

[ISACA, 2012] ISACA. Cobit 5, a business framework for governance and management of enterprise it. Report, ISACA, 2012.

[Kleer de and Williams, 1987] J. Kleer de and B.C. Williams. Diagnosing multiple faults. *Artificial Intelligence*, 32:97–130, 1987.

[Kluckhohn, 1951] C. Kluckhohn. *The study of culture*. The Policy Sciences. Stanford University Press, 1951.

[Kolko, 2010] J. Kolko. Abductive thinking and sensemaking: The drivers of design synthesis. *Design Issues*, 26(1):14, 2010.

[Laffont and Martimort, 2002] J-J. Laffont and D. Martimort. *Theory of Incentives: the principal - agent model*. Princeton University Press, Princeton, 2002.

[Macintyre, 1999] A Macintyre. Social structures and their threads to moral agency. *Philosophy*, 74(289):311 – 329, 1999.

[Mason, 2015] Mason, 2015.

[Mauss, 1950] M. Mauss. *The Gift*. Presses Universitaires de France in Sociologie et Anthropologie, 1950.

[Merchant, 1982] K.A. Merchant. The control function of management. *Sloan Management Review*, (Summer 1982):43–55, 1982.

[merriam webster, 2016] merriam webster. definition of a norm, 2016.

[Mische and Ann, 1998] Mustafa Emirbayer Mische and Ann. What is agency? *American Journal of Sociology*, 3(4):9621023, 1998.

[Myerson, 1982] R Myerson. Optimal coordination mechanisms in generalized principal-agent problems. *Journal of Mathematical Economics*, 10:67–81, 1982.

[Neuman, 1945] J. Neuman. First draft of a report on the edvac. Report, 1945.

[NIKE, 2016a] NIKE. Supply chain, 2016.

[NIKE, 2016b] NIKE. Transform manuafactuering, 2016.

[Norman, 1993] D.A. Norman. *Things that make us smart*. Basic Books, 1993.

[OECD, 2015] OECD. G20/oecd priciples of corporate governance. Report, OECD, 2015.

[Ouchi, 1978] W.G. Ouchi. An organisational failures framework. Report 461, Stanford, June, 1978 1978.

[Ouchi, 1980] W.G. Ouchi. Markets, bureaucraties, and clans. *Administrative Science Quarterly*, 25(1):129–141, 1980.

[Paape, 2007] L Paape. *Corporate Governance: the impact on the role, Position, and Scope of services of Internal Audit*. Thesis, 2007.

[Poivet, 2006] R. Poivet. *Le realism esthetique*. Presses Universitaire de France, 2006.

[Poivet, 2010] R. Poivet. Moral and epistemic virtues: A thomistic and analytical perspective. *International Journal for Philosophy*, 15(1):1–15, 2010.

[Reiter, 1987] R. Reiter. The theory of diagnosis from first principles. *Artificial Intelligence*, 32:57–95, 1987.

[Senge, 1990] P.M. Senge. *The fifth discipline: the art and practise of learning organization*. Doubleday, New york, first edition, 1990.

[Simmel, 1900] G Simmel. A chapter in the philosophy of value. *Journal of Sociology*, V(5):577–603, 1900.

[Simons, 1990] R. Simons. The role of management controlsystems in creating competitive advantage: new perspectives. *Accounting, Organizations and Society*, 15(no 1/2):16, 1990.

[Simons, 1995] R. Simons. *Levers of Control: How Managers Use Innovative Control Systems to Drive Strategic Renewal*. Harvard Business School Press., 1995.

[Sowa, 2000] J.F. Sowa. *Knowlegde representation: logical, philosophical, and computational foundations*. Thomson, Pacific Groove, 2000.

[Stanford, 2015] Stanford. Ethics-computer, 2015.

[Strawson, 1974] P.F. Strawson. *Freedom and resentment and other Essays*. Routledge, 2008 edition, 1974.

[Thomson, 1985] Jarvis Thomson. The trolly problem. *The Yale Law Journal*, 94(6):1395–1415, 1985.

[Williamson, 1975] O.E. Williamson. *Markets and hierarchies, analysis and antitrust implications*. The Free Press, 1975.

[Woolridge, 2000] M. Woolridge. *Intelligent Agents*, volume 1, book section 1, page 609. MIT press, London, 2000.

[Woolridge, 2011] M. Woolridge. *An introduction to multiagent systems*, volume 1. John Wiley & Sons, Chichester, 2011.

PART III

ENGINEERING

7
Interaction Protocols
MATTEO BALDONI, CRISTINA BAROGLIO, AMIT K. CHOPRA, AKIN GÜNAY

1 Introduction

A highly promising application of multiagent systems (MAS) is in domains that involve interaction between autonomous social *principals*—typically, humans and organizations. Business and finance, healthcare, scientific collaboration, and social media are all examples of such domains. Such systems are properly *sociotechnical*, reflecting the use of technology by social principals to carry out social processes [Chopra and Singh, 2016b]. Sociotechnical systems are naturally *decentralized*: each principal is a locus of autonomy and, therefore, decision-making. A central challenge facing the software engineering of sociotechnical systems is accommodating the autonomy of principals in interacting flexibly with each other while at the same time also supporting notions of what it means to interact correctly.

Multiagent systems (MAS) are ideally suited for modeling sociotechnical systems. In the basic model of a MAS, each agent represents an autonomous social principal. The agents interact with each other (on behalf of their principals) to carry out social processes. However, not all interactions are legal. Interaction protocols specify the *rules of encounter* between agents [Singh, 1998]. If an agent follow the rules, it is *compliant* with the protocol; otherwise it is noncompliant. Thus, a protocol specification serves as a standard of correctness for agent behavior in a multiagent system. Protocols also promote autonomy and flexibility because they leave principals free to design the agents as they wish, in accordance with their goals, potentially in a manner that will ensure protocol compliance but not necessarily.

Two kinds of protocols are relevant to MAS:

Messaging protocols describe constraints on message flow between agents, typically in terms of ordering and occurrence constraints. For example, a protocol for scientific collaboration may specify that a request for a resource (message) must precede any response. Such protocols are violated when a messaging constraint is violated. Agent UML [Huget, 2004], BSPL [Singh, 2011a], HAPN [Winikoff *et al.*, 2018] are all proposals along this line. In certain contexts, compliance with such protocols may be enforced that allows only legal messages to be sent, e.g. [Singh, 2011b]. In other cases, deviations will be automatically identified. Messaging protocols are being formalized in many business domains and led to the development of industry-supported standards. These efforts include Intel-led *RosettaNet* (e-business) [RosettaNet, 1998], ABN-AMRO led *TWIST* (foreign exchange transactions) [TWIST, 2006], and *HL7* (healthcare) [HL7, 2002], among many others. The efforts vary in their level of sophistication: some specify only message formats, others specify request-response protocols, whereas others specify only some possible *sequences* of interaction.

Meaning-based protocols describe the social *meanings* of messages in terms of normative expectations such as commitments, prohibitions, and so on. For example, the meaning of the request may be expressed as a commitment by the requester that if the request is granted, then the resource will be released by a specified deadline. In such protocols, noncompliance results from violation of commitments. For example, if the request is granted, but the resource is not released by the deadline. We will consider *commitment protocols* as exemplar of this style of protocols. Compliance of an agent with norms cannot, in general, be guaranteed: it would depend on the design of the agent.

A protocol is an abstraction for interaction. The key benefit is that it captures application-level logic pertaining to interactions in a reusable manner. In principle, protocols may be refined and composed. Naturally, we may want to verify that a protocol has desirable properties. A protocol may be statically verified for properties such as deadlock freedom, fairness, and safety before the interaction starts [El-Menshawy et al., 2011]. Agents may be verified for behavior in accordance with the specification (*conformance*) [Baldoni et al., 2006]. At runtime, the compliance of agents with protocols can be monitored [Chesani et al., 2013]. If the interacting agents conform to the specification of the role they play, the interaction inherits the properties verified on the protocol. For instance, if in a protocol specification roles are proved interoperable, any agent that will conform to the protocol is guaranteed that its interaction with any other agents, playing the other roles and compliant with their specification, will succeed [Rajamani and Rehof, 2002; Bravetti and Zavattaro, 2009].

In this chapter, we give an overview of messaging and commitment protocols with the help of examples specified in important languages. We highlight some of the key features and subtleties of these protocols and discuss the relevant interesting properties and their verification. We also highlight important open challenges in interaction protocol research.

The rest of the chapter is organized as follows. Section 2 introduces messaging protocols with the help of two different ways of specifying them, one in terms of control flow and the other in terms of information flow. Section 3 introduces the notions of commitments and commitment protocols. Section 4 introduces the various kinds of reasoning one can do with protocol specifications. Section 5 discusses challenges in interaction protocol specification.

Section 6 discusses some of the relevant literature.

2 Messaging Protocols

Languages for messaging protocols typically specify ordering and occurrence constraints between messages (more generally, message types). Typically these languages are operational in nature, meaning that the constraints are specified via control flow constructs such as sequence, choice, parallel, and so on, though there are also some declarative approaches [Montali et al., 2010]. Important theoretical challenges here concern correct enactment in the face of distribution (general multiparty settings and asynchrony) and minimal message delivery assumptions. For example, we would not want to assume FIFO delivery of messages between agents; such an assumption would be inadequate anyway in settings of more than two agents. Correctness is usually expressed in terms of liveness (e.g., deadlock freedom) and safety properties (e.g., agents making compatible choices). Representations vary in formality and sophistication, ranging from Agent UML (AUML) [Odell et al., 2000; Bauer et al., 2001] and state machines to Petri Nets [El Fallah-Seghrouchni et al., 2001], and pi-calculus. WS-CDL [WS-CDL, 2005] is a protocol language for modeling interactions among Web services. The Blindingly Simple Protocol Language (BSPL) is different from the above approaches for specifying protocols. Instead of specifying constraints directly between messages, in BSPL, one specifies constraints between the information items in a message.

Below we discuss AUML and BSPL as exemplar languages for specifying messaging protocols—AUML because it is based on UML notations that are widely used in software engineering, and BSPL because of (1) its basis in information, (2) its contrast with AUML, and (3) its support of fully decentralized asynchronous protocol enactments.

2.1 An Operational Approach: AUML

A AUML Sequence Diagram is an enhancement of a UML interaction diagram, which is an informal graphical notation for specifying interactions protocols. One specifies a protocol in AUML by specifying the control flow between messages using abstractions such as sequence, alternative, loop, and so on.

Figure 1 shows the FIPA (Foundation of Intelligent Physical Agents) Request Interaction Protocol in AUML. This protocol involves two roles, an INITIATOR and a PARTICIPANT. The INITIATOR sends a *request* to the PARTICIPANT, who either responds with a *refuse* or an *agree*. In the latter case, it follows up with a detailed response, which could be a *failure*, an *inform-done*, or an *inform-result*. The PARTICIPANT will omit the *agree* message unless the INITIATOR asked for a notification.

AUML sequence diagrams are a kind of automata which prescribes the legal sequencing of the protocol messages. The choice of relying on automata of some kind is well-supported in the literature concerning interaction protocols (see also [Cabac et al., 2003; Dunn-Davies et al., 2005]) but, as [Yolum and Singh, 2002; Winikoff et al., 2004] point out, such protocol specifications show

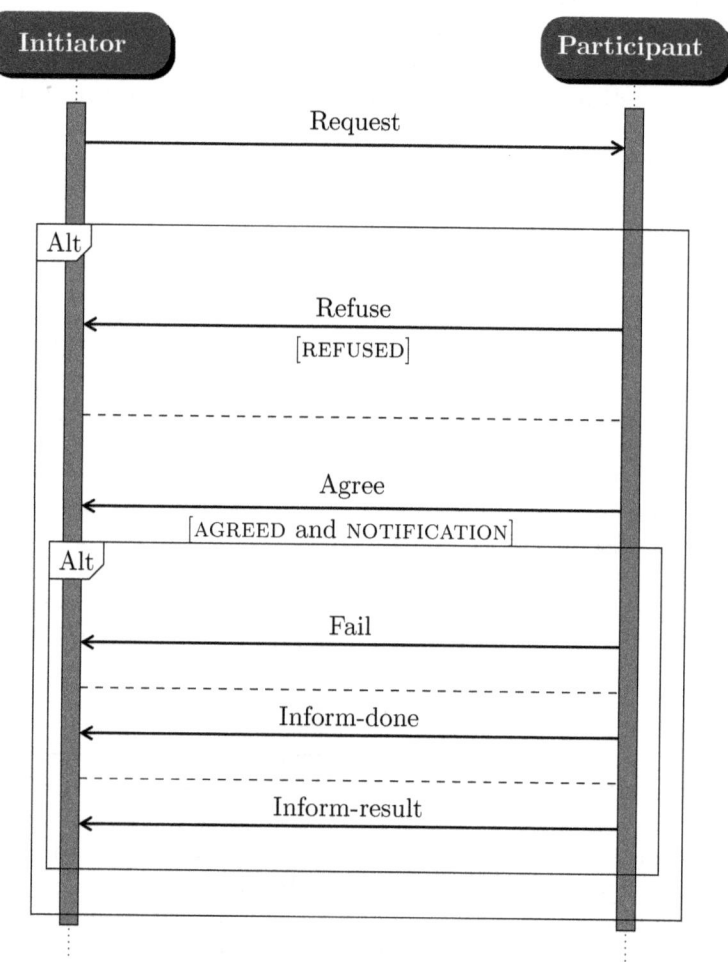

Figure 1. FIPA Request Interaction Protocol, from the FIPA specification [FIPA, 2003], expressed as a Agent UML (AUML) Sequence Diagram.

a rigidity that prevents agents from taking opportunities and handling exceptions in a dynamic and uncertain multiagent environment. This limitation can be overcome by commitment-based interaction protocols as we describe in Section 3.

2.2 Information-Oriented Approach: BSPL

The Blindingly Simple Protocol Language (BSPL) [Singh, 2011a] is a declarative, information-oriented approach for specifying protocols. Specifically, in a BSPL protocol specification, one eschews explicitly specifying ordering and occurrence relations between messages in favor of specifying the dependencies between the data communicated by the messages.

Listing 1.1. FIPA Request protocol in BSPL.

```
FIPA-Request {
  roles I, P // Initiator, Participant
  parameters out ID key, out job, out conf, out finished

  I ↦ P: Request[out ID, out job]
  P ↦ I: Agree[in ID, out conf]
  P ↦ I: Refuse[in ID, out conf, out finished]
  P ↦ I: Fail[in ID, in conf, out finished]
  P ↦ I: Done[in ID, in conf, out finished]
  P ↦ I: Result[in ID, in conf, out finished]
}
```

Listing 1.1 shows a protocol in BSPL that is analogous to the FIPA Request protocol. A BSPL protocol starts with declarations of the protocol's name, roles of the participants, and a set of protocol parameters. In Listing 1.1, the protocol's name is FIPA-Request. It includes the roles I (initiator) and P (participant), and the parameters ID, job, $conf$, and $finished$, from which ID is declared key. Hence, bindings of ID identify different enactments of the protocol. All protocol parameters are functionally dependent on the key parameters, which ensures integrity. A protocol enactment is said to be *complete* when all its parameters are bound. Each protocol parameter may be adorned out or in. In Listing 1.1, all protocol parameters are adorned out, meaning that their bindings are produced during the protocol's enactment. On the other hand, an in adorned parameter, which does not exist in Listing 1.1, indicates that the binding of that parameter must be provided externally. Overall, roles and parameters facilitate composition with other protocols. (BSPL also supports private roles and parameters, and an additional nil adornment, but we omit them here for brevity.)

The rest of Listing 1.1 declare the messages of the protocol, for which the declaration order is not significant. Each message declaration includes the sender and receiver, the name of the message, and its parameters. For example, *Request* is from I to P, and ID and job are its parameters. All key protocol parameters that appear in a message declaration are also key parameters for a message. Further, each message parameter is adorned *out* or *in* to capture causality between messages. For instance, *ID* and *job* are adorned *out* for *Request*, meaning that their bindings must be produced by *I* when sending *Request*. In *Agree*, *ID* and *job* are both *in*, meaning that their bindings must be known by *I* from the emission or reception of a prior message (e.g., reception of *Request*) to be able to send *Agree*. Beside causality, *out* adornment also ensures mutual exclusion. That is if a parameter is adorned *out* it more than one messages (e.g., *Agree* and *Refuse*), only one of these message can be sent in an enactment. Otherwise, the common parameter could have different bindings for the same key and violate integrity.

BSPL is formal and is more expressive than AUML. BSPL explicitly supports parameters and, through keys, the identification of protocol instances. BSPL is declarative and is designed to support composition. Intuitively, a message declaration is an atomic protocol; any other protocol (e.g. *FIPA Request*) is a composition of protocols. BSPL specifications may be verified for liveness and safety [Singh, 2012]. Implementations of BSPL in middleware can ensure that only correct messages (that preserve integrity and respect the information flow constraints) may be sent by any party [Singh, 2011b]. Winikoff et al. [2018] discuss the differences between AUML and BSPL in detail.

3 Commitment-Based Interaction Protocols

Commitment-based interaction protocols are meaning-based. They capture interaction patterns in a declarative way given in terms of commitments, involving a set of predefined roles. By executing protocol actions agents can create new commitments or manipulate existing commitments, e.g. they can release another agent from some commitment. Any agent who knows the meaning of the protocol actions, and is aware of a sequence of executions of such actions, will be able to deduce which commitments hold and their state. This section introduces social commitments and commitment-based interaction protocols, including both their basic definition (Section 3.1) and more recent evolutions that allow tackling temporal regulations inside commitments as well as dialectical commitments (Section 3.2).

3.1 Fundamentals of Social Commitments

Social Commitments. Since the late 1980s, many studies on distributed artificial intelligence, on formal theories of collective activity, on team work, and on cooperation [Castelfranchi, 1995; Singh, 1997; Norman et al., 2004] implicitly identified *commitment*, the glue of group activity. Commitments link the actions of the group members and the group members with each other. The value of commitments as fundamental building blocks of interaction is recognized also outside artificial intelligence, like in sociology, where commitments are fundamental to the definition of organizations [Elder-Vass, 2010], and in economics where models based on commitment and trust were, for instance, proposed to understand the functioning of relational marketing [Morgan and Hunt, 1994].

This chapter relies on the notion of *social commitment*, as defined by Singh [1999]. *C(debtor, creditor, antecedent, consequent)* denotes a social commitment, meaning that the *debtor* debtor is committed to the *creditor* creditor to bring about the *consequent* consequent condition, when the antecedent *antecedent* condition holds. For instance, *C(merchant, customer, goods-paid, goods-delivered)* means that if the merchant is paid for goods, then merchant is committed to the customer to deliver the goods. Hence, commitments let agents have *expectations* on the behavior of their counterparts. In particular, the creditor of a detached commitment expects that its debtor will sooner or later bring about the consequent condition—debtors are, indeed, *liable* in case their commitments are violated.

With reference to [Telang *et al.*, 2012] each social commitment can be in one of the following states. A commitment is *Violated* when its antecedent is true but its consequent will forever be false, or it is canceled when *Detached*; *Satisfied*, meaning that the engagement is accomplished; *Expired*, meaning that it is no longer in effect and therefore the debtor would not fail to comply even if does not accomplish the consequent. Typically, a commitment should be *Active* when it is initially created. Active has two substates: *Conditional* (as long as the antecedent does not occur) and *Detached* (the antecedent has occurred).

It is typically assumed that only the debtor has control over the creation and discharge of commitments and the creditor has control over detachment. Further, the literature discusses commitment operations such as delegation (to a new debtor) and assignment (to a new creditor) that can only be done by the debtor and creditor, respectively [Singh, 1999].

Commitment-based interaction protocols. *Commitment protocols* were introduced in the seminal works by Yolum and Singh [2001a; 2001b]. Agents share a social state that contains commitments and other literals that are relevant to their interaction. Every agent can affect the social state by executing actions, whose definition is given in terms of updates to the social state (e.g. add a new commitment, release another agent from some commitment, satisfy a commitment). Hence, in its basic interpretation, a commitment protocol is made of a *set of actions*, involving the foreseen roles and whose semantics is agreed upon by all of the participants [Yolum and Singh, 2001a; Yolum and Singh, 2001b; Chopra, 2009]. The only constraint that commitment protocols include, to say that an interaction is successful, is that all commitments are discharged. These characteristics give commitment-based protocols great flexibility and give to the involved agents great autonomy. In fact, they are free to apply the social actions in any order they wish if, in the end, commitments are discharged.

Several languages have been developed in the recent years to specify commitment protocols (and other normative constructs). Listing 1.2 shows an example commitment protocol specification in the Custard language [Chopra and Singh, 2016a]. The protocol models the interaction between two agents, namely a *merchant* and a *customer*, in a purchase scenario. It involves a single commitment, namely *Purchase*, where the *merchant* is the debtor and the *customer* is the creditor. The state of the commitment is determined with respect to a set of events. It is created when a *Quote* event occurs. It becomes detached if the *Goods-Paid* event occurs within ten time units after *Quote*. Otherwise, the commitment expires. Finally, the commitment becomes discharged if the *Goods-Delivered* event occurs within five time units after *Goods-Paid*.

Listing 1.2. A commitment protocol in Custard.
```
commitment Purchase merchant to customer
  create Quote
  detach Goods–Paid [ , Quote + 10]
  discharge Goods–Delivered [ , Goods–Paid + 5]
```

The events of the protocol are defined in an information schema as we show in Listing 1.3, where each event corresponds to a relation. For instance, a *Quote* event has three attributes, namely, *purchaseID*, *goods*, and *price*, which capture the information associated with the event. Furthermore, the attribute *purchaseID* is defined *key* for this event. Instances of the events in a schema are stored in the information stores of the agents in a distributed manner. The event instances should satisfy the integrity constraints of the schema (e.g., each event instance must have a unique key). This information-oriented approach enables management of multiple protocol instances (e.g., each *Quote* event initiates a new protocol with a unique *purchaseID*). States of protocol instances can be determined by querying the information store, which is generated automatically by Custard from the protocol specification.

Listing 1.3. A commitment protocol in Custard.
```
schema
  Quote(purchaseID, goods, price) key purchaseID time t
  Goods–Paid(purchaseID, goods, paymentInfo) key purchaseID time t
  Goods–Delivered(purchaseID, goods, deliveryInfo) key purchaseID time t
```

Realistic scenarios often involves multiple agents and multiple commitments between them. For instance, in our purchase example there could be a third courier agent to realize delivery of goods. Listing 1.4 extends the previous protocol with a new commitment, namely *Delivery*, from the courier to the merchant to capture the delivery of goods. Intuitively, the commitment states that, if there is a delivery order by the merchant, the courier should do the delivery within three time units. We omit the extended event schema for the new events *Delivery-Agreement* and *Delivery-Order* for brevity.

Listing 1.4. A commitment protocol in Custard.
```
commitment Purchase merchant to customer
  create Quote
  detach Goods–Paid [ , Quote + 10]
  discharge Goods–Delivered [ , Goods–Paid + 5]
```

```
commitment Delivery courier to merchant
  create Delivery-Agreement
  detach Delivery-Order
  discharge Goods-Delivered[, Delivery-Order + 3]
```

It is straightforward to extend the above protocol further with additional agents and commitments. Listing 1.5 shows two more commitments that can be added to the above protocol. The first commitment, namely *Payment*, captures the interaction of the customer with a new *bank* agent to realize the payment. The second commitment, namely *Refund* extends the interaction of the customer and merchant for the refund of returned goods. Note that detachment of *Refund* occurs when *Purchase* is discharged, which is a meta-level commitment state event.

Listing 1.5. A commitment protocol in Custard.
```
// Purchase and Delivery commitments as defined above

commitment Payment bank to customer
  create Open-Account
  detach Payment-Order
  discharge Goods-Paid[, Payment-Order + 2]

commitment Refund merchant to customer
  create Quote
  detach discharged Purchase and Goods-Returned[, Goods-Delivered + 5]
  discharge Refunded[, Goods-Returned + 5]
```

These examples show several strengths of commitment protocols. First, commitment protocols support autonomy of agents. That is, a commitment protocol defines the expectations of agents, but it does not define how these expectations should be satisfied. Hence, agents are free to fulfill (or not) these expectations as they see fit. For instance, the customer could realize the payment either herself or through the bank. Similarly, the merchant could herself deliver the goods or make a delivery order to the courier. Note that a procedural specification approach, such as AUML, can also represent such situations. However, in most cases, the number of possible realizations of an interaction grows rapidly and makes the procedural approach impractical. Second, commitment protocols are modular, which makes them easy to extend with new agents and commitments, and also compose with existing protocols as our examples demonstrate.

3.2 Advanced Concepts

Temporal Regulations inside Commitment Protocols. Some works explored the possibility to introduce temporal regulations inside commitment protocols to capture constraints on the coordination between agents. 2CL [Baldoni et al., 2013] is a first proposal that aims capturing temporal regulations on the flow of events in commitment-based protocols, decoupling *constitutive* rules from *regulative* rules. Following Searle [1995], constitutive rules, by identifying certain behaviors as foundational of a certain type of activity, create that activity. Regulative rules, instead, contingently constrain a previously constituted activity. The decoupling of constitutive and regulative components is motivated by desired properties such easier re-use of actions in different contexts, easier customization of protocols, and easier composition of protocols. A multiagent system that clearly separates a constitutive component from a regulative component in its protocols gains greater openness, interoperability, and modularity of design.

While 2CL enriches commitment-based protocols with temporal regulations an agent is said to accept when enacting a protocol role, the proposal of Marengo et al. [2011] improves the idea by giving regulations a real normative power. It does so by allowing the specification of commitments to temporal regulations. Consider the following example. A physician commits to a patient that if the patient has any sign of heart trouble after signing up with him, then the patient will be immediately referred to a laboratory for tests, the results of which will be evaluated by a specialist. Temporal constraints such as those in the example are traditionally captured as procedural workflows. Marengo *et al.* capture such constraints more broadly as regulations and express them more flexibly in a logical notation. The commitments among autonomous parties capture their business relationships naturally. Regulations are incorporated into the contents of commitments. Thus by reifying regulations into business relationships, normative force is brought to the specification, thereby providing a clear basis for the participants to guide their actions locally and to judge the compliance of their counterparties.

Commitment antecedent and consequent conditions are given in precedence logic. Precedence logic is an event-based logic [Singh, 2003]. It has three primary operators for specifying requirements about the occurrence of events: '∨' (choice), '∧' (concurrence), and '·' (before). The *before* operator enables one to express specifications such as *approve · perform*: both *approve* and *perform* must occur and in the specified order. The specifications are interpreted over runs. Each run is a sequence of events. The transitions correspond to event occurrences.

However, placing regulations inside commitments, as in the above example, leads to new challenges, among which the need to formalize the progression (the life cycle) of commitments, bearing in mind the events that have occurred. For example, we say that an active commitment $C(x, y, r, u)$ progresses to *satisfied* when u occurs. Analogously, we would like to say that $C(x, y, r, e \cdot f)$ progresses to $C(x, y, r, f)$ when e occurs. The challenge is to formalize general progression rules for an expressive event language.

Type Checking. While a main research issue in concurrency programming, including the actor model [Agha, 1986], is the definition of formal models, and type systems analogously to the sequential case, for what concerns agents, just a few studies were carried on aspects such as typing and type safety. In programming languages, type systems are used to help designers and developers to avoid code errors that can produce unpredictable results. In MAS, and in particular for what concerns interaction inside a MAS, typing systems would provide a means for answering questions like: does an agent have the means for carrying out the encoded interaction? Is an agent compliant to a role specification?

Baldoni et al. [2014a; 2018] propose an agent-based, dynamic (i.e., performed at role enactment time), and declarative type checking system for agent interactions that are modeled by means of commitment-based protocols. The typing system enjoys an important progression property: when an agent joins an interaction, it is guaranteed that it possesses all the required behaviors to carry out its part, as far as one of the interaction acceptance states. One of the reasons to express agent types and role requirements based on commitments is that the agent system developers are better supported if the typing system relies on abstractions that are typical of MASs, rather than relying on abstractions from other programming paradigms, as done by other proposals. Clearly, a type system allows only a lightweight check of the behavior of the involved agents, being more concerned with a safe usage rather than a full behavioral compatibility. It does not imply that an agent which has the same type of another agent will display the same behavior. This does not exclude the possibility to integrate deeper checks, for instance based on model checking such as [Bentahar et al., 2009]. Type checking, as a form of lightweight verification, adopts notions (e.g., substitutability), that are used also for coping with issues of interoperability and conformance, discussed in [Baldoni et al., 2006; Baldoni et al., 2009]. The conformance verification aims at guaranteeing that when an agent plays a role, or substitutes another agent in an on-going interaction, the interoperability of the system is preserved.

The explained commitment-based typing proposal overcomes the limitations of approaches to the typing of interaction protocols that are based on global session types by Ancona et al. [2013]. Briefly, Ancona et al. use global session types as an abstraction tool, which enables automatically generating monitor agents. A monitor agent has the aim of verifying the correctness of a multi-party interaction; to this aim, it intercepts all the exchanged messages and verifies whether the protocol is respected. Similarly to finite state automata, global session types have a prescriptive, procedural nature, whose drawbacks have been explained [Yolum and Singh, 2002; Winikoff et al., 2004]. Moreover, as the reader will observe, in this proposal typing is used to specify the interaction within a system from a global perspective, rather than provide an actual typing of the agents. Thus, questions concerning whether an agent can enact a role, or has some abilities required by the interaction remain unanswered.

4 Reasoning over Commitments

Commitment-based interaction protocols enable agents to reason about their interactions with others without knowing their internals. There are several reasoning tasks that should be performed by an agent to ensure its effective operation when enacting a protocol. Here we identify three such tasks: (i) whether a protocol *supports* an agent to achieve its goals, (ii) whether a protocol is *feasible* for an agent to enact, and (iii) whether a protocol is *verifiable*.

4.1 Goal Support

goal support Goal support [Günay et al., 2013; Günay et al., 2015b] generalizes the ideas of control and safety to ensure that an agent achieves its goals with respect to its commitments. That is, a protocol supports the goals of an agent, if the commitments of the protocol provide sufficient control to the agent for achieving its goals. For instance, suppose that a protocol involves the commitment $C(merchant, customer, goods\text{-}paid, goods\text{-}delivered)$, and the goal of the customer is to achieve *goods-delivered*. If we assume that the merchant fulfills its commitments, we can say that the commitment supports the customer's goal, since it makes the merchant committed for delivering the goods. However, note that the antecedent of the commitment requires a payment to occur first. Hence, the customer should adopt an intermediary goal to bring about *goods-paid* [Telang et al., 2012], and the protocol should also support this intermediary goal. If the customer controls the payment, we can conclude that the customer's goal is supported by the above commitment, since no matter what happens, the customer can make a payment and the merchant becomes committed to deliver the goods. If the customer does not control payment, the protocol supports the customer's goal, only if there is another commitment in the protocol that ensures occurrence of the payment. Kafalı et al. [2014] address computation of goal support using event calculus. Günay et al. [2015a; 2016] extend computation of goal support into a probabilistic setting considering uncertainty in the behaviors of agents.

4.2 Feasibility

feasibility In many settings agents need resources to perform actions. In this context, a commitment protocol is feasible for an agent, if either the agent has sufficient resources before the enactment of the protocol to fulfill its commitments, or the agent can acquire the necessary resources from other agents during the enactment of the protocol, which is subject to the commitments of the protocol and their feasibility [Günay and Yolum, 2013]. For instance, a merchant's commitment to deliver certain goods is feasible, if the goods are already in the merchant's stock. If this is not the case, there should be other feasible commitments which ensure acquisition of the goods by the merchant (e.g., a commitment from a supplier to provide the goods). Note that the definition of feasibility is recursive (e.g., the merchant should have sufficient resources to detach the supplier's commitment). Time constraints (e.g., deadlines) should also be taken into account for feasibility. If the merchant was committed to deliver the goods in three days, but the supplier is committed to provide the goods in five days, than the merchant's commitment might not be feasible. Günay and Yolum [2011; 2013] use constraint satisfaction techniques

4.3 Verifiability

verifiability We have seen that it is possible, by applying proper kinds of reasoning, to assess whether playing a protocol role is safe for an agent, and whether the agent has sufficient control. A related property concerns whether agents can or cannot judge the compliance, to the agreed protocol, of their parties in the interaction, as the interaction proceeds. This problem is studied in [Baldoni et al., 2015b], where an agent is said to be *compliant* if and only if it discharges all commitments of which it is the debtor. The research question is to characterize commitment protocols that, when enacted, support each participant verifying the compliance of the other participants. Verification requires each participant to be in condition to observe "relevant" events so that it can determine the progress of the commitment along its lifecycle. The relevant events are all those that affect commitments, e.g. by creating, detaching, violating them. An interaction is *verifiable* when relevant events can be observed by all participants. Thus, each of them knows which commitments are active (or expired, detached, or violated), and consequently knows about its own and the other principal's compliance. The problem is that, as in general, not all the concerned parties can observe all of the events relevant to their commitments. A simple example is that of a purchase, where delivery is made by a courier. As such, its completion is not directly observable by the seller who, however, is the one who has a commitment towards the client for delivery. So, for instance, the seller will not know, by direct observation, that the client received the goods (event which satisfies the commitment). The client's and the seller's views of the state of the interaction cannot be "aligned" alignment [Chopra and Singh, 2009; Chopra and Singh, 2015], hence yielding non-verifiability.

A solution is suggested by the way in which similar situations are tackled in the real world. When relevant events cannot be observed, the participants of the interaction enrich the protocol with new actions (like delivery tracking), whose meaning is a claim about the state of affairs (a position). Such claims imply the taking of responsibility of the truth of what is declared. So, the courier will let the seller know the state of the delivery by declaring it through a tracking service. Claims act as "bridges" between different interaction contexts. Baldoni et al. [2015b] observe that such claims raise expectations by some principals of others, and that each principal is accountable for the expectations it creates in others. Consequently, they propose to rely on *dialectical* commitments [Singh, 2008] to represent agents' claims. They also introduce a pragmatic rule because of which the participants can use dialectical commitments as equal to the conditions they concern. In other words, a dialectical commitment by the courier saying delivery occurred will satisfy the seller's commitment to delivery. By realizing claims as commitments, debtors become *accountable* for their declarations, and are held *liable* when the commitment is violated.

5 Challenges

Although foundations of commitment-based protocols are well studied, there are still several challenges to address to realize their use in practice. Below, we discuss two of these challenges in detail.

5.1 Run-time Protocol Synthesis

Typically, commitment protocols are defined at design-time, and embedded into the implementations of agents [Winikoff, 2007]. This approach simplified the development of a MAS. However, it also limits flexibility and adaptation of a MAS, since it requires protocols and agents to be tightly coupled. The alternative of design-time protocols is to enable agents in a MAS to create their own protocols at run-time according to their needs [Artikis, 2009]. synthesis We call this approach *run-time protocol synthesis*. The specific problem of run-time protocol synthesis is the following [Günay et al., 2015b]: given a set of agents, their goals, preferences, and capabilities, automatically synthesize a commitment protocol to regulate the interactions of the agents, which enables the agents to achieve their goals with respect to their preferences and capabilities. This problem is relatively straightforward to solve (at least conceptually) in a centralized manner for a simple propositional representation. However, realistic applications (e.g., IoT, healthcare) need a decentralized synthesis approach that supports an expressive representation, such as first-order logic. Decentralization is essential for efficiency and reliability. Besides, it is also necessary to ensure privacy of the agents. That is, the synthesis mechanism should not force agents to reveal their goals, preferences, and capabilities to others.

There are two recent approaches that address run-time synthesis of protocols. Telang et al. [2013; 2013] develop a centralized planning approach that uses hierarchical task network (HTN) planning to synthesize commitment protocols. Günay et al. [2013; 2015b] develop a distributed approach for run-time synthesis of protocols that is based on the goal support property we have discussed earlier. These two proposals establish the foundations to solve the run-time protocol synthesis challenge. However, both methods have some shortcomings. Telang et al. achieve a desirable degree of expressiveness by using a first-order formalization. However, they consider a fully centralized setting, in which all the information about the agents (e.g., private goals) are known to a central planner. On the other hand, Günay et al. do not require agents to share their private information. However, they use a propositional formalization, which suffers from limited expressive power. Besides, neither of these methods consider constraints such as time and resources. In the light of our discussion and the proposals, we identify the properties of an ideal run-time protocol synthesis method as follows: (i) The run-time protocol synthesis problem should be solved in a fully automated manner at run-time by the agents without human intervention. (ii) The synthesis method should be capable of creating commitment protocols in a sufficiently expressive representation to capture various aspects of practical problems, such as fine-grained time requirements and resource constraints. (iii) The method should be decentralized and preserve privacy of agents. There is a significant body of work both in multiagent planning [Durfee and Zilberstein, 2013] and distributed constraint optimization [Yokoo et al., 1998] techniques. We believe that utilization of these techniques is a promising direction for the solution of run-time protocol synthesis challenge.

5.2 Methodologies and Tools

Effective methodologies and tools for designing commitment protocols are required to integrate commitment protocols into the engineering processes of MAS. There are several proposals in this direction. Winikoff [2006; 2007] discusses implementation of commitment protocols in conventional agent-oriented programming languages, and develops a mapping from commitment protocols to BDI-style plans with extensions to semantics of BDI languages. Yolum [2007] formalizes design requirements of commitment protocols, such as effectiveness, consistency, and robustness and develops tools to analyze commitment protocols with respect to these requirements. Amoeba [Desai et al., 2009] is a methodology for designing business processes in which commitment protocols capture the business meaning of interactions among autonomous agents. Amoeba describes guidelines to specify cross-organizational processes using commitment protocols and uses composition to handle the evolution of requirements. 2CL Methodology [Baldoni et al., 2014b] extends Amoeba with temporal constraints. JaCaMo+ [Baldoni et al., 2015a] is a programming framework where Jason agents interact according to commitment-based protocols, which are modeled as special CArtAgO artifacts. JaCaMo+ handles enactment and monitoring of commitment-based protocols, including the social state of the interaction, and enables Jason agents both to be notified about the social events and to perform practical reasoning about the other agents' actions. Torroni and colleagues [Chesani et al., 2013] address several issues about monitoring of commitment protocols including exceptions [Kafalı and Torroni, 2012] and delegations [Kafalı and Torroni, 2011], and develop tools using reactive event calculus. Kafalı and Yolum [2016] also develop a methodology to create agents that can monitor interactions of their users in business settings. They use a flexible as-good-as relation to compare the state of the interaction with the expectations of the users.

Despite these methodologies and tools, there are still several issues to address for better integration of commitment protocols into the engineering of practical MAS. First, there is no standardization on the representation and interpretation of commitments. While all the methodologies and tools share a common basic model of commitments, they use different approaches to handle details such as representation of lifecycle, information, and time. Hence, development of standard models of commitments is necessary for the interoperability of different methodologies and tools. Second, most methodologies and tools are general purpose and inadequate to address domain specific issues of modern systems such Internet of Things and healthcare. Finally, there is no formal approach for the operationalization of commitment protocols (i.e., mapping of a commitment protocol to operational messages), which is essential for the efficient development of correct commitment-based systems, by eliminating manual transformations from commitment protocols to operational messages.

6 Related Approaches

In the literature there different proposals for specifying meaning-based interaction protocols. An approach that lies between automata-based approaches (like AUML [Bauer et al., 2001]) and commitment protocols is the one proposed in [Fornara, 2003; Fornara and Colombetti, 2004] which provides a commitment-based semantics for the speech acts of an agent communication language, and relies on interaction diagrams to regulate the interaction.

Proposals concerning Electronic Institutions, like the seminal papers by Esteva et al. [2004] and Tinnemeier et al. [2009], take a different perspective. Esteva et al. use an interaction protocol to regulate the dialogic interactions of a group of agents, constituting a scene. Scenes can be composed resulting into complex interactions with transitions from a scene to another. *Normative meaning* is associated to the actions (in [Tinnemeier et al., 2009] to facts) leading to the generation of obligations (prohibitions, etc.) that will then shape the interaction within the MAS. Once again in this kind of work, protocols often amount to automata prescribing specific courses of interaction.

Torroni et al. [2009] propose the idea of relying on *social expectations* as an alternative to commitments for providing a social semantics to the interaction. An expectation represents a desired behavior by describing an event and the time at which the event is expected to happen. Positive expectations have the form $E(p,t)$, meaning that the event p is expected to happen at time t. Negative expectations have the form $EN(p,t)$ capturing that event p should not happen at time t. A main difference of expectations from commitments is that expectations are not directed from a debtor to a creditor. Therefore, even if they have a normative aspect, expectations are not associated to a notion of accountability. This approach is at the basis of the SCIFF logical framework [Alberti et al., 2008] that supports both interaction specification and verification. A SCIFF specification is an abductive logic program, where social integrity constraints are used to model relations among events as well as expectations about events. Based on such integrity constraints, it is possible to define patterns of interactions by relating the time at which different events are expected to happen.

Interaction protocols bear similarities with business processes. In recent years, as Chesani et al. [2009] reports, workflow management systems for business processes address trade-offs between flexibility and expressiveness, introducing ways for deviating from the standardized process and even modifying the process. Verifying the compliance of complex and flexible processes to regulations is a hard task that cannot be performed a priori, due to the fact that the involved agents can stray from the original process or even modify it. Compliance verification, in such cases, is usually performed afterwards. An analyst observes traces of (possibly modified) process executions to understand if any violation occurred. The expectation-based approach was used to implement a tool that performs this kind of check in an automatic way.

The extensive work on interactions has led Chopra and Singh [Chopra and Singh, 2016b] to formulate Interaction-Oriented Software Engineering (IOSE) as a distinct paradigm from Agent-Oriented Software Engineering (AOSE). Whereas in AOSE, the focus is typically on specifying and composing agents and interaction is seen as instrumental to achieving the goals of agents, in IOSE the focus is on the specification and composition of interaction protocols without reference to agents. This separation of agent concerns from interaction concerns is likely to be of crucial importance going forward.

Acknowledgments: We thank Michael Winikoff and Neil Yorke-Smith for their extensive comments on an earlier draft. Chopra and Günay were supported by EPSRC grant EP/N027965/1 (Turtles).

BIBLIOGRAPHY

[Agha, 1986] Gul Agha. *Actors: A Model of Concurrent Computation in Distributed Systems*. MIT Press, Cambridge, MA, USA, 1986.

[Alberti et al., 2008] Marco Alberti, Federico Chesani, Marco Gavanelli, Evelina Lamma, Paola Mello, and Paolo Torroni. Verifiable agent interaction in abductive logic programming: The SCIFF framework. *ACM Transactions on Compututational Logic*, 9(4):29:1–29:43, 2008.

[Ancona et al., 2013] Davide Ancona, Sophia Drossopoulou, and Viviana Mascardi. Automatic generation of self-monitoring MASs from multiparty global session types in Jason. In *Declarative Agent Languages and Technologies X*, volume 7784 of *LNCS*, pages 76–95. Springer Berlin Heidelberg, 2013.

[Artikis, 2009] Alexander Artikis. Dynamic protocols for open agent systems. In *Proceedings of the 8th International Conference on Autonomous Agents and Multiagent Systems*, pages 97–104, 2009.

[Baldoni et al., 2006] Matteo Baldoni, Cristina Baroglio, Alberto Martelli, and Viviana Patti. A priori conformance verification for guaranteeing interoperability in open environments. In *Service-Oriented Computing – ICSOC 2006*, volume 4294 of *LNCS*, pages 339–351. Springer Berlin Heidelberg, 2006.

[Baldoni et al., 2009] Matteo Baldoni, Cristina Baroglio, Amit K. Chopra, Nirmit Desai, Viviana Patti, and Munindar P. Singh. Choice, interoperability, and conformance in interaction protocols and service choreographies. In *Proceedings of the 8th International Conference on Autonomous Agents and Multiagent Systems*, pages 843–850, 2009.

[Baldoni et al., 2013] Matteo Baldoni, Cristina Baroglio, Elisa Marengo, and Viviana Patti. Constitutive and regulative specifications of commitment protocols: a decoupled approach. *ACM Transactions on Intelligent Systems and Technology, Special Issue on Agent Communication*, 4(2):22:1–22:25, 2013.

[Baldoni et al., 2014a] Matteo Baldoni, Cristina Baroglio, and Federico Capuzzimati. Typing multi-agent systems via commitments. In *Engineering Multi-Agent Systems*, volume 8758 of *LNCS*, pages 388–405. Springer Berlin Heidelberg, 2014.

[Baldoni et al., 2014b] Matteo Baldoni, Cristina Baroglio, Elisa Marengo, Viviana Patti, and Federico Capuzzimati. Engineering commitment-based business protocols with the 2CL methodology. *Autonomous Agents and Multi-Agent Systems*, 28(4):519–557, 2014.

[Baldoni et al., 2015a] Matteo Baldoni, Cristina Baroglio, Federico Capuzzimati, and Roberto Micalizio. Leveraging commitments and goals in agent interaction. In *Proceedings of the 30th Italian Conference on Computational Logic*, volume 1459 of *CEUR Workshop Proceedings*, pages 85–100. CEUR-WS.org, 2015.

[Baldoni et al., 2015b] Matteo Baldoni, Cristina Baroglio, Amit K. Chopra, and Munindar P. Singh. Composing and verifying commitment-based multiagent protocols. In *Proceedings of the 24th International Joint Conference on Artificial Intelligence*, pages 10–17, 2015.

[Baldoni et al., 2018] Matteo Baldoni, Cristina Baroglio, Federico Capuzzimati, and Roberto Micalizio. Type checking for protocol role enactments via commitments. *Journal of Autonomous Agents and Multi-Agent Systems*, 32(3):349–386, 2018.

[Bauer et al., 2001] Bernhard Bauer, Jörg P. Müller, and James Odell. Agent UML: A formalism for specifying multiagent software systems. *International Journal of Software Engineering and Knowledge Engineering*, 11(3):207–230, 2001.

[Bentahar et al., 2009] Jamal Bentahar, John-Jules Ch. Meyer, and Wei Wan. Model checking communicative agent-based systems. *Knowledge-Based Systems*, 22(3):142–159, 2009.

[Bravetti and Zavattaro, 2009] Mario Bravetti and Gianluigi Zavattaro. A theory of contracts for strong service compliance. *Mathematical Structures in Computer Science*, 19(3):601–638, 2009.

[Cabac et al., 2003] Lawrence Cabac, Daniel Moldt, and Heiko Rölke. A proposal for structuring petri net-based agent interaction protocols. In *Applications and Theory of Petri Nets 2003*, volume 2679 of *LNCS*, pages 102–120. Springer Berlin Heidelberg, 2003.

[Castelfranchi, 1995] Cristiano Castelfranchi. Commitments: From individual intentions to groups and organizations. In *Proceedings of the 1st International Conference on Multiagent Systems*, pages 41–48, 1995.

[Chesani et al., 2009] Federico Chesani, Paola Mello, Marco Montali, Fabrizio Riguzzi, Maurizio Sebastianis, and Sergio Storari. Checking compliance of execution traces to business rules. In *Business Process Management Workshops*, volume 17 of *Lecture Notes in Business Information Processing*, pages 134–145. Springer Berlin Heidelberg, 2009.

[Chesani et al., 2013] Federico Chesani, Paola Mello, Marco Montali, and Paolo Torroni. Representing and monitoring social commitments using the event calculus. *Autonomous Agents and Multi-Agent Systems*, 27(1):85–130, 2013.

[Chopra and Singh, 2009] Amit K. Chopra and Munindar P. Singh. Multiagent commitment alignment. In *Proceedings of the 8th International Conference on Autonomous Agents and Multiagent Systems*, pages 937–944, 2009.

[Chopra and Singh, 2015] Amit K. Chopra and Munindar P. Singh. Generalized commitment alignment. In *Proceedings of the 14th International Conference on Autonomous Agents and Multiagent Systems*, pages 453–461, 2015.

[Chopra and Singh, 2016a] Amit K. Chopra and Munindar P. Singh. Custard: Computing norm states over information stores. In *Proceedings of the 15th International Conference on Autonomous Agents and Multiagent Systems*, pages 1096–1105, 2016.

[Chopra and Singh, 2016b] Amit K. Chopra and Munindar P. Singh. From social machines to social protocols: Software engineering foundations for sociotechnical systems. In *Proceedings of the 25th International World Wide Web Conference*, pages 903–914, 2016.

[Chopra, 2009] Amit K. Chopra. *Commitment Alignment: Semantics, Patterns, and Decision Procedures for Distributed Computing*. PhD thesis, North Carolina State University, Raleigh, NC, 2009.

[Desai et al., 2009] Nirmit Desai, Amit K. Chopra, and Munindar P. Singh. Amoeba: A methodology for modeling and evolving cross-organizational business processes. *ACM Transactions on Software Engineering Methodolgy*, 19(2), 2009.

[Dunn-Davies et al., 2005] H. R. Dunn-Davies, R. J. Cunningham, and S. Paurobally. Propositional statecharts for agent interaction protocols. *Electronic Notes in Theoretical Computer Science*, 134:55–75, 2005.

[Durfee and Zilberstein, 2013] Ed Durfee and Shlomo Zilberstein. Multiagent planning, control, and execution. In Gerhard Weiss, editor, *Multiagent Systems*, chapter 11, pages 485–546. MIT Press, 2013.

[El Fallah-Seghrouchni et al., 2001] Amal El Fallah-Seghrouchni, Serge Haddad, and Hamza Mazouzi. A formal study of interactions in multi-agent systems. *International Journal of Computers and Their Applications*, 8(1), 2001.

[El-Menshawy et al., 2011] Mohamed El-Menshawy, Jamal Bentahar, and Rachida Dssouli. Model checking commitment protocols. In *Modern Approaches in Applied Intelligence*, volume 6704 of *LNCS*, pages 37–47. Springer Berlin Heidelberg, 2011.

[Elder-Vass, 2010] Dave Elder-Vass. *The causal power of social structures: emergence, structure and agency*. Cambridge Univ Press, 2010.

[Esteva et al., 2004] Marc Esteva, Bruno Rosell, Juan A. Rodríguez-Aguilar, and Josep Lluís Arcos. AMELI: an agent-based middleware for electronic institutions. In *Proceedings of the 3rd International Joint Conference on Autonomous Agents and Multiagent Systems*, pages 236–243, 2004.

[FIPA, 2003] FIPA. FIPA interaction protocol specifications, 2003. FIPA: The Foundation for Intelligent Physical Agents.

[Fornara and Colombetti, 2004] Nicoletta Fornara and Marco Colombetti. A commitment-based approach to agent communication. *Applied Artificial Intelligence*, 18(9-10):853–866, 2004.

[Fornara, 2003] Nicoletta Fornara. *Interaction and Communication among Autonomous Agents in Multiagent Systems*. PhD thesis, Università della Svizzera italiana, Facoltà di Scienze della Comunicazione, 2003.

[Günay and Yolum, 2011] Akın Günay and Pınar Yolum. Detecting conflicts in commitments. In *Declarative Agent Languages and Technologies IX*, volume 7169 of *LNCS*, pages 51–66. Springer Berlin Heidelberg, 2011.

[Günay and Yolum, 2013] Akın Günay and Pınar Yolum. Constraint satisfaction as a tool for modeling and checking feasibility of multiagent commitments. *Applied Intelligence*, 39(3):489–509, 2013.

[Günay et al., 2013] Akın Günay, Michael Winikoff, and Pınar Yolum. Commitment protocol generation. In *Declarative Agent Languages and Technologies X*, volume 7784 of *LNCS*, pages 136–152. Springer Berlin Heidelberg, 2013.

[Günay et al., 2015a] Akın Günay, Song Songzheng, Yang Liu, and Jie Zhang. Automated analysis of commitment protocols using probabilistic model checking. In *Proceedinds of the 29th AAAI Conference on Artificial Intelligence*, pages 2060–2066, 2015.

[Günay et al., 2015b] Akın Günay, Michael Winikoff, and Pınar Yolum. Dynamically generated commitment protocols in open systems. *Autonomous Agents and Multi-Agent Systems*, 29(2):192–229, 2015.

[Günay et al., 2016] Akın Günay, Yang Liu, and Jie Zhang. Promoca: Probabilistic modeling and analysis of agents in commitment protocols. *Journal of Artificial Intelligence Research*, 57(1):465–508, 2016.

[HL7, 2002] HL7. HL7 reference information model, version 1.19. http://www.hl7.org/Library/data-model/RIM/C30119/Graphics/RIM_billboard.pdf, 2002.

[Huget, 2004] Marc-Philippe Huget. Agent UML notation for multiagent system design. *IEEE Internet Computing*, 8(4):63–71, 2004.

[Kafalı and Torroni, 2011] Özgür Kafalı and Paolo Torroni. Social commitment delegation and monitoring. In *Computational Logic in Multi-Agent Systems*, volume 6814 of *LNCS*, pages 171–189. Springer Berlin Heidelberg, 2011.

[Kafalı and Torroni, 2012] Özgür Kafalı and Paolo Torroni. Exception diagnosis in multiagent contract executions. *Annals of Mathematics and Artificial Intelligence*, 64(1):73–107, 2012.

[Kafalı and Yolum, 2016] Özgür Kafalı and Pınar Yolum. PISAGOR: A proactive software agent for monitoring interactions. *Knowledge Information Systems*, 47(1):215–239, 2016.

[Kafalı et al., 2014] Özgür Kafalı, Akın Günay, and Pınar Yolum. GOSU: Computing goal support with commitments in multiagent systems. In *Proceedinds of the 21st European Conference on Artificial Intelligence*, pages 477–482, 2014.

[Marengo et al., 2011] Elisa Marengo, Matteo Baldoni, Cristina Baroglio, Amit K. Chopra, Viviana Patti, and Munindar P. Singh. Commitments with regulations: reasoning about safety and control in REGULA. In *Proceedings of the 10th International Conference on Autonomous Agents and Multiagent Systems*, pages 467–474, 2011.

[Meneguzzi et al., 2013] Felipe Meneguzzi, Pankaj R. Telang, and Munindar P. Singh. A first-order formalization of commitments and goals for planning. In *Proceedings of the 27th AAAI Conference on Artificial Intelligence*, AAAI, pages 697–703, 2013.

[Montali et al., 2010] Marco Montali, Maja Pesic, Wil M. P. van der Aalst, Federico Chesani, Paola Mello, and Sergio Storari. Declarative specification and verification of service choreographies. *ACM Transactions on the Web*, 4(1):3:1–3:62, 2010.

[Morgan and Hunt, 1994] Robert M. Morgan and Shelby D. Hunt. The commitment-trust theory of relationship marketing. *Journal of Marketing*, 58(3):20–38, 1994.

[Norman et al., 2004] Timothy J. Norman, D. V. Carbogim, Eric C. W. Krabbe, and C. Douglas Walton. Argument and multi-agent systems. In *Argumentation Machines: New Frontiers in Argument and Computation*, pages 15–54. Springer Netherlands, 2004.

[Odell et al., 2000] James Odell, H. Van Dyke Parunak, and Bernhard Bauer. Representing agent interaction protocols in UML. In *Agent-Oriented Software Engineering*, volume 1957 of *LNCS*, pages 121–140. Springer Berlin Heidelberg, 2000.

[Rajamani and Rehof, 2002] Sriram K. Rajamani and Jakob Rehof. Conformance checking for models of asynchronous message passing software. In *Computer Aided Verification*, volume 2404 of *LNCS*, pages 166–179. Springer Berlin Heidelberg, 2002.

[RosettaNet, 1998] RosettaNet. http://www.rosettanet.org, 1998.

[Searle, 1995] John R. Searle. *The construction of social reality*. Free Press, New York, NY, US, 1995.

[Singh, 1997] Munindar P. Singh. Commitments among autonomous agents in information-rich environments. In *Proceedings of the 8th European Workshop on Modelling Autonomous Agents in a Multi-Agent World: Multi-Agent Rationality*, pages 141–155, 1997.

[Singh, 1998] Munindar P. Singh. Agent communication languages: Rethinking the principles. *IEEE Computer*, 31(12):40–47, 1998.

[Singh, 1999] Munindar P. Singh. An ontology for commitments in multiagent systems. *Artificial Intelligence and Law*, 7(1):97–113, 1999.

[Singh, 2003] Munindar P. Singh. Distributed enactment of multiagent workflows: temporal logic for web service composition. In *Proceedings of the 2nd International Joint Conference on Autonomous Agents and Multiagent Systems*, pages 907–914, 2003.

[Singh, 2008] Munindar P. Singh. Semantical considerations on dialectical and practical commitments. In *Proceedings of the 23rd AAAI Conference on Artificial Intelligence*, pages 176–181, 2008.

[Singh, 2011a] Munindar P. Singh. Information-driven interaction-oriented programming: BSPL, the blindingly simple protocol language. In *Proceedings of the 10th International Conference on Autonomous Agents and MultiAgent Systems*, pages 491–498, 2011.

[Singh, 2011b] Munindar P. Singh. LoST: Local state transfer—an architectural style for the distributed enactment of business protocols. In *Proceedings of the 9th IEEE International Conference on Web Services*, pages 57–64, 2011.

[Singh, 2012] Munindar P. Singh. Semantics and verification of information-based protocols. In *Proceedings of the 11th International Conference on Autonomous Agents and MultiAgent Systems*, pages 1149–1156, 2012.

[Telang et al., 2012] Pankaj R. Telang, Munindar P. Singh, and Neil Yorke-Smith. Relating goal and commitment semantics. In *Programming Multi-Agent Systems*, LNCS, pages 22–37. Springer Berlin Heidelberg, 2012.

[Telang et al., 2013] Pankaj R. Telang, Felipe Meneguzzi, and Munindar P. Singh. Hierarchical planning about goals and commitments. In *Proceedings of the 12th International Joint Conference on Autonomous Agents and Multiagent Systems*, pages 877–884, 2013.

[Tinnemeier et al., 2009] Nick A. M. Tinnemeier, Mehdi Dastani, John-Jules Ch. Meyer, and Leendert W. N. van der Torre. Programming normative artifacts with declarative obligations and prohibitions. In *Proceedings of the 2009 IEEE/WIC/ACM International Conference on Intelligent Agent Technology*, pages 145–152, 2009.

[Torroni et al., 2009] Paolo Torroni, Federico Chesani, Pınar Yolum, Marco Gavanelli, Munindar P. Singh, Evelina Lamma, Marco Alberti, and Paola Mello. *Modelling Interactions via Commitments and Expectations*, chapter 11, pages 263–284. IGI Global, 2009.

[TWIST, 2006] TWIST. Transaction workflow innovation standards team, February 2006. http://www.twiststandards.org.

[Winikoff et al., 2004] Michael Winikoff, Wei Liu, and James Harland. Enhancing commitment machines. In *Declarative Agent Languages and Technologies II*, volume 3476 of *LNCS*, pages 198–220. Springer Berlin Heidelberg, 2004.

[Winikoff et al., 2018] Michael Winikoff, Nitin Yadav, and Lin Padgham. A new hierarchical agent protocol notation. *Autonomous Agents and Multi-Agent Systems*, 32(1):59–133, 2018.

[Winikoff, 2006] Michael Winikoff. Implementing flexible and robust agent interactions using distributed commitment machines. *Multiagent and Grid Systems*, 2(4):365–381, 2006.

[Winikoff, 2007] Michael Winikoff. Implementing commitment-based interactions. In *Proceedings of the 6th International Joint Conference on Autonomous Agents and Multiagent Systems*, pages 128:1–128:8, 2007.

[WS-CDL, 2005] WS-CDL. Web services choreography description language version 1.0, November 2005. www.w3.org/TR/ws-cdl-10/.

[Yokoo et al., 1998] Makoto Yokoo, Edmund H. Durfee, Toru Ishida, and Kazuhiro Kuwabara. The distributed constraint satisfaction problem: Formalization and algorithms. *IEEE Transactions on Knowledge and Data Engineering*, 10(5):673–685, 1998.

[Yolum and Singh, 2001a] Pınar Yolum and Munindar P. Singh. Commitment machines. In *Intelligent Agents VIII*, volume 2333 of *LNCS*, pages 235–247, 2001.

[Yolum and Singh, 2001b] Pınar Yolum and Munindar P. Singh. Designing and executing protocols using the event calculus. In *Proceedings of the 5th International Conference on Autonomous Agents*, pages 27–28, 2001.

[Yolum and Singh, 2002] Pınar Yolum and Munindar P. Singh. Flexible protocol specification and execution: applying event calculus planning using commitments. In *Proceedings of the 1st International Joint Conference on Autonomous Agents and Multiagent Systems*, pages 527–534, 2002.

[Yolum, 2007] Pınar Yolum. Design time analysis of multiagent protocols. *Data and Knowledge Engineering*, 63(1):137–154, 2007.

8
Norm-aware and Norm-oriented Programming

MATTEO BALDONI, CRISTINA BAROGLIO, OLIVIER BOISSIER, JOMI F. HÜBNER, AND ROBERTO MICALIZIO

1 Introduction

Multiagent systems involve different abstractions. At the individual level, agents are situated in an environment, in which they act. Quoting [Weyns et al., 2007] "the environment is a first-class abstraction that provides the surrounding conditions for agents to exist and that mediates both the interaction among agents and the access to resources." At the system level, it is widely recognized that further abstractions become handy, like organizations and interactions [Demazeau, 1995], aimed at enabling a meaningful and fruitful coordination of the autonomous and heterogeneous agents in the system. Thus, agents are not only situated in a physical environment, they are also situated in a social environment [Lindblom and Ziemke, 2003] where they have relationships with other agents and are subject to the regulations of the society they belong to. Lopez and Scott [Lopez and Scott, 2000] identified two components in social situatedness. On the one hand, an institutional component where the "... social structure is seen as comprising those cultural or normative patterns that define the expectations agents hold about each other's behavior and that organize their enduring relations with each other." On the other hand, a relational component where social structure amounts to "the relationships themselves, understood as patterns of causal interconnection and interdependence among agents and their actions, as well as the positions that they occupy." Norms, and normative reasoning, are at the basis of both components of social situatedness. Then, at the system level, norms produce obligations that drive the agents' behavior, while at the agent level, commitments oblige agents towards each other.

Only a few proposals in the literature tackle in an integrated way agents, environment, and norms. In order to explain the value of considering computational systems as based on these three abstractions, this chapter positions such kind of systems in the landscape of modularization of software, ranging from functional decomposition to business artifacts, using as a touchstone Meyer's forces of computation [Meyer, 1997]. Then, it introduces some state-of-the-art tools that integrate agents, environment, and norms. Such tools show the versatility of norms as programming components that can be used to specify how agents are coordinated at a system level, how they act on the environment, how the environment impacts on the normative state, and how agents can autonomously and directly create a coordination of their activities on an agent-to-agent basis.

2 Software Engineering Perspective

In order to highlight the role of norms in system programming, we resort to *Meyer's forces of computation*. They provide a neutral touchstone, unrelated to any specific programming approach or modularization mechanism. According to Meyer, three forces are at play when we use software to perform some computations [Meyer, 1997, Ch. 5, p. 101]: *processors*, *actions*, and *objects*. A processor can be a *process* or a *thread* (we use both the terms processor and process to refer to this force); actions are the *operations* that make the computation; objects are the *data* to which actions are applied.

A software system, in order to execute, uses processes to apply certain actions to certain objects. The form of the actions depends on the considered level of granularity: they can be instructions of a programming language, as well as they can be major steps of a complex algorithm. Moreover, the form of actions conditions the way in which processes operate on objects. Some objects are built by a computation for its own needs and exist only while the computation proceeds; others (e.g., files or databases) are external and may outlive individual computations. In the following, we analyze the most important proposals concerning software modularization, showing how they, sometimes implicitly, give more or less strength to Meyer's forces, and the drawbacks that follow. We, then, devote a special attention to the abstractions of agent, environment, and norm, explaining how altogether they provide a programming paradigm which is much more balanced, with respect to the use of the three forces, than all the previously considered ones.

Top-down functional decomposition puts in the center the notion of process, where a process is seen as implementing a given function. A system is built by stepwise refinement, each refinement step decreases the abstraction of the specification. The approach disregards objects, just considered as data structures that are instrumental to the function specification and internal to processes. Actions are defined only in terms of the instructions provided by the programming language and of other functions built on top of them, into which a process is structured. As a consequence, this approach, though intuitive and suitable to the development of individual algorithms, does not scale up well when data are shared among concurrent processes.

The *Object-Oriented approach* to modularization results from an effort aimed at showing the limits of the functional approach [Meyer, 1997]. Objects (data) often have a life of their own, independent from the processes that use them. Objects become, then, the fundamental notion of the model. They provide the actions by which–and only by which–it is possible to operate on them (data operations). Objects have a static nature: actions are invoked on objects; the process that invokes an operation is the one that causes the evolution of the object. Thus, there is no decoupling between the use of an object and the management of that object. Moreover, the model does not supply conceptual notions for representing tasks, in particular when concurrency is involved.

The key concept in the *actor model* [Hewitt et al., 1973] (by which *active objects* are largely inspired) is that everything is an actor. Interaction between actors occurs only through direct asynchronous message passing, with no restriction on the order in which messages are received; recipients of messages are identified by opaque addresses. The decoupling between the sender of a message and the communications sent makes it possible both to tackle asynchronous communication and to define actors' control structures as patterns of passing messages. Many authors, such as [Mitchell, 2002; Tasharofi et al., 2013; Neykova and Yoshida, 2014], note that the actor model does not address the issue of *coordination*. Coordination requires the possibility for an actor to have expectations about another actor's behavior, but asynchronous message passing alone gives no means to foresee how a message receiver will behave. For example, in the object-paradigm, methods return the computed results to their callers. In the actor model this is not guaranteed because this simple pattern requires the exchange of two messages: no way for specifying patterns of message exchanges between actors is provided. The lack of such mechanisms hinders the verification of properties of a system of interacting actors. Similar problems are well-known also in the area that studies enterprise application integration [Alonso et al., 2004] and service-oriented computing [Singh and Huhns, 2005], that can be considered as heirs to the actor model. The above problem can better be understood by referring to Meyer's forces. The actor model supports the realization of object/data management processes (these are the internal behaviors of the actors, that rule how the actor evolves), but it does not support the design and the modularization of processes that perform the object use, which would be *external* to the actors.

Business processes have been increasingly adopted by enterprises and organizations to conceptually describe their dynamics, and those of the sociotechnical systems they live in. More specifically, a business process describes how a set of interrelated activities can lead to a precise and measurable result (a product or a service) in response to an *external event* (e.g., a new order) [Weske, 2007]. Among the main advantages of this process-centric view, we have the fact that it enables analysis of an enterprise functioning, it enables comparison of business processes, it enables the study of compliance to norms (e.g., [Governatori, 2010]), and also to identify critical points like bottlenecks by way of simulations. On the negative side, business processes show the same limits as functional decomposition. Specifically, they are typically represented in an activity-centric way; i.e., by emphasizing which flows of activities are acceptable, without providing adequate abstractions to capture the data that are manipulated along such flows.

The *artifact-centric approach* [Bhattacharya et al., 2007; Cohn and Hull, 2009; Calvanese et al., 2013] counterposes a data-centric vision to the activity-centric vision described above. *Artifacts* are concrete, identifiable, self-describing chunks of information, the basic building blocks by which business models and operations are described. They include an *information model* of the data, and a *lifecycle model*, that contains the key states through which the data evolve, together with their transitions. The lifecycle model is not only used at runtime but also at design time to understand who is responsible of which transitions. On the negative side, like in the case of the actor model, business artifacts disregard the design and the modularization of those processes that operate on them. Moreover, verification problems are much harder to tackle because the explicit presence of data, together with the possibility of incorporating

2.1 Agents, Environments, and Norms

In [Russell and Norvig, 2003; Wooldridge, 2009], *agents* are defined as entities that observe their environment and act upon it so as to achieve their own goals. Two fundamental characteristics of agents are *autonomy* and *situatedness*. Agents are autonomous in the sense that they have a sense-plan-act deliberative cycle, which gives them control of their internal state and behavior; autonomy, in turn, implies proactivity; i.e., the ability of an agent to take action towards the achievement of its (delegated) objectives without being solicited to do so. Agents are situated because they can sense and manipulate the environment in which operate. The environment can be physical or virtual, and is understood by agents in terms of relevant data. The difference between the agent paradigm and the previously cited paradigms is that the agent paradigm introduces two equally important abstractions: the *agent* and the *environment* [Weyns et al., 2007]. Such a dichotomy does not find correspondence in the other models and gives a first-class role to both Meyer's process and object forces. Processes realize algorithms aimed at achieving objectives. This is exactly the gist of the agent abstraction and the rationale behind its proactivity: agents exploit their deliberative cycle (as control flow), possibly together with the key abstractions of belief, desire, and intention (as logic), in order to realize algorithms (i.e., processes), for acting in their environment and pursue their goals[1]. The manifestation of the object force is the environment abstraction. The environment does not exhibit the kind of autonomy explained for agents even when its definition includes a process. Being reactive, rather than active, makes the environment more similar to an actor whose behavior is triggered by the messages it receives, that are all served equally.

Actions are the capabilities agents have to modify their environment. The process force is mapped onto a cycle in which the agent observes the world (updating its beliefs), deliberates which intentions to achieve, plans how to achieve them, and finally executes the plan [Bratman, 1990]. Beliefs and intentions are those components of the process abstraction that create a bridge respectively towards the object/data force (i.e., the environment) and the action force. Beliefs concern the environment. Intentions lead to action [Wooldridge, 2009], meaning that if an agent has an intention, then the expectation is that the agent will make a reasonable attempt to pursue it. In this sense, intentions play a central role in the selection and the execution of action. Consequently, instead of being subordinate to the process force the action force is put in relation to it by means of intentions. This is a difference with respect to functional decomposition (see previous section), where actions are produced as modular component processes (e.g. procedures) by refining a given goal through a top-down strategy.

The action force is better considered by *normative multiagent systems* [Jones and Carmo, 2001; Boella et al., 2007], which take inspiration from mechanisms

---

[1] Summarizing, objects "do it" for free because they are data, agents are processes and "do it" because it is functional to their objectives.

that are typical of human communities, and have been widely studied in the research area on multiagent systems. According to [Boella et al., 2007] a normative multiagent system is: "a multiagent system together with normative systems in which agents on the one hand can decide whether to follow the explicitly represented norms, and on the other the normative systems specify how and to which extent the agents can modify the norms". The deliberative cycle of agents is affected by the norms and by the obligations these norms generate as a consequence of the agents' actions. Each agent is free to adapt its behavior to (local or coordination) changing conditions, e.g., by re-ranking its goals based on the context or by adopting new goals. Institutions and organizations set the ground for coordination and cooperation among agents. Intuitively, an institution is an organizational structure for coordinating the activities of multiple interacting agents, that typically embodies some rules (norms) that govern participation and interaction. In general, an organization adds to this societal dimension a set of organizational goals and powers to create institutional facts or to modify the norms and obligations of the normative system [Boella and van der Torre, 2004]. Agents, playing one or more roles, must accomplish the organizational goals respecting the norms. Institutions and organizations are, thus, a way to realize functional decomposition in an agent setting.

In the normative multiagent systems domain, several proposals focus on *regulative norms* that, through obligations, permissions, and prohibitions, specify the patterns of actions and interactions agents should adhere to, even though deviations can still occur and have to be properly considered [Jones and Carmo, 2001]. These regulative norms have been combined with *constitutive norms* [Jones and Sergot, 1997; Noriega, 1997; Boella and van der Torre, 2004; Grossi, 2007; Chopra and Singh, 2008; Criado et al., 2013], which support the creation of institutional realities by defining institutional actions that make sense only within the institutions they belong to. A typical example is that of "raising a hand", which counts as "make a bid" in the context of an auction. Institutional actions allow agents to operate within an institution. Citing [Criado et al., 2013], the impact on the agent's deliberative cycle is that agents can "reason about the social consequences of their actions." In this light, going back to Meyer's forces, if agents are abstractions for processes and environments for objects, then *norms* are abstractions of the *action force* because norms model actions and, thus, condition the way in which processes operate on objects. In fact, norms specify either institutional actions, or the conditions for the use of such actions, consequently regulating the acceptable behavior of the agents in a system. This view is also supported by the fact that norms concern "doing the right thing" rather than "doing what leads to a goal" [Therborn, 2002].

The development of software systems using the agent-oriented approach requires tools that properly integrate the programming of the agents, the environment and the norms (both regulative and constitutive), all of them as first class entities. Although there are many tools to independently program each of these parts, we have few tools that consider their integration. One example is the integration of 2APL (an agent programming language) and 2OPL (an normative programming language) [Dastani, 2008; Dastani et al., 2009; Dastani, 2015]. An example that also considers the environment as a first class entity is JaCaMo [Boissier et al., 2013]. JaCaMo integrates Jason [Bordini et al., 2007] (an agent programming language), CArtAgO [Ricci et al., 2011] (an environment programming language) and Moise [Hübner et al., 2007] (an organization and normative programming language) and is considered in a later section to illustrate normative programming and awareness.

3 Norm-Oriented Programming

This section presents an implemented programming language for norms to further illustrate how the action force is realized by norms in the case of coordination. The coordination of the agents in JaCaMo can be conceived at two levels: goals and actions. In the former, global plans are used to define dependencies between collective goals[2] that agents can commit to. Due to their commitments, agents are then *obliged* to achieve those goals. They are however free to decide which actions they will perform to achieve them. In the latter, interaction protocols are used to define the exact sequence of actions that are expected from participant agents and the resources they use. Once an agent participates in a protocol, from a system perspective, it is *obliged* to perform the expected actions as specified by the protocol. At both levels, norms are used to express obligations and mechanisms are used to monitor whether agents are compliant, since they are autonomous and might violate norms. In the rest of this section, we briefly present a language to program the norms and then focus on the action level to illustrate the language.

3.1 Normative Programming Language

An important feature in modularizing norms in a dynamic MAS is to keep them independent from both the agents and the environment, so that the developer does not need to update the normative program whenever new agents arrive or the environment changes. While the independence from the agents is usually achieved by the notion of role, the independence from the environment is managed via constitutive rules. For instance, a norm like `the auction winner is obliged to pay` can be applied to several agents (those that play the role of `auction winner`) and environments (those that have concrete elements to be interpreted as `to pay`).

In the case of JaCaMo, the Normative Programming Language (NPL) has this property [Hübner et al., 2011]. The language is quite simple and based on just two constructs: *obligation* and *regimentation*[3]. With these two primitives, others constructions are possible:

1. Prohibitions are represented either via regimentation or via an obligation for someone else to decide how to handle the situation (e.g., to impose some sanction). For example, consider the norm "it is prohibited to submit a paper with more than 16 pages". In the case of regimentation of this norm, attempts to submit a paper with more than 16 pages will fail (i.e., they will be prevented from taking place). In case this norm

---

[2]While collective goals are created by the overall system and possibly shared between agents, individual goals are created by the agents.

[3]Regimentation is a preventive strategy to enforce norms whereby agents are not capable of violation [Jones and Sergot, 1993].

is not to be regimented, the designer could handle the prohibition by defining an obligation for another agent as in "when a paper with more than 16 pages is submitted, the chair must decide whether to accept the submission or not".

2. Permissions are defined by omission, as in [Grossi et al., 2007]; that is, if something is not obligatory nor prohibited it is simply permitted.

3. Sanctions are represented as obligations (i.e., some agent is *obliged* to apply the sanction).

NPL norm syntax has the general form:

$$\text{norm } id : \varphi \rightarrow \psi$$

where id is a unique *identifier* of the norm; φ is a formula that determines the *activation condition* for the norm; and ψ is the *consequence* of the activation of the norm. Two types of norm consequences ψ are possible:

- *fail* – `fail(r)`: represents the case where the norm is regimented; the argument r represents the reason for the failure;

- *obl* – `obligation(a,r,g,d)`: represents the case where an obligation for some agent a is created. Argument r is the reason for the obligation; g is the formula that represents the obligation itself (either an action or a state of the world that the agent must bring about); and d is the deadline to fulfil the obligation.

A simple example to illustrate the language is given below:

```
// the auction winner has 4 hours to pay the product
norm n1: winner(A) & bid(A,V)
  -> obligation(A,n1,pay(V),'now'+'4 hours').

// example of a regimented norm; bids should be greater than zero
norm n2: bid(_,V) & V <= 0
  -> fail(n2(bid(V))).
```

The interpretation of such programs creates obligations for the participating agents. An obligation has a run-time life-cycle as defined in Figure 1 and explained below (the formal semantics is presented in [Hübner et al., 2011]).

1. An obligation is created when the activation condition φ of some norm n holds. The activation condition formula is used to instantiate the values of variables a, r, g, and d of the obligation to be created.

2. Once created, the initial state of an obligation is *active*.

3. The state changes to *fulfilled* when agent a fulfils the norm's obligation g before the deadline d.

4. The obligation state changes to *unfulfilled* when agent a does not fulfil the norm's obligation g before the deadline d.

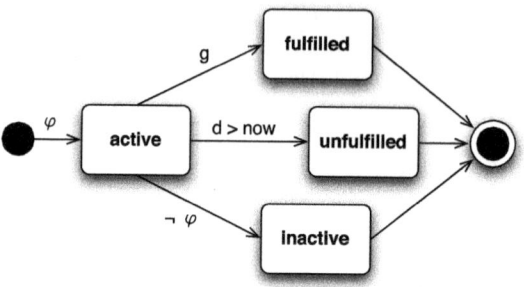

Figure 1. Life-cycle of obligations in NPL

5. As soon as the activation condition (φ) of the norm that created the obligation ceases to hold, the state changes to *inactive*.

The evaluation of the formulas φ and g are based on institutional facts and actions (as introduced in Sec. 2.1). For instance the fact that Alice has transferred some money to a bank account is considered as satisfying a norm only if this action is institutional. In the case of JaCaMo, all institutional facts and actions are defined by a *count-as* program as proposed in [de Brito et al., 2015; de Brito et al., 2017]. A count-as program links the environment and the institution by defining how concrete facts and actions in the environment are interpreted from an institutional point of view. For instance the following count-as program defines the institutional facts and actions used in the norms presented earlier in the auction scenario.

```
tell(A,bid(V)) count-as bid(A,V)
            // tell is a speech act, a brute fact in the environment
            // and bid/2 is an institutional fact

A count-as winner(A)
   while bid(A,V) & not (bid(A2,V2) & V2 > V1 & A2 \= A1) &
       auction(closed)
            // A is the name of concrete agent
            // and winner/1 is an institutional fact

bank_deposit(A,V) count-as pay(V)
   if winner(A)
            // bank_deposit is a concrete action in the environment
            // pay/1 is an institutional action
            // the back deposit is considered by the institution only if
            // performed by the winner
```

3.2 Interaction Protocols

Interaction protocols are a common tool to define how the agents should behave at the action level. In this section we adopt the proposal presented in [Zatelli and Hübner, 2014; Zatelli et al., 2016] since it has the following main property: it is integrated with both the environment and norms. Regarding the integration with the environment, the protocol language considers both the agent-to-artifact interaction and the agent-to-agent interaction based on speech actions.

```
1. protocol voting {
2.   participants:
3.     initiator   agent;
4.     voter       agent;
5.     ballotBox   artifact;
6.   states: k1 initial; k2; k3; k4 final;
7.   transitions:
8.     k1 - k2 # initiator -- message[tell] "object(X)" -> voter;
9.     k2 - k3 # voter     -- action "vote(Y)" -> ballotBox;
10.    k2 - k3 # timeout 30000;
11.    k3 - k4 # ballotBox -- event "winner(Y)" -> initiator;
12. }
```

Figure 2. Example of a Simple Voting Protocol

Regarding the integration with norms, a protocol specification is translated to a set of norms for its regulation.

The Zatelli et al. proposal considers an interaction protocol as composed of a set of participants (agents or artifacts), transitions, and states. Each transition links a source state to a target state and it can be fired by concrete facts from the environment like messages sent, events produced by artifacts, and the execution of actions on artifacts. When some transition is fired, a new state is achieved and the protocol execution evolves. For example, the protocol in Fig. 2 specifies the coordination required for a very simple voting process where agents and artifacts are participating. By this protocol (line 8) we expect a behavior where an initiator agent announces (by a tell message) the object of the election to voters participating in the protocol. Voter should then perform the vote action in a ballot artifact (line 9). When all votes have being performed or some timeout has achieved (i.e., the state k3 is achieved), the ballot artifact produces an event announcing the winner (line 11).

Agents are free to join a particular a protocol, but when they do so (by indicating which kind of participant they will play), the system managing the execution of the protocol will produce obligations for them. These obligations are based on NPL norms automatically created from the protocol specification. The creation of these norms is quite simple: one norm is created for each transition where the source is an agent.[4] For instance, the norms for the protocol of Fig. 2 are:

```
norm k1_k2_a: state(k1) & play(A,initiator) & play(B,voter)
  -> obligation(A, voting_protocol, send(A,B,tell,object(_)), 'now'+'1 hour')

norm k2_k3_a: state(k2) & play(A,voter) & play(B,ballotBox)
  -> obligation(A, voting_protocol, do(A,B,vote(_)), 'now'+'1 hour')
```

Most of the institutional facts used by these norms are produced by count-as rules: play(A,P) is based on the action of agent A joining the protocol as participant P; send(A,B,P,C) is based on the agent A sending a P message to B with content that match C; and do(A,B,O) is based on the agent A doing

---
[4]Timeout transitions and those triggered by artifacts do not require regulation by norms, since there is no autonomy involved in the transition.

the action O on artifact B. However, the institutional fact state(S) is internally produced by the protocol management system. For instance, when the obligation of the initiator is fulfilled by sending the voting object message, the protocol evolves from state k1 to k2 and thus the institutional fact state(k2) holds.

An agent playing voter is obliged to vote not when it joins the protocol, but *after* receiving the message with the voting object, since the norm k2_k3_a that obliges it to vote is activated if the protocol is in state k2.

By means of norms derived from a protocol specification, we are thus regulating and coordinating the agent at the *action* level. Of course it is a design decision to regulate the agents at that level. For some application it is preferable to regulate and coordinate the agents with norms referring to goals, letting agents to select the proper actions to achieve them.

This section has focused on the norms independently of how the agents will handle them: obligations are created by the normative system towards the agents, setting the expected behavior from a system perspective. The system has monitoring mechanisms to verify whether the agents are following the norms, despite their internal architectures. Nevertheless, better results can be achieved if agents are able to reason about norms. The next section explores how the agents can internally handle what is expected from them to do.

4 Norm-aware Interaction

Since the late '80s, studies on distributed artificial intelligence, studies on formal theories of collective activity, team, or group work, and studies on cooperation implicitly identified in *commitment* the glue of group actvity: commitments link the actions of the group members and the group members with each other [Castelfranchi, 1995; Singh, 1997; Norman *et al.*, 2003]. In particular, *social commitments* [Singh, 1999] are a kind of social relationship with a normative value, that makes it possible for the agents to have expectations on one another and coordinate their activities. On this foundation, works like [Baldoni *et al.*, 2015b; Baldoni *et al.*, 2015a; Baldoni *et al.*, 2018a] propose to complement the interaction protocol in [Zatelli and Hübner, 2014], and more in general organizational approaches, with a relational representation of interaction, where agents, by their own action, directly create normative bonds (represented by social commitments) with one another, and use such bonds to coordinate their activities.

A social commitment models the directed relation between two agents: a *debtor* and a *creditor*, that are both aware of the existence of such a relation and of its current state: A commitment $C(x, y, s, u)$ captures that agent x (debtor) commits to agent y (creditor) to bring about the consequent condition u when the antecedent condition s holds. Antecedent and consequent conditions are conjunctions or disjunctions of events and commitments. Since debtors are expected to satisfy their engagements, commitments have a normative value, providing social expectations on the agents' behaviors.

A commitment is autonomously taken by a debtor towards a creditor on its own initiative and is manipulated by agents through the standard operations *create, cancel, release, discharge, assign, delegate* [Singh, 1999]. Commitment

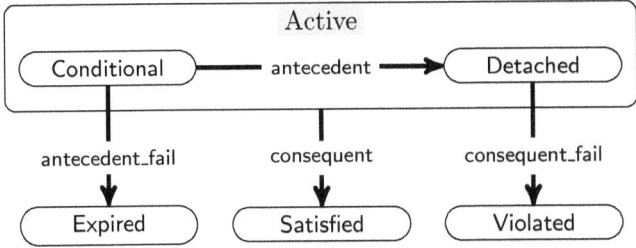

Figure 3. Commitment life cycle.

evolution follows the lifecycle formalized in [Telang *et al.*, 2011], which is reported in Figure 3. A commitment should be *Active* when it is initially created. Active has two substates: *Conditional* as long as the antecedent does not occur, and *Detached* when the antecedent has occurred. A commitment is *Violated* either when its antecedent is true but its consequent will forever be false, or when it is canceled when Detached. It is *Satisfied*, when the engagement is accomplished. It is *Expired*, when it is no longer in effect and therefore the debtor would not fail to comply even if does not accomplish the consequent.

A social commitment, whose antecedent condition is true, amounts to a directed obligation with an important difference. In essence, an obligation is a system level norm while a commitment is an agent level constraint. At system level, something happens and an obligation is created on some agent. At the agent level, an agent creates a conditional social commitment towards some other agent, based on its own beliefs and goals [Telang *et al.*, 2012]. In the most typical case, such a commitment binds the debtor agent to bring about the consequent condition, in the context in which the antecedent condition holds.[5] The creditor agent will detach the conditional commitment if and when it deems it useful to its own purposes, thus activating the obligation of the debtor agent. The motive that leads the two agents to behave in this way is that they have goals they are not able to achieve on their own, so they seek for cooperation by way of interaction. Such agents do so also because they do not have behavioral guidelines, provided by an organization. The interaction, i.e. the causal relationship between the actions of the two concerned agents, is an effect of the presence of the conditional commitment. The creditor performs some kind of normative reasoning on such a commitment, inferring how to act in order to activate the obligation for its debtor to make the consequent condition true.

The choice of commitments is, thus, motivated by the fact that they are taken by an agent as a result of an internal deliberative process. This preserves the autonomy of the agents and is fundamental to harmonize deliberation with goal achievement. The agent does not just react to some obligations, but it rather includes a deliberative capacity by which it creates engagements towards other agents while it is trying to achieve its goals (or to the aim of achieving

---

[5] In general the debtor is not requested to make the condition true by its own actions but will be liable in case of violation.

its goals). Citing Singh [Singh, 2011], an agent would become a debtor of a commitment based on the agent's own communications: either by directly saying something or having another agent communicate something in conjunction with a prior communication of the debtor. That is, there is a causal path from the establishment of a commitment to prior communications by the debtor of that commitment.

Commitment-based interaction protocols assume that a (notional) *social state* is available and inspectable by all the involved agents [Baldoni et al., 2015c]. The social state traces which commitments currently exist between any two agents, and the states of these commitments according to the commitments life-cycle. The explicit representation of the life-cycle of commitments can be used as an interface towards agent behaviors, in order to tackle those state transitions that are of interest for the achievement of the agent's goals. The social state is, thus, a concrete piece of information that belongs to the environment (object force), and that, by evolving according to a known lifecycle, recalls the business artifacts as defined by data-centric approaches [Nigam and Caswell, 2003; Cohn and Hull, 2009; Baldoni et al., 2016]. The structure and lifecycle of such a piece of information are pivotal in harmonizing the object force with the action force through the commitments, and offer to the agents a precious element that can be used both at design time (for programming the agents), and at run-time to allow agents to take into account also the current commitments and their expected evolution in the process of deciding how to operate. Agents will act upon the social state to achieve their goals by creating new commitments or by detaching/discharging the currently active commitments.

Commitments can, thus, be used by agents in their practical reasoning together with beliefs, intentions, and *goals*. In particular, Telang et al. [Telang et al., 2012] point out that goals and commitments are complementary: a commitment specifies how an agent relates to another one, and hence describes what an agent is willing to bring about for another agent. On the other hand, a goal denotes an agent's attitude towards some condition; that is, a state of the world that the agent should achieve. An agent can create a commitment towards another agent to achieve one of its goals; but at the same time, an agent determines the goals to be pursued relying on the commitments it has towards others.

4.1 Programming Interaction with Commitments

JaCaMo+ is an extension of JaCaMo that allows Jason agents to engage commitment-based interactions which are reified as CArtAgO artifacts. In JaCaMo+ an artifact represents the social state of an interaction and provides the roles agents enact. The use of artifacts enables the implementation of monitoring functionalities for verifying that the on-going interactions respect the commitments and for detecting violations and violators. Specifically, a JaCaMo+ artifact encodes a commitment protocol, that is structured into a set of roles. By enacting a role, an agent gains the rights to perform social actions, whose execution has public social consequences, expressed in terms of commitments. If an agent tries to perform an action which is not associated with the role it is enacting, the artifact raises an exception that is notified to the violator. On the other hand, when an agent performs an action that pertains to its role, the social state is updated accordingly by adding new commitments, or by modifying the state of existing commitments.

JaCaMo+ extends the *Jason* component of JaCaMo by allowing the specification of plans whose triggering events involve commitments. JaCaMo+ represents a commitment as a term *cc(debtor, creditor, antecedent, consequent, status)* where *debtor* and *creditor* identify the involved agents (or agent roles), while *antecedent* and *consequent* are the commitment conditions. *Status* is the commitment state (the set being defined in the commitments life-cycle). Commitments operations (e.g. create) are realized as CArtAgO internal operations. Thus, commitment operations cannot be invoked directly by the agents, but the commitment protocol actions will use them as primitives to modify the social state. In a Jason plan specification, commitments can be used wherever beliefs can be used. In contrast to beliefs, their assertion/deletion can only occur through the artifact, in consequence to a social state change. This template shows a Jason plan triggered by the addition of a commitment in the social state:

$$+cc(debtor, creditor, antecedent, consequent, status) : \langle context \rangle \leftarrow \langle body \rangle.$$

More precisely, the plan is triggered when a commitment, that unifies with the one in the plan head, appears in the social state. The syntax is the standard for Jason plans. *Debtor* and *creditor* are to be substituted by the proper roles. The plan may be devised so as to change the commitment status (e.g. the debtor will try to satisfy the comment), or it may be devised so as to allow the agent to react to the commitment presence (e.g., collecting information). Similar schemas can be used for commitment deletion. Further, commitments can also be used in contexts and in plans as test goals (In Jason syntax: ?cc(...)), or achievement goals (!cc(...)). Addition or deletion of such goals can as well be managed by plans, for example:

$$+!cc(debtor, creditor, antecedent, consequent, status) : \langle context \rangle \leftarrow \langle body \rangle.$$

The plan is triggered when the agent creates an achievement goal concerning a commitment. Consequently, the agent will act upon the artifact so as to create the desired social relationship. After the execution of the plan, the commitment *cc(debtor, creditor, antecedent, consequent, status)* will hold in the social state, and will be projected onto the belief bases of all agents focusing on the artifact.

Let us see how Dijktra's dining philosophers can be programmed in JaCaMo+. Intuitively, this example shows the advantage of a separation of concerns between the coordination logic in terms of norms (i.e., the action force, as explained in Section 2.1) and the agent's logic (i.e., the process force). When these two forces are kept separate, a better software modularization is possible. In fact, on one hand, it becomes possible to implement and verify the interaction artifact independently of the agents that will use it. On the other hand, it becomes possible to implement the agents' plans in different ways as long as they keep on addressing the commitments' state changes that may occur along the interaction.

```
1  counter(0).
2  !start.
3  +!start : true
4       <- focusWhenAvailable("philoArtifact"); enact("philosopher").
5  +enacted(Id,"philosopher",Role_Id)
6       <- +enactment_id(Role_Id); .my_name(Me);
7          in("philo_init",Me,Left,Right);
8          +my_left_fork(Left); +my_right_fork(Right); !!living.
9  +!living :
10      <- !thinking; !eating.
11 +!eating : my_left_fork(Left) & my_right_fork(Right) & counter(C)
12      <- ?enactment_id(Role_Id);
13         askForks(Left, Right, C).
14 +cc(My_Role_Id, "philosopher", available(Left,Right,C),
15     returnForks(Left,Right,C),"DETACHED")
16     :   enactment_id(My_Role_Id) & my_left_fork(Left) &
17         my_right_fork(Right) & counter(C)
18      <- !eat(Left, Right, C); returnForks(Left, Right, C).
19 +cc(My_Role_Id, "philosopher", available(Left,Right,C),
20     returnForks(Left,Right,C), "SATISFIED")
21     :   enactment_id(My_Role_Id) & my_left_fork(Left)
22      <- ?counter(C); -+counter(C+1); !living.
23 +!eat(Left, Right, C): my_left_fork(Left) & my_right_fork(Right)
24     & available(Left, Right, C) & counter(C)
25      <- .my_name(Me); ?enactment_id(Role_Id);
26         println(Me, " ", Role_Id, " eating").
27 +!thinking : counter(C)
28      <- .my_name(Me); ?enactment_id(Role_Id);
29         println(Me, " ", Role_Id, " thinking, time ",C).
```

Listing 1.1. The philosopher agent program in JaCaMo+.

An agent has a !*living* main cycle (ln. 9) that alternates the goals !*thinking* and !*eating*. Coordination is needed just for eating: to this aim, forks must be available. The interaction artifact provides the role *philosopher*, empowered with an operation *askForks*. When an agent wants to eat, it invokes such an operation which in turn creates a commitment to return the forks, whose antecedent condition is to have the forks assigned. Eventually, forks will be ready and the commitment will be detached, so the agent can use the forks (i.e., eat) before discharging the commitment by returning them. The agent who executes the operation is the debtor of such a commitment, any other philosopher is the creditor. The antecedent condition is that forks are available and the consequent is that forks will be returned. Note that fork assignment is decided by way of a coordination policy that is implemented in the artifact. The *askForks* operation hides the synchronization for using forks. The only concern on the agent side is to address the meaningful state changes of the commitment that was created by means of *askForks*. In our case, only *Detached* and *Satisfied* are meaningful. When the commitment is detached, the agent eats and then executes *returnForks*, thus satisfying its commitment. When the commitment is satisfied, the agent can re-start its main cycle (!*living*). Knowing the social meanings of artifact operations is sufficient for coordinating with others correctly. The connection between the event "commitment detached" and the associated plan is not only causal, but rather the plan has the aim of satisfying the consequent condition of the commitment (*returnForks*).

5 Challenges

The development of software systems with integrated agent, environment, and norms can still be improved in several aspects. We mention here three challenges related to the environment dimension:

- While the impact of the environment on the normative state is addressed by the constitutive rules, the proper design and instrumentalization of the environment to achieve organizational goals still deserves further investigation.

- Agent coordination should not only concern agent activities but it should also account for, and in some cases be driven by, the environment and its evolution. One major challenge is that this kind of coordination calls for a declarative specification of the environment life cycle, upon which the normative system should be based. The advantage would be twofold. Agents would be capable of reasoning on the system as a whole (because they would have expectations on both other agents and on the environment) with an understanding of the implications on their own behavior, in terms of duties, prohibitions and such like. On the other hand, it would also be possible to perform property analysis at the level of specification and norms rather than on the system as a whole.

- A third challenge concerns tackling accountability through a proper formalization and to support agent programming through typing. Accountability is a fundamental concept at the basis of interaction which is still little explored in the MAS literature [Chopra and Singh, 2016; Baldoni et al., 2018c]. It concerns the identification of who should give account for some situation of interest. In organizational contexts it is often used when some undesired situation occurs, with the aim of improving performance. Typing concerns capturing those requirements that agent programs should satisfy, for instance, to play an organizational role [Baldoni et al., 2018b]. It is a feature at the heart of software engineering.

Acknowledgments: We thank Viviana Mascardi and Julian Padget for their extensive comments on an earlier draft.

BIBLIOGRAPHY

[Alonso et al., 2004] Gustavo Alonso, Fabio Casati, Harumi Kuno, and Vijay Machiraju. *Web Services*. Springer, 2004.

[Baldoni et al., 2015a] Matteo Baldoni, Cristina Baroglio, Federico Capuzzimati, and Roberto Micalizio. Empowering Agent Coordination with Social Engagement. In M. Gavanelli, E. Lamma, and F. Riguzzi, editors, *AI*IA 2015: Advances in Artificial Intelligence, XIV International Conference of the Italian Association for Artificial Intelligence*, volume 9336 of *LNAI*, pages 89–101, Ferrara, Italy, September 2015. Springer.

[Baldoni et al., 2015b] Matteo Baldoni, Cristina Baroglio, Federico Capuzzimati, and Roberto Micalizio. Exploiting Social Commitments in Programming Agent Interaction. In Q. Chen, P. Torroni, S. Villata, J. Y. Hsu, and A. Omicini, editors, *PRIMA 2015: Principles and Practice of Multi-Agent Systems, 18th International Conference*, number 9387 in Lecture Notes in Computer Science, pages 566–574, Bertinoro, Italy, October 26th–30th 2015. Springer.

[Baldoni et al., 2015c] Matteo Baldoni, Cristina Baroglio, Amit K. Chopra, and Munindar P. Singh. Composing and Verifying Commitment-Based Multiagent Protocols. In

Q. Yang and M. Wooldridge, editors, *Proc. of 24th International Joint Conference on Artificial Intelligence, IJCAI 2015*, pages 10–17, Buenos Aires, Argentina, July 25th-31th 2015. AAAI Press / International Joint Conferences on Artificial Intelligence.

[Baldoni et al., 2016] Matteo Baldoni, Cristina Baroglio, Diego Calvanese, Roberto Micalizio, and Marco Montali. Towards Data- and Norm-Aware Multiagent Systems. In *Engineering Multi-Agent Systems - 4th International Workshop, EMAS 2016, Singapore, Singapore, May 9-10, 2016, Revised, Selected, and Invited Papers*, pages 22–38, 2016.

[Baldoni et al., 2018a] Matteo Baldoni, Cristina Baroglio, Federico Capuzzimati, and Roberto Micalizio. Commitment-based Agent Interaction in JaCaMo+. *Fundamenta Informaticae*, 159:1–33, 2018.

[Baldoni et al., 2018b] Matteo Baldoni, Cristina Baroglio, Federico Capuzzimati, and Roberto Micalizio. Type Checking for Protocol Role Enactments via Commitments. *Journal of Autonomous Agents and Multi-Agent Systems*, 2018. In press, DOI: https://doi.org/10.1007/s10458-018-9382-3.

[Baldoni et al., 2018c] Matteo Baldoni, Cristina Baroglio, Katherine M. May, Roberto Micalizio, and Stefano Tedeschi. Computational Accountability in MAS Organizations with ADOPT. *Journal of Applied Sciences, special issue "Multi-Agent Systems"*, 8(4):489, March 2018.

[Bhattacharya et al., 2007] Kamal Bhattacharya, Nathan S. Caswell, Santhosh Kumaran, Anil Nigam, and Frederick Y. Wu. Artifact-centered operational modeling: Lessons from customer engagements. *IBM Systems Journal*, 46(4):703–721, 2007.

[Boella and van der Torre, 2004] Guido Boella and Leendert W. N. van der Torre. Regulative and constitutive norms in normative multiagent systems. In Didier Dubois, Christopher A. Welty, and Mary-Anne Williams, editors, *Principles of Knowledge Representation and Reasoning: Proceedings of the Ninth International Conference (KR2004), Whistler, Canada, June 2-5, 2004*, pages 255–266. AAAI Press, 2004.

[Boella et al., 2007] Guido Boella, Leendert W. N. van der Torre, and Harko Verhagen. Introduction to normative multiagent systems. In Guido Boella, Leendert W. N. van der Torre, and Harko Verhagen, editors, *Normative Multi-agent Systems, 18.03. - 23.03.2007*, volume 07122 of *Dagstuhl Seminar Proceedings*. Internationales Begegnungs- und Forschungszentrum für Informatik (IBFI), Schloss Dagstuhl, Germany, 2007.

[Boissier et al., 2013] Olivier Boissier, Rafael H. Bordini, Jomi F. Hbner, Alessandro Ricci, and Andrea Santi. Multi-agent oriented programming with JaCaMo. *Science of Computer Programming*, 78(6):747 – 761, 2013.

[Bordini et al., 2007] Rafael H. Bordini, Jomi Fred Hübner, and Michael Wooldrige. *Programming Multi-Agent Systems in AgentSpeak using Jason*. Wiley Series in Agent Technology. John Wiley & Sons, 2007.

[Bratman, 1990] Michael E. Bratman. What is intention? In P. Cohen, J. Morgan, and M. Pollack, editors, *Intensions in Communication*, pages 15–31. MIT Press, Cambridge, MA, 1990.

[Calvanese et al., 2013] Diego Calvanese, Giuseppe De Giacomo, and Marco Montali. Foundations of data-aware process analysis: a database theory perspective. In Richard Hull and Wenfei Fan, editors, *Proceedings of the 32nd ACM SIGMOD-SIGACT-SIGART Symposium on Principles of Database Systems, PODS 2013, New York, NY, USA - June 22 - 27, 2013*, pages 1–12. ACM, 2013.

[Castelfranchi, 1995] Cristiano Castelfranchi. Commitments: From Individual Intentions to Groups and Organizations. In Victor R. Lesser and Les Gasser, editors, *Proceedings of the First International Conference on Multiagent Systems (ICMAS)*, pages 41–48, San Francisco, California, USA, June 1995. The MIT Press.

[Chopra and Singh, 2008] Amit K. Chopra and Munindar P. Singh. Constitutive interoperability. In *Proceedings of the 7th Int. J. Conf. on Autonomous agents and multiagent systems, Volume 2*, pages 797–804. International Foundation for Autonomous Agents and Multiagent Systems, 2008.

[Chopra and Singh, 2016] Amit K. Chopra and Munindar P. Singh. From social machines to social protocols: Software engineering foundations for sociotechnical systems. In *Proc. of the 25th Int. Conf. on WWW*, 2016.

[Cohn and Hull, 2009] David Cohn and Richard Hull. Business Artifacts: A Data-centric Approach to Modeling Business Operations and Processes. *IEEE Data Eng. Bull.*, 32(3):3–9, 2009.

[Criado et al., 2013] Natalia Criado, Estefania Argente, Pablo Noriega, and Vicent Botti. Reasoning about constitutive norms in BDI agents. *Logic Journal of IGPL*, 2013.

[Dastani et al., 2009] Mehdi Dastani, Nick A. M. Tinnemeier, and John-Jules Ch. Meyer. A programming language for normative multi-agent systems. In V. Dignum, editor, *Multi-Agent Systems: Semantics and Dynamics of Organizational Models*, chapter XVI, pages 397–417. Information Science Reference, Hershey, PA, USA, 2009.

[Dastani, 2008] Mehdi Dastani. 2APL: a practical agent programming language. *Autonomous Agent and Multi-Agent Systems*, 16:241–248, 2008.

[Dastani, 2015] Mehdi Dastani. Programming multi-agent systems. *Knowledge Eng. Review*, 30(4):394–418, 2015.

[de Brito et al., 2015] Maiquel de Brito, Jomi F. Hübner, and Olivier Boissier. Bringing constitutive dynamics to situated artificial institutions. In *Proc. of 17th Portuguese Conference on Artificial Intelligence (EPIA 2015)*, volume 9273 of *LNCS*, pages 624–637. Springer, 2015.

[de Brito et al., 2017] Maiquel de Brito, Jomi Fred Hübner, and Olivier Boissier. Situated artificial institutions: stability, consistency, and flexibility in the regulation of agent societies. *Autonomous Agents and Multi-Agent Systems*, pages 1–33, 2017.

[Demazeau, 1995] Yves Demazeau. From interactions to collective behaviour in agent-based systems. In *Proceedings of the 1st. European Conference on Cognitive Science*, pages 117–132, Saint-Malo, 1995.

[Governatori, 2010] Guido Governatori. Law, logic and business processes. In *Third International Workshop on Requirements Engineering and Law, RELAW 2010, Sydney, NSW, Australia, September 28, 2010*, pages 1–10. IEEE, 2010.

[Grossi et al., 2007] Davide Grossi, Huib Aldewered, and Frank Dignum. *Ubi Lex, Ibi Poena*: Designing norm enforcement in e-institutions. In P. Noriega, J. Vázquez-Salceda, G. Boella, O. Boissier, V. Dignum, N. Fornara, and E. Matson, editors, *Coordination, Organizations, Institutions, and Norms in Agent Systems II*, volume 4386 of *LNAI*, pages 101–114. Springer, 2007. Revised Selected Papers.

[Grossi, 2007] Davide Grossi. *Designing Invisible Handcuffs, Formal Investigations in Institutions and Organizations for Multi-agent Systems*. PhD thesis, University of Utrecht, 2007.

[Hewitt et al., 1973] Carl Hewitt, Peter Bishop, and Richard Steiger. A universal modular ACTOR formalism for artificial intelligence. In Nils J. Nilsson, editor, *Proceedings of the 3rd International Joint Conference on Artificial Intelligence. Standford, CA, August 1973*, pages 235–245. William Kaufmann, 1973.

[Hübner et al., 2007] Jomi Fred Hübner, Jaime Simão Sichman, and Olivier Boissier. Developing organised multi-agent systems using the MOISE+ model: Programming issues at the system and agent levels. *International Journal of Agent-Oriented Software Engineering*, 1(3/4):370–395, 2007.

[Hübner et al., 2011] Jomi F. Hübner, Olivier Boissier, and Rafael H. Bordini. A normative programming language for multi-agent organisations. *Annals of Mathematics and Artificial Intelligence*, 62(1-2):27–53, 2011.

[Jones and Carmo, 2001] Andrew J.I. Jones and José Carmo. Deontic logic and contrary-to-duties. In Dov Gabbay, editor, *Handbook of Philosophical Logic*, pages 203–279. Kluwer, 2001.

[Jones and Sergot, 1993] Andrew J. I. Jones and Marek Sergot. On the characterization of law and computer systems: the normative systems perspective. In *Deontic logic in computer science: normative system specification*, pages 275–307. John Wiley and Sons Ltd., Chichester, UK, 1993.

[Jones and Sergot, 1997] Andrew J. I. Jones and Marek J. Sergot. A formal characterisation of institutionalised power. *Journal of the IGPL*, 4:429–445, 1997.

[Lindblom and Ziemke, 2003] Jessica Lindblom and Tom Ziemke. Social Situatedness of Natural and Artificial Intelligence: Vygotsky and Beyond. *Adaptive Behaviour*, 11(2):79–96, 2003.

[Lopez and Scott, 2000] Jose Lopez and John Scott. *Social Structure*. Open University Press, 2000.

[Meyer, 1997] Bertrand Meyer. *Object-oriented Software Construction (2Nd Ed.)*. Prentice-Hall, Inc., Upper Saddle River, NJ, USA, 1997.

[Mitchell, 2002] John C. Mitchell. *Concepts in programming languages*. Cambridge University Press, Cambridge, New York (N. Y.), 2002.

[Neykova and Yoshida, 2014] Rumyana Neykova and Nobuko Yoshida. Multiparty Session Actors. In eva Kühn and Rosario Pugliese, editors, *Coordination Models and Languages - 16th IFIP WG 6.1 International Conference, COORDINATION 2014, Held as Part of the 9th International Federated Conferences on Distributed Computing Techniques,*

DisCoTec 2014, Berlin, Germany, June 3-5, 2014, Proceedings, volume 8459 of *Lecture Notes in Computer Science*, pages 131–146. Springer, 2014.

[Nigam and Caswell, 2003] Anil Nigam and Nathan S. Caswell. Business artifacts: An approach to operational specification. *IBM Systems Journal*, 42(3):428–445, 2003.

[Noriega, 1997] Pablo Noriega. *Agent-Mediated Auctions: The Fishmarket Metaphor*. PhD thesis, Institut d'Investigatió en Intelligència Artificial, 1997.

[Norman et al., 2003] Timothy J. Norman, D. V. Carbogim, Eric C. W. Krabbe, and C. Douglas Walton. Argument and multi-agent systems. In *Argumentation Machines: New Frontiers in Argument and Computation*, volume 9 of *Argumentation Library*, pages 15–54. 2003.

[Ricci et al., 2011] Alessandro Ricci, Michele Piunti, and Mirko Viroli. Environment programming in multi-agent systems: an artifact-based perspective. *Autonomous Agents and Multi-Agent Systems*, 23(2):158–192, 2011.

[Russell and Norvig, 2003] Stuart J. Russell and Peter Norvig. *Artificial Intelligence: A Modern Approach*. Pearson Education, 2 edition, 2003.

[Singh and Huhns, 2005] Munindar P. Singh and Michael N. Huhns. *Service-oriented computing - semantics, processes, agents*. Wiley, 2005.

[Singh, 1997] Munindar P. Singh. Commitments Among Autonomous Agents in Information-Rich Environments. In *Proceedings of the 8th European Workshop on Modelling Autonomous Agents in a Multi-Agent World: Multi-Agent Rationality*, pages 141–155, London, UK, UK, 1997. Springer-Verlag.

[Singh, 1999] Munindar P. Singh. An ontology for commitments in multiagent systems. *Artif. Intell. Law*, 7(1):97–113, 1999.

[Singh, 2011] Munindar P. Singh. Commitments in multiagent systems some controversies, some prospects. In Fabio Paglieri, Luca Tummolini, Rino Falcone, and Maria Miceli, editors, *The Goals of Cognition. Essays in Honor of Cristiano Castelfranchi*, chapter 31, pages 601–626. College Publications, London, 2011.

[Tasharofi et al., 2013] Samira Tasharofi, Peter Dinges, and Ralph E. Johnson. Why Do Scala Developers Mix the Actor Model with Other Concurrency Models? In *Proceedings of the 27th European Conference on Object-Oriented Programming*, ECOOP'13, pages 302–326, Berlin, Heidelberg, 2013. Springer-Verlag.

[Telang et al., 2011] Pankaj R. Telang, Munindar P. Singh, and Neil Yorke-Smith. Relating goal and commitment semantics. In *ProMAS*, volume 7217 of *Lecture Notes in Computer Science*, pages 22–37. Springer, 2011.

[Telang et al., 2012] Pankaj R. Telang, Neil Yorke-Smith, and Munindar P. Singh. Relating Goal and Commitment Semantics. In *Proc. of ProMAS*, volume 7212 of *LNCS*, pages 22–37. Springer, 2012.

[Therborn, 2002] Göran Therborn. Back to norms! on the scope and dynamics of norms and normative action. *Current Sociology*, 50:863–880, 2002.

[Weske, 2007] Mathias Weske. *Business Process Management: Concepts, Languages, Architectures*. Springer, 2007.

[Weyns et al., 2007] Danny Weyns, Andrea Omicini, and James Odell. Environment as a first class abstraction in multiagent systems. *JAAMAS*, 14(1):5–30, 2007.

[Wooldridge, 2009] Michael J. Wooldridge. *Introduction to multiagent systems, 2nd edition*. Wiley, 2009.

[Zatelli and Hübner, 2014] Maicon R. Zatelli and Jomi F. Hübner. The interaction as an integration component for the JaCaMo platform. In Fabiano Dalpiaz, Jürgen Dix, and M. Birna van Riemsdijk, editors, *Proc. 2nd International Workshop on Engineering Multi-agent Systems (EMAS @ AAMAS 2014)*, volume 8758 of *LNCS*, pages 431–450. Springer, 2014.

[Zatelli et al., 2016] Maicon Rafael Zatelli, Alessandro Ricci, and Jomi F. Hbner. Integrating interaction with agents, environment, and organisation in JaCaMo. *International Journal of Agent-Oriented Software Engineering (IJAOSE)*, 5(2,3):266 – 302, 2016.

PART IV

LOGICAL ANALYSIS

9
Multiagent Deontic Logic and its Challenges from a Normative Systems Perspective

GABRIELLA PIGOZZI AND LEENDERT VAN DER TORRE

Introduction

Deontic logic [von Wright, 1951a; Gabbay *et al.*, 2013] is the field of logic that is concerned with normative concepts such as obligation, permission, and prohibition. Alternatively, a deontic logic is a formal system capturing the essential logical features of these concepts. Typically, a deontic logic uses Op to mean that it is obligatory that p, (or it ought to be the case that p), and Pp to mean that it is permitted, or permissible, that p. The term 'deontic' is derived from the ancient Greek *déon*, meaning that "which is binding or proper".

Deontic logic can be used for reasoning about normative multiagent systems, i.e. about multiagent systems with normative systems in which agents can decide whether to follow the explicitly represented norms, and the normative systems specify how and to which extent agents can modify the norms [Boella *et al.*, 2006; Andrighetto *et al.*, 2013]. Normative multiagent systems need to combine normative reasoning with agent interaction, and thus raise the challenge to relate the logic of normative systems to game theory [van der Torre, 2010].

Traditional (or "standard") deontic logic is a normal propositional modal logic of type KD, which means that it extends the propositional tautologies with the axioms $K : O(p \rightarrow q) \rightarrow (Op \rightarrow Oq)$ and $D : \neg(Op \wedge O\neg p)$, and it is closed under the inference rules *modus ponens* $p, p \rightarrow q/q$ and *generalization* or *necessitation* p/Op. Prohibition and permission are defined by $Fp = O\neg p$ and $Pp = \neg O\neg p$. Traditional deontic logic is an unusually simple and elegant theory. An advantage of its modal-logical setting is that it can easily be extended with other modalities such as epistemic or temporal operators and modal accounts of action. In this chapter we illustrate the combination of deontic logic with a modal logic of action, called STIT logic [Horty, 2001].

Not surprisingly for such a highly simplified theory, there are many features of actual normative reasoning that traditional deontic logic does not capture. Notorious are the so-called 'paradoxes of deontic logic', which are usually dismissed as consequences of the simplifications of traditional deontic logic. For example, Ross's paradox [Ross, 1941] is the counterintuitive derivation of "you ought to mail or burn the letter" from "you ought to mail the letter." It is typically viewed as a side effect of the interpretation of 'or' in natural language.

In this chapter we discuss also an example of norm based semantics, called input/output logic, to discuss challenges related to norms and detachment. Maybe the most striking feature of the abstract character of traditional deontic logic is that it does not explicitly represent the norms of the system, only the obligations and permissions which can be detached from the norms in a given context. This is an obvious limitation when using deontic logic to reason about normative multiagent systems, in which norms are represented explicitly.

In this chapter we consider the following fifteen challenges for multiagent deontic logic. The list of challenges is by no means final. Other problems may be considered equally important, such as how a hierarchy of norms (or of the norm-giving authorities) is to be respected, how general abstract norms relate to individual concrete obligations, how norms can be interpreted, or how various kinds of imperatives can be distinguished. We do not consider deontic logics for specification and verification of multiagent systems [Broersen et al., 2003; Ågotnes et al., 2010], but we focus on normative reasoning within multiagent systems. The three central concepts in these challenges are preference, agency, and norms. Regarding agency, we consider individual agent action as well as agent interaction in games.

1. Contrary-to-duty reasoning, preference and violation — preference
2. Non-deterministic actions: ought-to-do vs ought-to-be — agency
3. Moral luck and the driving example — agency
4. Procrastination: actualism vs possibilism — agency
5. Jørgensen's dilemma and the problem of detachment — norms
6. Multiagent detachment — norms
7. Coherence of a normative system — norms
8. Normative conflicts and dilemmas — preference & norms
9. Descriptive dyadic obligations and norms — preference & norms
10. Permissive norms — preference & norms
11. Meaning postulates and intermediate concepts — norms
12. Constitutive norms — norms
13. Revision of a normative system — norms
14. Merging normative systems — norms
15. Games, norms and obligations — norms & agency

To discuss these challenges, we repeat the basic definitions of so-called standard deontic logic, dyadic standard deontic logic, deontic STIT logic, and input/output logic. The chapter thus contains several definitions, but these are not put to work in any theorems or propositions, for which we refer to the handbook of deontic logic and normative systems [Gabbay et al., 2013]. The point of introducing formal definitions in this chapter is just to have a reference for the interested reader. Likewise, the interested reader should consult the handbook of deontic logic and normative systems for a more comprehensive description of the work done on each challenge, as in this chapter we can mention only a few references for each challenge. Moreover, the challenge of detachment is considered in more detail in the chapter by Parent and van der Torre [2017a], and the challenge of norm interpretation is discussed in the chapter by da Costa Pereira et al. [2018].

1 Contrary-to-duty reasoning, preference and violation

In this section we discuss how the challenge of the contrary-to-duty paradoxes leads to traditional modal deontic logic introduced at the end of the sixties, based on dyadic operators and preference based semantics. Moreover, we contrast this use of preference in deontic logic with the use of preference in decision theory.

1.1 Chisholm's paradox

Suppose we are given a code of conditional norms, that we are presented with a condition (input) that is unalterably true, and asked what obligations (output) it gives rise to. It may happen that the condition is something that should not have been true in the first place. But that is now water under the bridge: we have to "make the best out of the sad circumstances" as B. Hansson [1969] put it. We therefore abstract from the deontic status of the condition, and focus on the obligations that are consistent with its presence. How to determine this in general terms, and if possible in formal ones, is the well-known problem of contrary-to-duty conditions as exemplified by the notorious contrary-to-duty paradoxes. Chisholm's paradox [Chisholm, 1963] consists of the following four sentences:

(1) It ought to be that a certain man go to the assistance of his neighbours.
(2) It ought to be that if he does go, he tell them he is coming.
(3) If he does not go then he ought not to tell them he is coming.
(4) He does not go.

Furthermore, intuitively, the sentences derive the following sentence (5):

(5) He ought not to tell them he is coming.

Chisholm's paradox is a contrary-to-duty paradox, since it contains both a primary obligation to go, and a secondary obligation not to tell if the agent does not go. Traditionally, the paradox was approached by trying to formalise each of the sentences in an appropriate language of deontic logic. However, in traditional (or "standard") deontic logic, i.e. the normal propositional modal logic of type KD, it turned out that either the set of formulas is inconsistent, or one formula is a logical consequence of another formula. Yet intuitively the natural-language expressions that make up the paradox are consistent and independent from each other: this is why it is called a paradox. The problem is thus:

Challenge 1 *How do we reason with contrary-to-duty obligations which are in force only in case of norm violations?*

There are various kinds of scenarios which are similar to Chisholm's scenario. For example, there is a key difference between contrary-to-duties proper, and reparatory obligations, because the latter cannot be atemporal [Prakken and Sergot, 1996]. Though Chisholm presented his challenge as essentially a single agent decision problem, we can as well reformulate it as a multiagent reasoning problem:

(1) It is obligatory that i sees to it that p (i should do p).
(2) It is obligatory that j sees to it that q if i does not see to it that p
    (j should sanction i if i does not do as told).
(3) It is obligatory that j does not see to it that q if i sees to it that p
    (j should not sanction i if i does as told).
(4) i does not do as told.

The logic may give us the paradoxical conclusion that j should see to it that q and he should see to it that not q. For example, van Benthem, Grossi and Liu [2014] give the following example, in the formulation proposed by Åqvist [1967]:

(1) It ought to be that Smith refrains from robbing Jones.
(2) Smith robs Jones.
(3) If Smith robs Jones, he ought to be punished for robbery.
(4) It ought to be that if Smith refrains from robbing Jones he is not punished for robbery.

As explained in detail in the following subsections, the development of dyadic deontic operators as well as the introduction of temporally relative deontic logic operators can be seen as a direct result of Chisholm's paradox. Since the robbing takes place before the punishment, the example can quite easily be represented once time is made explicit [van der Torre and Tan, 1998]. If you make time explicit or you direct obligations to different agents, then the paradox disappears, in a way. However, both the fact that time and agency are present may distract from the key point behind the example. Therefore also atemporal, non-agency version of the paradox allow to address to the core challenge of the issue. For example, Prakken and Sergot [1996] consider the following variant of Chisholm's scenario:

(1) It ought to be that there is no dog.
(2) If there is a dog, there should be a sign.
(3) If there is no dog, there should be no sign.
(4) There is a dog.

When a new deontic logic is proposed, the traditional contrary-to-duty examples are always the first benchmark examples to be checked. It may be observed here that some researchers in deontic logic doubt that contrary-to-duties can still be considered a challenge, because due to extensive research by now we know pretty much everything about them. The deontic logic literature is full of (at least purported) solutions. In other words, these researchers doubt that deontic logic still needs more research on contrary-to-duties. Indeed, it appears to be difficult to make an original contribution to this vast literature, but new twists are still identified [Parent and van der Torre, 2017b].

1.2 Monadic deontic logic

Traditional or 'standard' deontic logic, often referred to as SDL, was introduced by Von Wright [1951a].

1.2.1 Language

Let Φ be a set of propositional letters. The language of traditional deontic logic $\mathfrak{L}_D$ is given by the following BNF:

$$\varphi := \bot \mid p \mid \neg\varphi \mid (\varphi \wedge \varphi) \mid \bigcirc\varphi \mid \Box\varphi$$

where $p \in \Phi$. The intended reading of $\bigcirc\varphi$ is "φ is obligatory" and the intended reading of $\Box\varphi$ is "φ is necessary". Moreover we use $P\varphi$, read as "φ is permitted", as an abbreviation of $\neg\bigcirc\neg\varphi$ and $F\varphi$, "φ is forbidden", as an abbreviation of $\bigcirc\neg\varphi$. Likewise, $\vee$, $\rightarrow$ and $\leftrightarrow$ are defined in the usual way.

1.2.2 Semantics

The semantics is based on an accessibility relation that gives all the ideal alternatives of a world.

Definition 1 *A deontic relational model $M = (W, R, V)$ is a structure where:*

- *W is a nonempty set of worlds.*
- *R is a serial relation over W. That is, $R \subseteq W \times W$ and for all $w \in W$, there exist $v \in W$ such that Rwv.*
- *V is a valuation function that assigns a subset of W to each propositional letter p. Intuitively, $V(p)$ is the set of worlds in which p is true.*

A formula $\bigcirc\varphi$ is true at world w when φ is true in all the ideal alternatives of w.

Definition 2 *Given a relational model M, and a world s in M, we define the satisfaction relation $M, s \models A$ ("world s satisfies A in M") by induction on A using the clauses:*

- *$M, s \models p$ iff $s \in V(p)$.*
- *$M, s \models \neg\varphi$ iff not $M, s \models \varphi$.*
- *$M, s \models (\varphi \wedge \psi)$ iff $M, s \models \varphi$ and $M, s \models \psi$.*
- *$M, s \models \bigcirc\varphi$ iff for all t, if Rst then $M, t \models \varphi$.*
- *$M, s \models \Box\varphi$ iff for all $t \in W$, $M, t \models \varphi$.*

For a set Γ of formulas, we write $M, s \models \Gamma$ iff for all $\varphi \in \Gamma$, $M, s \models \varphi$. For a set Γ of formulas and a formula φ, we say that φ is a consequence of Γ (written as $\Gamma \models \varphi$) if for all models M and all worlds $s \in W$, if $M, s \models \Gamma$ then $M, s \models \varphi$.

1.2.3 Limitations

The following example is a variant of the scenario originally phrased by Chisholm in 1963. There is widespread agreement in the literature that, from the intuitive point of view, this set of sentences is consistent, and its members are logically independent of each other.

(A) It ought to be that Jones does not eat fast food for dinner.

(B) It ought to be that if Jones does not eat fast food for dinner, then he does not go to McDonald's.

(C) If Jones eats fast food for dinner, then he ought to go to McDonald's.

(D) Jones eats fast food for dinner.

Below are three ways to formalise this example. The first attempt is inconsistent. The second attempt is redundant due to $\bigcirc \neg f \models \bigcirc(f \to m)$. The third attempt is redundant due to $f \models \neg f \to \bigcirc \neg m$.

$$
\begin{array}{llllll}
(A_a) & \bigcirc \neg f & (A_b) & \bigcirc \neg f & (A_c) & \bigcirc \neg f \\
(B_a) & \bigcirc(\neg f \to \neg m) & (B_b) & \bigcirc(\neg f \to \neg m) & (B_c) & \neg f \to \bigcirc \neg m \\
(C_a) & f \to \bigcirc m & (C_b) & \bigcirc(f \to m) & (C_c) & f \to \bigcirc m \\
(D_a) & f & (D_b) & f & (D_c) & f
\end{array}
$$

However, it is not very hard to meet the two requirements of consistency and logical independence. The following representation is an example. It comes with apparently strong assumptions, because B_1/C_1 seem to say that my (conditional) obligations are necessary. For instance, Anderson argued that norms are contingent, because we make our rules; they are not (logical) necessities. However, we could also say that the $\square$ is just part of the definition of a strict conditional. Also, we could represent the first obligation as $\square \bigcirc \neg f$.

(A_1) $\bigcirc \neg f$
(B_1) $\square(\neg f \to \bigcirc \neg m)$
(C_1) $\square(f \to \bigcirc m)$
(D_1) $\neg f$

More seriously, a drawback of the SDL representation $A_1 - D_1$ is that it does not represent that ideally, the man does not eat fast food and does not go to McDonald's. In the ideal world, Jones goes to McDonald, yet he does not eat fast food. Moreover, there does not seem to be a similar solution for the following variant of the scenario. It is a variant of Forrester's paradox [Forrester, 1984], also known as the gentle murderer paradox: You should not kill, but if you kill, you should do it gently.

(AB) It ought to be that Jones does not eat fast food and does not go to McDonald's.

(C) If Jones eats fast food, then he ought to go to McDonald's.

(D) Jones eats fast food for dinner.

Moreover, SDL uses a binary classification of worlds into ideal/non-ideal, whereas many situations require a trade-off between violations. The challenge is to extend the semantics of SDL in order to overcome this limitation. For example, one can add distinct modal operators for primary and secondary obligations, where a secondary obligation is a kind of reparational obligation. From $A_2 - D_2$ we can derive only $\bigcirc_1 m \land \bigcirc_2 \neg m$, which is perfectly consistent.

(A_2) $\bigcirc_1 \neg f$
(B_2) $\bigcirc_1(\neg f \to \neg m)$
(C_2) $f \to \bigcirc_2 m$
(D_2) f

However, it may not always be easy to distinguish primary from secondary obligations, because it may depend on the context whether an obligation is primary or secondary. For example, if we leave out **A**, then **C** would be a primary obligation instead of a secondary one. Carmo and Jones [2002] therefore put as an additional requirement for a solution of the paradox that **B** and **C** are represented in the same way (as in A_1-D_1). Also, the distinction between $\bigcirc_1$ and $\bigcirc_2$ is insufficient for extensions of the paradox that seem to need also operators like $\bigcirc_3$, $\bigcirc_4$, etc, such as the following **E** and **F**.

(E) If Jones eats fast food but does not go to McDonald's, then he should go to Quick.

(F) If Jones eats fast food but does not go to McDonald's or to Quick, then he should
...

1.2.4 SDL proof system

The proof system of traditional deontic logic Λ_D is the smallest set of formulas of $\mathfrak{L}_D$ that contains all propositional tautologies, together with the following axioms:

K $\bigcirc(\varphi \to \psi) \to (\bigcirc\varphi \to \bigcirc\psi)$

D $\bigcirc\varphi \to P\varphi$

and is closed under *modus pones*, and *generalization* (that is, if $\varphi \in \Lambda_D$, then $\bigcirc\varphi \in \Lambda_D$).

For every $\varphi \in \mathfrak{L}_D$, if $\varphi \in \Lambda_D$ then we say φ is a theorem and write $\vdash \varphi$. For a set of formulas Γ and formula φ, we say φ is deducible form Γ (write $\Gamma \vdash \varphi$) if $\vdash \varphi$ or there are formulas $\psi_1, \ldots, \psi_n \in \Gamma$ such that $\vdash (\psi_1 \wedge \ldots \wedge \psi_n) \to \varphi$.

1.3 Dyadic deontic logic

Inspired by rational choice theory in the sixties, preference-based semantics for traditional deontic logic was used by, for example, Danielsson [1968], Hansson [1969], van Fraassen [1972], Lewis [1973], and Spohn [1975]. The obligations of Chisholm's paradox can be represented by a preference ordering, like:

$$\neg f \wedge \neg m > \neg f \wedge m > f \wedge m > f \wedge \neg m$$

Extensions like **E** and **F** can be incorporated by further refining the preference relation. The language is extended with dyadic operators $\bigcirc(p|q)$, which is true iff the preferred q worlds satisfy p. The class of logics is called Dyadic 'Standard' Deontic Logic or DSDL. The notation is inspired by the representation of conditional probability.

1.3.1 Language

Given a set Φ of propositional letters. The language of DSDL $\mathcal{L}_D$ is given by the following BNF:

$$\varphi := \bot \mid p \mid \neg\varphi \mid (\varphi \wedge \varphi) \mid \Box\varphi \mid \bigcirc(\varphi|\varphi)$$

The intended reading of $\Box\varphi$ is "necessarily φ", $\bigcirc(\varphi|\psi)$ is "It ought to be φ, given ψ". Moreover we use $P(\varphi|\psi)$, read as "φ is permitted, given ψ", as an abbreviation of $\neg \bigcirc (\neg\varphi|\psi)$, and $\Diamond\varphi$, read as "possibly φ", as an abbreviation of $\neg\Box\neg\varphi$.

Unconditional obligations are defined in terms of the conditional ones: $\bigcirc p = \bigcirc(p|\top)$, where $\top$ stands for any tautology.

1.3.2 Semantics

The semantics is based on an accessibility relation that gives all better alternatives of a world.

Definition 3 *A preference model $M = (W, \geq, V)$ is a structure where:*

- W *is a nonempty set of worlds.*
- $\geq$ *is a reflexive, transitive relation over W satisfying the following limitedness requirement: if $||\varphi|| \neq \emptyset$ then $\{x \in ||\varphi|| : (\forall y \in ||\varphi||) x \geq y\} \neq \emptyset$. Here $||\varphi|| = \{x \in W : M, x \vDash \varphi\}$.*
- V *is a standard propositional valuation such that for every propositional letter p, $V(p) \subseteq W$.*

Definition 4 *Formulas of $\mathcal{L}_D$ are interpreted in preference models.*

- $M, s \vDash p$ *iff* $s \in V(p)$.
- $M, s \vDash \neg\varphi$ *iff not* $M, s \vDash \varphi$.
- $M, s \vDash (\varphi \wedge \psi)$ *iff* $M, s \vDash \varphi$ *and* $M, s \vDash \psi$.
- $M, s \vDash \Box\varphi$ *iff* $\forall t \in W, M, t \vDash \varphi$.
- $M, s \vDash \bigcirc(\psi|\varphi)$ *iff* $\forall t(((M, t \vDash \varphi) \& \forall u(M, u \vDash \varphi) \Rightarrow t \geq u) \Rightarrow M, t \vDash \psi)$.

Intuitively, $\bigcirc(\psi|\varphi)$ holds whenever the best φ-worlds are ψ-worlds.

The Chisholm's scenario can be formalised in DSDL as follows:

$(A_3) \bigcirc \neg f$

$(B_3) \bigcirc (\neg m | \neg f)$

$(C_3) \bigcirc (m | f)$

$(D_3) f$

A challenge of both the multiple obligation solution using $\bigcirc_1$, $\bigcirc_2$, ... and the preference based semantics is to combine preference orderings, for example combining the Chisholm preferences with preferences originating from the Good Samaritan paradox:

(**AB'**) A man should not be robbed.
(**C'**) If he is robbed, he should be helped.
(**D'**) A man is robbed.

$$\neg r \wedge \neg h > r \wedge h > r \wedge \neg h$$

The main drawback of DSDL is that in a monotonic setting, we cannot detach the obligation $\bigcirc m$ from the four sentences. In fact, the preference based solution represents **A**, **B** and **C**, but has little to say about **D**. So the dyadic representation $A_3 - D_3$ highlights the dilemma between factual detachment (FD) and deontic detachment (DD). We cannot have both FD and DD, as we derive a dilemma $\bigcirc \neg m \wedge \bigcirc m$.

$$\frac{\bigcirc(m|f), f}{\bigcirc m} FD \qquad \frac{\bigcirc(\neg m|\neg f), \bigcirc \neg f}{\bigcirc \neg m} DD$$

1.3.3 DSDL proof system

The proof system of traditional deontic logic Λ_D, also referred as Aqvist's system G, is the smallest set of formulas of $\mathcal{L}_D$ that contains all propositional tautologies, the following axioms. The names of the labels are taken from Parent [2008]:

S5   S5-schemata for $\square$
COK $\bigcirc(B \to C|A) \to (\bigcirc(B|A) \to \bigcirc(C|A))$
Abs $\bigcirc(B|A) \to \square\bigcirc(B|A)$
CON $\square B \to \bigcirc(B|A)$
Ext $\square(A \leftrightarrow B) \to (\bigcirc(C|A) \leftrightarrow \bigcirc(C|B))$
Id $\bigcirc(A|A)$
C $\bigcirc(C|(A \wedge B)) \to \bigcirc((B \to C)|A)$
D* $\Diamond A \to (\bigcirc(B|A) \to P(B|A))$
S $(P(B|A) \wedge \bigcirc((B \to C)|A)) \to \bigcirc(C|(A \wedge B))$

and is closed under *modus ponens*, and *generalization* (that is, if $\varphi \in \Lambda_D$, then $\square\varphi \in \Lambda_D$).

1.3.4 The use of preferences in decision theory

Arrow's condition of rational choice theory says that if C are the best alternatives of A, and $B \cap C$ is nonempty, then $B \cap C$ are the best alternatives of $A \cap B$. This principle is reflected by the S axiom of DSDL:

$$(P(B|A) \wedge \bigcirc((B \to C)|A)) \to \bigcirc(C|(A \wedge B))$$

Moreover, we may represent a preference or comparative operator $\succ$ in the language, and define the dyadic operator in terms of the preference logic:

$$O(\psi \mid \phi) =_{def} (\phi \wedge \psi) \succ (\phi \wedge \neg\psi)$$

One may wonder whether the parallel between deontic reasoning and rational choice can be extended to utility theory, decision theory, game theory, planning, and so on. First, consider a typical example from Prakken and Sergot's Cottage Regulations [Prakken and Sergot, 1996]: there should be no fence, if there is a fence there should be a white fence, if there is a non-white fence, it should be black, if there is a fence which is neither white nor black, then .... This part of the cottage regulations is related to Forrester's paradox [Forrester, 1984]. However, note the following difference between Forrester's paradox and the cottage regulations. Once you kill someone, it can no longer be undone, whereas if you build a fence, you can still remove it. The associated preferences of the fence example are:

$$no\ fence > white\ fence > black\ fence > \ldots$$

If this represents a utility ordering over states, then we miss the representation of action [Pearl, 1993]. For example, it may be preferred that the sun shines, but we do not say that the sun should shine. As a simple model of action, one might distinguish controllable from uncontrollable propositions [Boutilier, 1994], and restrict obligations to controllable propositions. Moreover, we may consider actions instead of states: we should remove the fence if there is one, we may paint the fence white, we may paint it black, etc.

$$remove > paint\ white > paint\ black > \ldots$$

We may interpret this preference ordering as an ordering of expected utility of actions. Alternatively, the ordering may be generated by another decision rule, such as maximin or minimal regret. Once we are working with a decision theoretic semantics, we may represent probabilities explicitly, or model causality. For example, let n stand for not doing homework and g for getting a good grade for a test. Then we may have the following preference order, which does not reflect that doing homework causes good grades:

$$n \wedge g > \neg n \wedge g > n \wedge \neg g > \neg n \wedge \neg g$$

1.3.5 The use of goals in planning and agent theory

We may interpret $O\phi$ or $O(\phi \mid \psi)$ as goals for ϕ, rather than obligations. This naturally leads to the distinction between maintenance and achievement goals, and to extensions of the logic with beliefs and intentions. Belief-Desire-Intention or BDI logics have been developed as formalizations of BDI theory.

BDI theory is developed in the theory of mind and has been based on folk psychology. In planning, more efficient alternatives to classical planning have been developed, for example based on hierarchical or graph planning.

The following example is a more challenging variant of Chisholm's scenario using anankastic conditionals [Condoravdi and Lauer, 2016], also known as hypothetical imperatives. The four sentences can be given a consistent interpretation, when the second sentence is interpreted as a classical conditional, and the third sentence is interpreted as an anankastic conditional.

(a) It ought to be that you do not smoke.

(b) If you want to smoke, then you should not buy cigarettes.

(c) If you want to smoke, then you should buy cigarettes.

(d) You want to smoke.

1.4 Defeasible Deontic Logic: detachment and constraints

Defeasible deontic logics (DDLs) use techniques developed in non-monotonic logic, such as constrained inference [Horty, 1997; Makinson and van der Torre, 2001]. Using these techniques, we can derive $\bigcirc m$ from only the first two sentences **A** and **B**, but not from all four sentences **A-D**. Consequently, the inference relation is not monotonic. For example, we may read $O(\phi|\psi)$ as follows: if the facts are exactly ψ, then ϕ is obligatory. This implies that we no longer have that $O(\phi)$ is represented by $O(\phi|\top)$.

In a similar fashion, in deontic update semantics [van der Torre and Tan, 1998; van der Torre and Tan, 1999; van der Torre and Tan, 1999] facts are updates that restrict the domain of the model. They make a fact 'settled' in the sense that it will never change again even after future updates of the same sort. Van Benthem et al. [2014] use dynamic logic to phrase such a dynamic approach within standard modal logic including reduction axioms and standard model theory. They rehabilitate classical modal logic as a legitimate tool to do deontic logic, and position deontic logic within the growing dynamic logic literature.

A drawback of the use of non-monotonic techniques is that we often have that violated obligations are no longer derived. This is sometimes referred to as the drowning problem. For example, in the cottage regulations, if it is no longer derived that there should be no fence once there is a fence, then how do we represent that a violation has occurred?

A second related drawback of this solution is that it does not give the cue for action that the decision maker should change his mind. For example, once there is a fence, it does not represent the obligation to remove the fence.

A third drawback of this approach is that the use of non-monotonic logic techniques like constraints should also be used to represent exceptions, and it thus raises the challenge how to distinguish violations from exceptions. This is highlighted by Prakken and Sergot's cottage regulations [Prakken and Sergot, 1996].

(A") It ought to be that there is no fence around the cottage.

(BC") If there is a fence around the cottage, then it ought to be white.

(G") If the cottage is close to a cliff, then there ought to be a fence.

(D") There is a fence around the cottage, which is close to a cliff.

We say more about defeasible deontic logic in Section 8.

1.5 Alternative approaches

Carmo and Jones [2002] suggest that the representation of the facts is challenging, instead of the representation of the norms. In their approach, depending on the formalisation of the facts various obligations can be detached.

Another approach to Chisholm's paradox is to detach both obligations of the dilemma $O\neg m \wedge Om$, and represent them consistently using some kind of minimal deontic logic, for example using techniques from paraconsistent logic. From a practical reasoning point of view, a drawback of this approach is that a dilemma is not very useful as a moral cue for action. Moreover, intuitively it is not clear that the example presents a true dilemma. We say more about dilemmas in Section 9.

A recent representation of Chisholm's paradox [Parent and van der Torre, 2014; Parent and van der Torre, 2014; Sun and van der Torre, 2014] is to replace deontic detachment by so-called aggregative deontic detachment (ADD), and to derive from **A-D** the obligation $O(\neg f \wedge \neg m)$ and Om, but not $O\neg m$.

$$\frac{O(m|f), f}{Om} FD \qquad \frac{O(\neg m|\neg f), O\neg f}{O(\neg m \wedge \neg f)} ADD$$

A possible drawback of these approaches is that we can no longer accept the principle of weakening (also known as inheritance).

$$\frac{O(\neg m \wedge \neg f | \top)}{O(\neg m | \top)} W$$

2 Non-deterministic actions: ought-to-do vs ought-to-be

We now turn to three specific challenges on agency and obligation, discussed in much more detail by Horty [Horty, 2001; Broersen and van der Torre, 2003]. His textbook is a prime reference for the use of deontic logic for multiagent systems. The central challenge Horty addresses is whether ought-to-do can be reduced to ought-to-be. A particular problem is the granularity of actions in case of non-deterministic effects, like flipping a coin or throwing a dice.

Challenge 2 *How to define obligations to perform non-deterministic actions?*

At first sight, we may define an obligation to do an action as an obligation that such an action is done, and we can thus reuse SDL or DSDL to define obligations regarding non-deterministic actions. In other words, it may seem that we can reduce ought-to-do to ought-to-be. However, as we discuss in Section 2.2, such a reduction is problematic. To explain this challenge, we first introduce a logic to express non-deterministic actions, so-called See-To-It-That or STIT logic.

2.1 Horty's STIT logic

We give a very brief overview of the main concepts of Horty's STIT logic. For more details and motivation we refer to Horty's textbook on obligation and agency [Horty, 2001]. As illustrated in Figure 1, a STIT model is a tree where each moment is a partitioning of traces or histories, where the partitioning $Choice_\alpha^m$ represents the choices of the agent at that moment. Each alternative of the choice is called an action K_1^m, K_2^m, etc. With each history a utility value is associated, and the higher the utility value, the better the history.

Formulas are evaluated with respect to moment-history pairs. Some typical formulas of Horty's utilitarian STIT-formalism are A, FA, $[\alpha\ cstit : A]$, and OA for 'the

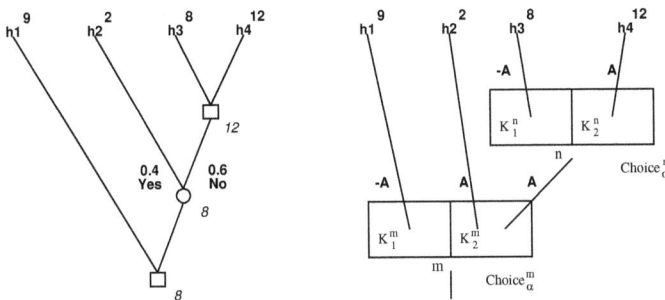

Figure 1. A decision tree and the corresponding utilitarian STIT-model

atomic proposition A', 'some time in the future A will be the case', 'agent α Sees To It That A', and 'it ought to be that A', respectively.

A is true at a moment-history pair m, h if and only if it is assigned the value true in the STIT-model, FA is true at a moment-history pair m, h if and only if there is some future moment on the history where A is true, $[\alpha\ cstit : A]$ is true at a moment history pair m, h if and only if A is true at all moment-history pairs through m that belong to the same *action* as m, h, and $\bigcirc A$ is true at a moment history pair m, h if and only if there is some history h' through m such that A is true at all pairs m, h'' for which the history h'' has a utility at least as high as h' ('moment determinate').

This semantic condition for the STIT-ought is a utilitarian generalisation of the standard deontic logic view (SDL) that 'it ought to be that A' means that A holds in all deontically optimal worlds.

On the STIT-model of Figure 1 we have $\mathcal{M}, m, h_3 \models A$ (directly from the valuation of atomic propositions on moment-history pairs), $\mathcal{M}, m, h_3 \models F\neg A$ (the proposition $\neg A$ is true later on, at moment n, on the history h_3 through m).

Also we have $\mathcal{M}, m, h_3 \models [\alpha\ cstit : A]$, because A holds for all histories through m that belong to the same action as h_3 (i.e. action K_2^m). Regarding ought-formulas we have: $\mathcal{M}, m, h_3 \models \bigcirc A$ and $\mathcal{M}, m, h_3 \models \bigcirc[\alpha\ cstit : A]$.

These two propositions are true for the same reason: the history h_4 through m has the highest utility (which means that we do not have to check conditions for histories with even higher utility) and satisfies both A and $[\alpha\ cstit : A]$ at m.

2.2 Gambling problem

Horty argues that ought-to-do statements are not just special kinds of ought-to-be statements. In particular, he claims that 'agent α ought to see to it that A' cannot be modelled by the formula $\bigcirc[\alpha\ cstit : A]$ ('it ought to be that agent α sees to it that A').

Justification of this claim is found in the 'gambling example'. This example concerns the situation where an agent faces the choice between gambling to double or lose five dollar (action K_1) and refraining from gambling (action K_2). This situation is sketched in the figure 2.

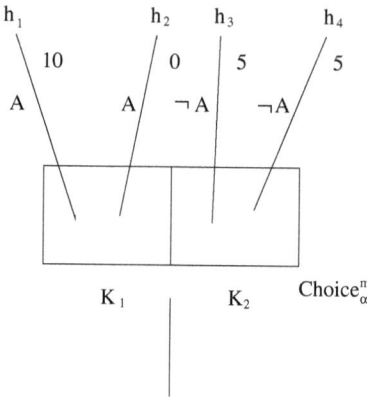

Figure 2. The gambling problem

The two histories that are possible by choosing action K_1 represent ending up with ten dollar by gaining five, and ending up with nothing by loosing all, respectively.

Also for action K_2, the game event causes histories to branch. But, for this action the two branches have equal utilities because the agent is not taking part in the game, thereby preserving his 5 dollar. Note this points to redundancy in the model representation: the two branches are logically indistinguishable, because there is no formula whose truth value would change by dropping one of them.

$\bigcirc[\alpha \; cstit : A]$ is true at m for history h_1 and for all histories with a higher utility (i.e. none), the formula $[\alpha \; cstit : A]$ is true. However, a reading of $\bigcirc[\alpha \; cstit : A]$ as 'agent α ought to perform action K_1' is counter-intuitive for this example. From the description of the gambling scenario it does not follow that one action is better than the other. In particular, without knowing the odds (the probabilities), we cannot say anything in favor of action K_1: by choosing it, we may either end up with more or with less utility than by doing K_2. The only thing one may observe is that action K_1 will be preferred by more adventurous agents. But that is not something the logic is concerned with.

This demonstrates that 'agent α ought to see to it that A' cannot be modelled by $\bigcirc[\alpha \; cstit : A]$. The cause of the mismatch can be explained as follows. Adapting and generalising the main idea behind SDL to the STIT-context, ought-to-be statements concern truth in a set of optimal histories ('worlds' in SDL). Optimality is directly determined by the utilities associated with individual histories. If ought-to-be is about optimal histories, then ought-to-do is about optimal actions. But, since actions are assumed to be non-deterministic, actions do not correspond with individual histories, but with *sets* of histories. This means that to apply the idea of optimality to the definition of ought-to-do operators, we have to generalise the notion of optimality such that it applies to *sets* of histories, namely, the sets that make up non-deterministic actions. More specifically, we have to *lift* the ordering of histories to an ordering of actions.

The ordering of actions suggested by Horty is very simple: an action is strictly better than another action if all of its histories are at least as good as any history of the other action, and not the other way around.

Having lifted the ranking of histories to a ranking of actions, the utilitarian ought conditions can now be applied to actions. Thus, Horty defines the new operator 'agent α ought to see to it that A (in formula form: $\odot[\alpha\ cstit : A]$)' as the condition that for all actions not resulting in A there is a higher ranked action that does result in A, plus that all actions that are ranked even higher also result in A. This 'solves' the gambling problem. We do not have $\odot[\alpha\ cstit : A]$ or $\odot[\alpha\ cstit : \neg A]$ in the gambling scenario, because in the ordering of actions, K_1 is not better or worse than K_2.

3 Moral luck and the driving example

The gambling problem may be seen as a kind of moral luck: whether we obtain the utility of 10 or 0 is not due to our actions, but due to luck. The issue of moral luck is even more interesting in the case of multiple agents, where it depends on the actions of other agents whether you get utility 10 or 0.

Challenge 3 *How to deal with moral luck in normative reasoning?*

The driving example [Horty, 2001, p.119-121] is used to illustrate the difference between so-called dominance act utilitarianism and orthodox perspective on the agent's ought. Roughly, dominance act utilitarianism is that α ought to see to it that A just in case the truth of A is guaranteed by each of the optimal actions available to the agent—formally, that $\odot[\alpha\ cstit : A]$ should be settled true at a moment m just in case $K \subseteq |A|_m$ for each $K \in Optimal_\alpha^m$. When we adopt the orthodox perspective, the truth or falsity of ought statements can vary from index to index. The orthodox perspective is that α should see to it that A at a certain index just in case the truth of A is guaranteed by each of the actions available to the agent that are optimal given the circumstances in which he finds himself at this index.

> "In this example, two drivers are travelling toward each other on a one-lane road, with no time to stop or communicate, and with a single moment at which each must choose, independently, either to swerve or to continue along the road. There is only one direction in which the drivers might swerve, and so a collision can be avoided only if one of the drivers swerves and the other does not; if neither swerves, or both do, a collision occurs. This example is depicted in Figure 3, where α and β represent the two drivers, K_1 and K_2 represent the actions available to α of swerving or staying on the road, K_3 and K_4 likewise represent the swerving or continuing actions available to β, and m represents the moment at which α and β must make their choice. The histories h_1 and h_3 are the ideal outcomes, resulting when one driver swerves and the other one does not; collision is avoided. The histories h_2 and h_4, resulting either when both drivers swerve or both continue along the road, represent non-ideal outcomes; collision occurs. The statement A, true at h_1 and h_2, expresses the proposition that α swerves." [Horty, 2001, p.119]

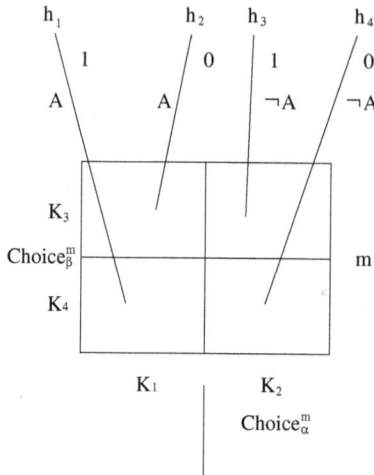

Figure 3. The driving example and moral luck

From the dominance point of view both actions available to α are classified as optimal, written as $Optimal_\alpha^m = \{K_1, K_2\}$. One of the optimal actions available to α guarantees the truth of A and the other guarantees the truth of $\neg A$. Consequently $M, m \not\models \odot[\alpha \; cstit : A]$ and $M, m \not\models \odot[\alpha \; cstit : \neg A]$. From the orthodox point of view, we have $M, m, h_1 \models \odot[\alpha \; cstit : A]$ and $M, m, h_2 \models \odot[\alpha \; cstit : \neg A]$. What α ought to do at an index depends on what β does.

Horty concludes that from the standpoint of intuitive adequacy, the contrast between the orthodox and dominance deontic operators provides us with another perspective on the issue of moral luck, the role of external factors in our moral evaluations [Horty, 2001, p.121]. The orthodox ought is the one who after the actual event looks back to it. For example, when there has been a collision then α might say—perhaps while recovering from the hospital bed—that he ought to have swerved. The dominance ought is looking forward. Though the agent may legitimately regret his choice, it is not one for which he can be blamed, since either choice, at the time, could have led to a collision.

4 Procrastination: actualism vs possibilism

Practical reasoning is intimately related to reasoning about time. For example, if you are obliged and willing to visit a relative, but you always procrastinate this visit, then we may conclude that you violated this obligation. In other words, each obligation to do an action should come with a deadline [Broersen et al., 2004; Boella et al., 2008].

Challenge 4 *How to deal with procrastination in normative reasoning?*

The example of Procrastinate's choices [Horty, 2001, p. 162] illustrates the notion of strategic oughts. A strategy is a generalized action involving a series of actions. Like an action, a strategy determines a subset of histories. The set of admissible histories for a strategy σ is denoted $Adh(\sigma)$.

A crucial new concept here is the concept of a *Field*, which is basically a subtree of the STIT model which denotes that the agent's reasoning is limited to this range. A strategic ought is defined analogous to dominance act utilitarianism, in which action is replaced by strategy in a field. α ought to see to it that A just in case the truth of A is guaranteed by each of the optimal strategies available to the agent in the field—formally, that $\odot[\alpha\ cstit : A]$ should be settled true at a moment m just in case $Adh(\sigma) \subseteq |A|_m$ for each $\sigma \in Optimal_\alpha^m$. Horty observes some complications, and that a 'proper treatment of these issues might well push us beyond the borders of the current representational formalism' [p.150].

Horty also uses the example of Procrastinate's choices to distinguish between actualism and possibilism, for which he uses the strategic oughts, and in particular the notion of a field. Roughly, actualism is the view that an agent's current actions are to be evaluated against the background of the actions he is actually going to perform in the future. Possibilism is the view that an agent's current actions are to be evaluated against the background of the actions that he might perform in the future, the available future actions.

The example is due to Jackson and Pargetter [1986].

> "Professor Procrastinate receives an invitation to review a book. He is the best person to do the review, has the time, and so on. The best thing that can happen is that he says yes, and then writes the review when the book arrives. However, suppose it is further the case that were to say yes, he would not in fact get around to writing the review. Not because of incapacity or outside interference or anything like that, but because he would keep on putting the task off. (This has been known to happen.) This although the best thing that can happen is for Procrastinate to say yes and then write, and he *can* do exactly this, what *would* happen in fact were he to say yes is that he would not write the review. Moreover, we may suppose, this latter is the worst thing which may happen.
>
> [...]
>
> According to possibilism, the fact that Procrastinate would not write the review were he to say yes is irrelevant. What matters is simply what is possible for Procrastinate. He can say yes and then write; that is best; that requires *inter alia* that he says yes; therefore, he ought to say yes. According to actualism, the fact that Procrastinate would not actually write the review were he to say yes is crucial. It means that to say yes would be in fact to realize the worst. Therefore, Procrastinate ought to say no."

Horty represents the example by the STIT model in Figure 4. Here, m_1 is the moment at which Procrastinate, represented as the agent α, chooses whether or not to accept the invitation: K_1 represents the choice of accepting, K_2 the choice of declining. If Procrastinate accepts the invitation, he then faces at m_2 the later choice of writing the review or not: K_3 represents the choice of writing the review, K_4 another choice that results in the review not being written. For convenience, Horty also supposes that at m_3 Procrastinate has a similar choice whether or not to write the review:

K_5 represents the choice of writing, K_6 the choice of not writing. The history h_1, in which Procrastinate accepts the invitation and then writes the review, carries the greatest value of 10; the history h_2, in which Procrastinate accepts the invitation and then neglects the task, the least value of 0; the history h_4, in which he declines, such that a less competent authority reviews the book, carries an intermediate value of 5; and the peculiar h_3, in which he declines the invitation but then reviews the book anyway, carries a slightly lower value of 4, since he wastes his time, apart from doing no one else any good. The statement A represents the proposition that he accepts the invitation; the statement B represents the proposition that Procrastinate will write the review.

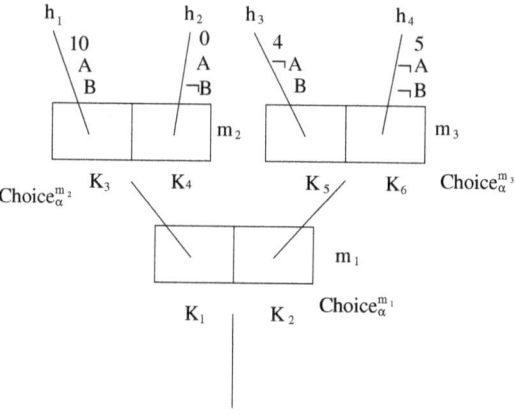

Figure 4. Procrastinate's choices

Now, in the possibilist interpretation, $M = \{m_1, m_2, m_3\}$ is the background field. In this interpretation, Procrastinate ought to accept the invitation because this is the action determined by the best available strategy—first accepting the invitation, and then writing the review. Formally, $Optimal_\alpha^M = \{\sigma_6\}$ with $\sigma_6 = \{\langle m_1, K_1 \rangle, \langle m_2, K_3 \rangle\}$. Since $Adh(\sigma_6) \subseteq |A|_m$, the strategic ought statement $\odot[\alpha \ cstit : A]$ is settled true in the field M. In the actualist interpretation, the background field may be narrowed to the set $M' = \{m_1\}$, which shifts from the strategic to the momentary theory of oughts. In this case, we have $\odot[\alpha \ cstit : A]$ is settled false. It is as if we choose to view Procrastinate as gambling on his own later choice in deciding whether to accept the invitation. However, from this perspective, this should not be viewed as a gamble; an important background assumption—and the reason that he should decline the invitation—is that he will not, in fact, write the review.

5 Jørgensen's dilemma and the problem of detachment

A philosophical problem that has had a major impact in the development of deontic logic is Jørgensen's dilemma. In a nutshell, given that norms cannot be true or false, the dilemma implies that deontic logic cannot be based on traditional truth functional semantics. In particular, building on a tradition of Alchourrón and Bulygin in the seventies, Makinson [1999] argues that norms need to be represented explicitly. SDL, DSDL and STIT logic represent logical relations between deontic operators, but they do not explicitly represent a distinction between norms and obligations. The explicit representation of norms is the basis of alternative semantics, that breaks with the idea of traditional semantics that norms and obligations have truth values, and most importantly, that discards the main technical and conceptual tool of traditional semantics, namely possible worlds. As an example, in this section we illustrate this alternative semantics using input/output logic.

5.1 Jørgensen's dilemma

While normative concepts are the subject of deontic logic, it is quite difficult to see how there can be a logic of such concepts at all. Norms like individual imperatives, promises, legal statutes, and moral standards are usually not viewed as being true or false. E.g. consider imperative or permissive expressions such as "John, leave the room!" and "Mary, you may enter now": they do not describe, but demand or allow a behavior on the part of John and Mary. Being non-descriptive, they cannot meaningfully be termed true or false. Lacking truth values, these expressions cannot—in the usual sense—be premise or conclusion in an inference, be termed consistent or contradictory, or be compounded by truth-functional operators. Hence, though there certainly exists a logical study of normative expressions and concepts, it seems there cannot be a logic of norms: this is Jørgensen's dilemma [Jørgensen, 1938; Makinson, 1999].

Though norms are neither true nor false, one may state that *according to the norms*, something ought to be done or is permitted: the statements "John ought to leave the room" and"Mary is permitted to enter" are then true or false descriptions of the normative situation. Such statements are sometimes called normative statements, as distinguished from norms. To express principles such as the principle of conjunction: $O(p \wedge q) \leftrightarrow (Op \wedge Oq)$, with Boolean operators having truth-functional meaning at all places, deontic logic has resorted to interpreting its formulas Op, Fp, Pp not as representing norms, but as representing such normative statements. A possible logic of normative statements may then reflect logical properties of underlying norms—thus logic may have a "wider reach than truth", as Von Wright [1957] famously stated.

Since the truth of normative statements depends on a normative situation, in the way in which the truth of the statement "John ought to leave the room" depends on whether some authority ordered John to leave the room or not, it seems that norms must be represented in a logical semantics that models such truth or falsity. However, semantics used to model the truth or falsity of normative statements mostly fail to include norms. Standard deontic semantics evaluates deontic formulas with respect to sets of worlds, in which some are ideal or better than others—Ox is then defined to be true if x is true in all ideal or the best reachable worlds. Alternatively, norms, not ideality, should provide the basis on which normative statements are evaluated. Thus the following question arises, asked by D. Makinson [1999]:

Challenge 5 *How can deontic logic be reconstructed in accord with the philosophical position that norms are neither true nor false?*

In the older literature on deontic logic there has been a veritable 'imperativist tradition' of authors that have, deviating from the standard approach, in one way or other, tried to give truth definitions for deontic operators with respect to given sets of norms. Cf. among others S. Kanger [1957], E. Stenius [1963], T. J. Smiley [1963], Z. Ziemba [1971], B. van Fraassen [1973], Alchourrón and Bulygin [1981] and I. Niiniluoto [1986]. The reconstruction of deontic logic as logic about imperatives has been the project of Jörg Hansen beginning with [Hansen, 2001]. Input/output logic [Makinson and van der Torre, 2000] is another reconstruction of a logic of norms in accord with the philosophical position that norms direct rather than describe, and are neither true nor false. We explain it in more detail in the next section below.

5.2 Input/output logic

To illustrate a possible answer to the dilemma, we use Makinson and van der Torre's input/output logic [2000; 2001; 2003a], and we therefore assume familiarity with this approach (cf. [Makinson and van der Torre, 2003b] for an introduction). Input/output logic takes a very general view at the process used to obtain conclusions (more generally: outputs) from given sets of premises (more generally: inputs). While the transformation may work in the usual way, as an 'inference motor' to provide logical conclusions from a given set of premises, it might also be put to other, perhaps non-logical uses. Logic then acts as a kind of secretarial assistant, helping to prepare the inputs before they go into the machine, unpacking outputs as they emerge, and, less obviously, coordinating the two. The process as a whole is one of logically assisted transformation, and is an inference only when the central transformation is so. This is the general perspective underlying input/output logic. It is one of logic at work rather than logic in isolation; not some kind of non-classical logic, but a way of using the classical one.

Suppose that we have a set G (meant to be a set of conditional norms), and a set A of formulas (meant to be a set of given facts). The problem is then: how may we reasonably define the set of propositions x making up the output of G given A, which we write $out(G, A)$? In particular, if we view the output as a collection of descriptions of states of affairs that ought to obtain given the norms G and the facts A, what is a reasonable output operation that enables us to define a deontic O-operator that returns the normative statements that are true given the norms and the facts—the normative consequences given the situation? One such definition is the following:

$$G, A \models Ox \quad \text{iff} \quad x \in out(G, A)$$

So Ox is true iff the output of G under A includes x. Note that this is rather a description of how we think such an output should or might be interpreted, whereas 'pure' input/output logic does not discuss such definitions. For a simple case, let G include a conditional norm that states that if a is the case, x should obtain (we write $(a, x) \in G$). An unconditional norm that commits the agent to realizing x is represented by a conditional norm $(\top, x)$, where $\top$ means an arbitrary tautology. If a can be inferred from A, i.e. if $a \in Cn(A)$, and z is logically implied by x, then z should be among the normative consequences of G given A. An operation that does this is simple-minded output out_1:

$$out_1(G, A) = Cn(G(Cn(A)))$$

where $G(B) = \{y \mid (b, y) \in G \text{ and } b \in B\}$. So in the given example, Oz is true given $(a, x) \in G$, $a \in Cn(A)$ and $z \in Cn(x)$.

Simple-minded output may, however, not be strong enough. Sometimes, legal argumentation supports reasoning by cases: if there is a conditional norm (a, x) that states that an agent must bring about x if a is the case, and a norm (b, x) that states that the same agent must also bring about x if b is the case, and $a \vee b$ is implied by the facts, then we should be able to conclude that the agent must bring about x. An operation that supports such reasoning is basic output out_2:

$$out_2(G, A) = \cap\{Cn(G(V)) \mid v(A) = 1\}$$

where v ranges over Boolean valuations plus the function that puts $v(b) = 1$ for all formulae b, and $V = \{b \mid v(b) = 1\}$. It can easily be seen that now Ox is true given $\{(a, x), (b, x)\} \subseteq G$ and $a \vee b \in Cn(A)$.

This definition of out_2 may give rise to a mere feeling of merely technical adequacy, because of its recourse to intersection and valuations, neither of which quite corresponds to our natural course of reasoning in such situations. However, this semantics makes explicit what is present but implicit in the use of possible worlds in conditional logics: if you want to reason by cases in the logic, you need to represent the cases explicitly in the semantics.

It is quite controversial whether reasoning with conditional norms should support 'normative' or 'deontic detachment', i.e. whether it should be accepted that if one norm (a, x) commands an agent to make x true in conditions a, and another norm (x, y) directs the agent to make y true given x is true, then the agent has an obligation to make y true if a is factually true. Some would argue that as long as the agent has not in fact realized x, the norm to bring about y is not 'triggered'; others would maintain that obviously the agent has an obligation to make $x \wedge y$ true given that a is true. Moreover, the inference can be restricted to cases where the agent ought to make x true instantly rather than eventually, see [Makinson, 1999; Boella et al., 2008] If such detachment is viewed as permissible for normative reasoning, then one might use reusable output out_3 that supports such reasoning:

$$out_3(G, A) = \cap\{Cn(G(B)) \mid A \subseteq B = Cn(B) \supseteq G(B)\}$$

An operation that combines reasoning by cases with deontic detachment is then reusable basic output out_4:

$$out_4(G, A) = \cap\{Cn(G(V)) : v(A) = 1 \text{ and } G(V) \subseteq V\}$$

It may turn out that further modifications of the output operation are required in order to produce reasonable results for normative reasoning. Also, the proposal to employ input/output logic to reconstruct deontic logic may lead to competing solutions, depending on what philosophical views as to what transformations should be acceptable one subscribes to. All this is what input/output logic is about. However, it should be noted that input/output logic succeeds in representing norms as entities that are neither true nor false, while still permitting normative reasoning about such entities.

5.3 Contrary to duty reasoning reconsidered

In the input/output logic framework, the strategy for eliminating excess output is to cut back the set of generators to just below the threshold of yielding excess. To do that, input/output logic looks at the maximal non-excessive subsets, as described by the following definition:

Definition (Maxfamilies) *Let G be a set of conditional norms and A and C two sets of propositional formulas. Then maxfamily(G, A, C) is the set of maximal subsets $H \subseteq G$ such that $out(H, A) \cup C$ is consistent.*

For a possible solution to Chisholm's paradox, consider the following output operation $out^\cap$:

$$out^\cap(G, A) \;=\; \bigcap\{out(H, A) \mid H \in \textit{maxfamily}(G, A, A)\}$$

So an output x is in $out^\cap(G, A)$ if it is in output $out(H, A)$ of all maximal norm subsets $H \subseteq G$ such that $out(H, A)$ is consistent with the input A. Let a deontic O-operator be defined in the usual way with regard to this output:

$$G, A \models O^\cap x \quad \text{iff} \quad x \in out^\cap(G, A)$$

Furthermore, tentatively, and only for the task of shedding light on Chisholm's paradox, let us define an entailment relation between norms as follows:

Definition (Entailment relation) *Let G be a set of conditional norms, and (a, x) be a norm whose addition to G is under consideration. Then (a, x) is entailed by G iff for all sets of propositions A, $out^\cap(G \cup \{(a, x)\}, A) = out^\cap(G, A)$.*

So a (considered) norm is entailed by a (given) set of norms if its addition to this set would not make a difference for any set of facts A. Finally, let us use the following cautious definition of 'coherence from the start' (also called 'minimal coherence' or 'coherence per se'), see Section 7:

A set of norms G is 'coherent from the start' iff $\bot \notin out(G, \top)$.

Now consider a 'Chisholm norm set' $G = \{(\top, x), (x, z), (\neg x, \neg z),\}$, where $(\top, x)$ means the norm that the man must go to the assistance of his neighbors, (x, z) means the norm that it ought to be that if he goes he ought to tell them he is coming, and $(\neg x, \neg z)$ means the norm that if he does not go he ought not to tell them he is coming. It can be easily verified that the norm set G is 'coherent from the start' for all standard output operations out_n, since for these either $out(G, \top) = Cn(\{x\})$ or $out(G, \top) = Cn(\{x, z\})$, and both sets $\{x\}$ and $\{x, z\}$ are consistent. Furthermore, it should be noted that all norms in the norm set G are independent from each other, in the sense that no norm $(a, x) \in G$ is entailed by $G \setminus \{(a, x)\}$ for any standard output operation $out_n^{(+)}$: for $(\top, x)$ we have $x \in out^\cap(G, \top)$ but $x \notin out^\cap(G \setminus \{(\top, x)\}, \top)$, for (x, z) we have $z \in out^\cap(G, x)$ but $z \notin out^\cap(G \setminus \{(x, z)\}, x)$, and for $(\neg x, \neg z)$ we have $\neg z \in out^\cap(G, \neg x)$ but $\neg z \notin out^\cap(G \setminus \{(\neg x, \neg z)\}, \top)$. Finally consider the 'Chisholm fact set' $A = \{\neg x\}$, that includes as an assumed unalterable fact the proposition $\neg x$, that the man will not go to the assistance of his neighbors: we have $\textit{maxfamily}(G, A, A) = \{G \setminus \{(\top, x)\}\} = \{\{(x, z), (\neg x, \neg z),\}\}$ and either $out(G \setminus \{(\top, x)\}, A) = Cn(\{\neg z\})$ or $out(G \setminus \{(\top, x)\}, A) = Cn(\{\neg x, \neg z\})$ for all standard output operations $out_n^{(+)}$, and so $O^\cap \neg z$ is true given the norm and fact sets G and A, i.e. the man must not tell his neighbors he is coming. Thus:

$$G, A \models O^\cap \neg z$$

6 Multiagent detachment

In Section 6.1 we introduce normative multiagent systems using agents and controllable propositions, and we introduce a challenge for detachment for multiagent systems. In Section 6.2 we give a solution for the challenge in these formalisms.

6.1 Challenge for multiagent detachment

Olde Loohuis [2009] argues that the assumption that other agents comply with their norms reflects that agents live in a responsible world. However, Makinson [1999] observes that if all we know is that "John owes Peter $1000" and "if John pays Peter $1000, then Peter is obliged to give John a receipt," then we cannot detach that Peter has to give John a receipt unconditionally based on the assumption that John will pay Peter the money.

We assume that the normative system is known to all agents, and in this section we assume that it does not change over time, and that each norm is directed to one agent only. The agents reason about the consequences of the normative system, that is, which obligations and permissions can be detached from it. With an explicit normative system, the agents should act such that they do not violate norms. Moreover, in this section we assume that each (instance of a) norm specifies the behavior of a single individual agent. For example, a norm may say that an agent should drive to the right hand side of the street, but we do not consider group norms saying that agents should live together in harmony.

We do not assume a full action theory as in STIT logic, but we assume a minimal action theory: the set of propositions is partitioned into parameters (uncontrollable propositions) and decision variables (controllable propositions). Boutilier [1994] traces this idea back to discrete event systems, see also Cholvy and Garion [2001]. It is an abstract and general approach, since we can instantiate the propositions with action descriptions like do(action) or done(action). Note that this generality is in line with game theory, which abstracts away sequential decisions in extensive games by representing conditional plans as strategic games. Boutilier observes that the theory can be extended to a full fledged action theory by, for example, introducing a causal theory. By convention, the proposition letters p, p_1, etc are parameters, a, a_1, ..., are decision variables for agent 1, b, b_1, ..., are decision variables for agent 2, etc. Norms are written as pairs of propositional formulas, where (p_1, p_2) is read as "if p_1 is the case, then p_2 ought to be the case," (a_1, a_2) is read as "if agent 1 does a_1, then he has to do a_2," and so on. We restrict the propositional language to conjunctions of literals (propositional atoms or their negations), so we do not consider disjunctions or material implications.

Definition 5 (Normative multi agent system, individual norms) *A normative multiagent system is a tuple NMAS= $\langle A, P, c, N \rangle$ where A is a set of agents, P is a set* of atomic propositions, $c: P \to A$ is a partial function which maps the propositions to the agents controlling them, and N is a set of pairs of conjunctions of literals built of P, such that if $(\phi, \psi) \in N$, then all propositional atoms in ψ are controlled by a single agent.

Our action theory may be seen as a simple kind of STIT theory, in the sense that an obligation for a proposition p controlled by agent α may be read as: "the agent α ought to see to it that p is the case." Though this abstracts away from the temporal issues of STIT operators, it still has the characteristic property of STIT logics that actions have a higher granularity than worlds.

Makinson [1999] illustrates the intricacies of temporal reasoning with norms, obligations and agents by discussing the iteration of detachment, in the sense that from the two conditional norms "if ϕ, then obligatory ψ" and "if ψ, then obligatory χ" together with the fact ϕ, we can derive not only that ψ is obligatory, but also that χ is obligatory. Makinson's challenge is how to detach obligations based on the principle that agents cannot assume that other agents comply with their norms, but they assume that they themselves comply with their norms. In other words, deontic detachment holds only for the single agent a-temporal case.

First, Makinson argues that iteration of detachment often appears to be appropriate. He gives the following example, based on instructions to authors preparing manuscripts.

Example 1 (Manuscript [Makinson, 1999]) *Let the set of norms be* $(25x15, 12) =$ *"if $25x15$, then obligatory 12" and* $(12, refs10) =$ *"if 12, then obligatory $refs10$", where $25x15$ is "The text area is 25 by 15 cm", 12 is "The font size for the main text is 12 points", and $refs10$ is "The font size for the list of references is 10 points". Moreover, consider a single agent controlling the three variables. If the facts contain $25x15$, then we want to detach not only that it is obligatory that 12, but also that it is obligatory that $refs10$.*

Second, he argues that iteration of detachment sometimes appears to be inappropriate by discussing the following example, which he attributes to Sven Ove Hansson.

Example 2 (Receipt [Makinson, 1999]) *Let instances of the norms be*

$(owe_{jp}, pay_{jp}) =$ *"if owe_{jp}, then obligatory pay_{jp}" and*
$(pay_{jp}, receipt_{pj}) =$ *"if pay_{jp}, then obligatory $receipt_{pj}$"*

where owe_{jp} is "John owes Peter $1000", pay_{jp} is "John pays Peter $1000", and $receipt_{pj}$ is "Peter gives John a receipt for $1000". Moreover, assume that the first variable is not controlled by an agent, the second is controlled by John, and the third is controlled by Peter. Intuitively Makinson would say that in the circumstance that John owes Peter $1000, considered alone, Peter has no obligation to write any receipt. That obligation arises only when John fulfils his obligation.

Makinson observes that there appear to be two principal sources of difficulty here. One concerns the passage of time, and the other concerns bearers of the obligations. Sven Ove Hansson's example above involves both of these factors.

"We recall that our representation of norms abstracts entirely from the question of time. Evidently, this is a major limitation of scope, and leads to discrepancies with real-life examples, where there is almost always an implicit time element. This may be transitive, as when we say "when b holds then a should eventually hold", or "... should simultaneously hold". But it may be intransitive, as when we say "when b holds then a should hold within a short time" or "... should be treated as a matter of first priority to bring about". Clearly, iteration of detachment can be legitimate only when the implicit time element is either nil or transitive. Our representation also abstracts from the question of bearer, that is, who (if anyone) is assigned responsibility for carrying out what is required. This too can lead to discrepancies. Iteration of detachment becomes questionable as soon as some promulgations have different bearers from others, or some are impersonal (i.e. without bearer) while others are not. Only when the locus of responsibility is held constant can such an operation take place." [Makinson, 1999]

Challenge 6 *How to define detachment for multiple agents?*

Broersen and van der Torre [2007] consider the temporal aspects of the example. In this section we consider the actions of the agents. The following example extends the discussion of the example to aggregative deontic detachment.

Example 3 (continued) *Consider again* (owe_{jp}, pay_{jp}) *and* $(pay_{jp}, receipt_{pj})$, *where the first variable is not controlled by an agent, the second is controlled by John, and the third is controlled by Peter. In the circumstance that John owes Peter $1000, considered alone, do we want to derive the obligation for* $pay_{jp} \wedge receipt_{pj}$, *that is, the obligation that "John pays Peter $1000", and "Peter gives John a receipt for $1000"? In many systems the obligation for* $pay_{jp} \wedge receipt_{pj}$ *implies the obligation for* $receipt_{pj}$, *such that the answer will be negative. However, if the obligation for* $pay_{jp} \wedge receipt_{pj}$ *does not imply the obligation for* $receipt_{pj}$, *then maybe the obligation for* $pay_{jp} \wedge receipt_{pj}$ *is not as problematic as the obligation for* $receipt_{pj}$. *Moreover, the obligation for* $pay_{jp} \wedge receipt_{pj}$ *is a compact representation of the fact that ideally, the exchange of money and receipt takes place.*

6.2 Deontic detachment for agents

As the iterative approaches seem most natural to most people, we define deontic detachment of agents using these iterative approaches. The question thus arises whether we consider sequential or iterated detachment. The following example illustrates this question, not discussed by Makinson [1999].

Example 4 $N = \{(p, a), (a, b_1), (a \wedge b_1, b_2)\}$ *where p is a parameter, a is a decision variable of agent 1, and b_1 and b_2 are decision variables of agent 2. In context* $F = \{p, a\}$, *do we want to detach only b_1, or both b_1 and b_2? If we can detach b_2, then this implies that despite the fact that a and b_1 are decision variable from distinct agents we can use* $(a \wedge b_1, b_2)$ *to detach b_2.*

In the above example, we believe that b_2 should be derivable, because only b_1 is reused when b_2 is detached, and both b_1 and b_2 are decision variables of the same agent. In other words, when considering the norm $(a \wedge b_1, b_2)$ to detach b_2, we should not consider the norm and reject it because there is a variable in the input which refers to another agent, but we should consider it since we have $a \in F$ as a fact, and b_1 already in the output, we can derive b_2 too.

If b_2 should not be derivable, then we could simply restrict the set of norms that we select from N to satisfy the syntactic criterion, just like we selected the set of norms N_0. However, if b_2 should be derivable, then we have to define detachment procedures for each agent, and combine them afterwards. This is formalized in the following detachment procedure for agents.

Definition 6 (Iterative detachment for agents.) *Agent $a \in A$ controls a propositional formula ϕ, written as $c(\phi) = a$, if and only if for all atoms $x \in \phi$ we have $c(x) = a$.*

$$N_0^a = \{(\phi, \psi) \in N \mid F \cup \{\phi\} \not\models \neg\psi, c(\psi) = a\}$$

$E_0^{ia} = \emptyset$. *For $n = 1$ to ∞ do $E_{n+1}^{ia} = \{\psi \mid (\phi, \psi) \in N_0^a, F \cup E_n^{ia} \models \phi\}$ if consistent with F, E_n^{ia} otherwise. $out^{ia}(N, F, a) = Cn(\cup E_i^{ia})$, and $out^{ia}(N, F) = \cup_{a \in A} out^{ia}(N, F, a)$.*

We leave the logical analysis of this ans related approaches to future work.

7 Coherence

Consider norms which at the same time require you to leave the room and not to leave the room. In such cases, we are inclined to say that there is something wrong with the normative system. This intuition is captured by the SDL axiom $D : \neg(Ox \wedge O\neg x)$ that states that there cannot be co-existing obligations to bring about x and to bring about $\neg x$, or, using the standard cross-definitions of the deontic modalities: x cannot be both, obligatory and forbidden, or: if x is obligatory then it is also permitted. However, what does this tell us about the normative system?

Since norms do not bear truth values, we cannot, in any usual sense, say that such a set of norms is inconsistent. All we can consider is the consistency of the output of a set of norms. We like to use the term *coherence* with respect to a set of norms with consistent output. For a start, consider the notion of minimal coherence in Section 5.3:

(0)   A set of norms G is minimal coherent   iff $\bot \notin out(G, \emptyset)$.

This is clearly very weak, as for example the norms $(a, x), (a, \neg x)$ would be coherent. Alternatively, we might try to define coherence as follows:

(1)   A set of norms G is coherent   iff $\bot \notin out(G, A)$.

However, this definition seems not quite sufficient: one might argue that one should be able to determine whether a set of norms G is coherent or not regardless of what arbitrary facts A might be assumed. A better definition would be $(1a)$:

(1a)   A set of norms G is coherent   iff   there exists a set of formulas A such that $\bot \notin out(G, A)$.

For $(1a)$ it suffices that there exists a situation in which the norms can be, or could have been, fulfilled. However, consider the set of norms $G = \{(a, x), (a, \neg x)\}$ that requires both x to be realized and $\neg x$ to be realized in conditions a: it is immediate that e.g. for all output operations out_n, we have $\bot \notin out_n(G, \neg a)$: no conflicting demands arise when $\neg a$ is factually assumed. Yet something seems wrong with a normative system that explicitly considers a fact a only to tie to it conflicting normative consequences. The dual of $(1a)$ would be

(1b) A set of norms G is coherent   iff   for all sets of formulas A, $\bot \notin out(G, A)$.

Now a set G with $G = \{(a, x), (a, \neg x)\}$ would no longer be termed coherent. $(1b)$ makes the claim that for no situation A, two norms $(a, x), (b, y)$ would ever come into conflict, which might seem too strong. We may wish to restrict A to sets of facts that are consistent, or that are not in violation of the norms. The question is, basically, how to distinguish situations that the norm-givers should have taken care of, from those that describe misfortune or otherwise unhappy circumstances. A weaker claim than $(1b)$ would be $(1c)$:

(1c) A set of norms G is coherent   iff   for all a with $(a, x) \in G$, $\bot \notin out(G, a)$.

By this change, consistency of output is required just for those factual situations that the norm-givers have foreseen, in the sense that they have explicitly tied normative consequences to such facts. Still, $(1c)$ might require further modification, since if a is a foreseen situation, and so is b, then also $a \vee b$ or $a \wedge b$ might be counted as foreseen situations for which the norms should be coherent.

As one anonymous reviewer suggested, another solution consists in combining elements of previous proposals:

(1d) A set of norms G is coherent   iff   for each $A \subseteq \{a \mid (a, x) \in G\}$, if A is non-empty and consistent, then $\bot \notin out(G, A)$.

However, there is a further difficulty: let G contain a norm $(a, \neg a)$ that, for conditions in which a is unalterably true, demands that $\neg a$ be realized. We then have $\neg a \in out_n(G, a)$ for the principal output operations out_n, but not $\bot \in out_n(G, a)$. Certainly the term 'incoherent' should apply to a normative system that requires the agent to accomplish what is—given the facts in which the duty arises—impossible. However, since not every output operation supports 'throughput', i.e. the input is not necessarily included in the output, neither (1) nor its variants implies that the agent can actually realize all propositions in the output, though they might be logically consistent. We might therefore demand that the output be not merely consistent, but consistent with the input:

(2) A set of norms G is coherent   iff $out(G, A) \cup A \not\models \bot$.

However, with definition (2) we obtain the questionable result that for any case of norm-violation, i.e. for any case in which $(a, x) \in G$ and $(a \wedge \neg x) \in Cn(A)$, G must be termed incoherent—Adam's fall would only indicate that there was something wrong with God's commands. One remedy would be to leave aside all those norms whose violation is entailed by the circumstances A, i.e. instead of $out(G, A)$ consider

$out(\{(a, x) \in G \mid (a \land \neg x) \notin Cn(A)\}, A)$—but then a set G such that $(a, \neg a) \in G$ would not be incoherent.[1] It seems it is time to formally state our problem:

Challenge 7 *When is a set of norms to be termed 'coherent'?*

As can be seen from the discussion above, input/output logic provides the tools to formally discuss this question, by rephrasing the question of coherence of the norms as one of consistency of output, and of output with input. Both notions have been explored in the input/output framework as 'output under constraints', see also the motivation regarding contrary-to-duty reasoning in Section 1.4.:

Definition (Output under constraints) *Let G be a set of conditional norms and A and C two sets of propositional formulas. Then G is coherent in A under constraints C when $out(G, A) \cup C$ is consistent.*

Future study must define an output operation, determine the relevant states A, and find the constraints C, such that any set of norms G would be appropriately termed coherent or incoherent by this definition.

8 Normative conflicts and dilemmas

There are essentially two views on the question of normative conflicts: in the one view, they do not exist. In the other view, conflicts and dilemmas are ubiquitous.

According to the view that normative conflicts are ubiquitous, it is obvious that we may become the addressees of conflicting normative demands at any time. My mother may want me to stay inside while my brother wants me to go outside with him and play games. I may have promised to finish a paper by the end of a certain day, while for the same day I have promised a friend to come to dinner—now it is late afternoon and I realize I will not be able to finish the paper if I visit my friend. Social convention may require me to offer you a cigarette when I am lighting one for myself, while concerns for your health should make me not offer you one. Legal obligations might collide - think of the case where the SWIFT international money transfer program was required by US anti-terror laws to disclose certain information about its customers, while under European law that also applied to that company, it was required not to disclose this information. Formally, let there be two conditional norms (a, x) and (b, y): unless we have that either $(x \rightarrow y) \in Cn(a \land b)$ or $(y \rightarrow x) \in Cn(a \land b)$ there is a possible situation $a \land b \land \neg(x \land y)$ in which the agent can still satisfy each norm individually, but not both norms collectively. But to assume this for any two norms (a, x) and (b, y) is clearly absurd. Nevertheless, as discussed extensively in Section 1 of this chapter, Lewis's [1973; 1974] and Hansson's [1969] deontic semantics imply that there exists a 'system of spheres', in our setting: a sequence of boxed contrary-to-duty norms $(\top, x_1), (\neg x_1, x_2), (\neg x_1 \land \neg x_2, x_3), \dots$ that satisfies this condition. So any logic about norms must take into account possible conflicts. But standard deontic logic SDL includes D: $\neg(Ox \land O\neg x)$ as one of its axioms, and it is not immediately clear how deontic reasoning could accommodate conflicting norms.

---

[1] Temporal dimensions are not considered here. In an approach that would consider dynamic norms, one may argue, throughput should not be included in a definition of coherence as any change involves an inconsistency between the way things were and the way they become.

Challenge 8a. *How can deontic logic accommodate possible conflicts of norms?*

The literature on normative conflicts and dilemmas is vast. As highlighted earlier in this chapter, here we do not aim at an exhausting literature review on the topic; for that, the interested reader is referred to Goble's [2013] chapter in the handbook of deontic logic and normative systems. If we accept the view that normative conflicts not only genuinely exist but are also ubiquitous, one classical way to deal with such conflicts consists in denying that 'ought' implies 'can', as done by Lemmon [1962]. Another common solution is to deny the principle of conjunction, that is, to deny that oughting to do x and y separately implies ought to do both [Marcus, 1980; van Fraassen, 1973; Goble, 2000]. However, this solution was challenged by Horty's example [1994; 1997; 2003; 2012] where, from "Smith ought to fight in the army or perform alternative national service" and "Smith ought not to fight in the army", we should be able to derive "Smith ought to perform alternative national service". By withdrawing the principle of conjunction, this argument is no longer valid. The distribution rule states that x necessitates y implies that, if one ought to do x, then one ought to do y. As Goble [2013] observes, although this principle has been often criticized for its role in many deontic paradoxes, its responsibility in connection with normative conflicts has rarely been discussed. Keeping the principle of conjunction while removing the distribution rule would validate Horty's argument [Goble, 2009]. For other systems that restrict the distribution principle, see [Goble, 2005; Goble, 2009].

In an input/output setting one could say that there exists a conflict whenever $\bot \in Cn(out(G, A) \cup A)$, i.e. whenever the output is inconsistent with the input: then the norms cannot all be satisfied in the given situation. There appear to be two ways to proceed when such inconsistencies cannot be ruled out. For the concepts underlying the 'some-things-considered' and 'all-things-considered' O-operators defined below cf. Horty [1997] and Hansen [2004; 2005a]. For both, it is necessary to recur to the the notion of a *maxfamily*(G, A, A), i.e. the family of all maximal $H \subseteq G$ such that $out(H, A) \cup A$ is consistent. On this basis, input/output logic defines the following two output operations $out^\cup$ and $out^\cap$:

$$out^\cup(G, A) = \bigcup\{out(H, A) \mid H \in \text{maxfamily}(G, A, A)\}$$
$$out^\cap(G, A) = \bigcap\{out(H, A) \mid H \in \text{maxfamily}(G, A, A)\}$$

Note that $out^\cup$ is a non-standard output operation that is not closed under consequences, i.e. we do not generally have $Cn(out^\cup(G, A)) = out^\cup(G, A)$. Finally we may use the intended definition of an O-operator

$$G, A \models Ox \quad \text{iff} \quad x \in out(G, A)$$

to refer to the operations $out^\cup$ and $out^\cap$, rather than the underlying operation $out(G, A)$ itself, and write $O^\cup x$ and $O^\cap x$ to mean that $x \in out^\cup(G, A)$ and $x \in out^\cap(G, A)$, respectively. Then the 'some-things-considered', or 'bold' O-operator $O^\cup$ describes x as obligatory given the set of norms G and the facts A if x is in the output of some $H \in \text{maxfamily}(G, A, A)$, i.e. if some subset of non-conflicting norms, or: some coherent normative standard embedded in the norms, requires x to be true. It is immediate that neither the SDL axiom $D : \neg(Ox \land O\neg x)$ nor the agglomeration principle $C : Ox \land Oy \to O(x \land y)$ holds for $O^\cup$, as there may be two competing standards demanding x and $\neg x$ to be realized, while there may be none that demands the impossible $x \wedge \neg x$. However, the 'all-things-considered', or 'sceptic', O-operator $O^\cap$ describes x as obligatory given the norms G and the facts A if x is in the outputs of all $H \in$ *maxfamily*(G, A, A), i.e. it requires that x must be realized according to all coherent normative standards. Note that by this definition, both SDL theorems D and C are validated.

The opposite view, that normative conflicts do not exist, appeals to the very notion of obligation: it is essential for the function of norms—to direct human behavior—that the subject of the norms is capable of following them. To state a norm that cannot be fulfilled is a meaningless use of language. To state two norms which cannot both be fulfilled is confusing the subject, not giving him or her directions. To say that a subject has two conflicting obligations is therefore a misuse of the term 'obligation'. So there cannot be conflicting obligations, and if things appear differently, a careful inspection of the normative situation is required that resolves the dilemma in favor of the one or other of what only appeared both to be obligations. In particular, this inspection may reveal that the apparent conflicts in reality comes from some ambiguities in the examples, for instance where a moral 'ought' is not compatible with a legal 'ought': thus, there is no real conflict, because the two 'oughts' refer to two different spheres, and each should be represented with a different operator [Castañeda, 1981; Castañeda, 1982]. Or again, a priority ordering of the apparent obligations may help resolving the conflict, e.g. in Ross [1930], von Wright [1963; 1968], and Hare [1981]. The problem that arises for such a view is then how to determine the 'actual obligations' in face of apparent conflicts, or, put differently, in the face of conflicting 'prima facie' obligations.

Challenge 8b. How can the resolution of apparent conflicts be semantically modeled?

Again, both the $O^\cup$ and the $O^\cap$-operator may help to formulate and solve the problem: $O^\cup$ names the conflicting *prima facie* obligations that arise from a set of norms G in a given situation A, whereas $O^\cap$ resolves the conflict by only telling the agent to do what is required by all maximal coherent subsets of the norms: so there might be conflicting 'prima facie' $O^\cup$-obligations, but no conflicting 'all things considered' $O^\cap$-obligations. The view that a priority ordering helps to resolve conflicts seems more difficult to model. A good approach appears to be to let the priorities help us to select a set $P(G, A, A)$ of preferred maximal subsets $H \in$ *maxfamily*(G, A, A). We may then define the $O^\cap$-operator not with respect to the whole of *maxfamily*(G, A, A), but only with respect to its selected preferred subsets $P(G, A, A)$. Ideally, in order to resolve all conflicts, the priority ordering should narrow down the selected sets to $card(P(G, A, A)) = 1$, but this generally requires a strict ordering of the norms in G. The demand that all norms can be strictly ordered is itself subject of philosophical dispute. Some moral requirements may be incomparable: this is Sartre's paradox, where the requirement that Sartre's student stays with his ailing mother conflicts with the requirement that the student joins the resistance against the German occupation [Sartre, 1946]. Other moral requirements may be of equal weight, e.g. two simultaneously obtained obligations towards identical twins, of which only one can be fulfilled [Marcus, 1980]. The difficult part is then to define a mechanism that determines the preferred maximal subsets by use of the given priorities between the norms. There have been several proposals to this effect, not all of them successful, and the reader is referred to the discussions in Boella and van der Torre [2003] and Hansen [2005b; 2008].

9 Descriptive dyadic obligations

Dyadic deontic operators, that formalize e.g. 'x ought to be true under conditions a' as $O(x|a)$, were introduced over 50 years ago by G. H. von Wright [1956]. Their introduction was due to Prior's paradox of derived obligation: often a primary obligation Ox is accompanied by a secondary, 'contrary-to-duty' obligation that pronounces y (a sanction, a remedy) as obligatory if the primary obligation is violated. At the time, the usual formalization of the secondary obligation would have been $O(\neg x \to y)$, but given Ox and the axioms of standard deontic logic SDL, $O(\neg x \to y)$ is derivable for any y. A bit later, Chisholm's paradox showed that formalizing the secondary obligation as $\neg x \to Oy$ produces similarly counterintuitive results. So to deal with such contrary-to-duty conditions, the dyadic deontic operator $O(x|a)$ was invented. For a historical account the reader is referred to Hilpinen and McNamara's chapter in the handbook of deontic logic and normative systems [Hilpinen and McNamara, 2013].

In Section 1.3 we have extensively discussed DSDL. The perhaps best-known semantic characterization of dyadic deontic logic is B. Hansson's [1969] system DSDL3, axiomatized by Spohn [1975]. Hansson's idea was that the circumstances (the conditions a) are something which has actually happened (or will unavoidably happen) and which cannot be changed afterwards. Ideal worlds in which $\neg a$ is true are therefore excluded. However, some worlds may still be better than others, and there should then be an obligation to make 'the best out of the sad circumstances". Consequently, Hansson presents a possible worlds semantics in which all worlds are ordered by a preference (betterness) relation. $O(x|a)$ is then defined true if x is true in the best a-worlds. Here, we intend to employ semantics that do not make use of any prohairetic betterness relation, but that model deontic operators with regard to given sets of norms and facts.

Challenge 9 *How to define dyadic deontic operators with regard to given sets of norms and facts?*

Input/output logic assumes a set of (conditional) norms G, and a set of unalterable facts A. The facts A may describe a situation that is inconsistent with the output $out(G, A)$: suppose there is a primary norm $(\top, a) \in G$ and a secondary norm $(\neg a, x) \in G$, i.e. $G = \{(\top, a), (\neg a, x)\}$, and $A = \{\neg a\}$. Though $a \in out(G, A)$, it makes no sense to describe a as obligatory since a cannot be realized any more in the given situation—no crying over spilt milk. Rather, the output should include only the consequent of the secondary obligation x—it is the best we can make out of these circumstances. To do so, we return to the definitions of $maxfamily(G, A, A)$ as the set of all maximal subsets $H \subseteq G$ such that $out(H, A) \cup A$ is consistent, and the set $out^\cap(G, A)$ as the intersection of all outputs from $H \in maxfamily(G, A, A)$, i.e. $out^\cap(G, A) = \bigcap\{out(H, A) \mid H \in maxfamily(G, A, A)\}$. We may then define:

$$G \models O(x|a) \quad \text{iff} \quad x \in out^\cap(G, \{a\})$$

Thus, relative to the set of norms G, $O(x|a)$ is defined true if x is in the output under a of all maximal sets H of norms such that their output under $\{a\}$ is consistent with a. In the example where $G = \{(\top, a), (\neg a, x)\}$ we therefore obtain $O(x|\neg a)$ but not $O(a|\neg a)$ as being true, i.e. only the consequent of the secondary obligation is described as obligatory in conditions $\neg a$.

In the above definition, the antecedent a of the dyadic formula $O(x|a)$ makes the inputs explicit: the truth definition does not make use of any facts other than a. This may be unwanted; one might consider an input set A of *given* facts, and employ the antecedent a only to denote an additional, *assumed* fact. Still, the output should contradict neither the given nor the assumed facts, and the output should include also the normative consequences x of a norm (a, x) given the assumed fact a. This may be realized by the following definition:

$$G, A \models O(x|a) \quad \text{iff} \quad x \in out^{\cap}(G, A \cup \{a\})$$

So, relative to a set of norms G and a set of facts A, $O(x, a)$ is defined true if x is in the output under $A \cup \{a\}$ of all maximal sets H of norms such that their output under $A \cup \{a\}$ is consistent with $A \cup \{a\}$.

Hansson's description of dyadic deontic operators as describing defeasible obligations that are subject to change when more specific, namely contrary-to-duty situations emerge, may be the most prominent view, but it is by no means the only one. Earlier authors like von Wright [1961; 1962] and Anderson [1959] have proposed more normal conditionals, which in particular support 'strengthening of the antecedent' SA $O(x|a) \rightarrow O(x|a \wedge b)$. From an input/output perspective, such operators can be accommodated by defining

$$G, A \models O(x|a) \quad \text{iff} \quad x \in out(G, A \cup \{a\})$$

It is immediate that for all standard output operations out_n this definition validates SA. The properties of dyadic deontic operators that are, like the above, semantically defined within the framework of input/output logic, have not been studied so far. The theorems they validate will inevitably depend on what output operation is chosen, cf. Hansen [2008] for some related conjectures.

10 Permissive norms

In formal deontic logic, permission is studied less frequently than obligation. For a long time, it was naively assumed that it can simply be taken as a dual of obligation, just as possibility is the dual of necessity in modal logic. Permission is then defined as the absence of an obligation to the contrary, and the modal operator P defined by $Px =_{def} \neg O \neg x$. Today's focus on obligations is not only in stark contrast how deontic logic began, for when von Wright [1951b] started modern deontic logic in 1951, it was the P-operator that he took as primitive, and defined obligation as an absence of a permission to the contrary. Rather, more and more authors have come to realize how subtle and multi-faceted the concept of permission is. Much energy was devoted to solving the problem of 'free choice permission', where one may derive from the statement that one is permitted to have a cup of tea or a cup of coffee that it is permitted to have a cup of tea, and it is permitted to have a cup of coffee, or for short, that $P(x \vee y)$

implies Px and Py (cf. Kamp [1973]). Von Wright, in his late work starting with [von Wright, 1983], dropped the concept of inter-definability of obligations and permissions altogether by introducing P-norms and O-norms, where one may call something permitted only if it derives from the collective contents of some O-norms and at most one P-norm. This concept of 'strong permission' introduced deontic 'gaps': whereas in standard deontic logic SDL, $O\neg x \lor Px$ is a tautology, meaning that any state of affairs is either forbidden or permitted, von Wright's new theory means that in the absence of explicit P-norms only what is obligatory is permitted, and that nothing is permitted if also O-norms are missing. Perhaps most importantly, Bulygin [1986] observed that an authoritative kind of permission must be used in the context of multiple authorities and updating normative systems: if a higher authority permits you to do something, a lower authority can no longer prohibit it. Summing up, the understanding of permission is still in a less satisfactory state than the understanding of obligation and prohibition. Indeed, a whole chapter in the handbook of deontic logic and normative systems is devoted to the various forms of permission [Hansson, 2013].

Challenge 10 *How to distinguish various kinds of permissions and relate them to obligations?*

From the viewpoint of input/output logic, one may first try to define a concept of negative permission in the line of the classic approach. Such a definition is the following:

$$G, A \models P^{neg} x \quad \text{iff} \quad \neg x \notin out(G, A)$$

So something is permitted by a code iff its negation is not obligatory according to the code and in the given situation. As innocuous and standard as such a definition seems, questions arise as to what output operation *out* may be used. Simple-minded output out_1 and basic output out_2 produce counterintuitive results: consider a set of norms G of which one norm $(work, tax)$ demands that if I am employed then I have to pay taxes. For the default situation $A = \{\top\}$ then $P^{neg}(work \land \neg tax)$ is true, i.e. it is by default permitted that I am employed and do not pay taxes. Stronger output operations out_3 and out_4 that warrant reusable output exclude this result, but their use in deontic reasoning is questionable due to contrary-to-duty reasoning, as discussed in Section 1.

In contrast to a concept of negative permission, one may also define a concept of 'strong' or 'positive permission'. This requires a set P of explicit permissive norms, just as G is a set of explicit obligations. As a first approximation, one may say that something is positively permitted by a code iff the code explicitly presents it as such. However, this leaves a central logical question unanswered as to how explicitly given permissive and obligating norms may generate permissions that—in some sense— follow from the explicitly given norms. Pursuing von Wright's later approach, we may define:

$$G, P \models P^{stat}(x/a) \quad \text{iff} \quad x \in out(G \cup \{(b, y)\}, a) \text{ for some } (b, y) \in P \cup \{(\top, \top)\}$$

So there is a permission to realize x in conditions a if x is generated under these conditions either by the norms in G alone, or the norms in G together with some explicit permission (b, y) in P. We call this a 'static' version of strong permission. For example, consider a set G consisting of the norm $(work, tax)$, and a set P consisting of the sole license $(18y, vote)$ that permits all adults to take part in political elections. Then all of the following are true: $P^{stat}(tax/work)$, $P^{stat}(vote/18y)$, $P^{stat}(tax/work \wedge male)$ and also $P^{stat}(vote/\neg work \wedge 18y)$ (so even unemployed adults are permitted to vote).

Where negative permission is liberal, in the sense that anything is permitted that does not conflict with one's obligations, the concept of static permission is quite strict, as nothing is permitted that does not explicitly occur in the norms. In between, one may define a concept of 'dynamic permission' that defines something as permitted in some situation a if forbidding it for these conditions would prevent an agent from making use of some explicit (static) permission. The formal definition reads:

$$G, P \models P^{dyn}(x/a) \quad \text{iff} \quad \neg y \in out(G \cup \{(a, \neg x)\}, b) \text{ for some } y \text{ and conditions } b \text{ such that } G, P \models P^{stat}(y/b)$$

Consider the above static permission $P^{stat}(vote/\neg work \wedge 18y)$ that even the unemployed adult populations is permitted to vote, generated by the sets $P = \{(18y, vote)\}$ and $G = \{(work, tax)\}$. We might also like to say, without reference to age, that the unemployed are protected from being forbidden to vote, and in this sense are permitted to vote, but $P^{stat}(vote/\neg work)$ is not true. And we might like to say that adults are protected from being forbidden to vote unless they are employed, and in this sense are permitted to be both unemployed and take part in elections, but also $P^{stat}(\neg work \wedge vote/18y)$ is not true. Dynamic permissions allow us to express such protections, and make both $P^{dyn}(vote/\neg work)$ and $P^{dyn}(\neg work \wedge vote/18y)$ true: if either $(\neg work, \neg vote)$ or $(18y, (\neg work \to \neg vote))$ were added to G we would obtain $\neg vote$ as output in conditions $(\neg work \wedge 18y)$ in spite of the fact that, as we have seen, $G, P \models P^{stat}(vote/\neg work \wedge 18y)$.

The relation of permission and obligation can also be studied from a multiagent perspective. Think of two brothers who are fighting for a toy, and the mother obliges the son who's playing with the toy to permit his brother to play as well.

There are, ultimately, a number of questions for all these concepts of permissions that Makinson and van der Torre have further explored [Makinson and van der Torre, 2003a]. Other kinds of permissions have been discussed from an input/output perspective in the literature, too, for example permissions as exceptions of obligations [Boella and van der Torre, 2003]. It seems input/output logic is able to help clarify the underlying concepts of permission better than traditional deontic semantics. One challenge is Governatori's paradox [Governatori, 2015], containing a conditional norm whose body and head are permissions: "the collection of medical information is permitted provided that the collection of personal information is permitted."

11 Meaning postulates and intermediate concepts

To define a deontic operator of individual obligation seems straightforward if the norm in question is an individual command or act of promising. For example, if you are the addressee α of the following imperative sentence

(1)   You, hand me that screwdriver, please.

and you consider the command valid, then what you ought to do is to hand the screwdriver in question to the person β uttering the request. In terms of input/output logic, let x be the proposition that α hands the screwdriver to β: with the set of norms $G = \{(\top, x)\}$, the set of facts $A = \{\top\}$, and the truth definition Ox iff $x \in out(A, G)$: then we obtain that Ox is true, i.e. it is true that it ought to be that α hands the screwdriver to β.

Norms that belong to a legal system are more complex, and thus more difficult to reason about. Consider, for example

(2) An act of theft is punished by a prison sentence not exceeding 5 years or a fine.

Things are again easy if you are a judge and you know that the accused in front of you has committed an act of theft—then you ought to hand out a verdict that commits the accused to pay a fine or to serve a prison sentence not exceeding 5 years. However, how does the judge arrive at the conclusion that an act of theft has been committed? 'Theft' is a legal term that is usually accompanied by a legal definition such as the following one:

(3) Someone commits an act of theft if that person has taken a movable object from the possession of another person into his own possession with the intention to own it, and if the act occurred without the consent of the other person or some other legal authorization.

It is noteworthy that (3) is not a norm in the strict sense—it does not prescribe or allow a behavior—but rather a stipulative definition, or, in more general terms, a *meaning postulate* that constitutes the legal meaning of theft. Such sentences are often part of the legal code. They share with norms the property of being neither true nor false: stipulative definitions are neither empirical statements nor descriptive statements. In this sense we say that they are neither true nor false. However, they are held to be true by definition. The significance of (3) is that it decomposes the complex legal term 'theft' into more basic legal concepts. These concepts are again the subject of further meaning postulates, among which may be the following:

(4) A person in the sense of the law is a human being that has been born.
(5) A movable object is any physical object that is not a person or a piece of land.
(6) A movable object is in the possession of a person if that person is able to control the uses and the location of the object.
(7) The owner of an object is—within the limits of the law—entitled to do with it whatever he wants, namely keep it, use it, transfer possession or ownership of the object to another person, and destroy or abandon it.

Not all of definitions (4)-(7) may be found in the legal statutes, though they may be viewed as belonging to the normative system by virtue of having been accepted in legal theory and judicial reasoning. They constitute 'intermediate concepts': they link legal terms (person, movable object, possession etc.) to words describing natural facts (human being, born, piece of land, keep an object etc.).

Any proper representation of legal norms must include means of representing meaning postulates that define legal terms, decompose legal terms into more basic legal terms, or serve as intermediate concepts that link legal terms to terms that describe natural facts. But for deontic logic, with its standard possible worlds semantics, a comprehensive solution to the problem of representing meaning postulates is so far lacking (cf. Lindahl [1997]).

Challenge 11 *How can meaning postulates and intermediate terms be modeled in semantics for deontic logic reasoning?*

The representation of intermediate concepts is of particular interest, since such concepts arguably reduce the number of implications required for the transition from natural facts to legal consequences and thus serve an economy of expression (cf. Lindahl and Odelstad [2006] and their recent overview chapter [Lindahl and Odelstad, 2013]). Lindahl and Odelstad use the term 'ownership' as an example to argue as follows: let $F_1, ..., F_p$ be descriptions of some situations in which a person α acquires ownership of an object γ, e.g. by acquiring it from some other person β, finding it, building it from owned materials, etc., and let $C_1, ..., C_n$ be among the legal consequences of α's ownership of γ, e.g. freedom to use the object, rights to compensation when the object is damaged, obligations to maintain the object or pay taxes for it etc. To express that each fact F_i has the consequence C_j, $p \times n$ implications are required. The introduction of the term $Ownership(x, y)$ reduces the number of required implications to $p + n$: there are p implications that link the facts $F_1, ..., F_p$ to the legal term $Ownership(x, y)$, and n implications that link the legal term $Ownership(x, y)$ to each of the legal consequences $C_1, ..., C_n$. The argument obviously does not apply to all cases: one implication $(F_1 \vee ... \vee F_p) \to (C_1 \wedge ... \wedge C_n)$ may often be sufficient to represent the case that a variety of facts $F_1, ..., F_p$ has the same multitude of legal consequences $C_1, ..., C_n$. However, things may be different when norms that link a number of factual descriptions to the same legal consequences stem from different normative sources, may come into conflict with other norms, can be overridden by norms of higher priority, or be subject to individual exemption by norms that grant freedoms or licenses: in these cases, the norms must be represented individually. So it seems worthwhile to consider ways to incorporate intermediate concepts into a formal semantics for deontic logic.

In an input/output framework, a first step could be to employ a separate set T of theoretical terms, namely meaning postulates, alongside the set G of norms. Let T consists of intermediates of the form (a, x), where a is a factual sentence (e.g. that β is in possession of γ, and that α and β agreed that α should have γ, and that β hands γ to α), and x states that some legal term obtains (e.g. that α is now owner of γ). To derive outputs from the set of norms G, one may then use $A \cup out(T, A)$ as input, i.e. the factual descriptions together with the legal statements that obtain given the intermediates T and the facts A.

It may be of particular interest to see that such a set of intermediates may help resolve possible conflicts in the law. Let $(\top, \neg dog)$ be a statute that forbids dogs on the premises, but let there also be a higher order principle that no blind person may be required to give up his or her guide dog. Of course the conflict may be solved by modifying the statute (e.g. add a condition that the dog in question is not a guide dog), but then modifying a statute is usually not something a judge, faced with such a norm, is allowed to do: the judge's duty is solely to consider the statute, interpret it according to the known or supposed will of the norm-giver, and apply it to the given facts. The judge may then come to the conclusion that a fair and considerate norm-giver would not have meant the statute to apply to guide dogs, i.e. the term "dog" in the statute is a theoretical term whose extension is smaller than the natural term. So the statute must be re-interpreted as reading $(\top, \neg tdog)$ with the additional intermediate $(dog \land \neg guidedog, tdog) \in T$, and thus no conflict arises for the case of blind persons that want to keep their guide dog. While this seems to be a rather natural view of how judicial conflict resolution works (the example is taken from an actual court case), the exact process of creating and modifying theoretical terms in order to resolve conflicts must be left to further study.

12 Constitutive norms

Constitutive norms like counts-as conditionals are rules that create the possibility of or define an activity. For example, according to Searle [1995], the activity of playing chess is constituted by action in accordance with these rules. Chess has no existence apart from these rules. The institutions of marriage, money, and promising are like the institutions of baseball and chess in that they are systems of such constitutive rules or conventions. They have been identified as the key mechanism to normative reasoning in dynamic and uncertain environments, for example to realize agent communication, electronic contracting, dynamics of organizations, see, e.g., Boella and van der Torre [2006a].

Challenge 12 *How to define counts-as conditionals and relate them to obligations and permissions?*

For Jones and Sergot [1996], the counts-as relation expresses the fact that a state of affairs or an action of an agent "is a sufficient condition to guarantee that the institution creates some (usually normative) state of affairs". They formalize this introducing a conditional connective $\Rightarrow_s$ to express the "counts-as" connection holding in the context of an institution s. They characterize the logic of $\Rightarrow_s$ as a conditional logic, with axioms for agglomeration $((x \Rightarrow_s y) \& (x \Rightarrow_s z)) \supset (x \Rightarrow_s (y \land z))$, left disjunction $((x \Rightarrow_s z) \& (y \Rightarrow_s z)) \supset ((x \lor y) \Rightarrow_s z)$ and transitivity $((x \Rightarrow_s y) \& (y \Rightarrow_s z)) \supset (x \Rightarrow_s z)$. The flat fragment can be phrased as an input/output logic as follows [Boella and van der Torre, 2006b].

Definition 7 *Let L be a propositional action logic with $\vdash$ the related notion of derivability and Cn the related consequence operation $Cn(x) = \{y \mid x \vdash y\}$. Let CA be a set of pairs of L, $\{(x_1, y_1), \ldots, (x_n, y_n)\}$, read as '$x_1$ counts as y_1', etc. Moreover, consider the following proof rules conjunction for the output (AND), disjunction of the input (OR), and transitivity (T) defined as follows:*

$$\frac{(x, y_1), (x, y_2)}{(x, y_1 \land y_2)} AND \qquad \frac{(x_1, y), (x_2, y)}{(x_1 \lor x_2, y)} OR \qquad \frac{(x, y_1), (y_1, y_2)}{(x, y_2)} T$$

For an institution s, the counts-as output operator out_{CA} is defined as the closure operator on the set CA using the rules above together with a tacit rule that allows replacement of logical equivalents in input and output. We write $(x, y) \in out_{CA}(CA, s)$. Moreover, for $X \subseteq L$, we write $y \in out_{CA}(CA, s, X)$ if there is a finite $X' \subseteq X$ such that $(\wedge X', y) \in out_{CA}(CA, s)$, indicating that the output y is derived by the output operator for the input X, given the counts-as conditionals CA of institution s. We also write $out_{CA}(CA, s, x)$ for $out_{CA}(CA, s, \{x\})$.

Example 5 *If for some institution s we have $CA = \{(a, x), (x, y)\}$, then we have $out_{CA}(CA, s, a) = \{x, y\}$.*

The recognition that statements like "X counts as Y in context c" may have different meanings in different situations lead Grossi et al. [2006; 2008] to propose a family of operators capturing four notions of counts-as conditionals. Starting from a simple modal logic of contexts, several logics are used to define the family of operators. All logics have been proven to be sound and strongly complete. By using a logic of acceptance, Lorini et al. [Lorini and Longin, 2008; Lorini et al., 2009] investigate another aspect of constitutive norms, that is, the fact that agents of a society need to accept such norms in order for them to be in force.

Considering the legal practice, Governatori and Rotolo [2008] propose a study of constitutive norms within the framework of defeasible logic. This allows them to capture de defeasibility of counts-as conditionals: even in presence of a constitutive norms like "X counts as Y in context c", the inference of Y from X can be blocked in presence of exceptions.

There is presently no consensus on the logic of counts-as conditionals, probably due to the fact that the concept is not studied in depth yet. For example, the adoption of the transitivity rule T for their logic is criticized by Artosi et al. [2004]. Jones and Sergot say that "we have been unable to produce any counter-instances [of transitivity], and we are inclined to accept it". Neither of these authors considers replacing transitivity by cumulative transitivity (CT): $((x \Rightarrow_s y) \& (x \wedge y \Rightarrow_s z)) \supset (x \Rightarrow_s z)$, that characterizes operations out_3, out_4 of input/output logic. For a more comprehensive overview on constitutive norms, the reader is referred to the chapter by Grossi and Jones [2013] in the handbook of deontic logic and normative systems.

The main issue in defining constitutive norms like counts-as conditionals is defining their relation to regulative norms like obligations and permissions. Boella and van der Torre [2006b] use the notion of a logical architecture combining several logics into a more complex logical system, also called logical input/output nets (or *lions*).

The notion of logical architecture naturally extends the input/output logic framework, since each input/output logic can be seen as the description of a 'black box'. In figure 5 there are boxes for counts-as conditionals (CA), institutional constraints (IC), obligating norms (O) and explicit permissions (P). The norm base (NB) component contains sets of norms or rules, which are used in the other components to generate the component's output from its input. The figure shows that the counts-as conditionals are combined with the obligations and permissions using iteration, that is, the counts-as conditionals produce institutional facts, which are input for the

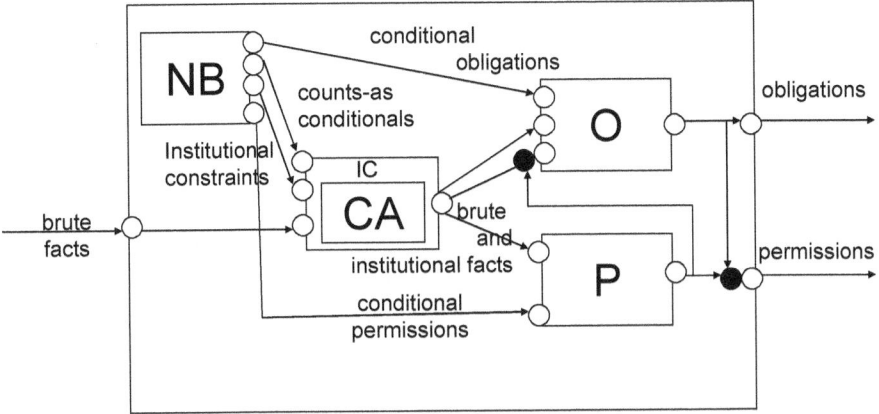

Figure 5. Logical Architecture of a Normative System.

norms. Roughly, if we write $out(CA, G, A)$ for the output of counts-as conditionals together with obligations, $out(G, A)$ for obligations as before, then $out(CA, G, A) = out(G, out_{CA}(CA, A))$.

There are many open issues concerning constitutive norms, since their logical analysis has not attracted much attention yet. How to distinguish among various kinds of constitutive norms? How are constitutive norms (x counts as y) distinguished from classifications (x is a y)? What is the relation with intermediate concepts?

13 Revision of a set of norms

In general, a code G of regulations is not static, but changes over time. For example, a legislative body may want to introduce new norms or to eliminate some existing ones. A different (but related) type of change is the one induced by the fusion of two (or more) codes—a topic addressed in the next section. A related but different issue not addressed here is that of how norms come about, how they propagate in the society, and how they change over time (cf. the chapter by Frantz and Pigozzi [forthcoming]).

Little work exists on the logic of the revision of a set of norms. To the best of our knowledge, Alchourrón and Makinson [1981; 1982] were the first to study the changes of a legal code. The addition of a new norm n causes an enlargement of the code, consisting of the new norm plus all the regulations that can be derived from n. Alchourrón and Makinson distinguish two other types of change. When the new norm is incoherent with the existing ones, we have an *amendment* of the code: in order to coherently add the new regulation, we need to reject those norms that conflict with n. Finally, *derogation* is the elimination of a norm n together with whatever part of G implies n.

Alchourrón and Makinson [1981] assume a "hierarchy of regulations". Alchourrón and Bulygin [1981] also considered the *Normenordnung* and the consequences of gaps in this ordering. For example, in jurisprudence the existence of precedents is an established method to determine the ordering among norms.

However, although Alchourrón and Makinson aim at defining change operators for a set of norms of some legal system, the only condition they impose on G is that it is a non-empty and finite set of propositions. In other words, a norm x is taken to be simply a formula in propositional logic. Thus, they suggest that "the same concepts and techniques may be taken up in other areas, wherever problems akin to inconsistency and derogation arise" ([Alchourrón and Makinson, 1981], p. 147).

This explains how their work (together with Gärdenfors's analysis of counterfactuals) could ground that research area that is now known as *belief revision*. Belief revision is the formal study of how a set of propositions changes in view of new information that may be inconsistent with the existing beliefs. Expansion, revision and contraction are the three belief change operations that Alchourrón, Gärdenfors and Makinson identified in their approach (called AGM) and that have a clear correspondence with the changes on a system of norms we mentioned above.

Challenge 13 *How to revise a set of regulations or obligations?*

Recently, AGM theory has been reconsidered as a framework for norm change. However, beside syntactic approaches where norm change is performed directly on the set of norms (as in AGM), there are also proposals that appeared in the dynamic logic literature and that could be described as semantic approaches.

One example of this is the dynamic context logic proposed by Aucher et al. [2009], where norm change is a form of model update. Point of depart is a dynamic variant of the logic of context used to study counts-as conditionals introduced by Grossi et al. [2008]. Context expansion and context contraction operators are defined. Context expansion and context contraction represent the promulgation and the derogation of constitutive norms respectively. One of the advantages of this approach is that it can be used for the formal specification and verification of computational models of interactions based on norms.

A formal account clearly rooted in the legal practice is the one proposed by Governatori and Rotolo [2010]. In particular, the removal of norms can be performed by annulment or by abrogation. The crucial difference between these two mechanisms is that annulment removes a norm from the code and all its effects (past and future) are cancelled. Abrogation, on the other hand, does not operate retroactively, and so it leaves the effects of an abrogated norm holding in the past.

It should then be clear that, in order to capture the difference between annulment and abrogation, the temporal dimension is pivotal. For this reason, Governatori and Rotolo's first attempt is to use theory revision in Defeasible Logic without temporal reasoning is unsuccessful as it cannot capture retroactivity. They the add a temporal dimension to Defeasible Logic to keep track of the changes in a normative system and to deal with retroactivity. Norms are represented along two temporal dimensions: the time of validity when the norm enters in the normative system and the time of effectiveness when the norm can produce legal effects. This leads to keep multiple versions of a normative system are needed. If Governatori and Rotolo [2010] manage to capture the temporal dimension that plays a role in legal modifications, the resulting formalisation is rather complex.

To overcome such complexity without losing hold on the legal practice, Governatori et al. [2013] explored three AGM-like contraction operators to remove rules, add exceptions and revise rule priorities.

Boella et al. [2016] also use AGM theory, where propositional formulas are replaced by pairs of propositional formulas to represent rules, and the classical consequence operator Cn is replaced by an input/output logic. Within this framework, AGM contraction and revision of rules are studied. It is shown that results from belief base dynamics can be transferred to rule base dynamics. However, difficulties arise in the transfer of AGM theory change to rule change. In particular, it is shown that the six basic postulates of AGM contraction are consistent only for some input/output logics but not for others. Furthermore, it is shown how AGM rule revision can be defined in terms of AGM rule contraction using the Levi identity.

When we turn to a proper representation of norms, as in the input/output logic framework, the AGM principles thus prove to be too general to deal with the revision of a normative system. For example, one difference between revising a set of beliefs and revising a set of regulations is the following: when a new norm is added, coherence may be restored by modifying some of the existing norms, not necessarily retracting some of them. The following example clarifies this point:

Example. If we have $\{(\top, a), (a, b)\}$ and we have that c is an exception to the obligation to do b, then we need to retract (c, b). Two possible solutions are $\{(\neg c, a), (a, b)\}$ or $\{(\top, a), (a \wedge \neg c, b)\}$.

Stolpe [2010] also combines input/output logic and AGM theory to propose an abstract model of norm change. Contraction is used to represent the derogation of a norm, that is, the elimination of a norm together with whatever part of the code that implies that norm. This is rendered as an AGM partial meet contraction with a selection function for a set of norms in input/output logic. Stolpe gives a complete AGM-style characterisation of the derogation operation. Revision, on the other hand, serves to study the amendment of a code, which happens when we wish to add a new norm which is incoherent with the existing ones. Amendment is defined as a norm revision obtained via the Levi identity.

Future research must investigate whether general patterns in the revision and contraction of norms exist and how to formalize them. Another open question is whether other logics can offer a general framework for modelling norm change. Finally, more case studies showing that formally defined operators serve for a conceptual analysis of normative change are needed.

14 Merging sets of norms

We now turn to another type of change, that is the aggregation of regulations. This problem has been only recently addressed in the literature and therefore the findings are still incomplete.

The first noticeable thing is the lack of general agreement about where the norms that are to be aggregated come from:

1. some papers focus on the merging of conflicting norms that belong to the same normative system [Cholvy and Cuppens, 1999];

2. other papers assume that the regulations to be fused belong to different systems [Booth et al., 2006]; and finally

3. some authors provide patterns of possible rules to be combined, and consider both cases 1. and 2. above [Grégoire, 2004].

The first situation seems to be more a matter of coherence of the whole system rather than a genuine problem of fusion of norms. However, such approaches have the merit to reveal the tight connections between fusion of norms, non-monotonic logics and defeasible deontic reasoning. The initial motivation for the study of belief revision was the ambition to model the revision of a set of regulations. In contrast to this, the generalization of belief revision to *belief merging* is primarily dictated by the goal to tackle the problem—arising in computer science—of combining information from different sources. The pieces of information are represented in a formal language and the aim is to merge them in an (ideally) unique knowledge base. See Konieczny and Grégoire [2006] for a survey on logic-based approaches to information fusion.

Challenge 14 *Can the belief merging framework deal with the problem of merging sets of norms?*

If, following Alchourrón and Makinson, we assume that norms are unconditional, then we could expect to use standard merging operators to fuse sets of norms. Yet once we consider conditional norms, as in the input/output logic framework, problems arise again. Moreover, most of the fusion procedures proposed in the literature seem to be inadequate for the scope.

To see why this is the case, we need to explain the merging approach in a few words. Let us assume that we have a finite number of belief bases $K_1, K_2, \ldots, K_n$ to merge. IC is the belief base whose elements are the integrity constraints (i.e., any condition that we want the final outcome to satisfy). Given a multi-set $E = \{K_1, K_2, \ldots, K_n\}$ and IC, a merging operator $\mathcal{F}$ is a function that assigns a belief base to E and IC. Let $\mathcal{F}_{IC}(E)$ be the resulting collective base from the IC fusion on E.

Fusion operators come in two types: model-based and syntax-based. The idea of a model-based fusion operator is that models of $\mathcal{F}_{IC}(E)$ are models of IC, which are preferred according to some criterion depending on E. Usually the preference information takes the form of a total pre-order on the interpretations induced by a notion of distance $d(w, E)$ between an interpretation w and E.

Syntax-based merging operators are usually based on the selection of some consistent subsets of $\bigcup E$ [Baral et al., 1992; Konieczny, 2000]. The bases K_i in E can be inconsistent and the result does not depend on the distribution of the well formed formulas over the members of the group. Konieczny [2000] refers the term 'combination' to the syntax-based fusion operators to distinguish them from the model-based approaches.

Finally, the model-based aggregation operators for bases of equally reliable sources can be of two sorts. On the one hand, there are majoritarian operators that are based on a principle of distance-minimization [Lin and Mendelzon, 1996]. On the other hand, there are egalitarian operators, which look at the distribution of the distances in E [Konieczny, 1999]. These two types of merging try to capture two intuitions that often guide the aggregation of individual preferences into a social one. One option is to let the majority decide the collective outcome, and the other possibility is to equally distribute the individual dissatisfaction.

Obviously, these intuitions may well serve in the aggregation of individual knowledge bases or individual preferences, but have nothing to say when we try to model the fusion of sets of norms. Hence, for this purpose, syntactic merging operators may be more appealing. Nevertheless, the selection of a coherent subset depends on additional information like an order of priority over the norms to be merged, or some other meta-principles.

The reader may wonder about the relationships between merging sets of norms and the revision of a normative system. In particular, one may speculate that Challenge 14 is not independent of Challenge 13, and that a positive answer to Challenge 14 implies an answer to 13. This is indeed an interesting question, but we believe that the answer to this question is not straightforward. Konieczny and Pino Pérez [2011] have shown that there are close links between belief merging operators and belief revision ones. In particular, they show that an IC merging operator is an extension of an AGM revision operator. However, as we have seen, it is not clear whether IC merging operators could be properly used to study the merging of norms.

An alternative approach is to generalize existing belief change operators to merging rules. This is the approach followed by Booth et al. [2006], where merging operators defined using a consolidation operation and possibilistic logic are applied to the aggregation of conditional norms in an input/output logic framework. However, at this preliminary stage, it is not clear whether such methodology is more fruitful for testing the flexibility of existing operators to tackle other problems than the ones they were created for, or if this approach can really shed some light on the new riddle at hand.

Grégoire [2004] takes a different perspective. Here, real examples from the Belgian-French bilateral agreement preventing double taxation are considered. These are fitted into a taxonomy of the most common legal rules with exceptions, and the combination of each pair of norms is analyzed. Moreover, both the situations in which the regulations come from the same system and those in which they come from different ones are contemplated, and some general principles are derived. Finally, a merging operator for rules with abnormality propositions is proposed. A limitation of Grégoire's proposal is that only the aggregation of rules with the same consequence is taken into account and, in our opinion, this neglects other sorts of conflicts that may arise, as we see now.

Cholvy and Cuppens [1999] also call for non-monotonic reasoning in the treatment of contradictions, and present a method for merging norms. The proposal assumes an order of priority among the norms to be merged and this order is used to resolve the incoherence. Even though this is quite a strong assumption, Cholvy and Cuppens's work takes into consideration a broader type of incoherence than Grégoire [2004]. In their example, an organization that works with secret documents has two rules. R_1 is "It is obligatory that any document containing some secret information is kept in a safe, when nobody is using this document". R_2 is "If nobody has used a given document for five years, then it is obligatory to destroy this document by burning it". As they observe, in order to deduce that the two rules are conflicting, we need to introduce the constraint that keeping a document and destroying it are contradictory actions. That is, the notion of coherence between norms can involve information not given by any norms.

15 Games, norms and obligations

Deontic logic has been developed as a logic for practical reasoning, and normative systems are used to guide, control, or regulate desired system behaviour. This raises a number of questions. For example, how are deontic logic and the logic of normative systems related to alternative decision and agent interaction models such as BDI theory, decision theory, game theory, or social choice theory? Moreover, how can deontic logic be extended with cognitive concepts such as beliefs, desires, goals, intentions, and commitments? Though there have been a few efforts to base deontic logic with a logic of knowledge to define knowledge based obligation [Pacuit *et al.*, 2006], or to extend deontic logic with BDI concepts [Broersen *et al.*, 2003], we believe that such extensions have not been fully explored yet. For example, Kolodny and MacFarlane [2010] describe a decision problem involving miners, as well as several dialogues scenarios, which highlight the problems of normative reasoning with agents.

Maybe the most fundamental challenge has become apparent in this chapter. We discussed how deontic STIT logics are based on interactions of agents in games, and we discussed how norm based deontic logics have been developed on the basis of detachment. However, these two approaches have not been combined yet. So this is our final challenge in this chapter.

Challenge 15 *How can deontic logic be based on both norm and detachment, as well as decision and game theory?*

Norms and games have been related before. Lewis [1979] introduced master-slave games and Bulygin [1986] introduced Rex-minister-subject games in a discussion on the role of permissive norms in normative systems and deontic logic. Moreover, deontic logic has been used as an element in games to partially influence the behavior of individual agents [Boella and van der Torre, 2007]. Van der Torre [2010] proposes games as the foundation of deontic logic. He illustrate the notion of a violation game using a metaphor from daily life. A person faces the parental problem of letting the son go to bed in time, or letting him make his homework. The mother is obliging her son to eat his vegetables. As illustrated in the first drawing of figure 6, the son did what his mother asked him to do.

However, in the second drawing his behavior has changed. The son does not like vegetables, and when the parents tell the boy to eat his vegetables, he just says "No!" At the third drawing, when the son's desire not to eat vegetables became stronger than

Violation Game 1: Conformance

Violation Games: Problem

Violation Game 2: Incentives

Violation Games: Problem

Violation Game 3: Negotiation

Figure 6. Conformance, violation, incentive, violation, negotiation (Drawings by Egberdien van der Torre), from [van der Torre, 2010].

his motivation to obey his parents, the parents adapted their strategy and introduced the use of incentives. They told their son, "if you empty your plate you will get a dessert", or sometimes, "if you don't finish your plate, you don't get a dessert." The boy has a desire to eat a dessert, and this desire is stronger than the desire not to eat vegetables, so he is eating his vegetables again. However, after some time we reach the fourth situation in figure 6 where the incentive no longer works. The boy starts to protest and to negotiate. In those cases, the parents sometimes decide that the son will get his dessert even without eating his vegetables, for example, because the child still has eaten at least some of them, or because it is his birthday, or simply because they are not in the mood to argue. As visualized in the fifth situation in figure 6, this makes the boy very happy. It is precisely this aspect that characterizes a violation game. The violation does not follow necessarily from the norm, but is subject to exceptions and negotiation.

Figure 7 models this example by a standard extensive game tree. Let's look first at one moment in time. The child decides first whether to eat his vegetables or not. But in this decision, he takes the response of his parents into account. In other words, he has a model of how the parents will respond to his behavior. In the deontic logic we propose here, based on a violation game, it is obligatory to empty the plate when the boy expects that not eating his vegetables leads to violation, not when a violation logically follows. By the way, we identify the recognition of violation and the sanction in the example for illustrative purposes, in reality usually two distinct steps can be

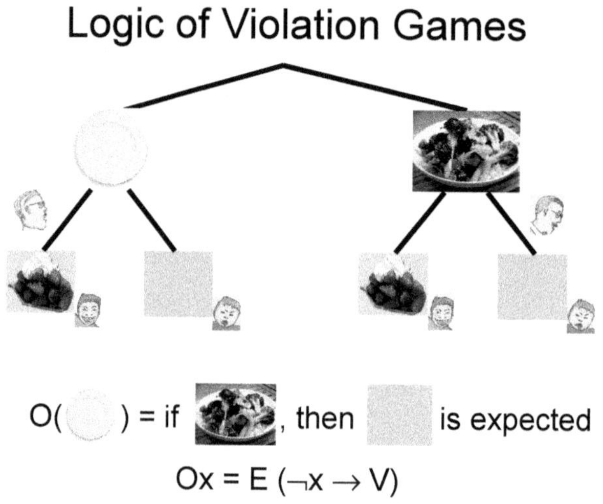

Logic of Violation Games

O( ) = if , then is expected

Ox = E (¬x → V)

Figure 7. Expectation, from [van der Torre, 2010]

The general definition of obligation based on violation games extends this basic idea to behavior over periods of time. Let's consider the three phases in the example. Borrowing from terminology from classical game theory, we say that it is obligatory to eat the vegetables, when not eating them and the strategy that this leads to a violation, is an equilibrium. In the first phase in which the son eats his vegetables, the violation is only implicit since it does not occur. In the second phase not eating the vegetables is identified with the absence of the dessert. In the third phase, the boy may sometimes eat his vegetables, and sometimes not. As long as the norm is in force, he will still believe to be sanctioned most of the time when he does not eat his vegetables. When the sanction is not applied most of the time we have reached a fourth phase, in which we say that the norm is no longer in force.

Summarizing, norms are rules defining a violation game.

Definition 8 (Violation games [van der Torre, 2010]) *Violation games are social interactions among agents to determine whether violations have occurred, and which sanctions will be imposed for such violations. A normative system is a specification of violation games.*

Since norms do not have truth values, we cannot say that two normative systems

Logic of Violation Games

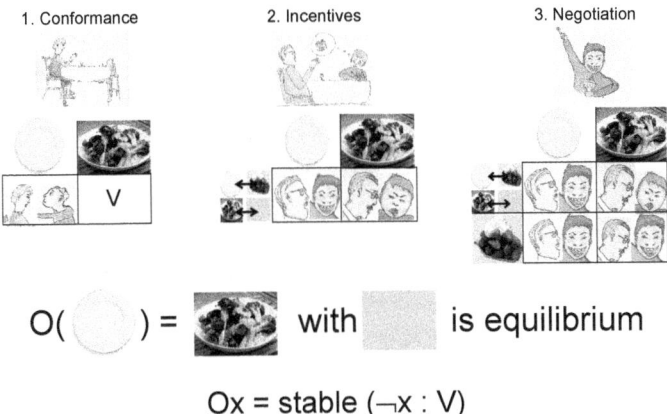

Figure 8. Equilibrium, from [van der Torre, 2010]

are logically equivalent, or that a normative system implies a norm. Therefore it has been proposed to take equivalence of normative systems as the fundamental principle of deontic logic. Implication is then replaced by acceptance and redundancy, which are defined in terms of norm equivalence: a norm is accepted by a normative system if adding it to the normative system leads to an equivalent normative system, and a norm is redundant in a normative system if removing it from the normative system leads to an equivalent normative system. The fundamental notion of equivalence of normative systems can be defined in terms of violation games.

Definition 9 (Equivalence of normative systems [van der Torre, 2010]) *Two normative systems are equivalent if and only if they define the same set of violation games.*

Finally, we can now give a more precise definition of an autonomous system. Remember that auto means self, and nomos means norm.

Definition 10 (Autonomy [van der Torre, 2010]) *A system is autonomous if and only if it can play violation games.*

Violation games are the basis of normative reasoning and deontic logic, but more complex games must be considered too. Consider for example the following situation. If a child is in the water and there is one bystander, chances are that the bystander will jump into the water and save the child. However, if there are one hundred bystanders, chances are that no-one jumps in the water and the child will drown. How to reason about such bystander effects?

Van der Torre suggests that an extension of violation games, called norm creation games [Boella and van der Torre, 2007], may be used to analyze the situation. An agent reasons as follows. What is the explicit norm I would like to adopt for such situations? Clearly, if I would be in the water and I could not swim, or it is my child drowning in the water, then I would like prefer that someone would jump in the water. To be precise, I would accept a norm that in such cases, the norm for each individual would be to jump into the water. Consequently, one should act according to this norm, and everyone should jump into the water. Norm creation games can be used to give a more general definition of a normative system.

Definition 11 (Norm creation games [van der Torre, 2010]) *Norm creation games are social interactions among agents to determine which norms are in force, whether norm violations have occurred, and which sanctions will be imposed for such violations. A normative system is a specification of norm creation games.*

There are many details to be further discussed here. For example, if there is a way to discriminate among the people and it may be assumed that all people would follow this discrimination, then only some people have to jump into the water (the men, the good swimmers if they can be identified, the tall people, and so on). In general, and as common in legal reasoning, the more that is known about the situation, the more can be said about the protocol leading to the norm.

For the semantics of the new deontic logic founded on violation games, one needs a way to derive obligations from norms, as in the iterative detachment approach, or input/output logic. The extension now is to represent the agents and their games into the semantic structures, and derive the norms from that using game theoretic methods. As the norm creation game illustrates, also protocols for norm creation must be represented to model more complex games.

The language of the new deontic logic founded on violation games will be richer than most of the deontic logics studied thus far. There will be formal statements referring to the regulative, permissive and constitutive norms, as in the input/output logic framework, but there will also be an explicit representation of the games the agents are playing. Many choices are possible here, and the area of game theory will lead the way.

We need other approaches that represent norms and obligations at the same time, since deontic logic founded on violation games has to built on it. We also also have to study time, actions, mental modalities, permissions and constitutive norms, since they all play a role in violation games. We also need a precise understanding of Anderson's idea of violation conditions which do not necessarily lead to sanctions, but to the more abstract notion of "a bad state," i.e. a state in which something bad has happened. Whereas many of these deontic problems have been studied in isolation in the deontic logic literature, we believe that violation games will work as a metaphor to bring these problems together, and study their interdependencies.

16 Summary

The aim of this chapter is to introduce readers of the handbook to the area of deontic logic and its challenges. The interested reader is advised to download the handbook of deontic logic and normative systems, and should not take our chapter only as its guidance. In particular, in this chapter we have not gone into the formal aspects of deontic logic. Deontic logicians have developed monadic modal logics, non-monotonic ones, rule based systems, and much more. The formalisms developed in deontic logic have also been adopted by a wider logic community, in particular the preference based deontic logics have been adopted in many areas [Makinson, 1993].

As far as open problems are concerned, in the context of the handbook this concerns mainly the problems of *multiagent* deontic logic and problems related to normative systems. We have addressed the following challenges.

How to reconstruct the history of traditional deontic logic as a challenge to deal with contrary to duty reasoning, violations and preference (Challenge 1)?

What are the challenges in game theoretic approach to normative reasoning (Section 2), which is based on non-deterministic actions (Challenge 2), moral luck (Challenge 3) and procrastination (Challenge 4)?

How to reconstruct the history of modern deontic logic as a challenge to deal with Jørgensen's dilemma and detachment (Challenge 5), and more generally to bridge the tradition of normative system with the tradition of modal deontic logic?

What is the challenge in multi agent detachment of obligations from norms? For example, when detaching obligations from norms, when do agents assume that other agents comply with their norms (Challenge 6)? In game theory, agents assume that other agents are rational in the sense of acting in their best interest. Analogously, multiagent deontic logic raises the question when agents assume that other agents comply with their norms. For answering the question, we assume that every norm is directed towards a single agent, and that the normative system does not change.

How do norm based semantics handle the traditional challenges in deontic logic? These problems are when a set of norms may be termed 'coherent' (Challenge 7), how to deal with normative conflicts (Challenge 8), how to interpret dyadic deontic operators that formalize 'it ought to be that x on conditions α' as $O(x/\alpha)$ (Challenge 9), how various concepts of permission can be accommodated (Challenge 10), how meaning postulates and counts-as conditionals can be taken into account (Challenge 11 and 12), and how sets of norms may be revised and merged (Challenge 13 and 14).

Finally, how can the two approaches of game based deontic logic and norm based deontic logic be combined? (Challenge 15)

Acknowledgement

The authors thank Jan Broersen and Jörg Hansen for their joint work on earlier versions of some sections of this chapter, and Davide Grossi and Xavier Parent for insighful and helpful comments on a preliminary version of this chapter.

The contribution of Gabriella Pigozzi was supported by the Deutsche Forschungsgemeinschaft (DFG) and the Czech Science Foundation (GACR) as part of the joint project From Shared Evidence to Group Attitudes (RO 4548/6-1).

BIBLIOGRAPHY

[Ågotnes *et al.*, 2010] Thomas Ågotnes, Wiebe van der Hoek, and Michael Wooldridge. Robust normative systems and a logic of norm compliance. *Logic Journal of the IGPL*, 18(1):4–30, 2010.

[Alchourrón and Bulygin, 1981] C. E. Alchourrón and E. Bulygin. The expressive conception of norms. In R. Hilpinen, editor, *New Studies in Deontic Logic*, pages pp 95–124. Reidel, Dordrecht, 1981.

[Alchourrón and Makinson, 1981] C. E. Alchourrón and D. Makinson. Hierarchies of regulations and their logic. In R. Hilpinen, editor, *New Studies in Deontic Logic*, pages 125–148. Reidel, Dordrecht, 1981.

[Alchourrón and Makinson, 1982] C. E. Alchourrón and D. Makinson. On the logic of theory change: Contraction functions and their associated revision functions. *Theoria*, 48:14–37, 1982.

[Anderson, 1959] A. R. Anderson. On the logic of commitment. *Philosophical Studies*, 19:23–27, 1959.

[Andrighetto et al., 2013] G. Andrighetto, G. Governatori, P. Noriega, and L. van der Torre, editors. *Normative Multi-Agent Systems*, volume 4 of *Dagstuhl Follow-Ups*. Schloss Dagstuhl–Leibniz-Zentrum fuer Informatik, Dagstuhl, Germany, 2013.

[Åqvist, 1967] L. Åqvist. Good Samaritans, contrary-to-duty imperatives and epistemic obligations. *Noûs*, 1:361–379, 1967.

[Artosi et al., 2004] A. Artosi, A. Rotolo, and S. Vida. On the logical nature of count-as conditionals. In *The Law of Electronic Agents. Proceedings of the LEA04 Workshop*, pages 9–33, Bologna, 2004.

[Aucher et al., 2009] G. Aucher, D. Grossi, A. Herzig, and E. Lorini. Dynamic context logic. In X. He, J. Horty, and E. Pacuit, editors, *Logic, Rationality, and Interaction: Second International Workshop, LORI 2009, Chongqing, China, October 8-11, 2009. Proceedings*, pages 15–26, Berlin, Heidelberg, 2009. Springer Berlin Heidelberg.

[Baral et al., 1992] C. Baral, S. Kraus, J. Minker, and V. S. Subrahmanian. Combining knowledge bases consisting of first-order theories. *Computational Intelligence*, 8:45–71, 1992.

[Boella and van der Torre, 2003] G. Boella and L. van der Torre. Permissions and obligations in hierarchical normative systems. In *Proceedings of the 9th International Conference on Artificial Intelligence and Law, ICAIL 2003, Edinburgh, Scotland, UK, June 24-28, 2003*, pages 109–118, Edinburgh, 2003.

[Boella and van der Torre, 2006a] G. Boella and L. van der Torre. Constitutive norms in the design of normative multiagent systems. In *Computational Logic in Multi-Agent Systems, 6th International Workshop, CLIMA VI*, LNCS 3900, pages 303–319. Springer, 2006.

[Boella and van der Torre, 2006b] G. Boella and L. van der Torre. A logical architecture of a normative system. In *Deontic Logic and Artificial Normative Systems, 8th International Workshop on Deontic Logic in Computer Science (DEON'06)*, volume 4048 of *LNCS*, pages 24–35, Berlin, 2006. Springer.

[Boella and van der Torre, 2007] Guido Boella and Leendert van der Torre. A game-theoretic approach to normative multi-agent systems. In *Normative Multi-agent Systems, 18.03. - 23.03.2007*, Dagstuhl Seminar Proceedings, 2007.

[Boella et al., 2006] G. Boella, L. van der Torre, and H. Verhagen. Introduction to normative multiagent systems. *Computation and Mathematical Organizational Theory, special issue on normative multiagent systems*, 12(2-3):71–79, 2006.

[Boella et al., 2008] G. Boella, J. M. Broersen, and L. van der Torre. Reasoning about constitutive norms, counts-as conditionals, institutions, deadlines and violations. In The Duy Bui, Tuong Vinh Ho, and Quang-Thuy Ha, editors, *Intelligent Agents and Multi-Agent Systems, 11th Pacific Rim International Conference on Multi-Agents, PRIMA 2008, Hanoi, Vietnam, December 15-16, 2008. Proceedings*, volume 5357 of *Lecture Notes in Computer Science*, pages 86–97. Springer, 2008.

[Boella et al., 2016] G. Boella, G. Pigozzi, and L. van der Torre. AGM contraction and revision of rules. *Journal of Logic, Language and Information*, 25(3-4):273–297, 2016.

[Booth et al., 2006] R. Booth, S. Kaci, and L. van der Torre. Merging rules: Preliminary version. In *Proceedings of the Eleventh International Workshop on Non-Monotonic Reasoning (NMR'06)*, 2006.

[Boutilier, 1994] C. Boutilier. Toward a logic for qualitative decision theory. In *Proceedings of the 4th International Conference on Principles of Knowledge Representation and Reasoning (KR'94). Bonn, Germany, May 24-27, 1994.*, pages 75–86, Bonn, 1994.

[Broersen and van der Torre, 2003] J. M. Broersen and L. van der Torre. What an agent ought to do. *Artif. Intell. Law*, 11(1):45–61, 2003.

[Broersen and van der Torre, 2007] J. Broersen and L. van der Torre. Reasoning about norms, obligations, time and agents. In *Agent Computing and Multi-Agent Systems, 10th Pacific Rim International Conference on Multi-Agents, PRIMA 2007, Bangkok, Thailand, November 21-23, 2007. Revised Papers*, pages 171–182, 2007.

[Broersen et al., 2003] J. Broersen, M. Dastani, and L. van der Torre. BDIOCTL: Obligations and the specification of agent behavior. In Georg Gottlob and Toby Walsh, editors, *IJCAI-03, Proceedings of the Eighteenth International Joint Conference on Artificial Intelligence, Acapulco, Mexico, August 9-15, 2003*, pages 1389–1390. Morgan Kaufmann, 2003.

[Broersen et al., 2004] J. M. Broersen, F. Dignum, V. Dignum, and J-J Ch. Meyer. Designing a deontic logic of deadlines. In A. Lomuscio and D. Nute, editors, *Deontic Logic in Computer Science, 7th International Workshop on Deontic Logic in Computer Science, DEON 2004, Madeira, Portugal, May*

26-28, 2004. Proceedings, volume 3065 of *Lecture Notes in Computer Science*, pages 43–56. Springer, 2004.

[Bulygin, 1986] E. Bulygin. Permissive norms and normative systems. In A. Martino and F. Socci Natali, editors, *Automated Analysis of Legal Texts*, pages 211–218. Publishing Company, Amsterdam, 1986.

[Carmo and Jones, 2002] J. Carmo and A.J.I. Jones. Deontic logic and contrary-to-duties. In D.Gabbay and F.Guenthner, editors, *Handbook of Philosophical Logic*, volume 8, pages 265–343. Kluwer, 2002.

[Castañeda, 1981] H-N Castañeda. The paradoxes of deontic logic: The simplest solution to all of them in one fell swoop. In R. Hilpinen, editor, *New Studies in Deontic Logic: Norms, Actions, and the Foundations of Ethics*, pages 37–85. Springer Netherlands, Dordrecht, 1981.

[Castañeda, 1982] H-N Castañeda. The logical structure of legal systems: A new perspective. In A. A. Martino, editor, *Deontic Logic, Computational Linguistics and Legal Information Systems, volume II*, page 21?37. North Holland Publishing, Dordrecht, 1982.

[Chisholm, 1963] R.M. Chisholm. Contrary-to-duty imperatives and deontic logic. *Analysis*, 24:33?36, 1963.

[Cholvy and Cuppens, 1999] L. Cholvy and F. Cuppens. Reasoning about norms provided by conflicting regulations. In P. McNamara and H. Prakken, editors, *Norms, Logics and Information Systems*, pages 247–264. IOS, Amsterdam, 1999.

[Cholvy and Garion, 2001] L. Cholvy and C. Garion. An attempt to adapt a logic of conditional preferences for reasoning with contrary-to-duties. *Fundam. Inform.*, 48(2-3):183–204, 2001.

[Condoravdi and Lauer, 2016] C. Condoravdi and S. Lauer. Anankastic conditionals are just conditionals. *Semantics and Pragmatics*, 9(8):1–69, November 2016.

[da Costa Pereira et al., 2018] C. da Costa Pereira, B. Liao, A. Malerba, A. Rotolo, A. Tettamanzi, and L. van der Torre. Handling norms in mas by means of argumentation. In *This volume*. 2018.

[Danielsson, 1968] S. Danielsson. *Preference and Obligation: Studies in the Logic of Ethics*. Filosofiska freningen, Uppsala, 1968.

[Forrester, 1984] J. Forrester. Gentle murder, or the adverbial Samaritan. *Journal of Philosophy*, 81:193–196, 1984.

[Frantz and Pigozzi, forthcoming] C. Frantz and G. Pigozzi. Modelling norm dynamics in MAS. In A. Chopra, L. van der Torre, H. Verhagen, and S. Villata, editors, *Handbook of Normative Multiagent Systems*. College Publications, forthcoming.

[Gabbay et al., 2013] D. Gabbay, J. Horty, X. Parent, R. van der Meyden, and L. van der Torre, editors. *Handbook of Deontic Logic and Normative Systems*. College Publications, London, UK, 2013.

[Goble, 2000] L. Goble. Multiplex semantics for deontic logic. *Nordic Journal of Philosophical Logic*, 5:113–134, 2000.

[Goble, 2005] L. Goble. A logic for deontic dilemmas. *Journal of Applied Logic*, 3:461–483, 2005.

[Goble, 2009] L. Goble. Normative conflicts and the logic of 'ought'. *Noûs*, 43:450–489, 2009.

[Goble, 2013] L. Goble. Prima facie norms, normative conflicts, and dilemmas. In D. Gabbay, J. Horty, X. Parent, R. van der Meyden, and L. van der Torre, editors, *Handbook of Deontic Logic*, pages 249–352. College Publications, 2013.

[Governatori and Rotolo, 2008] G. Governatori and A. Rotolo. A computational framework for institutional agency. *Artificial Intelligence and Law*, 16(1):25–52, 2008.

[Governatori and Rotolo, 2010] G. Governatori and A. Rotolo. Changing legal systems: legal abrogations and annulments in defeasible logic. *Logic Journal of IGPL*, 18(1):157–194, 2010.

[Governatori et al., 2013] G. Governatori, A. Rotolo, F. Olivieri, and S. Scannapieco. Legal contractions: A logical analysis. In E. Francesconi and B. Verheij, editors, *Proceedings of the 9th International Conference on Artificial Intelligence and Law, ICAIL 2003, Edinburgh, Scotland, UK, June 24-28, 2003*, pages 63–72. ACM, 2013.

[Governatori, 2015] G. Governatori. Thou shalt is not you will. In *Proceedings of the 15th International Conference on Artificial Intelligence and Law, ICAIL 2015, San Diego, CA, USA, June 8-12, 2015*, pages 63–68, 2015.

[Grégoire, 2004] E. Grégoire. Fusing legal knowledge. In *Proceedings of the 2004 IEEE Int. Conf. on Information Reuse and Integration (IEEE-IRI'2004)*, pages 522–529, 2004.

[Grossi and Jones, 2013] D. Grossi and A. Jones. Constitutive norms and counts-as conditionals. In D. Gabbay, J. Horty, X. Parent, R. van der Meyden, and L. van der Torre, editors, *Handbook of Deontic Logic and Normative Systems*, pages 407–441. College Publications, 2013.

[Grossi et al., 2006] D. Grossi, J-J Ch. Meyer, and F. Dignum. Classificatory aspects of counts-as: An analysis in modal logic. *Journal of Logic and Computation*, 16(5):613–643, 2006.

[Grossi et al., 2008] D. Grossi, J-J Ch. Meyer, and F. Dignum. The many faces of counts-as: A formal analysis of constitutive-rules. *Journal of Applied Logic*, 6(2):192–217, 2008.

[Hansen, 2001] J. Hansen. Sets, sentences, and some logics about imperatives. *Fundamenta Informaticae*, 48:205–226, 2001.
[Hansen, 2004] J. Hansen. Problems and results for logics about imperatives. *Journal of Applied Logic*, 2:39–61, 2004.
[Hansen, 2005a] J. Hansen. Conflicting imperatives and dyadic deontic logic. *Journal of Applied Logic*, 3:484–511, 2005.
[Hansen, 2005b] J. Hansen. Deontic logics for prioritized imperatives. *Artificial Intelligence and Law*, 3(3-4):484–511, 2005.
[Hansen, 2008] J. Hansen. Prioritized conditional imperatives: problems and a new proposal. *Autonomous Agents and Multi-Agent Systems*, 17(1):11–35, 2008.
[Hansson, 1969] B. Hansson. An analysis of some deontic logics. *Nôus*, 3:373–398, 1969. Reprinted in [Hilpinen, 1971] 121–147.
[Hansson, 2013] S. O. Hansson. The varieties of permission. In D. Gabbay, J. Horty, X. Parent, R. van der Meyden, and L. van der Torre, editors, *Handbook of Deontic Logic and Normative Systems*, pages 195–240. College Publications, 2013.
[Hare, 1981] R. M. Hare. *Moral Thinking*. Clarendon Press, Oxford, 1981.
[Hilpinen and McNamara, 2013] R. Hilpinen and P. McNamara. Deontic logic: A historical survey and introduction. In D. Gabbay, J. Horty, X. Parent, R. van der Meyden, and L. van der Torre, editors, *Handbook of Deontic Logic and Normative Systems*, pages 3–136. College Publications, 2013.
[Hilpinen, 1971] R. Hilpinen, editor. *Deontic Logic: Introductory and Systematic Readings*. Reidel, Dordrecht, 1971.
[Horty, 1994] J. F. Horty. Moral dilemmas and nonmonotonic logic. *Journal of Philosophical Logic*, 23:35–65, 1994.
[Horty, 1997] J. F. Horty. Nonmonotonic foundations for deontic logic. In D. Nute, editor, *Defeasible Deontic Logic*, pages 17–44. Kluwer, Dordrecht, 1997.
[Horty, 2001] J. Horty. *Agency and deontic logic*. Oxford University Press, 2001.
[Horty, 2003] J. F. Horty. Reasoning with moral conflicts. *Noûs*, 37:557–605, 2003.
[Horty, 2012] J. F. Horty. *Reasons and Defaults*. Oxford University Press, 2012.
[Jackson and Pargetter, 1986] F. Jackson and R. Pargetter. Oughts, options, and actualism. *Philosophical Review*, 99:233–255, 1986.
[Jones and Sergot, 1996] A. Jones and M. Sergot. A formal characterisation of institutionalised power. *Journal of IGPL*, 3:427–443, 1996.
[Jørgensen, 1938] J. Jørgensen. Imperatives and logic. *Erkenntnis*, 7:288–296, 1938.
[Kamp, 1973] H. Kamp. Free choice permission. *Proceedings of the Aristotelian Society*, 74:57–74, 1973.
[Kanger, 1957] S. Kanger. New foundations for ethical theory: Part 1. duplic., 42 p., 1957. Reprinted in [Hilpinen, 1971] 36–58.
[Kolodny and MacFarlane, 2010] N. Kolodny and J. MacFarlane. Ifs and oughts. *Journal of Philosophy*, 107(3):115–143, 2010.
[Konieczny and Grégoire, 2006] S. Konieczny and E. Grégoire. Logic-based approaches to information fusion. *Information Fusion*, 7:4–18, 2006.
[Konieczny and P. Pérez, 2011] S. Konieczny and Ramón P. Pérez. Logic based merging. *Journal of Philosophical Logic*, 40(2):239–270, 2011.
[Konieczny, 1999] S. Konieczny. *Sur la Logique du Changement: Révision et Fusion de Bases de Connaissance*. PhD thesis, University of Lille, France, 1999.
[Konieczny, 2000] S. Konieczny. On the difference between merging knowledge bases and combining them. In *KR 2000, Principles of Knowledge Representation and Reasoning Proceedings of the Seventh International Conference, Breckenridge, Colorado, USA, April 11-15, 2000.*, volume 8, pages 135–144. Morgan Kaufmann, 2000.
[Lemmon, 1962] E. J. Lemmon. Moral dilemmas. *The Philosophical Review*, 70:139–158, 1962.
[Lewis, 1973] D. Lewis. *Counterfactuals*. Basil Blackwell, Oxford, 1973.
[Lewis, 1974] D. Lewis. Semantic analyses for dyadic deontic logic. In S. Stenlund, editor, *Logical Theory and Semantic Analysis*, pages 1 – 14. Reidel, Dordrecht, 1974.
[Lewis, 1979] D. Lewis. A problem with permission. In E. Saarinen, R. Hilpinen, I. Niiniluoto, and M. P. Hintikka, editors, *Essays in Honour of Jaako Hintikka: On the Occasion of His Fiftieth Birthday on January 12, 1979*, pages 163–175. Reidel, Dordrecht, 1979.
[Lin and Mendelzon, 1996] J. Lin and A. Mendelzon. Merging databases under constraints. *International Journal of Cooperative Information Systems*, 7:55–76, 1996.

[Lindahl and Odelstad, 2006] L. Lindahl and J. Odelstad. Intermediate concepts in normative systems. In L. Goble and J-J Ch. Meyer, editors, *Deontic Logic and Artificial Normative Systems: 8th International Workshop on Deontic Logic in Computer Science, DEON 2006, Utrecht, The Netherlands, July 12-14, 2006. Proceedings*, pages 187–200. Springer Berlin Heidelberg, Berlin, Heidelberg, 2006.

[Lindahl and Odelstad, 2013] L. Lindahl and J. Odelstad. The theory of joining-systems. In D. Gabbay, J. Horty, X. Parent, R. van der Meyden, and L. van der Torre, editors, *Handbook of Deontic Logic and Normative Systems*, pages 545–634. College Publications, 2013.

[Lindahl, 1997] L. Lindahl. Norms, meaning postulates, and legal predicates. In E. Garzón Valdés, editor, *Normative Systems in Legal and Moral Theory. Festschrift for Carlos E. Alchourrón and Eugenio Bulygin*, pages 293–307. Duncker & Humblot, Berlin, 1997.

[Lorini and Longin, 2008] E. Lorini and D. Longin. A logical account of institutions: From acceptances to norms via legislators. In *Proceedings of the Eleventh International Conference on Principles of Knowledge Representation and Reasoning, KR'08*, pages 38–48. AAAI Press, 2008.

[Lorini et al., 2009] E. Lorini, D. Longin, B. Gaudou, and A. Herzig. The logic of acceptance: Grounding institutions on agents' attitudes. *Journal of Logic and Computation*, 19(6):901–940, 2009.

[Makinson and van der Torre, 2000] D. Makinson and L. van der Torre. Input-output logics. *Journal of Philosophical Logic*, 29(4):383–408, 2000.

[Makinson and van der Torre, 2001] D. Makinson and L. van der Torre. Constraints for input-output logics. *Journal of Philosophical Logic*, 30(2):155–185, 2001.

[Makinson and van der Torre, 2003a] D. Makinson and L. van der Torre. Permissions from an input-output perspective. *Journal of Philosophical Logic*, 32(4):391–416, 2003.

[Makinson and van der Torre, 2003b] D. Makinson and L. van der Torre. What is input/output logic? In B. Löwe, W. Malzkorn, and T. Räsch, editors, *Foundations of the Formal Sciences II : Applications of Mathematical Logic in Philosophy and Linguistics (Papers of a conference held in Bonn, November 10-13, 2000)*, Trends in Logic, vol. 17, pages 163–174, Dordrecht, 2003. Kluwer. Reprinted in this volume.

[Makinson, 1993] D. Makinson. Five faces of minimality. *Studia Logica*, 52:339–379, 1993.

[Makinson, 1999] D. Makinson. On a fundamental problem of deontic logic. In P. McNamara and H. Prakken, editors, *Norms, Logics and Information Systems. New Studies on Deontic Logic and Computer Science*, pages 29–54. IOS Press, 1999.

[Marcus, 1980] R. B. Marcus. Moral dilemmas and consistency. *Journal of Philosophy*, 77:121?136, 1980.

[Niiniluoto, 1986] I. Niiniluoto. Hypothetical imperatives and conditional obligation. *Synthese*, 66:111–133, 1986.

[Olde Loohuis, 2009] Loes Olde Loohuis. Obligations in a responsible world. In *LORI*, pages 251–262, 2009.

[Pacuit et al., 2006] E. Pacuit, R. Parikh, and E. Cogan. The logic of knowledge based obligation. *Synthese*, 149(2):311–341, 2006.

[Parent and van der Torre, 2014] X. Parent and L. van der Torre. "sing and dance!" - input/output logics without weakening. In F. Cariani, D. Grossi, J. Meheus, and X. Parent, editors, *Deontic Logic and Normative Systems - 12th International Conference, DEON 2014, Ghent, Belgium, July 12-15, 2014. Proceedings*, volume 8554 of *Lecture Notes in Computer Science*, pages 149–165. Springer, 2014.

[Parent and van der Torre, 2017a] X. Parent and L. van der Torre. Detachment in normative systems: examples, inference patterns, properties. In *This volume*. 2017.

[Parent and van der Torre, 2017b] X. Parent and L van der Torre. The pragmatic oddity in norm-based deontic logics. In *Proceedings of The 16th International Conference on Artificial Intelligence and Law*, 2017.

[Parent, 2008] X. Parent. On the strong completeness of åqvist's dyadic deontic logic G. In *Deontic Logic in Computer Science, 9th International Conference, DEON 2008, Luxembourg, Luxembourg, July 15-18, 2008. Proceedings*, pages 189–202, 2008.

[Pearl, 1993] J. Pearl. From conditional oughts to qualitative decision theory. In D. Heckerman and E. H. Mamdani, editors, *UAI '93: Proceedings of the Ninth Annual Conference on Uncertainty in Artificial Intelligence, The Catholic University of America, Providence, Washington, DC, USA, July 9-11, 1993*, pages 12–22. Morgan Kaufmann, 1993.

[Prakken and Sergot, 1996] H. Prakken and M. Sergot. Contrary-to-duty obligations. *Studia Logica*, 57:91–115, 1996.

[Ross, 1930] W. D. Ross. *The Right and the Good*. Clarendon Press, Oxford, 1930.

[Ross, 1941] A. Ross. Imperatives and logic. *Theoria*, 7:53–71, 1941. Reprinted in *Philosophy of Science* 11:30–46, 1944.

[Sartre, 1946] J.-P. Sartre. *L'Existentialisme est un Humanisme*. Nagel, Paris, 1946.

[Searle, 1995] J.R. Searle. *The Construction of Social Reality*. The Free Press, New York, 1995.

[Smiley, 1963] T. J. Smiley. The logical basis of ethics. *Acta Philosophica Fennica*, 16:237–246, 1963.

[Spohn, 1975] W. Spohn. An analysis of Hansson's dyadic deontic logic. *Journal of Philosophical Logic*, 4:237–252, 1975.

[Stenius, 1963] E. Stenius. The principles of a logic of normative systems. *Acta Philosophica Fennica*, 16:247–260, 1963.

[Stolpe, 2010] A. Stolpe. Norm-system revision: Theory and application. *Artificial Intelligence and Law*, 18:247–283, 2010.

[Sun and van der Torre, 2014] X. Sun and L. van der Torre. Combining constitutive and regulative norms in input/output logic. In F. Cariani, D. Grossi, J. Meheus, and X. Parent, editors, *Deontic Logic and Normative Systems - 12th International Conference, DEON 2014, Ghent, Belgium, July 12-15, 2014. Proceedings*, volume 8554 of *Lecture Notes in Computer Science*, pages 241–257. Springer, 2014.

[van Benthem *et al.*, 2014] J. van Benthem, D. Grossi, and F. Liu. Priority structures in deontic logic. *Theoria*, 80(2):116–152, 2014.

[van der Torre and Tan, 1998] L. van der Torre and Y. Tan. An update semantics for prima facie obligations. In *Proceedings of The 17th European Conference on Artificial Intelligence*, pages 38–42, 1998.

[van der Torre and Tan, 1999] L. van der Torre and Y. Tan. An update semantics for defeasible obligations. In K. B. Laskey and H. Prade, editors, *UAI '99: Proceedings of the Fifteenth Conference on Uncertainty in Artificial Intelligence, Stockholm, Sweden, July 30 - August 1, 1999*, pages 631–638. Morgan Kaufmann, 1999.

[van der Torre, 2010] L. van der Torre. Violation games: a new foundation for deontic logic. *Journal of Applied Non-Classical Logics*, 20(4):457–477, 2010.

[van Fraassen, 1972] B. C. van Fraassen. The logic of conditional obligation. *Journal of Philosophical Logic*, 1:417–438, 1972.

[van Fraassen, 1973] B. van Fraassen. Values and the heart's command. *Journal of Philosophy*, 70:5–19, 1973.

[von Wright, 1951a] G. H. von Wright. Deontic logic. *Mind*, 60:1–15, 1951.

[von Wright, 1951b] G. H. von Wright. *An Essay in Modal Logic*. North-Holland, Amsterdam, 1951.

[von Wright, 1956] G. H. von Wright. A note on deontic logic and derived obligation. *Mind*, 65:507–509, 1956.

[von Wright, 1957] G.H von Wright. *Logical Studies*. Routledge and Kegan, London, 1957.

[von Wright, 1961] G. H. von Wright. A new system of deontic logic. *Danish Yearbook of Philosophy*, 1:173–182, 1961. Reprinted in [Hilpinen, 1971] 105–115.

[von Wright, 1962] G. H. von Wright. A correction to a new system of deontic logic. *Danish Yearbook of Philosophy*, 2:103–107, 1962. Reprinted in [Hilpinen, 1971] 115–119.

[von Wright, 1963] G. H. von Wright. *Norm and Action*. Routledge & Kegan Paul, London, 1963.

[von Wright, 1968] G. H. von Wright. *An Essay in Deontic Logic and the General Theory of Action*. North Holland, Amsterdam, 1968.

[von Wright, 1983] G. H. von Wright. Norms, truth and logic. In G. H. von Wright, editor, *Practical Reason: Philosophical Papers vol. I*, pages 130–209. Blackwell, Oxford, 1983.

[Ziemba, 1971] Z. Ziemba. Deontic syllogistics. *Studia Logica*, 28:139–159, 1971.

10
Detachment in Normative Systems: Examples, Inference Patterns, Properties

XAVIER PARENT AND LEENDERT VAN DER TORRE

1 Introduction

The *Handbook of Deontic Logic and Normative Systems* [Gabbay et al., 2013] describes a debate between the traditional or standard semantics for deontic logic and alternative approaches. The traditional semantics is based on possible world models, whereas many alternative approaches refer to foundations in normative systems, algebraic methods, or non-monotonic logic. In particular, whereas Anderson [1956] argued to refer explicitly to normative systems and also Åqvist [2002] builds on it, various alternative approaches such as input/output logic [Makinson, 1999; Makinson and van der Torre, 2000] represent norms explicitly in the semantics.

Proponents of alternative approaches typically refer to limitations in the traditional approach, although the traditional approach has been generalised or extended to handle many of these limitations [Horty, 2014]. The development of formal and conceptual bridges between traditional and alternative approaches is one of the main current challenges in the area of normative systems and deontic logic. The following three limitations are frequently discussed.

Dilemmas. Examples discussed in the literature are those of van Fraassen [1973], Makinson's Möbius strip [Makinson, 1999], Prakken and Sergot's cottage regulations [Prakken and Sergot, 1996], and Horty's priority examples [Horty, 2007].

Defeasibility. The traditional approach does not distinguish various kinds of defeasibility. Legal norms are often assumed to be defeasible, and there is an increasing interest in philosophy in defeasibility, such as the defeasibility of moral reasons [Horty, 2007; Parent, 2011].

Identity. Many traditional deontic logics validate the formula $\bigcirc(\alpha|\alpha)$, read as "α is obligatory given α," "whose intuitive standing is open to question" [Makinson, 1999]. This has been dismissed as a harmless borderline case by proponents of the traditional semantics, but it hinders the representation of fulfilled obligations and violations, playing a central role in normative reasoning. Consider a logic validating identity: the formula $\bigcirc(\alpha|\neg\alpha)$, which represents explicitly that there is a violation, is not satisfiable; the obligation of α disappears, in context $\neg\alpha$. (See Section 2 in this chapter.)

Different disciplines and applications have put forward different requirements for the development of formal methods for normative systems and deontic logic. For example, in linguistics compositionality is an important requirement, as deontic statements must be integrated into a larger theory of language. In legal informatics, constitutive and permissive norms play a central role, and legal norms may conflict. It is an open problem whether there can be a unique formal method which can be widely applied across disciplines, or even whether there is a single framework of formal methods which can be used. In this sense, there may be an important distinction between classical and normative reasoning, since there is a unique first order logic for classical logic reasoning about the real world using sets, relations and functions. The situation for normative reasoning may be closer to the situation for non-monotonic reasoning, where also a family of reasoning methods have been proposed, rather than a unique method.

In this chapter we do not want to take a stance on these discussions, but we want to provide techniques and ideas to compare traditional and alternative approaches. We focus on inference patterns and proof-theory instead of semantical considerations. In particular, in this paper we are interested in the question:

> Which obligations can be detached from a set of rules or conditional norms in a context?

Our angle is different from the more traditional one in terms of inference rules.

There are many frameworks for reasoning about rules and norms, and there are many examples about detachment from normative systems, many of them problematic in some sense. However, there are few properties to compare and analyse ways to detach obligations from rules and norms, and they are scattered over the literature. We are not aware of a systematic overview of these properties. We address our research question by surveying examples, inference patterns and properties from the deontic logic literature.

Examples: Van Fraassen's paradox, Forrester's paradox, Prakken and Sergot's cottage regulations, Jeffrey's disarmament example, Chisholm's paradox, Makinson's möbius strip, and Horty's priority examples. They illustrate challenges for normative reasoning with deontic dilemmas, contrary-to-duty reasoning, defeasible obligations, reasoning by cases, deontic detachment, prioritised obligations, and combinations of these.

Inference Patterns: Conjunction, weakening of the consequent, forbidden conflict, factual detachment, strengthening of the antecedent, violation detection, compliance detection, reinstatement, deontic detachment, transitivity, and various variants of these patterns.

Framework: We develop a *framework* for deontic logics representing and resolving conflicts. By framework we mean that we do not develop a single logic, but many of them. This reflects that there is not a single logic of obligation and permission, but many of them, and which one is to be used depends on the application.

Properties: Factual detachment, violation detection, substitution, replacements of equivalents, implication, paraconsistency, conjunction, factual monotony, norm monotony, and norm induction.

The term "property" is more general than the term "inference pattern". An inference pattern describes a property of a certain form. The inference patterns listed above appear also in the list properties. For instance, factual monotony echoes strengthening of the antecedent. In some cases, we use the same name for both the inference pattern and the corresponding property.

A formal framework to compare formal methods should make as little assumptions as possible, so it is widely applicable. We only assume that the context is a set of facts $\{a, b, \ldots\}$ and that the conditional norms are of the type "if a is the case, then it ought to be the case that b" where a and b are sentences of a propositional language. This is more general than some rule-based languages based on logic programming, where a is restricted to a conjunction of literals and b is a single literal. However, it is less expressive than many other languages, that contain, for example, modal or first order sentences, constitutive and permissive norms, mixed norms such as "if a is permitted, then b is obligatory," nested operators, time, actions, knowledge, and so on. There are few benchmark examples discussed in the literature for such an extended language (see [Governatori and Hashmi, 2015] for a noteworthy exception) and we are not aware of any properties specific for such extended languages. Extending our formal framework and properties to such extended languages is therefore left to further research.

Our framework is built around a notion of detachment. In traditional approaches "if a, then it ought that b" is typically written as either $a \rightarrow \bigcirc b$ or as $\bigcirc(b|a)$, and in alternative approaches it is sometimes written as (a, b). To be able to compare the different reasoning methods, we will not distinguish between these ways to represent normative systems. The challenge for comparing the formal approaches is that traditional methods typically derive conditional obligations, whereas alternative methods typically do not, maybe because they assume norms do not have truth values and thus they cannot be derived from other norms. Instead, they derive only unconditional obligations. To compare these approaches, one may assume that the derivation of a conditional obligation "if a, then it ought that b" is short for "if the context is exactly $\{a\}$, then the obligation $\bigcirc b$ is detached." Alternatively, the detachment of an obligation for b in context a in alternative systems may be written as the derivation of a pair (a, b), as it is done in the proof theory of input/output logics [Makinson, 1999;

Makinson and van der Torre, 2000]. These issues are discussed in more detail in Section 3 of this chapter.

A remark on notation and terminology. We use Greek letters $\alpha, \beta, \gamma, \ldots$ for propositional formulas, and roman letters $a, b, c, \ldots, p, q, \ldots$ for (distinct) propositional atoms. Throughout this chapter the terms "rule" and "conditional norm" will be used interchangeably. The term "rule" is most often used in computer science (with reference to so-called rule-systems and expert systems), and the term "conditional norm" in philosophy and linguistics. Readers should feel free to use the term they prefer. The unconditional obligation for α will be written as $\bigcirc\alpha$, while the conditional obligation for α given β will be written as $O(\alpha|\beta)$, or as (β, α). We do not assume a specific semantics for these constructs.

We give two examples below.

Example 1.1 (Deontic explosion) *The deontic explosion requirement says that we should not derive all obligations from a dilemma. Now consider a dilemma with obligations for $\alpha \wedge \beta$ and $\neg\alpha \wedge \gamma$. It may be tempting to think that an obligation for $\beta \wedge \gamma$ should follow:*

$$\frac{\bigcirc(\alpha \wedge \beta) \quad \bigcirc(\neg\alpha \wedge \gamma)}{\bigcirc\beta \qquad \bigcirc\gamma}$$
$$\overline{\bigcirc(\beta \wedge \gamma)}$$

Assuming that we have replacements by logical equivalents, if we substitute a for α, $a \vee b$ for β, and $\neg a \vee b$ for γ, then we would derive from the obligations for a and $\neg a$ the obligation for c: deontic explosion. We should not derive the obligation for $\beta \wedge \gamma$, because $\alpha \wedge \beta$ and $\neg\alpha \wedge \gamma$ are classically inconsistent. As we show in Section 2.1, the obligation for $\beta \wedge \gamma$ should be derived only under suitable assumptions.

Example 1.2 (Aggregation) *Consider an iterative approach deriving from the two norms "obligatory c given $a \wedge b$" and "obligatory b given a" that in some sense we have in context a that c is obligatory. This derivation of the obligation for c is made by so-called deontic detachment, because it is derived from the fact a together with the obligation for b. However, if the input is a together with the negation of b, then (intuitively) c should not be derived. However, we can (still intuitively) make the following two derivations. First, we can derive "obligatory a and b given c," a norm which is accepted by the two norms (Parent and van der Torre [2014a; 2014b]).*

$$\frac{\bigcirc(\alpha|\beta \wedge \gamma), \bigcirc(\beta|\gamma)}{\bigcirc(\alpha \wedge \beta|\gamma)} \qquad \frac{(\gamma, \beta), (\gamma \wedge \beta, \alpha)}{(\gamma, \beta \wedge \alpha)}$$

Second, we can also derive the ternary norm "given α, and assuming β, γ is obligatory." However, we would need to extend the language with such expressions as done by van der Torre [2003] and Xin & van der Torre [2014]. Different motivations for *using a ternary operator can be given. For instance, one may want to reason about exceptions to norms. This is the approach taken by van der Torre [2003], who works with expressions of the form "given* α, γ *is obligatory unless* β."

This chapter is organised as follows. In Section 2 we introduce benchmark examples of deontic logic, and discuss them using inference patterns. In Section 3, we introduce the formal framework and its properties. Our approach is general and conceptual, and we abstract away from any specific system from literature. The reader will find in the *Handbook of Deontic Logic and Normative Systems* sample systems which can serve to exemplify the general considerations offered in this chapter.

The present chapter does not cover the notion of permission nor does it cover the notion of counts-as conditional. These topics will be a subject for future research. The reader is referred to the chapter by S. O. Hansson and to the chapter by A. Jones and D. Grossi in the aforementioned handbook for an overview of the state-of-the-art and perspectives for future research regarding these notions.

The present chapter complements the chapter "Multiagent Deontic Logic and its Challenges from a Normative Systems Perspective", by G. Pigozzi and L. van der Torre, contained in this volume. There is an inevitable overlap in coverage between the two chapters. They both discuss topics that have played a prominent role in the development of the field, like the topic of contrary-to-duty (CTD) reasoning and the topic of conflicts. However, overlapping topics are discussed from a different perspective. The present chapter focuses on the single agent case, which must be understood first before moving on to the multiagent case, which is dealt with in the other chapter. Our aim is to present a critical review of benchmark examples from literature, which may serve as a reference base for subsequent research. We discuss these benchmark examples with reference to a number of inference patterns, rather than in the context of some specific systems. We extract from these benchmark examples a number of core properties, which may serve as a tool for comparing existing and future systems.

2 Benchmark examples and inference patterns

In this section we discuss benchmark examples of deontic logic. The analysis in this section is based on a number of inference patterns. We do not consider ways in which deontic statements can be given a semantics. These principles must be understood as expressing strict rules. For future reference, we list the inference patterns in Table 1, in the order they are discussed in this section.

pattern	name
$\bigcirc \alpha_1, \bigcirc \alpha_2 / \bigcirc(\alpha_1 \wedge \alpha_2)$	AND
$\bigcirc \alpha_1, \bigcirc \alpha_2, \Diamond(\alpha_1 \wedge \alpha_2) / \bigcirc(\alpha_1 \wedge \alpha_2)$	RAND
$\bigcirc \alpha_1 / \bigcirc(\alpha_1 \vee \alpha_2)$	W
$\bigcirc(\alpha_1|\beta), \bigcirc(\alpha_2|\beta), \Diamond(\alpha_1 \wedge \alpha_2) / \bigcirc(\alpha_1 \wedge \alpha_2|\beta)$	RANDC
$\bigcirc(\alpha_1|\beta) / \bigcirc(\alpha_1 \vee \alpha_2|\beta)$	WC
$\bigcirc(\alpha_1|\beta), \bigcirc(\alpha_2|\beta), \Diamond(\alpha_1 \wedge \alpha_2 \wedge \beta) / \bigcirc(\alpha_1 \wedge \alpha_2|\beta)$	RANDC2
$\bigcirc(\alpha_1 \wedge \alpha_2|\beta_1), \bigcirc(\neg \alpha_1 \wedge \alpha_3|\beta_1 \wedge \beta_2) / \bigcirc(\neg \beta_2|\beta_1)$	FC
$\bigcirc(\alpha|\beta), \beta / \bigcirc \alpha$	FD
$\bigcirc(\alpha|\beta_1) / \bigcirc(\alpha|\beta_1 \wedge \beta_2)$	SA
$\bigcirc(\alpha|\beta_1), \Diamond(\alpha \wedge \beta_1 \wedge \beta_2) / \bigcirc(\alpha|\beta_1 \wedge \beta_2)$	RSA
$\bigcirc(\alpha|\beta) / \bigcirc(\alpha|\beta \wedge \neg \alpha)$	VD
$\bigcirc(\alpha|\beta \wedge \neg \alpha) / \bigcirc(\alpha|\beta)$	VD$^-$
$\bigcirc(\alpha|\beta_1), C / \bigcirc(\alpha|\beta_1 \wedge \beta_2)$	RSA$_C$
$\bigcirc(\alpha|\beta) / \bigcirc(\alpha|\beta \wedge \alpha)$	CD
$\bigcirc(\alpha|\beta \wedge \alpha) / \bigcirc(\alpha|\beta)$	CD$^-$
$\bigcirc(\alpha_1|\beta_1), \bigcirc(\neg \alpha_1 \wedge \alpha_2|\beta_1 \wedge \beta_2) / \bigcirc(\alpha_1|\beta_1 \wedge \beta_2 \wedge \neg \alpha_2)$	RI
$\bigcirc(\alpha_1|\beta_1), \bigcirc(\neg \alpha_1 \wedge \alpha_2|\beta_1 \wedge \beta_2),$ $\bigcirc(\neg \alpha_2|\beta_1 \wedge \beta_2 \wedge \beta_3) / \bigcirc(\alpha_1|\beta_1 \wedge \beta_2 \wedge \beta_3)$	RIO
$\bigcirc(\alpha|\beta_1), \bigcirc(\alpha|\beta_2) / \bigcirc(\alpha|\beta_1 \vee \beta_2)$	ORA
$\bigcirc(\alpha|\beta), \bigcirc \beta / \bigcirc \alpha$	DD
$\bigcirc(\alpha|\beta), \bigcirc(\beta|\gamma) / \bigcirc(\alpha|\gamma)$	T
$\bigcirc(\alpha|\beta \wedge \gamma), \bigcirc(\beta|\gamma) / \bigcirc(\alpha|\gamma)$	CT
$\bigcirc(\alpha|\beta \wedge \gamma), \bigcirc(\beta|\gamma) / \bigcirc(\alpha \wedge \beta|\gamma)$	ACT

Table 1. Inference patterns

The letter C in RSA$_C$ stands for the condition: there is no premise $\bigcirc(\alpha'|\beta')$ such that $\beta_1 \wedge \beta_2$ logically implies β', β' logically implies β_1 and not vice versa, α and α' are contradictory and $\alpha \wedge \beta'$ is consistent. RSA$_C$ is not a rule in the usual proof-theoretic sense. For it has a statement that quantifies over all other premises as an auxiliary condition. Thus the rule is not on a par with the other rules, like for instance weakening of the output.

2.1 Van Fraassen's paradox

We first discuss deontic explosion in van Fraassen's paradox, then the trade-off between on the one hand "ought implies can" and on the other hand the representation of violations in the violation detection problem, whether it is forbidden to put oneself into a dilemma, and finally the use of priorities to resolve conflicts.

2.1.1 Deontic explosion: conjunction versus weakening

It is a well-known problem from paraconsistent logic that the removal of all inconsistent formulas from the language is insufficient to reason in the presence of a contradiction, because there may still be explosion in the sense that all formulas of the language are derived from a contradiction. The following derivation illustrates how we can derive q from p and $\neg p$ in propositional logic, where all formulas in the derivation are classically consistent.

$$\frac{\dfrac{p}{q \vee p} \quad \neg p}{\dfrac{q \wedge \neg p}{q}}$$

This derivation involves the following rules: replacements of logical equivalents, $\vee$-introduction, $\wedge$-introduction, and $\wedge$-elimination.

A similar phenomenon occurs in deontic logic, if we reason about deontic dilemmas or conflicts, that is situations where $\bigcirc p$ and $\bigcirc \neg p$ both hold. Van der Torre and Tan [2000] call this deontic explosion problem "van Fraassen's paradox," because van Fraassen [1973] gave the following (informal) analysis of dilemmas in deontic logic. He rejects the conjunction pattern AND:

$$\text{AND:} \frac{\bigcirc \alpha_1, \bigcirc \alpha_2}{\bigcirc(\alpha_1 \wedge \alpha_2)}$$

This is because AND warrants the move from $\bigcirc p \wedge \bigcirc \neg p$ to $\bigcirc(p \wedge \neg p)$, and such a conclusion is not consistent with the principle 'ought implies can', formalised as $\neg \bigcirc (p \wedge \neg p)$. However, he does not want to reject the conjunction pattern in all cases. In particular, he wants to be able to derive $\bigcirc(p \wedge q)$ from $\bigcirc p \wedge \bigcirc q$ when p and q are distinct propositional atoms. His suggestion is that a restriction should be placed on the conjunction pattern: one derives $\bigcirc(\alpha_1 \wedge \alpha_2)$ from $\bigcirc \alpha_1$ and $\bigcirc \alpha_2$ only if $\alpha_1 \wedge \alpha_2$ is consistent. He calls the latter inference pattern *Consistent Aggregation*, renamed to restricted conjunction (RAND) by van der Torre and Tan in their following variant of van Fraassen's suggestion.

Example 2.1 (Van Fraassen's paradox [van der Torre and Tan, 2000]) *Consider a deontic logic without nested modal operators in which dilemmas like $\bigcirc p \wedge \bigcirc \neg p$ are consistent, but which validates $\neg \bigcirc \bot$, where $\bot$ stands for any contradiction like $p \wedge \neg p$. Moreover, assume that it satisfies replacement of logical equivalents and at least the following two inference patterns* Restricted Conjunction (RAND), *also called consistent aggregation, and* Weakening (W), *where $\Diamond \phi$ can be read as "ϕ is possible" (possibility is not necessarily the same as consistency).*

$$\text{RAND:} \frac{\bigcirc \alpha_1, \bigcirc \alpha_2, \Diamond(\alpha_1 \wedge \alpha_2)}{\bigcirc(\alpha_1 \wedge \alpha_2)} \qquad \text{W:} \frac{\bigcirc \alpha_1}{\bigcirc(\alpha_1 \vee \alpha_2)}$$

Moreover, assume the two premises 'Honor thy father or thy mother!' $\bigcirc(f \vee m)$ and 'Honor not thy mother!' $\bigcirc \neg m$. The left derivation of Figure 1 illustrates how the desired conclusion 'thou shalt honor thy father' $\bigcirc f$ can be derived from the premises. Unfortunately, the right derivation of Figure 1 illustrates that we cannot accept restricted conjunction and weakening in a monadic deontic logic, because we can derive every $\bigcirc \beta$ from $\bigcirc \alpha$ and $\bigcirc \neg \alpha$.

$$\dfrac{\dfrac{\bigcirc(f \vee m) \quad \bigcirc \neg m}{\bigcirc(f \wedge \neg m)} \text{ RAND}}{\bigcirc f} \text{ W} \qquad \dfrac{\dfrac{\dfrac{\bigcirc \alpha}{\bigcirc(\alpha \vee \beta)} \text{ W} \quad \bigcirc \neg \alpha}{\bigcirc(\neg \alpha \wedge \beta)} \text{ RAND}}{\bigcirc \beta} \text{ W}$$

Figure 1. Van Fraassen's paradox

Van Fraassen's paradox has a counterpart in dyadic deontic logic. The paradox consists in deriving $\bigcirc(\gamma|\beta)$ from $\bigcirc(\alpha|\beta)$ and $\bigcirc(\neg\alpha|\beta)$ using the following rules of *Restricted Conjunction for the Consequent* (RANDC) and *Weakening of the Consequent* (WC).

$$\text{RANDC} : \dfrac{\bigcirc(\alpha_1|\beta), \bigcirc(\alpha_2|\beta), \Diamond(\alpha_1 \wedge \alpha_2)}{\bigcirc(\alpha_1 \wedge \alpha_2|\beta)} \qquad \text{WC} : \dfrac{\bigcirc(\alpha_1|\beta)}{\bigcirc(\alpha_1 \vee \alpha_2|\beta)}$$

2.1.2 Violation detection problem: unrestricted versus restricted conjunction

Whereas $p \wedge \neg p$ can not be derived in a paraconsistent logic, we can consistently represent the formula $\bigcirc(p \wedge \neg p)$ in a modal logic, and we can block deontic explosion using a minimal modal logic [Chellas, 1980]. This raises the question whether we should accept the conjunction pattern unrestrictedly or in its restricted form.

The choice between the two can be illustrated as follows. Suppose we can derive the obligation $\bigcirc(p \wedge \neg p)$ from $\bigcirc(p)$ and $\bigcirc(\neg p)$ without deriving $\bigcirc f$, or any other counterintuitive consequence. In that case, is $\bigcirc(p \wedge \neg p)$ by itself a consequence we want to block? This presents us with a choice. On the one hand we would like to block $\bigcirc(p \wedge \neg p)$, because it contradicts the "ought implies can" principle. On the other hand, we would like to allow the derivation of $\bigcirc(p \wedge \neg p)$, because such a formula represents explicitly the fact that there is a dilemma.

This choice is even more subtle in dyadic deontic logic. There is the extra question as to whether the "ought implies can" reading implies that the obligation in the consequent must only be consistent in itself, or consistent with the antecedent too. The latter requirement is represented by the following variant of the *Restricted Conjunction for the Consequent* pattern, which we call RANDC2.

$$\text{RANDC2} : \dfrac{\bigcirc(\alpha_1|\beta), \bigcirc(\alpha_2|\beta), \Diamond(\alpha_1 \wedge \alpha_2 \wedge \beta)}{\bigcirc(\alpha_1 \wedge \alpha_2|\beta)}$$

On the one hand we would like to block the derivation of $\bigcirc(p \wedge q \mid \neg p \vee \neg q)$ from $\bigcirc(p \mid \neg p \vee \neg q)$ and $\bigcirc(q \mid \neg p \vee \neg q)$ because "ought implies can". On the other hand we would like to be able to derive it in order to make explicit that $\neg p \vee \neg q$ gives rise to a dilemma, and is not consistent with the fulfillment of the two obligations appearing as premises.

The alternative restricted conjunction pattern RANDC2 highlights the distinction between what we call the violability and the temporal interpretation of dyadic deontic logic. The former interprets the obligation $O(\alpha \mid \beta)$ as "given that β has been settled beyond repair, we should do α to make the best out of the sad circumstances" [Hansson, 1969] and the latter as "if α is the case now, what should be the case next?" The violability interpretation says that $O(\neg \alpha \mid \alpha)$ represents that α is a violation. For example, if you are going to kill, then do it gently. The temporal interpretation says that the present situation must be changed—which may or may not indicate a violation. For example, the temporal interpretation may be used to express a conditional obligation like "if the light is on, turn it off!"

We would like to point out that the violability interpretation is more expressive, in the sense that the temporal interpretation can be represented by introducing distinct propositional letters for what is the case now, and what is the case in the next moment. For example, "if the light is on, turn it off" can be represented by $\bigcirc(\neg on_2 \mid on_1)$, where on_1 represents that the light is on now, and on_2 that it is on at the next moment in time. In the temporal interpretation, however, it seems impossible to represent all violations in a natural way. Thus, a temporal interpretation with future directed obligations only seems to be a strong limitation.

We use the name "violation detection problem" to refer to the phenomenon that with the restricted conjunction pattern the representation (and hence the detection) of violations is made impossible. We continue the discussion on the violation detection problem in Section 2.2, where we discuss restricted inference patterns formalising contrary-to-duty reasoning.

2.1.3 Forbidden conflicts

Here is another question raised by dilemmas: is it forbidden to create a dilemma? The following inference pattern is called *Forbidden Conflict* (FC). If the inference pattern is accepted, then it is not allowed to bring about a conflict, because a conflict is sub-ideal.

$$\text{FC}: \frac{\bigcirc(\alpha_1 \wedge \alpha_2 \mid \beta_1), \bigcirc(\neg \alpha_1 \wedge \alpha_3 \mid \beta_1 \wedge \beta_2)}{\bigcirc(\neg \beta_2 \mid \beta_1)}$$

Here is an example, taken from van der Torre and Tan [1997]. Assume the premises $\bigcirc k$ and $\bigcirc(p \wedge \neg k \mid d)$, where k can be read as 'keeping a promise', p as 'preventing a disaster' and d as 'a disaster will occur if nothing is done to prevent it'. (FC) yields $\bigcirc \neg d$. There are situations where this is the right outcome. Consider a person having the obligation to keep a promise to show up at a birthday party. We have $\bigcirc k$, but also

$\bigcirc(p \wedge \neg k \,|\, d)$. She does not want to go, and so before leaving she does something that might result in a disaster later on, like leaving the coffee machine on. During the party, she leaves and goes home, using her second obligation as an excuse. Nobody will contest that leaving the machine on (on purpose) was a violation already, viz. $\bigcirc \neg d$.

An instance of this inference pattern has been discussed in defeasible deontic logic, and we return to it in Section 2.3.

2.1.4 Resolving dilemmas

To resolve a conflict between an obligation for p and an obligation for $\neg p$, we need additional information. For example, a total preference order on sets of propositions can resolve all dilemmas by picking the preferred set of obligations among the alternatives of the dilemma, and weaker relations on sets of propositions such as a total pre-order or a partial order leaves some dilemmas unresolved.

The most studied source for a preference order over sets of propositions is a preference order over propositions, which is then lifted to an order on sets of propositions. For example, an ordering on obligations can be derived from an ordering on the authorities who created the obligations, or the moment in time they were created. The level of preference of an obligation may reflect its priority.

Consider three obligations with priority 3, 2 and 1, and a dilemma between the first and the latter two. To represent the priority of an obligation, we write it in the $\bigcirc$ notation. A higher number reflects a higher priority.

$$\{\textcircled{3}(p \wedge q), \textcircled{2}\neg p, \textcircled{1}\neg q\}$$

In other words, we can either satisfy the most important obligation $\textcircled{3}(p \wedge q)$, or two less important obligations $\textcircled{2}\neg p$ and $\textcircled{1}\neg q$. Can this dilemma be resolved? There are various well known possibilities in the area of non-monotonic logic. Whether they can be used depends on the origin of the priorities and the application.

The issue of lifting priorities from obligations to sets of them gets more challenging when we consider conditional obligations and deontic detachment, as discussed later on in Section 2.7.

2.2 Forrester's paradox

We first discuss factual detachment in Forrester's paradox, then the problematic derivation of secondary obligations from primary ones, and finally what we call the violation detection problem for Forrester's paradox.

2.2.1 Factual detachment versus conjunction

Forrester's paradox consists of the four sentences 'Smith should not kill Jones,' 'if Smith kills Jones, then he should do it gently,' 'Smith kills Jones', and 'killing someone gently logically implies killing him.' The preference based models of dyadic deontic logic give a natural representation of the two obligations: not killing is preferred to gentle killing, and both are preferred to other forms of killing. However, the following example illustrates that it is less clear how to combine dyadic obligation with factual detachment, deriving unconditional obligations from conditional ones.

Example 2.2 (Forrester's paradox) *Assume a dyadic deontic logic without nested modal operators that has at least replacement of logical equivalents, the Conjunction pattern* AND *and the following inference pattern called factual detachment* FD.

$$\text{FD} : \frac{\bigcirc(\alpha|\beta), \beta}{\bigcirc\alpha}$$

Furthermore, assume the following premise set with background knowledge that gentle murder implies murder $\vdash g \to k$.

$$S = \{\bigcirc(\neg k|\top), \bigcirc(g|k), k\}$$

The set S represents the Forrester paradox when k is read as 'Smith kills Jones' *and g as* 'Smith kills Jones gently.' *We say that the last obligation is a contrary-to-duty obligation with respect to the first obligation, because its antecedent is contradictory with the consequent of the first obligation. Figure 2 visualizes how we can represent the concept of contrary-to-duty as a binary relation among dyadic obligations: the obligation $\bigcirc(\alpha_2|\beta_2)$ is a contrary-to-duty with respect to $\bigcirc(\alpha_1|\beta_1)$ if and only if $\beta_2 \wedge \alpha_1$ is inconsistent.*

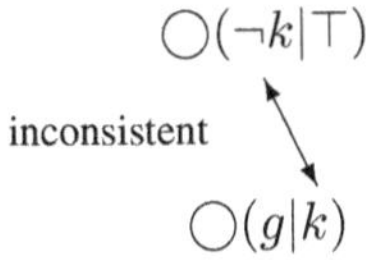

$$\bigcirc(\neg k|\top)$$
$$\text{inconsistent} \searrow$$
$$\bigcirc(g|k)$$

Figure 2. $\bigcirc(g|k)$ is a contrary-to-duty obligation with respect to $\bigcirc(\neg k|\top)$

The derivation in Figure 3 illustrates how the obligation $\bigcirc(\neg k \wedge g)$, i.e. $\bigcirc(\bot)$, can be derived from S by FD *and* AND.

$$\frac{\dfrac{\bigcirc(\neg k|\top) \quad \top}{\bigcirc(\neg k)}\text{FD} \quad \dfrac{\bigcirc(g|k) \quad k}{\bigcirc(g)}\text{FD}}{\bigcirc(\neg k \wedge g)}\text{AND}$$

Figure 3. Forrester's paradox

Forrester's paradox can be given two interpretations. First, the dilemma interpretation says that the two obligations give rise to a dilemma, just like the obligations $\bigcirc p$ and $\bigcirc \neg p$ in van Fraassen's paradox. Consequently, according to the dilemma interpretation, there is no problem, the derivation of $\bigcirc(\bot)$ just reflects the fact that there is a dilemma.

The coherent interpretation appeals to the independent and seemingly plausible principle 'ought implies can', $\neg \bigcirc (\bot|\alpha)$. According to this interpretation, the Forrester set is intuitively consistent with the 'ought implies can' principle, and so there is no dilemma, just an obligation to act as good as possible in the sub-ideal situation where the primary obligation has been violated.

There is a consensus in the literature that the example should be given a coherent interpretation, and that the dilemma interpretation is wrong.

2.2.2 Deriving secondary obligations from primary ones: Strengthening of the antecedent versus weakening of the consequent

The following example shows that Forrester's paradox can be used also to illustrate that combining the desirable inference patterns strengthening of the antecedent and weakening of the consequent is problematic in dyadic deontic logic. For example, strengthening of the antecedent is used to derive 'Smith should not kill Jones in the morning' $\bigcirc(\neg k|m)$ from the obligation 'Smith should not kill Jones' $\bigcirc(\neg k|\top)$ and weakening of the consequent is used to derive 'Smith should not kill Jones' $\bigcirc(\neg k|\top)$ from the obligation 'Smith should drive on the right side of the street and not kill Jones' $\bigcirc(r \wedge \neg k|\top)$.

Example 2.3 (Forrester's paradox, cont'd [van der Torre and Tan, 2000]) *Assume a dyadic deontic logic without nested modal operators that has at least replacement of logical equivalents and the following inference patterns* Strengthening of the Antecedent *(SA), the* Conjunction pattern for the Consequent *(ANDC) and* Weakening of the Consequent *(WC).*

$$\text{SA}: \frac{\bigcirc(\alpha|\beta_1)}{\bigcirc(\alpha|\beta_1 \wedge \beta_2)} \qquad \text{ANDC}: \frac{\bigcirc(\alpha_1|\beta), \bigcirc(\alpha_2|\beta)}{\bigcirc(\alpha_1 \wedge \alpha_2|\beta)} \qquad \text{WC}: \frac{\bigcirc(\alpha_1|\beta)}{\bigcirc(\alpha_1 \vee \alpha_2|\beta)}$$

The derivation in Figure 4 illustrates how the obligation $\bigcirc(\neg k \wedge g|k)$*, i.e.* $\bigcirc(\bot|k)$*, can be derived from S by* SA *and* ANDC. *Note that the dyadic obligation* $\bigcirc(\neg k|k)$ *can be given only a violability interpretation in this example, not a temporal interpretation, because it is impossible to undo a killing. That is, this dyadic obligation can be read only as "if Smith kills Jones, then this is a violation."*

The derivation is blocked when SA *is replaced by the following inference pattern* Restricted Strengthening of the Antecedent *(RSA).*

$$\text{RSA}: \frac{\bigcirc(\alpha|\beta_1), \Diamond(\alpha \wedge \beta_1 \wedge \beta_2)}{\bigcirc(\alpha|\beta_1 \wedge \beta_2)}$$

However, the obligation $\bigcirc(\bot|k)$ *can still be derived from S by* WC, RSA *and* ANDC. *This derivation from the set of obligations is represented on the right hand side of Figure 4. Like in Example 2.2, we can give the set a dilemma or a coherent interpretation.*

Detachment in Normative Systems: Examples, Inference Patterns, Properties

$$\dfrac{\dfrac{\bigcirc(\neg k|\top)}{\bigcirc(\neg k|k)} \text{ SA} \quad \bigcirc(g|k)}{\bigcirc(\neg k \wedge g|k)} \text{ ANDC} \qquad \dfrac{\dfrac{\dfrac{\bigcirc(\neg k|\top)}{\bigcirc(\neg g|\top)} \text{ WC}}{\bigcirc(\neg g|k)} \text{ RSA} \quad \bigcirc(g|k)}{\bigcirc(\neg g \wedge g|k)} \text{ ANDC}$$

Figure 4. Forrester's paradox

The underlying problem of the counterintuitive derivation in Figure 4 is the derivation of $\bigcirc(\neg g|k)$ from the first premise $\bigcirc(\neg k|\top)$ by WC and RSA, because it derives a contrary-to-duty obligation from its own primary obligation.

Since there is consensus that Forrester's paradox should be given a coherent interpretation, Forrester's paradox in Example 2.3 shows that combining strengthening of the antecedent and weakening of the consequent is problematic for *all* deontic logics.

2.2.3 Violation detection problem: restricted versus unrestricted strengthening of the antecedent

The choice between the unrestricted version and the restricted version of the law of strengthening of the antecedent has some similarity with the choice between the unrestricted version and the restricted version of the law of conjunction. This can be illustrated as follows. Suppose we have the obligation $\bigcirc(\neg k \mid \top)$. In that case, is $\bigcirc(\neg k|k)$ a consequence we want to block? This presents us with a choice. On the one hand, we would like to block $\bigcirc(\neg k|k)$, because it contradics the "ought implies can" principle. On the other hand, we would like to allow the derivation of $\bigcirc(\neg k|k)$, because this formula represents explicitly that there is a violation. (Cf. our explanatory comments on the violability interpretation, on p. 313.)

The following inference pattern *Violation Detection* (VD) formalizes the intuition that an obligation cannot be defeated by only violating it, and represents a solution to the violation detection problem. The VD pattern models the intuition that after violation the obligation to do α is still in force. Even if you drive too fast, you are still obliged to obey the speed limit.

$$\text{VD}: \dfrac{\bigcirc(\alpha|\beta)}{\bigcirc(\alpha|\beta \wedge \neg\alpha)} \qquad \text{VD}^-: \dfrac{\bigcirc(\alpha|\beta \wedge \neg\alpha)}{\bigcirc(\alpha|\beta)}$$

The inverse pattern VD⁻ says that violations do not come out of the blue. Although this inference pattern may seem intuitive at first sight, it appears too strong on further inspection.

Example 2.4 (Metro) *Consider the following derivation.*

$$\frac{\bigcirc(\alpha|\beta)}{\bigcirc(\alpha|\beta \wedge \neg\alpha)} \text{VD}$$
$$\overline{\bigcirc(\alpha|\alpha \vee \beta)} \text{VD}^-$$

For example, assume that if you travel by metro, you must have a ticket. We can derive that traveling by metro without a ticket is a violation. The two inference patterns together would derive that if you travel by metro or you buy a ticket, then you must buy a ticket. This is counterintuitive, because buying a ticket without traveling by metro does not involve any obligations. The example illustrates how reasoning about violations only can lead to the wrong conclusions.

Normative systems typically associate sanctions with violations, as an incentive for agents to obey the norms. Such sanctions can sometimes be expressed as contrary-to-duty obligations: the sanction to pay a fine if you do not return the book to the library in time, can be modelled as a contrary-to-duty obligation to pay the fine. By symmetry, though this is less often implemented in normative systems, rewards can be associated with compliance of obligations. In modal logic, an obligation for α is fulfilled if we have $\alpha \wedge \bigcirc \alpha$.

The following inference pattern *Compliance Detection* (CD) formalizes the intuition that an obligation cannot be defeated by only complying with it, analogous to the *Violation Detection* (VD) pattern.

$$\text{CD}: \frac{\bigcirc(\alpha|\beta)}{\bigcirc(\alpha|\beta \wedge \alpha)} \qquad \text{CD}^-: \frac{\bigcirc(\alpha|\beta \wedge \alpha)}{\bigcirc(\alpha|\beta)}$$

The following example illustrates that the inference pattern CD should not be confused with the inverse of CD$^-$, which seems to say that fulfilled obligations do not come out of the blue. Although this inference pattern may seem intuitive at first sight, it is highly counterintuitive on further inspection.

Example 2.5 (Forrester, continued) *Consider the following derivation.*

$$\frac{\bigcirc(\alpha \wedge \beta|\alpha)}{\bigcirc(\alpha \wedge \beta|\alpha \wedge \beta)} \text{CD}$$
$$\overline{\bigcirc(\alpha \wedge \beta|\top)} \text{CD}^-$$

You should kill gently, if you kill $\bigcirc(k \wedge g|k)$. Hence, by CD, you should kill gently, if you kill gently $\bigcirc(k \wedge g|k \wedge g)$ (a fulfilled obligation). However, this does not mean that there is an unconditional obligation to kill gently $\bigcirc(k \wedge g|\top)$. Hence, the inference pattern CD$^-$ should not be valid.

Without the CD pattern, we say that the fulfilled obligation "disappears," analogous to violations. A fulfilled obligation also disappears when we have as an axiom of the logic that $\bigcirc(\alpha|\beta) \leftrightarrow \bigcirc(\alpha \wedge \beta|\beta)$, because in that case $\bigcirc(\alpha \wedge \beta|\beta)$ does not hold because β is compliant with a norm.

2.3 Prakken and Sergot's cottage regulations

We first discuss the extension of Forrester's paradox with defeasible obligations, then we return to the violation detection problem, and finally we discuss reinstatement.

2.3.1 Violations and exceptions

The so-called cottage regulations are introduced by Prakken and Sergot [1996] to illustrate the distinction between contrary-to-duty reasoning and defeasible reasoning based on exceptional circumstances. It is an extended version of the Forrester or gentle murderer paradox discussed in Section 2.2. The following example is an alphabetic variant of the original example, because we replaced s, to be read as 'the cottage is by the sea,' by d, to be read as 'there is a dog.' Moreover, as is common, instead of representing background knowledge that w implies f, Prakken and Sergot represent a white fence by $w \wedge f$.

Example 2.6 (Cottage regulations [van der Torre and Tan, 1997]) *Assume a deontic logic that validates at least replacement of logical equivalents and the inference pattern* RSA$_C$.

$$\text{RSA}_C : \frac{\bigcirc(\alpha|\beta_1), C}{\bigcirc(\alpha|\beta_1 \wedge \beta_2)}$$

C: there is no premise $\bigcirc(\alpha'|\beta')$ such that $\beta_1 \wedge \beta_2$ logically implies β', β' logically implies β_1 and not vice versa, α and α' are contradictory and $\alpha \wedge \beta'$ is consistent. [van der Torre, 1994]

RSA$_C$ *formalises a principle of specificity to deal with exceptional circumstances. It is illustrated with Figure 5 (a). Suppose we are given these rules: you ought not to eat with your fingers; if you are served asparagus, you ought to eat with your fingers. One does not want to be able to strengthen the first obligation into: if you are served asparagus, you ought not to eat with your fingers. Such a strengthening is blocked by* RSA$_C$.

Now, assume the obligations

$$S = \{\bigcirc(\neg f|\top), \bigcirc(w \wedge f|f), \bigcirc(w \wedge f|d)\},$$

where f can be read as 'there is a fence around your house,' $w \wedge f$ as 'there is a white fence around your house' and d as 'you have a dog.' Notice that $\bigcirc(w \wedge f|f)$ is a contrary-to-duty obligation with respect to $\bigcirc(\neg f|\top)$ and $\bigcirc(w \wedge f|d)$ is not. If all we know is that there is a fence and a dog ($f \wedge d$), then the first obligation in S is intuitively overridden, and therefore it cannot be violated. Hence, the obligation $\bigcirc(\neg f | f \wedge d)$ should not be derivable. However, if all we know is that there is a *fence without a dog (f), then the first obligation in S is intuitively not overridden, and therefore it is violated. Hence, the obligation* $\bigcirc(\neg f|f)$ *should be derivable.*

One should be careful not to treat both $\bigcirc(w \wedge f|f)$ and $\bigcirc(w \wedge f|d)$ as more specific obligations that override the obligation $\bigcirc(\neg f \mid \top)$, because this is not correct for $\bigcirc(w \wedge f|f)$. The latter obligation should be treated as a contrary-to-duty obligation, i.e. as a case of violation. This interference of specificity and contrary-to-duty is represented in Figure 5. This figure should be read as follows. Each arrow is a condition: a two-headed arrow is a consistency check, and a single-headed arrow is a logical implication. For example, the condition C formalizes that an obligation $\bigcirc(\alpha|\beta)$ is overridden by $\bigcirc(\alpha'|\beta')$ if the conclusions are contradictory (a consistency check, the double-headed arrow) and the condition of the overriding obligation is more specific (β' logically implies β). Case (a) represents criteria for overridden defeasibility, and case (b) represents criteria for contrary-to-duty. Case (c) shows that the pair $\bigcirc(\neg f|\top)$ and $\bigcirc(w \wedge f|f)$ can be viewed as overridden defeasibility as well as contrary-to-duty.

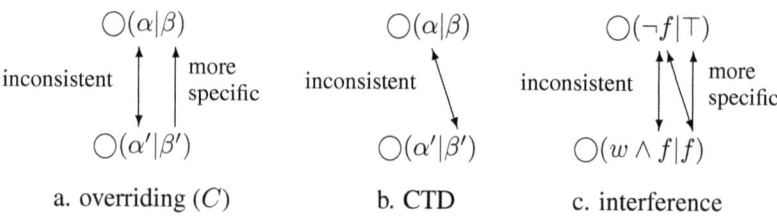

Figure 5. Specificity and CTD

2.3.2 Violation detection problem for defeasible obligations

What is most striking about the cottage regulations is the observation that when the premise $\bigcirc(\neg f \mid \top)$ is violated by f, then the obligation for $\neg f$ should be derivable, but not when $\bigcirc(\neg f|\top)$ is overridden by $f \wedge d$. In other words, we have to distinguish violations from exceptions.

In approaches where $\bigcirc(\alpha|\beta)$ implies that $\alpha \wedge \beta$ is consistent, we cannot represent this difference by deriving $\bigcirc(\neg f|f)$ and not deriving $\bigcirc(\neg f|d \wedge f)$. In this sense, this is again an example of the violation detection problem.

We can use priorities to represent the specificity example, by giving the more specific obligation a higher priority. Many conditional logics have specificity built in, but this must be combined with other conflict resolution methods, for example based on time or authority. This is an issue of reasoning about uncertainty, default reasoning, and nonmonotonic logic.

2.3.3 Reinstatement

The question raised by the inference pattern *Reinstatement* (RI) is whether an obligation can be overridden by an overriding obligation that itself is violated. The obligation $O(\alpha_1|\beta_1)$ is overridden by $O(\neg\alpha_1 \wedge \alpha_2|\beta_1 \wedge \beta_2)$ for $\beta_1 \wedge \beta_2$, but is it also overridden for $\beta_1 \wedge \beta_2 \wedge \neg\alpha_2$? If the last conclusion is not accepted, then the first obligation α_1 should be in force again. Hence, the original obligation is reinstated.

$$\text{RI}: \frac{O(\alpha_1|\beta_1), O(\neg\alpha_1 \wedge \alpha_2|\beta_1 \wedge \beta_2)}{O(\alpha_1|\beta_1 \wedge \beta_2 \wedge \neg\alpha_2)}$$

Suppose you are in the street, and see a child's bike unattended. As a general rule, you should not take the bike, viz. $O\neg t$ where t is for taking the bike. Now, suppose you also observe an elderly neighbor collapse with what might be a heart attack. You are a block away from the nearest phone from which you could call for help. In that more specific situation, you should take the bike and go call for help, $O(t \wedge h|e)$, where e and h are for an elderly neighbor collapses and go call for help, respectively. The obligation $O\neg t$ is overriden by $O(t \wedge h|e)$ for e. But it is not overriden for $e \wedge \neg g$. Of course, if you do not go for help, then the prohibition of t remains.

The following inference pattern RIO is a variant of the previous inference pattern RI, in which the overriding obligation is not factually defeated but overridden. The obligation $O(\alpha_1|\beta_1)$ is overridden by $O(\neg\alpha_1 \wedge \alpha_2|\beta_1 \wedge \beta_2)$ for $\beta_1 \wedge \beta_2$, and the latter is overridden by $O(\neg\alpha_2|\beta_1 \wedge \beta_2 \wedge \beta_3)$ for $\beta_1 \wedge \beta_2 \wedge \beta_3$. The inference pattern RIO says that an obligation cannot be overridden by an obligation that is itself overridden. Hence, an overridden obligation becomes reinstated when its overriding obligation is itself overridden.

$$\text{RIO}: \frac{O(\alpha_1|\beta_1), O(\neg\alpha_1 \wedge \alpha_2|\beta_1 \wedge \beta_2), O(\neg\alpha_2|\beta_1 \wedge \beta_2 \wedge \beta_3)}{O(\alpha_1|\beta_1 \wedge \beta_2 \wedge \beta_3)}$$

Example: you should not kill; if you find yourselves in a situation of self-defence, you should kill; if you find yourselves in a situation of self-defence, but your opponent is weak, you should not kill.

Van der Torre and Tan [1997] argue that Reinstatement does not hold in general, for example it does not hold for obligations under uncertainty. However, they argue also that these patterns hold for so-called prima facie obligations. The notion of prima facie obligation was introduced by Ross [1930]. He writes: 'I suggest '*prima facie* duty' or 'conditional duty' as a brief way of referring to the characteristic (quite distinct from that of being a duty proper) which an act has, in virtue of being of a certain kind (e.g. the keeping of a promise), of being an act which would be a duty proper if it were not at the same time of another kind which is morally significant' [Ross, 1930, p.19]. A prima facie duty is a duty proper when it is not overridden by another prima facie duty. When a prima facie obligation is overridden, it is not a proper duty but it is still in force: 'When we think ourselves justified in breaking, and indeed morally obliged to break, a promise [...] we do not for the moment cease to recognize a prima facie duty to keep our promise' [Ross, 1930, p.28].

Van der Torre and Tan argue also that the inference pattern Forbidden Conflict, discussed in Section 2.1.3, does not hold in general, but it holds for prima facie obligations. If the inference pattern is accepted, then it is not allowed to bring about a conflict, because a conflict is sub-ideal, even when it can be resolved.

2.4 Jeffrey's disarmament paradox

In general, reasoning by cases is a desirable property of reasoning with conditionals. In this reasoning scheme, a certain fact is proven by proving it for a set of mutually exclusive and exhaustive circumstances. For example, assume that you want to know whether you want to go to the beach. If you desire to go to the beach when it rains, and you desire to go to the beach when it does not rain, then you may conclude by this scheme 'reasoning by cases' that you desire to go to the beach under all circumstances. The two cases considered here are rain and no rain. This kind of reasoning schemes can be formalized by the following derivation: *If 'α if β' and 'α if not β,' then 'α regardless of β.'* Formally, if we write the conditional 'α if β' by $\beta > \alpha$, then it is represented by the following disjunction pattern for the antecedent.

$$\text{ORA:} \frac{\beta > \alpha, \neg\beta > \alpha}{\top > \alpha}$$

The following example illustrates that the disjunction pattern for the antecedent combined with strengthening of the antecedent derives counterintuitive consequences in dyadic deontic logic. Example 2.7 is based on the following classic illustration of Jeffrey [1983], see also the discussion by Thomason and Horty [1996].

Example 2.7 (Disarmament paradox [van der Torre and Tan, 2000]) *Assume a deontic logic that validates at least replacement of logical equivalents and the two inference patterns* RSA *and the* Disjunction pattern for the Antecedent *(ORA),*

$$\text{ORA :} \frac{O(\alpha|\beta_1), O(\alpha|\beta_2)}{O(\alpha|\beta_1 \vee \beta_2)}$$

and assume as premises the obligations 'we ought to be disarmed if there will be a nuclear war', 'we ought to be disarmed if there will be no war' *and* 'we ought to be armed if we have peace if and only if we are armed'. *They may be formalized as* $O(d|w)$, $O(d|\neg w)$ *and* $O(\neg d|d \leftrightarrow w)$, *respectively. The derivation in Figure 6 shows how we can derive the counterintuitive* $O(d \wedge \neg d|d \leftrightarrow w)$. *The derived obligation is inconsistent in most deontic logics, whereas intuitively the set of premises is consistent. The derivation of* $O(d|d \leftrightarrow w)$ *is counterintuitive, because it is not possible to fulfill this obligation together with the obligation* $O(d|\neg w)$ *it is derived from. The contradictory fulfillments are respectively* $d \wedge w$ *and* $d \wedge \neg w$.

$$\frac{\bigcirc(d|w) \quad \bigcirc(d|\neg w)}{\bigcirc(d|\top)} \text{ ORA}$$

$$\frac{\overline{\bigcirc(d|d \leftrightarrow w)} \text{ RSA} \quad \bigcirc(\neg d|d \leftrightarrow w)}{\bigcirc(d \land \neg d|d \leftrightarrow w)} \text{ AND}$$

Figure 6. The disarmament paradox

In other words, in this derivation the obligation $\bigcirc(d|d \leftrightarrow w)$ is considered to be counterintuitive, because it is not grounded in the premises. If $d \leftrightarrow w$ and w (the antecedent of the first premise) are true then d is trivially true, and if $d \leftrightarrow w$ and $\neg w$ (the antecedent of the second premise) are true then d is trivially false. In other words, if $d \leftrightarrow w$ then the first premise cannot be violated and the second premise cannot be fulfilled. Hence, the two premises do not ground the conclusion that for arbitrary $d \leftrightarrow w$ we have that $\neg d$ is a violation.

The example is difficult to interpret, because it makes use of a bi-implication. An alternative set of premises, also based on bi-implications, with analogous counterintuitive conclusions is $\{\bigcirc(d|d \leftrightarrow w), \bigcirc(d|\neg d \leftrightarrow w), \bigcirc(\neg d|w)\}$.

ORA also plays a role in the so-called miners' scenario introduced recently by Kolodny and MacFarlane [2010].

2.5 Chisholm's paradox

The second contrary-to-duty paradox we consider is Chisholm [1963]'s paradox. We first discuss the choice between deontic versus factual detachment, and then the representation of deontic detachment. We discuss the violation detection problem for deontic detachment only in Section 2.6 after we have introduced Makinson's Möbius strip example.

2.5.1 Deontic versus factual detachment

Chisholm's paradox consists of the three obligations of a certain man 'to go to his neighbours assistance,' 'to tell them that he comes if he goes,' and 'not to tell them that he comes if he does not go,' together with the fact 'he does not go.' The preference-based models of dyadic deontic logic again give a natural representation of the three sentences, just like for Forrester's paradox. For example, going to the assistance and telling is preferred to all the other possibilities, and not going to the assistance and not telling is preferred to not going and telling. It seems that the going and not telling and not going and telling may be ordered in various ways. However, the following example illustrates that it is difficult to combine factual with deontic detachment, and to derive unconditional obligations from conditional and unconditional ones.

Example 2.8 (Chisholm's paradox) *Assume a dyadic deontic logic without nested modal operators that has at least replacement of logical equivalents, the Conjunction pattern* AND *factual detachment* FD *and the following inference pattern deontic detachment* DD.

$$\text{DD}: \frac{O(\alpha|\beta), O\beta}{O\alpha}$$

Furthermore, consider the following premise set S.

$$S = \{O(a|\top), O(t|a), O(\neg t|\neg a), \neg a\}$$

The set S formalizes Chisholm's paradox when a is read as 'a certain man goes to the assistance of his neighbors' *and t as* 'the man tells his neighbors that he will come.' *Chisholm's paradox is more complicated than Forrester's paradox, because it also contains an* According-To-Duty (ATD) *obligation. We can represent the notion of according-to-duty as a binary relation among conditional obligations, just like the notion of contrary-to-duty. A conditional obligation $O(\alpha|\beta)$ is an ATD obligation of $O(\alpha_1 | \beta_1)$ if and only if β logically implies α_1. The condition of an ATD obligation is satisfied only if the primary obligation is fulfilled. The definition of ATD is analogous to the definition of CTD in the sense that an ATD obligation is an obligation conditional upon the fulfilment of an obligation and a CTD obligation is an obligation conditional upon a violation. The second obligation is an ATD obligation and the third obligation is a CTD obligation with respect to the first obligation, see Figure 7.*

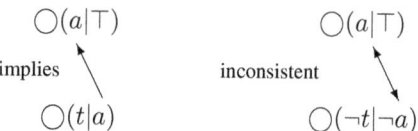

Figure 7. $O(t|a)$ is an ATD of $O(a|\top)$ and $O(\neg t|\neg a)$ is a CTD of $O(a|\top)$

The derivation in Figure 8 shows how the counterintuitive obligation $O(t \wedge \neg t)$, or $O\bot$, can be derived from S by FD, DD *and* AND. *Just like in Forrester's paradox, we can give a dilemma and a coherent interpretation to the scenario, and there is consensus that the latter one is preferred. This is not surprising, as Forrester's paradox shows that factual detachment and conjunction are problematic in themselves.*

2.5.2 Deriving secondary obligations from primary ones: three kinds of transitivity

Deontic detachment is related to the following three variants of transitivity: plain transitivity T, cumulative transitivity CT, and what Parent and van der Torre [2014a; 2014b] call aggregative cumulative transitivity ACT.

$$\text{T}: \frac{O(\alpha|\beta), O(\beta|\gamma)}{O(\alpha|\gamma)} \qquad \text{CT}: \frac{O(\alpha|\beta \wedge \gamma), O(\beta|\gamma)}{O(\alpha|\gamma)} \qquad \text{ACT}: \frac{O(\alpha|\beta \wedge \gamma), O(\beta|\gamma)}{O(\alpha \wedge \beta|\gamma)}$$

$$\frac{\bigcirc(t|a) \quad \dfrac{\bigcirc(a|\top) \quad \top}{\bigcirc(a)} \text{ FD}}{\bigcirc t} \text{ DD} \qquad \dfrac{\bigcirc(\neg t|\neg a) \quad \neg a}{\bigcirc(\neg t)} \text{ FD}$$
$$\bigcirc(t \wedge \neg t) \qquad \text{AND}$$

Figure 8. Chisholm's paradox

The left derivation illustrates that T can be derived from ACT together with SA and WC, and likewise CT can be derived from T and SA, and T can be derived from CT and SA. The right derivation illustrates how ANDC can be derived from SA and ACT. RANDC can be derived analogously from RSA and ACT.

$$\dfrac{\dfrac{\bigcirc(\alpha|\beta)}{\bigcirc(\alpha|\beta \wedge \gamma)} \text{SA} \quad \bigcirc(\beta|\gamma)}{\dfrac{\bigcirc(\alpha \wedge \beta|\gamma)}{\bigcirc(\alpha|\gamma)} \text{WC}} \text{ACT} \qquad \dfrac{\dfrac{\bigcirc(\alpha_1|\beta)}{\bigcirc(\alpha_1|\beta \wedge \alpha_2)} \text{SA} \quad \bigcirc(\alpha_2|\beta)}{\bigcirc(\alpha_1 \wedge \alpha_2|\beta)} \text{ACT}$$

The following variant of Chisholm's paradox illustrates that only ACT can be combined with restricted strengthening of the antecedent.

Example 2.9 (Chisholm's paradox, continued) *Assume a dyadic deontic logic that validates at least replacement of logical equivalents and the (intuitively valid) inference patterns* RSA *(or* SA*),* T *(or* CT*), and* ANDC*.*

The left derivation in Figure 9 illustrates how the counterintuitive $\bigcirc(\bot|\neg a)$ *can be derived from S. Again we can give a dilemma and a coherent interpretation, and there is consensus in the literature that it should get a coherent interpretation. The underlying problem is the derivation of* $\bigcirc(t|\neg a)$*, which seems counterintuitive since it derives a contrary-to-duty obligation from the primary* $\bigcirc(a|\top)$*. If we accept* RSA*, then we cannot accept* T *or* CT*.*

$$\dfrac{\dfrac{\bigcirc(t|a) \quad \bigcirc(a|\top)}{\bigcirc(t|\top)} \text{T/CT}}{\bigcirc(t|\neg a)} \text{RSA} \quad \bigcirc(\neg t|\neg a)}{\bigcirc(t \wedge \neg t|\neg a)} \text{AND} \qquad \dfrac{\dfrac{\dfrac{\bigcirc(t|a) \quad \bigcirc(a|\top)}{\bigcirc(a \wedge t|\top)} \text{ACT}}{\dfrac{\bigcirc(t|\top)}{\bigcirc(t|\neg a)} \text{RSA}} \quad \bigcirc(\neg t|\neg a)}{\bigcirc(t \wedge \neg t|\neg a)} \text{AND}$$

Figure 9. Chisholm's paradox

Assume a dyadic deontic logic that validates at least replacement of logical equivalents and the (intuitively valid) inference patterns RSA, ANDC, WC *and* ACT*. The right* derivation of Figure 9 illustrates how the counterintuitive $\bigcirc(\bot|\neg a)$ can be derived from S. However, without WC the counterintuitive obligation cannot be derived.

When we compare the two derivations of the contrary-to-duty paradoxes in dyadic deontic logic, we find the following similarity. The underlying problem of the counterintuitive derivations is the derivation of the obligation $\bigcirc(\alpha_1|\neg\alpha_2)$ from $\bigcirc(\alpha_1 \wedge \alpha_2|\top)$ by WC and RSA. It is respectively the derivation of $\bigcirc(\neg g|k)$ from $\bigcirc(\neg k|\top)$ in Figure 3 and $\bigcirc(t|\neg a)$ from $\bigcirc(a \wedge t|\top)$ in Figure 9. The underlying problem of the contrary-to-duty paradoxes is that a contrary-to-duty obligation can be derived from its primary obligation. It is no surprise that this derivation causes paradoxes. The derivation of a secondary obligation from a primary obligation confuses the different contexts found in contrary-to-duty reasoning. The context of primary obligation is the ideal state, whereas the context of a contrary-to-duty obligation is a violation state. Preference-based deontic logics were developed to semantically distinguish the different violation contexts in a preference ordering, but it appears more challenging to represent these contexts in derivations.

2.6 Makinson's Möbius strip

Makinson [1999]'s Möbius strip illustrates that dilemmas and deontic detachment can also be combined, leading to new challenges and distinctions. We discuss also the violation detection problem for deontic detachment.

2.6.1 Iterated deontic detachment

The so-called Möbius strip (whose name comes from the shape of the example in Figure 10) arises when we allow for deontic detachment to be iterated. We give the version of the example presented by Makinson and van der Torre in their input/output logic, though we use the dyadic representation.

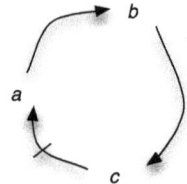

Figure 10. Möbius strip

Example 2.10 (Möbius strip) *Consider three conditional obligations stating $\neg a$ is obligatory given c, that c is obligatory given b, and that b is obligatory given a, to-*

gether with the fact that a is true.

$$\bigcirc(\neg a|c), \bigcirc(c|b), \bigcirc(b|a), a$$

For instance, a, b, c could represent "Alice (respectively Bob, Carol) is invited to dinner." The obligation $\bigcirc(b|a)$ says that if Alice is invited then Bob should be, and so on.

Makinson [1999] gives what we call here the coherent interpretation. He mentions that "intuitively, we would like to have" that under condition a, each of b and c is obligatory, even though we may not want to conclude for $\neg a$ under the same condition. He also indicates that "an approach inspired by maxi choice in AGM theory change" (like the one described in the paper in question) leads to three possible outcomes: both b and c are obligatory; only b is obligatory; neither of b and c is obligatory. The three sets of obligations corresponding to these outcomes are linearly ordered under set-theoretical inclusion.

In their input/output logic framework, Makinson and van der Torre [2001] present what we call here the dilemma interpretation of the example. They change the definitions such that precisely the dilemma among these three alternatives is the desired outcome of the example.

There does not seem to be consensus in the literature on which interpretation is the intuitive answer for this example. Deontic detachment has been severely criticised in the literature, so it may be questioned whether full transitivity is natural. However, the choice between coherent and dilemma interpretation is general and can be found in other examples, such as the following variant of Chisholm's paradox.

Example 2.11 (Chisholm's paradox, continued) *Consider this variant of the Möbius strip:*

$$\{\bigcirc(d|c), \bigcirc(c|b), \bigcirc(b|a), a, \neg d\}$$

By symmetry with the dilemma interpretation of Möbius strip, the dilemma interpretation gives three alternatives, $\{\bigcirc b, \bigcirc c\}$, $\{\bigcirc b\}$ and $\emptyset$. Now consider deontic detachment in Chisholm's paradox, together with the fact that we do not tell.

$$\bigcirc(t|a), \bigcirc(a|\top), \neg t$$

Again by symmetry, the dilemma interpretation gives two alternatives, $\{\bigcirc a\}$ and $\emptyset$.

The following example has been introduced by Horty [2007] in a prioritised setting, and we will consider it again in the section that comes next. Again the question is raised whether one solution can be a subset of another solution.

Example 2.12 (Order) *Consider the following set of obligations. a is for putting the heating on, and b is for opening the window.*

$$\bigcirc(a|\top), \bigcirc(b|\top), \bigcirc(\neg b|a)$$

The example is a dilemma, but the question is whether there are two or three alternatives. According to the first interpretation, the only two alternatives are the obligations for a and b, and the obligations for a and ¬b. According to the second interpretation, there is also the alternative of an obligation for b, without an obligation for a. The latter alternative is a subset of another alternative, analogous to the dilemma interpretation of the Möbius strip example.

2.6.2 Violation detection problem and transitivity

In the previous subsections, like most authors we have assumed that in the Möbius strip the derivation of the obligation for ¬a is intuitively not desirable. However, one can also view it as being intuitively desirable, for the following reason.

Example 2.13 (Möbius strip, continued) *Consider first the coherent interpretation of the Möbius strip, deriving obligations for b and c, but not for ¬a. With the transitivity* T *pattern, one may consider the derivation of the obligation for ¬a. This represents that a was actually a violation. With* ACT, *the violation can be represented by an obligation for* $b \wedge c \wedge \neg a$.

Consider now the dilemma interpretation, presenting three possible outcomes, either $\{\bigcirc b, \bigcirc c\}$, *or* $\{\bigcirc b\}$, *or* $\emptyset$. *In that case, a leads to a choice, and we may thus have an instance of the forbidden conflict pattern* FC *that derives that a is forbidden.*

2.7 Priority

We are given a set S of conditional obligations along with a priority relation defined on them.

Example 2.14 (Order [Horty, 2007], continued from Example 2.12) *Numbers represent the priority of the obligation, as in Section 2.1.4. Consider*

$$\{③(\neg b|a), ②(b|\top), ①(a|\top)\}$$

①, ②, *and* ③ *can be thought of as expressing commands uttered by a priest, a bishop, and a cardinal, respectively. There are three interpretations. The greedy interpretation derives obligations for a and b. It looks strange, because complying with* $①(a|\top)$ *triggers the most important norm* $③(\neg b|a)$, *which in turn cancels* $②(b|\top)$. *To put it another way, complying with* $①(a|\top)$ *and* $②(b|\top)$ *results in violating* $③(\neg b|a)$.

The last link interpretation derives $\bigcirc a$ *and* $\bigcirc \neg b$. *This looks strange too, because* $②(b|\top)$ *takes precedence over* $①(a|\top)$, *and* $③(\neg b|a)$ *will not be triggered (and* $②(b|\top)$ *cancelled) unless* $①(a|\top)$ *is fulfilled.*

The weakest link interpretation derives $\bigcirc b$ *only. In order not to trigger* $③(\neg b|a)$, *and avoid being in a violation state with respect to it, the agent goes for* $②(b|\top)$ *only.*

The idea underpinning Parent [2011]'s next example is similar. Parent argues that different outcomes are expected depending on whether the example is instantiated in the deontic or epistemic domain.

Example 2.15 (Cancer [Parent, 2011]) *Assume we have*

$$\{ ③(c|b), ②(b|a), ①(\neg b|a) \}$$

a is for the set of data used to set up a treatment against cancer, b is for receiving chemo as per the protocol, and c is for keeping WBCs (White Blood Cells) count to a safe level using a drug. In a diagram:

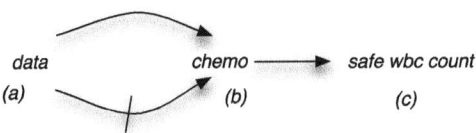

Figure 11. Cancer

Assume the input is a. In that case, we get $②(b|a)$ and $③(c|b)$, which derives $\bigcirc b$ and $\bigcirc c$. Given a, both $①(\neg b|a)$ and $②(b|a)$ are triggered. These two conflict. The stronger obligation takes precedence over the weaker one.

Assume the input is $\{a, \neg c\}$. In that case, we get $①(\neg b|a)$ which derives $\bigcirc \neg b$. The reason why may be explained as follows. Following one of Hansson [1969]'s suggestions, one might think of the input as someting settled as true. The question is: shall the agent do b or not? The ordering $② > ①$ says that b has priority over $\neg b$. So it would seem to follow that he should do b. But, in reply, it can be said that the ordering $③ > ②$ tells us that compliance with the stronger of the two conflicting norms triggers an obligation of even higher rank, namely the obligation to do c. Furthermore, c is already (settled as) false. Hence if the agent goes for b he will put himself in a violation state with respect to a norm with an even higher rank. The only way to avoid the violation of the most important norm is to go for $\neg b$. This is fully in line with what practitioners do: if the WBCs count cannot be maintained at a safe level, chemo is postponed.

In the epistemic domain, a different outcome is expected. This can be seen using the reliability interpretation discussed by Horty [2007, p. 391] among others. Under the latter interpretation, an epistemic conditional indicates something like a high conditional probability that its conclusion is satisfied, and the priority ordering measures relative strength of these conditional probabilities. For illustration purposes, assume that these conditional probabilities encode statistical assertions about some population groups, and instantiate a, b and c into (this is the example often used to illustrate the non-transitivity of default patterns) *being a student, being an adult,* and

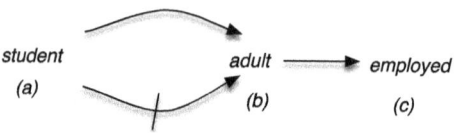

Figure 12. Student example

being employed. This is shown in Figure 12. Given input $\{a, \neg c\}$, the expected output remains b.

3 Formal framework

We extract ten basic properties from the examples, falling in three groups. We believe that the properties of factual detachment and violation detection, the logical properties of substitution, replacement by logical equivalents, implication and paraconsistency are desirable for methods to reason with normative systems, and that the properties of aggregation, factual and norm monotony, and norm induction are optional.

In this section we use the detachment terminology instead of the inference rules terminology.

3.1 Norms, obligations and factual detachment

The distinction between norms and obligations is fundamental in the modern approach to deontic logic. They are related via factual detachment, the detachment of an obligation from a norm.

3.1.1 Representing norms and imperatives explicitly

There are two traditions in normative reasoning, as witnessed by the two historical chapters in the *Handbook on Deontic Logic and Normative Systems* [Gabbay et al., 2013]. The first tradition of deontic logic is concerned with logical relations between obligations and permissions, or between the actual and the ideal. The second tradition of normative systems is concerned with normative reasoning, including reasoning about imperatives. Many people suggested a more comprehensive approach, by bringing the two traditions closer to each other, or proposing a uniform approach. For example, when van Fraassen [1973] is asking himself whether restricted conjunction can be formalized to reason about dilemmas, he suggests to represent imperatives explicitly.

> "But can this happy circumstance be reflected in the logic of the ought-statements alone? Or can it be expressed only in a language in which we can talk directly about the imperatives as well? This is an important

question, because it is the question whether the inferential structure of the 'ought' language game can be stated in so simple a manner that it can be grasped in and by itself. Intuitively, we want to say: there are simple cases, and in the simple cases the axiologist's logic is substantially correct even if it is not in general—but can we state precisely when we find ourselves in such a simple case? These are essentially technical questions for deontic logic, and I shall not pursue them here." [van Fraassen, 1973]

The distinction between norms and obligations was most clearly put forward by Makinson [1999], and we follow his notational conventions. To detach an obligation from a norm, there must be a context, and the norms must be conditional. Consequently, norms are a particular kind of rules.

3.1.2 Formal representation

In this section, a set of norms is represented by a set of pairs of formulae from a base logic, $(a_1, x_1), \ldots, (a_n, x_n)$. A norm (a, x) can be read as "if a is the case, then x ought to be the case." A normative system contains at least one set of norms, the regulative norms from which obligations and prohibitions can be detached. It often contains also permissive norms, from which explicit permissions can be detached, and constitutive norms, from which institutional facts can be detached.

The context is represented by a set of formulae of the same logic. A deontic operator $\bigcirc$ factually detaches obligations, represented by a set of formulae of the base logic, from a set of norms N in a context A, written as $\bigcirc(N, A)$. Unless there is a need for it, we adopt the convention that we do not prefix the detached formula with a modal operator. For example, from a norm that if you travel by metro, you must have a valid ticket (*metro*, *ticket*) in the context where you travel by metro, we derive *ticket* $\in \bigcirc(\{(metro, ticket)\}, \{metro\})$, but *ticket* itself is not prefixed with a deontic modality. Note that there is no risk of confusing facts and obligations. We know that *ticket* represents an obligation for *ticket*, because it is factually detached by the $\bigcirc$ operator.

To facilitate presentation and proofs, in this chapter we assume propositional logic as the base logic. We write $\beta \in \bigcirc(N, \alpha)$ for $\beta \in \bigcirc(N, \{\alpha\})$, and $\gamma \in \bigcirc((\alpha, \beta), A)$ for $\gamma \in \bigcirc(\{(\alpha, \beta)\}, A)$.

3.1.3 Arguments

Maybe the most important technical innovation of the modern approach is the following convention of writing an argument for α supported by A, traditionally written as $A \therefore \alpha$, as a pair (A, α):

$$(A, \alpha) \in \bigcirc(N) = \alpha \in \bigcirc(N, A)$$

We can move between $\bigcirc(N)$ and $\bigcirc(N, A)$ as we move between $\vdash$ and Cn in classical logic.

It is crucial to understand that the representation of arguments by a pair (A, α) is just a technical method to develop logical machinery: we use it to give more compact representations, to provide proof systems, and to make relations with other branches of logic. However, if you want to know what the argument $(A, \alpha) \in \bigcirc(N)$ means, then you always have to translate it back to $\alpha \in \bigcirc(N, A)$.

We reserve the term "norms" to explicit norms, in N. Obviously, one does not derive norms from norms.

In this section we give both the long and the short version of the properties we discuss, to prevent misreading.

3.1.4 Factual detachment

Factual detachment says that if there is a norm with precisely the context as antecedent, then the output contains the consequent. On the one hand this is relatively weak, as we require the context to be *precisely* the antecedent. A much stronger detachment principle imposes detachment when the antecedent is *implied* by the context. Between these two extremes, we can have that most obligations are detached, or in the most normal cases the obligation is detached. On the other hand the factual detachment principle is also quite strong, as in context a from the norm $(a, \bot)$ the contradiction $\bot$ is detached, and in case of a dilemma of (a, x) and $(a, \neg x)$, in context a both x and $\neg x$ are detached.

Definition 3.1 (Factual detachment) *A deontic operator $\bigcirc$ satisfies the factual detachment property if and only if for all sets of norms N and all sentences α and β we have:*

$$\frac{(\alpha, \beta) \in N}{\beta \in \bigcirc(N, \alpha)} \text{FD} \qquad \frac{(\alpha, \beta) \in N}{(\alpha, \beta) \in \bigcirc(N)} \text{FD} \qquad \frac{(\alpha, \beta) \in N}{(\alpha, \beta)} \text{FD}$$

3.2 Violation detection

The distinctive feature of norms and obligations with respect to other types of rules and modalities is that they can be violated. Obligations which cannot be violated are not real obligations, but obligations of a degenerated kind. It is not only that ought implies can, but more importantly, ought implies can-be-violated. Issues concerning violations can be found in most deontic examples. For example, dilemma examples arise because some obligation has to be violated, and contrary-to-duty examples arise because some obligation has been violated.

Modal logic offers a simple representation for violations. An obligation for α has been violated if and only we have $\neg \alpha \wedge \bigcirc \alpha$. In our notation with explicit norms, this is $\alpha \in \bigcirc(N, A)$ with $\neg \alpha \in Cn(A)$.

To make sure that violated obligations do not drown, we use the violation detection inference pattern, which we already discussed in Section 2.2.3.

Definition 3.2 (Violation detection) *A deontic operator $\bigcirc$ satisfies the violation detection property if and only for all sets of norms N, all sets of sentences A and all sentences α we have:*

$$\frac{\alpha \in \bigcirc(N, A)}{\alpha \in \bigcirc(N, A \cup \{\neg\alpha\})}\text{VD} \qquad \frac{(A, \alpha)}{(A \cup \{\neg\alpha\}, \alpha)}\text{VD}$$

Consequently, the restricted strengthening of the antecedent pattern is too weak.

3.3 Substitution

Whereas the first two properties define what is special about *deontic* logic, namely factual detachment and violation detection, the next four properties of substitution, replacements of logical equivalence, implication and paraconsistency say something about *logic*.

The first logical requirement is substitution, well known from classical propositional logic. It says that we can uniformly replace propositional letters by propositional formulae.

Definition 3.3 (Substitution) *Let a uniform substitution map each proposition letter to a propositional formula. A deontic operator $\bigcirc$ satisfies substitution if and only for all sets of norms N, all sets of formulae A, all sentences α and all uniform substitutions σ we have:*

$$\frac{\alpha \in \bigcirc(N, A)}{\alpha[\sigma] \in \bigcirc(N[\sigma], A[\sigma])}\text{SUB}$$

For example, it allows to replace propositional letters by distinct new letters, thus renaming them. This is an example of irrelevance of syntax, a core property of logic.

3.4 Replacement of logical equivalents

The following definition introduces two stronger types of irrelevance of syntax.

Definition 3.4 (Irrelevance of Syntax) *Let Cn be closure under logical consequence, and Eq closure under logical equivalence: $\alpha \in Eq(S)$ if and only if there is a β in S such that $Cn(\alpha) = Cn(\beta)$. We write $Eq(a_1, \ldots, a_n)$ for $Eq(\{a_1, \ldots, a_n\})$, and $Cn(a_1, \ldots, a_n)$ for $Cn(\{a_1, \ldots, a_n\})$. Here Cn is the consequence operation of the base logic on top of which the deontic operator $\bigcirc$ operates.*

A deontic operator $\bigcirc$ satisfies formula input (output) irrelevance of syntax if and only for all sets of norms N and all sets of formulae A we have:

$$\bigcirc(N, A) = \bigcirc(N, Eq(A)) \qquad (\bigcirc(N, A) = Eq(\bigcirc(N, A)))$$

and it satisfies set input (output) irrelevance of syntax if and only if for all sets of norms N and all sets of formulae A we have:

$$\bigcirc(N, A) = \bigcirc(N, Cn(A)) \qquad (\bigcirc(N, A) = Cn(\bigcirc(N, A)))$$

The following example illustrates the various types of irrelevance of syntax.

Example 3.5 (Irrelevance of syntax) Let $N = \{(a, x), (a, y)\}$ and $A = \{a\}$. The following table lists some possibilities for $\bigcirc(N, A)$:

$\emptyset$ $\qquad$ $\{x, y\}$ $\qquad$ $\{x, y, x \wedge y\}$
$\{x \wedge y, y \wedge x\}$ $\quad$ $\{x \wedge y, y \wedge x, x, y\}$ $\quad$ $\{x \wedge y, y \wedge x, x, y, x \vee y, y \vee x\}$
$Eq(x \wedge y)$ $\qquad$ $Eq(x \wedge y, x, y)$ $\qquad$ $Eq(x \wedge y, x, y, x \vee y)$
$Cn(x) \cup Cn(y)$ $\quad$ $Cn(x \wedge y)$

The first row gives some deontic operators which do not satisfy basic properties. For example, $\emptyset$ does not satisfy factual detachment, $\{x, y\}$ does not satisfy conjunction, and $\{x, y, x \wedge y\}$ does not satisfy variable renaming. That is, if we replace x and y in N, then we end up with the same set, but if we replace x and y in the output, we obtain $y \wedge x$. This violates the most basic property of irrelevance of syntax.

The second row gives some examples satisfying variable renaming for x and y. The set of obligations $\{x \wedge y, y \wedge x\}$ does not satisfy factual detachment again, and the set $\{x \wedge y, y \wedge x, x, y, x \vee y, y \vee x\}$ satisfies besides closure under conjunction also closure under disjunction. Whether this is desired depends on the application. However, all three examples do not satisfy formula output irrelevance of syntax. For example, they all three derive $x \wedge y$, but they do not derive the logically equivalent $x \wedge x \wedge y$.

The third and fourth row close the output under logical equivalence and logical consequence, respectively. $Cn(x \wedge y)$ in the last row satisfies set output irrelevance of syntax.

Input irrelevance is analogous to output irrelevance. For example, when the input is $a \wedge a$ rather than a, it may or may not derive again the same output. If it does not, then the operator violates formula input irrelevance of syntax. Moreover, if it does not treat $\{a, b\}$ and $\{a \wedge b\}$ the same, then it violates input set irrelevance of syntax.

The following example illustrates that output set irrelevance of syntax is too strong in the context of dilemmas, because it may lead to deontic explosion.

Example 3.6 (Irrelevance of syntax, continued) Let

$$N = \{(a, x \wedge y), (a, \neg x \wedge y)\}$$

and $A = \{a\}$. The following table lists some possibilities for $\bigcirc(N, A)$. We only list options closed under logical equivalence, i.e. which satisfy output formula irrelevance of syntax.

$Eq(x \wedge y, \neg x \wedge y)$ $\qquad$ $Eq(x \wedge y, \neg x \wedge y, x \wedge \neg x \wedge y)$
$Eq(x \wedge y, \neg x \wedge y, y)$ $\qquad$ $Eq(x \wedge y, \neg x \wedge y, y, x \wedge \neg x \wedge y)$
$Cn(x \wedge y) \cup Cn(\neg x \wedge y)$ $\quad$ $Cn(x \wedge y) \cup Cn(\neg x \wedge y) \cup Eq(x \wedge \neg x \wedge y)$
$Cn(x \wedge y, \neg x \wedge y)$

The last set $Cn(x \wedge y, \neg x \wedge y)$ derives the whole language, and thus gives rise to explosion. Hence we cannot accept it. The example illustrates that we cannot accept set output irrelevance of syntax.

The difference between the left and right column is that the right column is closed under conjunction, and represents with inconsistent formulae that there is a dilemma.

The difference between the first and the second row is that the second row is closed under disjunction. The difference between the second and the third row is that consistent formulae are closed under logical consequence.

$Cn(x \wedge y) \cup Cn(\neg x \wedge y) \cup Eq(x \wedge \neg x \wedge y))$ has the feature that violations and other obligations are treated in a distinct way.

In this chapter we require set input irrelevance of syntax, and formula output irrelevance of syntax. In addition, along the same lines we require that we can replace formulae within the norms by logically equivalent ones. All together, it corresponds to the following property of replacement of logical equivalents.

Definition 3.7 (Replacement of logically equivalent expressions) We say that two norms ar similar, written as $(\alpha_1, \beta_1) \approx (\alpha_2, \beta_2)$, if and only if $Cn(\alpha_1) = Cn(\alpha_2)$, and $N \approx M$ if and only if for all $(\alpha_1, \beta_1) \in N$ there is a $(\alpha_2, \beta_2) \in M$ such that $(\alpha_1, \beta_1) \approx (\alpha_2, \beta_2)$, and vice versa. A deontic operator $\bigcirc$ satisfies the replacement of Logical Equivalents property if and only if for all sets of norms N and M, all sets of formulae A and B, and all sentences α and β we have:

$$\frac{N \approx M, Cn(A) = Cn(B), Cn(\alpha) = Cn(\beta), \alpha \in \bigcirc(N, A)}{\beta \in \bigcirc(M, B)} \text{RLE}$$

The examples illustrate that there are other options in between formula and set output irrelevance of syntax, such as requiring that the output is closed under conjunction, or under disjunction, or both. We consider them in Section 3.7.

The principle of irrelevance of syntax has been criticized in belief revision theory. It is discussed by [Stolpe, 2010] in the context of a study of the notion of revision of a normative system. This notion falls outside the scope of the present chapter, and must be left as a topic for future research.

3.5 Implication

The four properties FD, VD, SUB and RLE defined thus far may be called positive properties, in the sense that they require something to be obligatory. That is why we could represent them as Horn rules: given a set of conditions, we require some obligation to be derivable. This contrasts with the examples in Section 2, where typically too much is derived.

The implication requirement in this section and the paraconsistency requirement in the following section may be called negative properties, in the sense that they forbid something to be obligatory. The first requirement makes use of the so-called materialisation of a normative system, which means that each norm (a, x) is interpreted as a material conditional $a \to x$, i.e. as the propositional sentence $\neg a \lor x$. The implication requirement says that if the materializations of N, written as $m(N)$, do not imply $a \to x$, then $(a, x) \notin \bigcirc(N)$. This represents the idea that we cannot derive more than we can derive in propositional logic. In general, implication in the base logic is the upper bound.

Definition 3.8 (Implication) *Let $m(N) = \{a \to x \mid (a, x) \in N\}$ be the set of materializations of N. A deontic operator $\bigcirc$ satisfies the implication property if and only if for all sets of norms N and all sets of sentences A we have $\bigcirc(N, A) \subseteq Cn(m(N) \cup A)$.*

The elements $(\{\alpha\}, \beta)$ of $\bigcirc(N)$ are a subset of $\{(\alpha, \beta) \mid \alpha \to \beta \in Cn(m(N))\}$. In most systems, the base logic is classical propositional logic, but it need not be so. For instance, Cn may be the consequence relation of intuitionistic propositional logic, as in [Parent et al., 2014]. Cn may also be what Makinson calls a pivotal consequence relation Cn_K, defined by $Cn_K(A) = C(A \cup K)$, where K is a set of formulas, and C is the consequence relation of classical propositional logic. [Stolpe, 2008] defines and studies two such input/output operations. They are aimed to model the interplay between norms and so-called material dependencies. We have $\bigcirc(N, A) \subseteq Cn_K(m(N) \cup A)$.

3.6 Paraconsistency

To prevent explosion we do not want to derive the whole language, unless maybe in pathological cases in which the normative system contains a norm for each propositional formula. A consequence relation may be said to be paraconsistent if it is not explosive, though there are various ways to make this formal.

To define our paraconsistency requirement, we distinguish obligations representing violations from other obligations. That is, we decompose an operator $\bigcirc(N, A)$ into two operators $V(N, A)$ and $\overline{V}(N, A)$, such that we have $V(N, A) = \{x \in \bigcirc(N, A) \mid \neg x \in Cn(A)\}$ and $\overline{V}(N, A) = \bigcirc(N, A) \setminus V(N, A)$. Trivially, we have

$$\bigcirc(N, A) = V(N, A) \cup \overline{V}(N, A)$$

The basic idea of our paraconsistency requirement is that obligations in $\overline{V}$ can be derived from a set of norms M in N, such that this set of norms M does not explode.

Definition 3.9 (Paraconsistency) *A deontic operator $\bigcirc$ satisfies the paraconsistency property if and only if for all sets of norms N, all sets of formulae A and all sentences α, if $\alpha \in \overline{V}(N, A)$, then there is a $M \subseteq N$ such that $\alpha \in \bigcirc(M, A)$ and $\bigcirc(M, A) \cup A$ is classically consistent.*

Implication and paraconsistency together imply that if $\alpha \in \overline{V}(N, A)$, then there is a $M \subseteq N$ such that $\alpha \in Cn(m(N) \cup A)$ and $\bigcirc(M, A) \cup A$ is classically consistent. This suggest an additional condition: if $\alpha \in \overline{V}(N, A)$, then there is a $M \subseteq N$ such that $\alpha \in Cn(m(N) \cup A)$ and $m(N) \cup A$ is classically consistent.

The underlying intuition to restrict to a set of norms was already raised in Example 1.1 in the introduction. There we observe that if we can derive $\bigcirc(\beta \wedge \gamma)$ from $\bigcirc(\alpha \wedge \beta)$ and $\bigcirc(\neg \alpha \wedge \gamma)$, and we have substitution and replacements of logical equivalents, then we also derive $\bigcirc(\beta)$ from $\bigcirc(\alpha)$ and $\bigcirc(\neg \alpha)$, in other words, we have deontic explosion. This can be verified by replacing β by $\alpha \vee \beta$ and γ by $\neg \alpha \vee \beta$. Therefore, we restrict the set of norms we use to a set of norms which is in some sense "consistent" with the input A.

3.7 Aggregation

The last four properties of aggregation, factual and norm monotony, and norm induction determine the kind of deontic logics we are going to study in our framework. We believe that other choices at this point may be of interest too, but we do not pursue them in this chapter.

Aggregation is a core issue in van Fraassen's paradox.

Definition 3.10 (Aggregation) *A deontic operator $\bigcirc$ satisfies the aggregation property if and only if for all sets of norms N, sets of sentences A and sentences α and β we have*

$$\frac{\alpha, \beta \in \bigcirc(N, A)}{\alpha \wedge \beta \in \bigcirc(N, A)} \text{AND} \qquad \frac{(A, \alpha), (A, \beta)}{(A, \alpha \wedge \beta)} \text{AND}$$

Van Fraassen's paradox shows that therefore we cannot accept weakening of the consequent. In the context of our present framework, we prefer to call it weakening of the output.

Definition 3.11 *A deontic operator $\bigcirc$ satisfies the weakening of the output property if and only if for all sets of norms N, sets of sentences A and sentences α and β we have*

$$\frac{\alpha \wedge \beta \in \bigcirc(N, A)}{\alpha, \beta \in \bigcirc(N, A)} \text{WO} \qquad \frac{(A, \alpha \wedge \beta)}{(A, \alpha), (A, \beta)} \text{WO}$$

Proposition 3.12 *There is no deontic operator $\bigcirc$ satisfying paraconsistency, aggregation, and weakening of the output.*

Proof. *Assume the statement does not hold, so there is a deontic $\bigcirc$ satisfying paraconsistency, aggregation and weakening of the output. Consider van Fraassen's paradox $N = \{(\top, p), (\top, \neg p)\}$. According to aggregation and weakening of the output, we have $(\top, q) \in \bigcirc(N)$. According to paraconsistency, $(\top, q) \notin \bigcirc(N)$. Contradiction.* ∎

3.8 Factual monotony

In this chapter we are interested in monotonic logics. Though non-monotonic logics may have their applications too, we believe they should be build on top of the monotonic ones.

Definition 3.13 (Factual monotony) *The factual monotony property holds for $\bigcirc$ if and only if for all sets of norms N, and all sets of sentences A and B, we have $\bigcirc(N, A) \subseteq \bigcirc(N, A \cup B)$.*

As this implies strengthening of the antecedent, Forrester's paradox illustrates that we cannot accept weakening of the consequent.

Proposition 3.14 *There is no deontic operator $\bigcirc$ satisfying paraconsistency, factual monotony, and weakening of the output.*

Proof. *Assume the statement does not hold, so there is a deontic $\bigcirc$ satisfying paraconsistency, factual monotony and weakening of the output. Consider the first norm of Forrester's paradox $N = \{(\top, \neg k)\}$. According to factual monotony and weakening of the output, we have $(k, \neg k \vee g) \in \bigcirc(N)$. According to paraconsistency, $(k, \neg k \vee g) \notin \bigcirc(N)$. Contradiction.* ∎

3.9 Norm monotony

Definition 3.15 (Norm monotony) *A deontic operator $\bigcirc$ satisfies the property of norm monotony if and only if for all sets of norms N and M we have $\bigcirc(N) \subseteq \bigcirc(N \cup M)$.*

A deontic operator $\bigcirc$ satisfies the property of monotony if and only if it satisfies those of factual and norm monotony, i.e. for all N, M, A, B we have $\bigcirc(N, A) \subseteq \bigcirc(N \cup M, A \cup B)$.

3.10 Norm induction

Norm induction says that if there is an output β for an input α, and we add the norm (α, β) to the normative system, then for all inputs, the output of the normative system stays the same. We call it norm induction, because the norm is induced from the relation between facts and obligations. The norm induction requirement considers a set M of such pairs (α, β).

Definition 3.16 (Norm induction) *A deontic operator $\bigcirc$ verifies the property of norm induction if and only if for all sets of norms N and M and all sets of sentences A we have $M \subseteq \bigcirc(N) \Rightarrow \bigcirc(N) = \bigcirc(N \cup M)$.*

The strong norm induction principle strengthens the norm induction principle to expansion of the normative system with new norms.

Definition 3.17 (Strong norm induction) *A deontic operator $\bigcirc$ satisfies the property of strong norm induction if and only if for all sets of norms N, N', M, and all sets of sentences A we have $M \subseteq \bigcirc(N) \Rightarrow \bigcirc(N \cup N') = \bigcirc(N \cup N' \cup M)$*

Clearly we have that the strong norm induction property implies the norm induction property.

Together, factual detachment, monotony and norm induction are equivalent to requiring that $\bigcirc$ is a closure operator.

Definition 3.18 (Closure operator) $\bigcirc$ *is a closure operator if and only if it satisfies the following three properties:*

INCLUSION $N \subseteq \bigcirc(N)$

MONOTONY $N \subseteq M$ implies $\bigcirc(N) \subseteq \bigcirc(M)$

IDEMPOTENCE $\bigcirc(N) = \bigcirc(\bigcirc(N))$

Their counterparts in terms of Cn are knowns as the "Tarskian" conditions, after A. Tarski. They can each be rephrased in terms of $\vdash$ ('proves') as follows.

REFLEXIVITY $A \vdash x$ for all $x \in A$

MONOTONY $A \vdash x$ implies $A \cup B \vdash x$

TRANSITIVITY $A \vdash x$ for all $x \in B$ and $B \vdash y$ imply $A \vdash y$

Inclusion for Cn translates into reflexivity of $\vdash$. Monotony for Cn translates into monotony of $\vdash$. Idempotence of Cn corresponds to the transitivity of $\vdash$.

4 Summary

Table 2 lists the examples we discussed in this chapter. Given that the world is full of conflicts, we have that normative systems are developed by humans and full of inconsistencies. We need to represent dilemmas consistently, if only to consider their resolution. Van Fraassen's paradox illustrates that doing so presents a basic dilemma: do we accept aggregation or closure under consequence? Forrester's paradox seems to indicate a dilemma too, as it presents two alternatives. In the cottage regulations, such a dilemma interpretation makes sense: either remove the fence, or paint it white. However, in Forrester's gentle murderer example, you cannot undo killing someone. So only the coherent interpretation makes sense. Dilemmas can be resolved by explicit priorities, for example reflecting the authority creating the obligation, or it can be derived from the specificity of the obligations. In the latter case, as illustrated by the cottage regulations, we have to be careful to distinguish violations from exceptions. Jeffrey's disarmament illustrates the problem of reasoning by cases in deontic reasoning. When conditions have an epistemic reading, reasoning by cases may not be valid. Deontic detachment and transitivity originate from Chisholm's paradox, though it is known in the literature as a contrary-to-duty paradox rather than a deontic detachment paradox. Chisholm's paradox illustrates that an alternative representation of the transitivity pattern makes it analogous to Forrester's paradox. Makinson's Möbius strip illustrates many of the problems of reasoning with transitivity. In particular, the dilemma interpretation highlights that we can have solutions being a strict subset of other solutions. More priority examples are introduced in the area of epistemic reasoning, and reasoning with defaults.

Ex.	obligations	patterns	
2.1	Fraassen	$\bigcirc p, \bigcirc \neg p$	AND, WC
2.2	Forrester	$\bigcirc(\neg k\|\top), \bigcirc(g\|k), \vdash g \to k$	FD, (R)AND
2.3	Forrester	$\bigcirc(\neg k\|\top), \bigcirc(g\|k), \vdash g \to k$	(R)SA, ANDC, WC
2.6	Cottage	$\bigcirc(\neg f\|\top), \bigcirc(w \wedge f\|f), \bigcirc(f\|d)$	RSA_o
2.7	Jeffrey	$\bigcirc(d\|w), \bigcirc(d\|\neg w), \bigcirc(\neg d\|d \leftrightarrow w)$	RSA, ORA
2.8	Chisholm	$\bigcirc(a\|\top), \bigcirc(t\|a), \bigcirc(\neg t\|\neg a), \neg a$	AND, FD, DD
2.9	Chisholm	$\bigcirc(a\|\top), \bigcirc(t\|a), \bigcirc(\neg t\|\neg a), \neg a$	T/ CT / ACT, ANDC
2.10	Möbius	$\bigcirc(\neg a\|c), \bigcirc(c\|b), \bigcirc(b\|a), a$	T/ CT
2.14	Priority	$③(\neg b\|a), ②(b\|\top), ①(a\|\top)$	T/ CT

Table 2. Summary of the examples

Maybe the most important technical innovation of our formal framework is the convention of writing an argument for α supported by A as a pair (A, α) with $(A, \alpha) \in \bigcirc(N)$, which means the same as $\alpha \in \bigcirc(N, A)$. We can move between $\bigcirc(N)$ and $\bigcirc(N, A)$ as we move between $\vdash$ and Cn in classical logic.

The ten properties of our formal framework listed in Table 3. We believe that all deontic logics have to satisfy the deontic properties of factual detachment and violation detection, and the logical properties of substitution, replacement by logical equivalents, implication and paraconsistency. Moreover, we discussed the optional properties of aggregation, factual and norm monotony, and norm induction.

There are two ways to look at the operator $\bigcirc$. First, given a set of norms, it derives sentences from sentences: $\alpha \in \bigcirc_N(A)$. This is the classical way deontic logics considered normative systems: facts go in, obligations go out. Secondly, it derives arguments from norms: $(A, \alpha) \in \bigcirc(N)$. These two views can be used to summarise our properties as follows.

First, the operator in $(A, \alpha) \in \bigcirc(N)$ must be a closure operator, which means that it satisfies factual detachment, norm monotony and norm induction. In addition, it must satisfy substitution and replacement of logical equivalents. Secondly, the operator in $\alpha \in \bigcirc_N(A)$ must satisfy violation detection, implication, paraconsistency,

FD	$(\alpha, \beta) \in N \Rightarrow \beta \in \bigcirc(N, \alpha)$	Factual detachment
VD	$(A, \beta) \Rightarrow (A \cup \{\neg\beta\}, \beta)$	Violation detection
SUB	$\alpha \in \bigcirc(N, A) \Rightarrow \alpha[\sigma] \in \bigcirc(N[\sigma], A[\sigma])$	Substitution
RLE	$N \approx M, Cn(A) = Cn(B), Cn(\alpha) = Cn(\beta),$	Replacement of
	$(A, \alpha) \in \bigcirc(N) \Rightarrow (B, \beta) \in \bigcirc(M)$	equivalents
IMP	$\bigcirc(N, A) \subseteq Cn(m(N) \cup A)$	Implication
PC	$\alpha \in \overline{V}(N, A) \Rightarrow \exists M \subseteq N : \alpha \in \bigcirc(M, A)$	Paraconsistency
	and $\bigcirc(M, A) \cup A$ consistent	
AND	$(A, \alpha)(A, \beta) \Rightarrow (A, \alpha \wedge \beta)$	Conjunction
FM	$(A, \alpha) \Rightarrow (A \cup B, \alpha)$	Factual monotony
NM	$\bigcirc(N) \subseteq \bigcirc(N \cup M)$	Norm monotony
NI	$M \subseteq O(N) \Rightarrow O(N) = O(N \cup M)$	Norm induction

Table 3. Properties

factual monotony, and aggregation.

The properties of norm monotony and norm induction have the effect that our logics will behave classically as Tarskian consequence operators. However, it is important to realise that the closure properties on $\bigcirc(N)$ are not as innocent as they are in other branches of philosophical logic. In particular norm induction is very strong, because it says that every argument (A, α) can itself be used as a norm. This may be true of some branches of case law, but it is probably too strong to be accepted as a universal law for norms. We therefore expect that future studies will first relax this requirement, before relaxing the others.

Finally, we may consider our ten properties as requirements for the further development of reasoning methods for normative systems and deontic logic. We have recently presented two logics satisfying all ten properties [Parent and van der Torre, 2014b], which shows that the ten properties are consistent in the sense that they can be satisfied simultaneously.

Acknowledgments. Thanks to an anonymous reviewer for valuable comments. This work is supported by the European Union's Horizon 2020 research and innovation programme under the Marie Curie grant agreement No: 690974 (Mining and Reasoning with Legal Texts, MIREL). Its support is gratefully acknowledged.

BIBLIOGRAPHY

[Anderson, 1956] A. R. Anderson. The formal analysis of normative systems. In N. Rescher, editor, *The Logic of Decision and Action*, pages 147–213. Univ. Pittsburgh, 1967, 1956.

[Åqvist, 2002] L. Åqvist. Deontic logic. In D. Gabbay and F. Guenthner, editors, *Handbook of philosophical logic*, volume 8, pages 147–264. Kluwer Academic publisher, 2002.

[Chellas, 1980] B.F. Chellas. *Modal Logic: An Introduction*. Cambridge University Press, 1980.

[Chisholm, 1963] R.M. Chisholm. Contrary-to-duty imperatives and deontic logic. *Analysis*, 24:33–36, 1963.

[Gabbay et al., 2013] D. Gabbay, J. Horty, R. van der Meyden, X. Parent, and L. van der Torre, editors. *Handbook of Deontic Logic and Normative Systems*, volume 1. College Publications, London, UK, 2013.

[Governatori and Hashmi, 2015] G. Governatori and M. Hashmi. Permissions in deontic event-calculus. In *Legal Knowledge and Information Systems - JURIX 2015: The Twenty-Eighth Annual Conference, Braga, Portual, December 10-11, 2015*, volume 279, pages 181–182. IOS Press, 2015.

[Hansson, 1969] B. Hansson. An analysis of some deontic logics. *Noûs*, 3:373–398, 1969. Reprinted in [Hilpinen, 1971, pp 121-147].

[Hilpinen, 1971] R. Hilpinen, editor. *Deontic Logic: Introductory and Systematic Readings*. Reidel, Dordrecht, 1971.

[Horty, 2007] J. Horty. Defaults with priorities. *Journal of Philosophical Logic*, 36:367–413, 2007.

[Horty, 2014] J. Horty. Deontic modals: why abandon the classical semantics? *Pacific Philosophical Quarterly*, 95:424–460, 2014.

[Jeffrey, 1983] R. Jeffrey. *The Logic of Decision*. University of Chicago Press, 2nd edition, 1983.

[Kolodny and MacFarlane, 2010] N. Kolodny and J. MacFarlane. Iffs and oughts. *Journal of Philosophy*, 107(3):115–143, 2010.

[Makinson and van der Torre, 2000] D. Makinson and L. van der Torre. Input/output logics. *Journal of Philosophical Logic*, 29(4):383–408, 2000.

[Makinson and van der Torre, 2001] D. Makinson and L. van der Torre. Constraints for input-output logics. *Journal of Philosophical Logic*, 30(2):155–185, 2001.

[Makinson, 1999] D. Makinson. On a fundamental problem in deontic logic. In P. Mc Namara and H. Prakken, editors, *Norms, Logics and Information Systems*, Frontiers in Artificial Intelligence and Applications, pages 29–54. IOS Press, Amsterdam, 1999.

[Parent and van der Torre, 2014a] X. Parent and L. van der Torre. Aggregative deontic detachment for normative reasoning (short paper). In T. Eiter, C. Baral, and G. De Giacomo, editors, *Principles of Knowledge Representation and Reasoning. Proceedings of the 14th International Conference - KR 14*. AAAI Press, 2014.

[Parent and van der Torre, 2014b] X. Parent and L. van der Torre. "Sing and dance!": Input/output logics without weakening. In F. Cariani, D. Grossi, J. Meheus, and X. Parent, editors, *Deontic Logic and Normative Systems - 12th International Conference, DEON 2014, Ghent, Belgium, July 12-15, 2014. Proceedings*, volume 8554 of *Lecture Notes in Computer Science*, pages 149–165. Springer, 2014.

[Parent et al., 2014] X. Parent, D. Gabbay, and L. van der Torre. Intuitionistic basis for input/output logic. In S. O. Hansson, editor, *David Makinson on Classical Methods for Non-Classical Problems*, pages 263–286. Springer Netherlands, Dordrecht, 2014.

[Parent, 2011] X. Parent. Moral particularism in the light of deontic logic. *Artificial Intelligence and Law*, 19(2-3):75–98, 2011.

[Prakken and Sergot, 1996] H. Prakken and M.J. Sergot. Contrary-to-duty obligations. *Studia Logica*, 57:91–115, 1996.

[Ross, 1930] D. Ross. *The Right and the Good*. Oxford University Press, 1930.

[Stolpe, 2008] A. Stolpe. Normative consequence: The problem of keeping it whilst giving it up. In G. Governatori and G. Sartor, editors, *Deontic Logic in Computer Science, 10th International Conference, DEON 2010. Proceedings*, volume 6181 of *Lecture Notes in Computer Science*, pages 174–188. Springer, 2008.

[Stolpe, 2010] A. Stolpe. Norm-system revision: Theory and application. *Artif. Intell. Law*, 18(3):247–283, 2010.

[Sun and van der Torre, 2014] X. Sun and L. van der Torre. Combining constitutive and regulative norms in input/output logic. In F. Cariani, D. Grossi, J. Meheus, and X. Parent, editors, *Deontic Logic and Normative Systems - 12th International Conference, DEON 2014. Proceedings*, volume 8554 of *Lecture Notes in Computer Science*, pages 241–257. Springer, 2014.

[Thomason and Horty, 1996] R. Thomason and R. Horty. Nondeterministic action and dominance: foundations for planning and qualitative decision. In *Proceedings of the Sixth Conference on Theoretical Aspects of Rationality and Knowledge (TARK'96)*, pages 229–250. Morgan Kaufmann, 1996.

[van der Torre and Tan, 1997] L. van der Torre and Y.-H. Tan. The many faces of defeasibility in defeasible deontic logic. In D. Nute, editor, *Defeasible Deontic Logic*, pages 79–121. Kluwer, 1997.

[van der Torre and Tan, 2000] L. van der Torre and Y.-H. Tan. Two-phase deontic logic. *Logique et analyse*, 43(171-172):411–456, 2000.

[van der Torre, 1994] L. van der Torre. Violated obligations in a defeasible deontic logic. In *Proceedings of the Eleventh European Conference on Artificial Intelligence (ECAI'94)*, pages 371–375. John Wiley & Sons, 1994.

[van der Torre, 2003] L. van der Torre. Contextual deontic logic: Normative agents, violations and independence. *Ann. Math. Artif. Intell.*, 37(1-2):33–63, 2003.

[van Fraassen, 1973] B.C. van Fraassen. Values and the heart command. *Journal of Philosophy*, 70:5–19, 1973.

11
Handling Norms in Multiagent Systems by Means of Formal Argumentation

CÉLIA DA COSTA PEREIRA, BEISHUI LIAO, ALESSANDRA MALERBA, ANTONINO ROTOLO, ANDREA G. B. TETTAMANZI, LEENDERT VAN DER TORRE, SERENA VILLATA

1 Introduction

Norms regulate our everyday life, and are used to assess conformance of behaviour with respect to regulations holding in multiagent systems. Agents undertake discussions about norms to assess their validity or applicability subject to particular conditions, to derive the obligations and permissions to be enforced, or to claim that a certain normative conclusion cannot be derived from the existing regulations. Given the profound importance of norms in multiagent systems, it is fundamental to understand, e.g., which norms are valid in certain environments, how to interpret them, and to determine the deontic conclusions of such norms. Some influential philosophers, such as Scott Shapiro [2011], argue that the law has an inherent teleological nature and that norms are plans, and in most existing normative multiagent systems, norms are like plans which aim at achieving the social goals the members of a society have decided to share [Boella *et al.*, 2009; Boella *et al.*, 2010]. However, it is not obvious that, for example, norms stating human rights can be considered as plans, and we therefore do not commit here to such philosophical claims.

Formal argumentation is typically based on logical arguments constructed from prioritised rules, and it is no surprise that the first applications of formal argumentation in the area of normative multiagent systems were concerned with the resolution of conflicting norms and norm compliance. Moreover, several frameworks have been proposed for normative and legal argumentation [Bench-Capon *et al.*, 2010], but no comprehensive formal model of normative reasoning from arguments has been proposed yet. In this chapter we discuss three challenges to illustrate the variety of applications of formal argumentation techniques in the field of normative multiagent systems.

- How can formal argumentation be used to explain existing approaches for reasoning about normative multiagent systems?

- How can new argumentation systems for reasoning about norms be developed, and how can these new argumentation systems be analysed?

- Which issues in the area of normative multiagent systems can be modelled and analysed using formal argumentation, besides the resolution of conflicting norms and checking compliance of a system with a set of norms?

First, we discuss how existing detachment procedures for prioritized norms can be represented in argumentation, by showing how the so-called Greedy and Reduction approaches can be represented in argumentation by applying the weakest link and the last link principles respectively [Liao et al., 2016]. Based on such representation results, formal argumentation can be used to explain the detachment of obligations and permissions from hierarchical normative systems in a new way.

Second, we discuss an instance of ASPIC$^+$ [Modgil and Prakken, 2013; Prakken and Sartor, 2013; van der Torre and Villata, 2014] capturing the inference schemes of arguments about norms like *legislative and interpretative* arguments. Moreover, we show how to adopt the input/output logic methodology [Makinson and van der Torre, 2000] for the analysis of these new argumentation systems [van der Torre and Villata, 2014].

Third, we discuss the model of da Costa Pereira et al. [2017], in which norm interpretation is a mechanism to deal with uncertainty, in contrast to existing models of norm interpretation in the context of Normative Multiagent Systems and AI&Law [Boella et al., 2009; Boella et al., 2010; Zurek and Araszkiewicz, 2013; Araszkiewicz and Zurek, 2015; Malerba et al., 2016; Araszkiewicz and Zurek, 2016]. This uncertainty reflects that, in legal theory, a definition of an empirical concept bounded in all now-foreseeable dimensions can break down in the face of unforeseen and unforeseeable events, and norms cannot anticipate all potential occurrences falling within the application scope of any legal norm [Hart, 1994; MacCormick and Summers, 1991]. In other words, it reflects that the interpretation of legal rules is often uncertain: legal language is vague, the concepts used to describe a legal rule are not always precise, and the purpose of the rule may be differently perceived [Heck, 1932; D'Amato, 1983; Liebwald, 2013]. The model uses fuzzy logic to measure the uncertainty of legal concepts, and argumentation is used to handle the conflicts between different interpretations of norms. More precisely, a fuzzy argumentation system [Tamani and Croitoru, 2014] to represent the interpretations, is combined with fuzzy labeling to evaluate the status of fuzzy arguments [da Costa Pereira et al., 2011]. As in many logical analyses of legal reasoning, the model is not purely descriptive and it is rather meant to offer a rational reconstruction for explaining and checking the robustness of interpretive arguments. A formal model for legal impreciseness must be cognitively sound, in the sense that it works on reliable cognitive assumptions.

The remainder of the chapter is organised as follows. Second 2 introduces how prioritized norms can be represented in argumentation. In Section 3, we discuss the logical properties of the static legal argumentation system proposed by Prakken and Sartor, and we reformulate it in a normative perspective. Section 4 motivates our adoption of graded categories as a tool to tackle the problem of open texture in legal interpretation. Section 5 introduces a model of fuzzy argumentation and fuzzy labeling, and Section 6 interprets a norm with flexibiity and conducts a case study by using an example from medically assisted reproduction. Second 7 discusses related work and Section 8 concludes.

2 Argumentation semantics for hierarchical normative systems

Consider the following benchmark example introduced by Hansen [2008], which we call here the *prioritised triangle* due to its graphical visualization in Figure 1.

Example 1 (Prioritised triangle [Hansen, 2008]) *Imagine you have been invited to a party. Before the event, you receive several imperatives, which we consider as the following set of norms.*
- *Your mother says: if you drink (p), then don't drive ($\neg x$).*
- *Your best friend says: if you go to the party (a), then you'll drive (x) us.*
- *An acquaintance says: if you go to the party (a), then have a drink with me (p).*

We assign numerical priorities to these norms, namely '3', '2' and '1' corresponding to the sources 'your mother', 'your best friend' and 'your acquaintance', respectively.

Let a, p and x respectively denote the propositions that you go to the party; you drink; and you drive. In terms of a hierarchical normative systems [Alchourron and Makinson, 1981], these norms are respectively represented as $(a,p)_1$, $(p,\neg x)_3$ and $(a,x)_2$. These three norms are visualized in Figure 1(a).

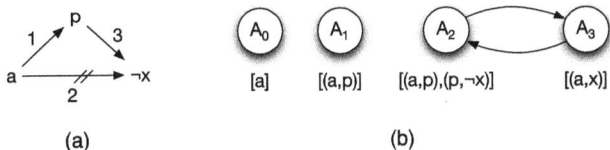

Figure 1. The prioritised normative system of the prioritised triangle example.

Consider the following two approaches resulting in different outcomes or *extensions* [Brewka and Eiter, 1999; Young et al., 2016; Liao et al., 2016].

Greedy approach Based on the context, a set of propositions that are known to hold, this approach always applies the norm with the highest priority that does not introduce inconsistency to an extension and the context. Here we say that a norm is applicable when its body is in the context or has been produced by other norms and added to the extension. In this example, we begin with the context $\{a\}$, and (a,x) is first applied. Then (a,p) is applied. Finally, $(p,\neg x)$ cannot be applied as this would result in a conflict, and so, by using the Greedy approach, we obtain the extension $\{p,x\}$.

Reduction approach In this approach, a candidate extension is identified. All norms which are applicable according to this candidate extension are selected and transformed into unconditional or body-free norms (i.e., a norm (a,b) selected in this way is transformed to a norm $(\top,b)$). The modified normative system, with the transformed norms is evaluated using the Greedy approach. The candidate extension is selected as an extension by the Reduction approach if it is identified as an extension according to this application of the Greedy approach. In this example, selecting a candidate extension $\{p, \neg x\}$, we get a set of body-free norms $\{(\top,p), (\top,\neg x), (\top,x)\}$. The priorities assigned to these norms are carried through from the original normative system, and are therefore respectively 1, 3 and 2. After applying the Greedy approach, we get $\{p, \neg x\}$, which is thus an extension of the Reduction approach. If on the other hand we had selected the candidate extension $\{p,x\}$, this new extension would not appear in the greedy evaluation, because $(\top, x)$ has a lower priority than $(\top, \neg x)$. Consequently $\{p,x\}$ is *not* an extension of the Reduction approach.

We now consider the prioritised triangle example in formal argumentation. Given a normative system, we may construct an argumentation framework as illustrated in Figure 1(b), which is a directed graph in which nodes denote arguments, and edges denote attacks between arguments. An argument is represented as a path of a directed graph starting from a node in the context. In this simple example, there are four arguments A_0, A_1, A_2 and A_3, represented as $[a], [a,p], [a,p,\neg x]$ and $[a,x]$, respectively. Since the conclusions of A_2 and A_3 are inconsistent, A_2 attacks A_3 and vice versa. Priorities allow us to transform these attacks into *defeats* according to different principles.

Last link ranks an argument based on the strength of its last inference, if the last link principle is applied, then $[a,p,\neg x]$ defeats $[a,x]$. As result, the principle allows us to conclude $\{p, \neg x\}$.

Weakest link ranks an argument based on the strength of its weakest inference. If the weakest link principle is used instead, $[a,x]$ defeats $[a,p,\neg x]$, and concludes $\{p,x\}$.

In this example, the last link principle thus gives the same result as the Reduction approach, and weakest link gives the same result as the Greedy approach. Liao *et al.* [2016] show that this is not a coincidence, but it holds for all totally ordered normative systems. This result addresses the challenge raised by Dung [1995] aiming at representing nonmonotonic logics through formal argumentation. In particular, argumentation is a way to exchange and communicate viewpoints, thus having an argumentation theory representing a nonmonotonic logic is desirable for such a logic, in particular when the argumentation theory is simple and efficient. Note that it is not helpful for the development of nonmonotonic logics themselves, but it helps when we want to apply such logics in distributed and multiagent scenarios.

Based on such representation results, formal argumentation can be used to explain the detachment of obligations and permissions from hierarchical normative systems in a new way. Moreover, many other challenges in normative reasoning have been expressed as inconsistent sets of formulas that are intuitively consistent, traditionally called deontic paradoxes. The most well known are the so-called contrary-to-duty paradoxes, which are concerned with handling norm violations. Techniques from non-monotonic reasoning have been applied to handle contrary-to-duty reasoning, and formal argumentation techniques can be applied in the same way [Pigozzi and van der Torre, to appear]. Finally, the most discussed practical problem in normative systems is norm conformance and compliance, which is a computational problem to check whether a business process is in accordance with a set of norms. Handling priorities among norms is again a central challenge for norm compliance, and formal argumentation techniques for resolving conflicts between norms can be extended with reasoning about business processes to reason about norm compliance [Tosatto et al., 2015].

3 New argumentation systems for normative reasoning

In the previous section, we used an argumentation system to explain the conclusions that are detached from a hierarchical normative system. The converse is done as well: new argumentation systems for normative reasoning have been developed for normative reasoning, for which detachment procedures have been defined to analyse these argumentation systems. We illustrate this by the argumentation system for legal reasoning proposed by Prakken and Sartor [2013], which has been analyzed and extended by van der Torre and Villata [2014].

Definition 1 (LAS-PS) *A legal argumentation system or LAS is a tuple $\langle \mathscr{L}, -, \mathscr{R} \rangle$ where $\mathscr{L}$ is the legal language of all sentences α, $-: \mathscr{L} \to 2^{\mathscr{L}}$ is a function given by $-(P) = \{\neg P\}$, $-(\neg P) = \{P\}$ and $-(N) = \emptyset$, and $\mathscr{R}$ contains the Defeasible modus ponens (DMP), rule for each possible norm N of the form $\phi_1 \wedge \ldots \wedge \phi_n \rightsquigarrow \psi$.*
DMP: $\phi_1, \ldots, \phi_n, \phi_1 \wedge \ldots \wedge \phi_n \rightsquigarrow \psi \Rightarrow \psi$;

In order to illustrate the legal argumentation framework, the running example proposed by Prakken and Sartor [2013] is adapted.

Example 2 (Smoking regulations) *Consider propositional atoms $P ::= a|b|c|d|e|f$ where a: "people want to smoke in a closed space", b: "the public place has special secluded smoking areas", c: "people need to smoke cannabis on medical grounds", d: "people are forbidden from smoking cannabis and tobacco in public places", e: "cannabis is allowed for medical treatment", f: "people are permitted to smoke cannabis in recreational cannabis establishments". $\mathscr{R}$ contains expressions for inference rules DMP of the form, for example: $a, a \rightsquigarrow b \Rightarrow b$ and $a, \neg c, a \wedge \neg c \rightsquigarrow d \Rightarrow d$.*

Prakken and Sartor [2013] follow Modgil and Prakken [2013], and do not consider a model theoretic semantics for this language. Instead, they define a set of arguments.

Definition 2 (LAS PS arguments) *A knowledge base K is a set of sentences of $\mathscr{L}$. The set of arguments A on the basis of a knowledge base K in a legal argumentation system LAS is called $Arg(LAS, K)$ and is the smallest set of expressions containing the literals in K and closed under the following rule:*
if $A_1, \ldots, A_n \subseteq Arg(LAS, K)$ and $concl(A_1), \ldots, concl(A_n) \Rightarrow L \in \mathscr{R}$ then we have also $(A_1, \ldots, A_n \Rightarrow L) \in Arg(LAS, K)$,
where $concl(A)$ is defined by $concl(L) = L$ and $concl(A_1, \ldots, A_n \Rightarrow L) = L$. We may leave out the brackets if there is no risk of confusion.

To study this notion of norm based argument, consequence is defined by considering only the conclusions of the arguments, in other words, by abstracting away the explicit arguments. Following input/output logic conventions, the consequence is called *Out*.

Definition 3 (Output PS) $Out(LAS, K) = \{concl(A) \mid A \in Arg(LAS, K)\}$.

Example 3 (Continued) *Consider the knowledge base of the smoking regulations $K_1 = \{a, b, c, e, a \wedge b \leadsto \neg d, c \wedge \neg d \wedge e \leadsto f\}$ where the norms state that*

- *if people want to smoke in a closed space and the public place has smoking special secluded areas, then people are not forbidden from smoking cannabis and tobacco in public places;*

- *if people need to smoke cannabis on medical grounds and it is not forbidden from smoking cannabis and tobacco in public places and cannabis is allowed for medical treatment, then people are permitted to smoke cannabis in recreational cannabis establishments;*

Arguments can be constructed combining DMP inference rules as follows:

- $A_1 : a, b, a \wedge b \leadsto \neg d \Rightarrow \neg d$;
- $A_2 : c, (a, b, a \wedge b \leadsto \neg d \Rightarrow \neg d), e, c \wedge \neg d \wedge e \leadsto f \Rightarrow f$.

Therefore, from arguments A_1, A_2, we have that $concl(A_1) = \neg d$ and $concl(A_2) = f$. We conclude that $Out(LAS_1, K_1) = \{a, b, c, \neg d, e, f\}$.

We now introduce a logical analysis. Van der Torre and Villata [2014] use a proof system with expressions $K \therefore L$. The proof system contains four rules, called Identity (ID), Strengthening of the input (SI), Factual Detachment (FD), and Deontic Detachment (DD). The former is sometimes called Monotonicity (Mon), and the latter two are sometimes called Modus Ponens (MP) or Cumulative Transitivity (CT). The notion of consequence is called simple-minded reusable throughput or out_3^+ by Makinson and van der Torre [2000].

Definition 4 (Derivations PS) *$der(LAS)$ is the smallest set of expressions $K \therefore L$ closed under the following four rules.*

ID: *$\{L\} \therefore L$ for a literal L*

SI: *from $K \therefore L$ derive $K \cup K' \therefore L$*

FD: *$\{L_1, \ldots, L_n, L_1 \wedge \ldots \wedge L_n \leadsto L\} \therefore L$ for a norm $L_1 \wedge \ldots \wedge L_n \leadsto L$*

DD: *from $K \therefore L_i$ for $1 \leq i \leq n$ and $K \cup \{L_1 \ldots, L_n\} \therefore L$ derive $K \therefore L$*

The close relation between arguments and derivations in a deontic logic or a logic of normative systems is illustrated by the following property:

$K \therefore L \in der(LAS)$ iff $L \in Out(LAS, K)$.

This is not surprising, as the similarity is quite clear from the structure of arguments. However, making the relation precise by framing the legal argument system into an input/output logic highlights a drawback of the legal argumentation system of

Prakken and Sartor: simple-minded reusable throughput is usually adopted for default logics and logic programs, not for the normative reasoning.

To establish the results with constrained input/output logic, only rebut is considered. Thus undercut is not considered. Moreover, they do not consider defeasible knowledge and undermining. So the only attack is the attack of an argument with an opposite literal. This is obviously a very simple notion of attack which is of little use in most applications, but it useful to establish the relation with logical approaches.

Definition 5 (Attack PS) *The set of sub-arguments of B is the smallest set containing B, and closed under the rule: if $A_1,\ldots,A_n \Rightarrow L$ is a sub-argument of B, then $A_1, \ldots, A_n$ are also sub-arguments of B.*

A attacks B iff there is a sub-argument B' of B such that $concl(A) \in -(concl(B'))$. We write $attack(AS,K)$ for the set of all attacks among $Arg(AS,K)$.

A semantics associates sets of extensions with an argumentation framework, where each extension consists of a set of arguments. For each extension, the output consists of the set of conclusions of the arguments, as for *Out* before. A semantics thus gives us a set of sets of conclusions, which is called an *Outfamily*.

Definition 6 (Outfamily PS) *An extension is a set of arguments, and an argumentation semantics $sem(arg, attack)$ is a function that takes as input a set of arguments and a binary attack relation among the arguments, and as output a set of extensions.*
$Outfamily(K, sem) = \{\{concl(A) \mid A \in S\} \mid S \in sem(arg(AS,K), attack(AS,K))\}$.

Constrained output in the input/output logic framework is defined as follows, being inspired by maximal consistent set constructions in belief revision and non-monotonic reasoning. *Maxf* takes the maximal sets of norms of K such that the output of K is consistent, and *Outf* takes the output of these maximal norm sets.

Definition 7 (Outf) *Let $K = K^L \cup K^N$ consist of literals K^L and norms K^N.*
$Conf(K) = \{N \subseteq K^N \mid Out(K^L \cup N) \text{ consistent}\}$
$Maxf(K) = \{N \subseteq K^N \mid N \text{ maximal w.r.t } \subseteq \text{ in } Conf(K)\}$
$Outf(K) = \{Out(K^L \cup N) \mid N \in Maxf(K)\}$

Theorem 1 (Characterization PS) *$Outfamily(KB, sem) = Outf(K)$ for sem is stable or preferred.*

Van der Torre and Villata [2014] add an additional modal operator O to the language. All norms are of the form $L_1 \wedge \ldots \wedge L_n \leadsto L$, as before, or $L_1 \wedge \ldots \wedge L_n \leadsto OL$. The body contains simple literals and the head contains either a literal or an obligation. They redefine the concepts or *LAS*, *Out*, *der*, etc. As there is no risk for confusion, we refer to them with the same names as in the previous sections.

Definition 8 (LAS O) *Given a set of propositional atoms. The literals, norms and legal language $\mathscr{L}$ are given by the following BNF.*
$L ::= P \mid \neg P$ with P in propositional atoms
$M ::= L \mid OL$

$$N ::= L \wedge \ldots \wedge L \rightsquigarrow M$$
$$\alpha ::= L \mid N$$

A legal argumentation system with obligations or LAS is as defined before, where the $-$ function is extended to obligations.

The definition of arguments is adapted in the obvious way. In the output, they consider only the obligatory propositions.

Definition 9 (Output O) $Out(LAS, K) = \{L \mid A \in Arg(LAS, K), concl(A) = OL\}$.

Example 4 We consider a revised version of the running example about smoking regulations. We have that the LAS_2 is based on propositional atoms $P ::= a|b|c|d|e$ where a: "the person wants to smoke in a closed space", b: "the person is in a private space", c: "the person needs to smoke on medical grounds", d: "the person is forbidden from smoking", e: "use electronic cigarettes". and $\mathcal{R}$ contains expressions for inference rules of the form:

- $a, a \rightsquigarrow b \Rightarrow b$;

- $a, \neg c, a \wedge \neg c \rightsquigarrow d \Rightarrow Od$;

Consider now the extended knowledge base of the smoking regulations represented by $K_2 = \{a, \neg b, \neg c, a \wedge \neg c \rightsquigarrow d, a \wedge b \rightsquigarrow \neg d, c \rightsquigarrow \neg d, a \wedge d \rightsquigarrow Oe\}$ where the norms state that

- if the person is in a closed space and she does not need to smoke on medical grounds, then the person is forbidden from smoking;

- if the person wants to smoke in a closed space and she is in a private space, then the person is not forbidden from smoking;

- if the person needs to smoke on medical grounds, then she is not forbidden from smoking;

- if the person wants to smoke in a closed space and she is forbidden from smoking, then it is obligatory to use electronic cigarettes;

We can construct the following arguments:

- $A_1 : a, \neg c, a \wedge \neg c \rightsquigarrow d \Rightarrow d$;

- $A_2 : a, (a, \neg c, a \wedge \neg c \rightsquigarrow d \Rightarrow d), a \wedge d \rightsquigarrow Oe \Rightarrow Oe$;

We have that $concl(A_1) = \{d\}$ and $concl(A_2) = \{Oe\}$, and we can thus conclude $Out(LAS_2, K_2) = \{e\}$ i.e., the conclusion is an obligation to use electronic cigarettes.

The constrained version can be defined analogously.

The proof system contains two rules, Strengthening of the Input (SI) and Factual Detachment (FD). The notion of consequence is called simple-minded output or out_1 by Makinson and van der Torre [2000].

Definition 10 (Derivations O) $der(LAS)$ *is the smallest set of expressions* $K \therefore L$ *closed under the following two rules.*

SI: *from* $K \therefore L$ *derive* $K \cup K' \therefore L$

FD: $\{L_1, \ldots, L_n, L_1 \wedge \ldots \wedge L_n \rightsquigarrow L\} \therefore L$ *for a norm* $L_1 \wedge \ldots \wedge L_n \rightsquigarrow OL$

Again we have $K \therefore L \in der(LAS)$ iff $L \in Out(LAS, K)$. The system does not satisfy deontic detachment, e.g. from $K = \{a, a \rightsquigarrow Ob, b \rightsquigarrow Oc\}$ we cannot derive Oc. This is reflected in the proof system by the lack of the DD rule.

Finally, van der Torre and Villata show how can to redefine the concepts of LAS, Out, der, etc., to re-introducing deontic detachment. This illustrates how the formal analysis can inspire the development of new argumentation systems.

4 From Open Texture to Graded Categories

4.1 Flexible legal interpretation based on graded categories

Legal systems are the product of human mind and are written in natural language. This implies that the basic processes of human cognition have to be taken into account when interpreting norms, and that, as natural languages are inherently vague and imprecise, so are norms.

The application of laws to a new situation is a metaphorical process: the new situation is mapped on to a situation in which applying law is obvious, by analogy. Here, by metaphor we mean using a well understood, prototypical situation to represent and reason about a less understood, novel situation. Metaphors are one of the basic building blocks of human cognition [Lakoff and Jonhson, 1980].

Norms are written with references to categories. As pointed out by Lakoff [1987], "Categorization is not a matter to be taken lightly. There is nothing more basic than categorization to our thought, perception, action, and speech." The "classical theory" that categories are defined by common properties is not entirely wrong, but it is only a small part of the story. It is now clear that categories may be based on prototypes. Some categories are vague or imprecise; some do not have gradation of membership, while others do. The category "US Senator" is well defined, but categories like "rich person" or "tall man" are graded, simply because there are different degrees of richness and tallness. However, it is important to notice that these degrees of membership depend both on the the context in which the norm will be applied and on the goal associated to the norm. To be considered tall in the Netherlands is not the same as to be considered tall in Portugal, for example. We have thus first to consider the context and then the goal associated to the norm.

We explore the use of fuzzy logic as a suitable technical tool to capture the imprecision related to categories. More precisely, a category may be represented as a fuzzy set: the membership of an element to a category is a graded notion.

As a result, we get that a norm may apply to a given situation only to a certain extent and different norms may apply to different extents to the same situation.

4.1.1 Fuzzy Logic

Fuzzy logic was initiated by Lotfi Zadeh [1965] with his seminal work on fuzzy sets. Fuzzy set theory provides a mathematical framework for representing and treating vagueness, imprecision, lack of information, and partial truth. Fuzzy logic is based on the notion of fuzzy set, a generalization of classical sets obtained by replacing the characteristic function of a set A, χ_A which takes up values in $\{0,1\}$, i.e. $\chi_A(x) = 1$ iff $x \in A$, $\chi_A(x) = 0$ otherwise, with a *membership function* μ_A, which can take up any value in $[0,1]$. The value $\mu_A(x)$ is the membership degree of element x in A, i.e., the degree to which x belongs in A. A fuzzy set is completely defined by its membership function. In fact, we can say that a fuzzy set *is* its membership function.

Operation on Fuzzy Sets The usual set-theoretic operations of union, intersection, and complement can be defined as a generalization of their counterparts on classical sets by introducing two families of operators, called triangular norms and triangular co-norms [Schweizer and Sklar, 1960; Schweizer and Sklar, 1983; Navara, 2007]. A triangular norm (or t-norm) is a binary operation $T : [0,1] \times [0,1] \to [0,1]$ satisfying the following conditions for $x, y, z \in [0,1]$:

- $T(x,y) = T(y,x)$ (commutativity);
- $T(x,T(y,z)) = T(T(x,y),z)$ (associativity):
- $y \leq z \Rightarrow T(x,y) \leq T(x,z)$ (monotonicity);
- $T(x,1) = x$ (neutral element 1).

A well-known property about t-norms is:

(1) $$T(x,y) \leq \min(x,y).$$

A triangular conorm (or t-conorm or s-norm), dual to a triangular norm, is a binary operation $S : [0,1] \times [0,1] \to [0,1]$, whose neutral element is 0 instead of 1, with all other conditions identical to those of a t-norm:

- $S(x,y) = S(y,x)$ (commutativity);
- $S(x,S(y,z)) = S(S(x,y),z)$ (associativity):
- $y \leq z \Rightarrow S(x,y) \leq S(x,z)$ (monotonicity);
- $S(x,0) = x$ (neutral element 0).

A well-known property about t-conorms is:

(2) $$S(x,y) \geq \max(x,y).$$

If T is a t-norm, then $S(x,y) \equiv 1 - T(1-x, 1-y)$ is a t-conorm and *vice versa*: T and S in this case form a *dual pair* of a t-norm and a t-conorm. Noteworthy examples of such dual pairs are:

- $T_M(x,y) = \min\{x,y\}$, $S_M(x,y) = \max\{x,y\}$ (minimum t-norm and maximum t-conorm or Gödel t-norm and t-conorm);

- $T_P(x,y) = xy$, $S_P(x,y) = x+y-xy$ (product t-norm and t-conorm or probabilistic product and sum);

- $T_L(x,y) = \max\{x+y-1, 0\}$, $S_L(x,y) = \min\{x+y, 1\}$ (Lukasiewicz t-norm and t-conorm or bounded sum);

For a given choice of a dual pair of a t-norm and a t-conorm (T,S), given two fuzzy sets A and B and an element x, the set-theoretic operations of union, intersection, and complement are thus defined as follows:

$$\begin{aligned}
(3) \quad \mu_{A \cup B}(x) &= S(\mu_A(x), \mu_B(x)); \\
(4) \quad \mu_{A \cap B}(x) &= T(\mu_A(x), \mu_B(x)); \\
(5) \quad \mu_{\bar{A}}(x) &= 1 - \mu_A(x).
\end{aligned}$$

4.2 Representing Norms

A norm r may be represented as a rule $b_1, \ldots, b_n \Rightarrow l$ such that l is the legal effect of r, such as an obligation linked to the norm [Sartor, 2005]. A norm then has a conditional structure such as $b_1, \ldots, b_n \Rightarrow l$ (if $b_1, \ldots, b_n$ hold, then l ought to be the case). An agent is compliant with respect to this norm if l is obtained whenever $b_1, \ldots, b_n$ is derived. Often, logical models of legal reasoning assume that conditions of norms give a complete description of their applicability [Sartor, 2005].

However, this assumption is too strong, due to the complexity and dynamics of the world. Norms cannot take into account all the possible conditions where they should or should not be applied, giving rise to the so called "penumbra": a core of cases which can clearly be classified as belonging to the concept. By a penumbra of hard cases, membership of the concept can be disputed. Moreover, not only does the world change as also pointed out in [Liebwald, 2013], giving rise to circumstances unexpected to the legislator who introduced the norm, but even the ontology of reality can change with respect to the one constructed by the law to describe the applicability conditions of norms. See, e.g., the problems concerning the application of existing laws to privacy, intellectual property or technological innovations in healthcare. To cope with unforeseen circumstances, the judicial system, at the moment in which a case concerning a violation is discussed in court, is empowered to interpret, i.e., to change norms, under some restrictions not to go beyond the purpose from which the norms stem.

The clauses of a norm often refer to imprecise concepts, which can take up different meanings depending on the purpose of the norm. The case for using fuzzy categories to account for such imprecise concepts has been made by da Costa Pereira et al. [2017]: those imprecise concepts are a product of the human mind and, more precisely, of a categorization process. According to prototype theory, which is one of the most prominent and influential accounts of the cognitive processes of categorization, each category is defined by one or more prototypes [Vanpaemel et al., 2005], which are typical exemplars of it. A prototype may be regarded as being represented by a property list which has salient properties of the objects that are classified into the concept.

We may formalize these notions in a way that is compatible with an underlying knowledge representation standard and technical infrastructure like the ones provided by the W3C for the Semantic Web, i.e. OWL based on description logics for the terminological part and RDF for the assertional part. This would allow a practical implementation of our proposal using state-of-the-art knowledge engineering technologies. Nevertheless, we keep our formalization abstract for the sake of clarity.

Definition 11 (Language) *Given a knowledge base* K, *an atom is a unary or binary predicate of the form* $C(s)$, $R(s_1, s_2)$, *where the predicate symbol C is a concept name in* K *and R is a role name in* K, s, s_1, s_2 *are terms. A term is either a variable (denoted by x,y,z) or a constant (denoted by a,b,c) standing for an individual name or data value.*

According to this formalisation, an individual object o is described by all the facts of the form $C(o)$, $R(o,y)$ and $R(y,o)$ such that $K \models C(o)$, $K \models R(o,y)$ and $K \models R(y,o)$, where $\models$ stands for entailment. We call these facts the *properties* of o.

Definition 12 (Graded Category) *A graded category* $\tilde{C}$ *is described by a non-empty set of prototypes* $\text{Prot}(\tilde{C}) = \{o_1, o_2, \ldots, o_n\}$, *where each* $o_i \in \text{Prot}(\tilde{C})$ *is an individual name in* K.

We can consider that the choice of the actual (more plausible) category with respect to a prototype may be seen as if the prototype represented a kind of generalisation, which applied deductively, will allow to "classify" (categorise) new "problems" (instances) [Ashley, 1991].

The membership of an instance to a category depends on its similarity to its prototype(s). Using a similarity measure with values in $[0, 1]$ allows us to represent graded categories as fuzzy sets. A similarity measure of that kind may be defined. Here, we adapt the contrast model of similarity proposed by Tversky [1977]. In such a model, an object is represented by means of a set of features and the similarity between two objects is defined as an increasing function of the features in common to the two objects, *common features*, and as a decreasing function of the features that are present in one object but not in the other, *distinctive features*.

Definition 13 (Number of Common Features) *Given two objects or individuals* a, b *in* K, *the number of their common features* $c(a, b)$ *is defined as*

$$\begin{aligned} c(a,b) &= \|\{C : K \models C(a) \wedge C(b)\}\| \\ &+ \|\{\langle R, c \rangle : K \models R(a,c) \wedge R(b,c)\}\| \\ &+ \|\{\langle c, R \rangle : K \models R(c,a) \wedge R(c,b)\}\|, \end{aligned}$$

where $\wedge$ *represents the* and *logical connective.*

Definition 14 (Number of Distinctive Features) *Given two objects or individuals a,b in K, the number of their distinctive features $\mathrm{dis}(a,b)$ is defined as*

$$\begin{aligned}
\mathrm{dis}(a,b) &= \|\{C : \mathrm{K} \models C(a) \oplus C(b)\}\| \\
&+ \|\{\langle R,c \rangle : \mathrm{K} \models R(a,c) \oplus R(b,c)\}\| \\
&+ \|\{\langle c,R \rangle : \mathrm{K} \models R(c,a) \oplus R(c,b)\}\|,
\end{aligned}$$

where $\oplus$ represents the exclusive or *logical connective.*

It might be the case, in a given application, that some features are more important than others. This might be taken into account by defining different weights for each feature, depending on the application. Let $w : \mathrm{Predicates} \to \mathbb{R}^+$ be a function associating a weight to each concept and role name in the language. The two functions c and dis might then be redefined as follows:

$$\begin{aligned}
c(a,b) &= \Sigma_{C:\mathrm{K} \models C(a) \wedge C(b)}\, w(C) \\
&+ \Sigma_R w(R) \cdot \|\{c : \mathrm{K} \models R(a,c) \wedge R(b,c)\}\| \\
&+ \Sigma_R w(R) \cdot \|\{c : \mathrm{K} \models R(c,a) \wedge R(c,b)\}\|; \\
\mathrm{dis}(a,b) &= \Sigma_{C:\mathrm{K} \models C(a) \oplus C(b)}\, w(C) \\
&+ \Sigma_R w(R) \cdot \|\{c : \mathrm{K} \models R(a,c) \oplus R(b,c)\}\| \\
&+ \Sigma_R w(R) \cdot \|\{c, : \mathrm{K} \models R(c,a) \oplus R(c,b)\}\|.
\end{aligned}$$

These boil down to Definitions 13 and 14 when $w(C) = 1$ for all C and $w(R) = 1$ for all R.

Definition 15 (Object Similarity) *Given two objects or individuals a,b in K, their similarity is defined as*

$$s(a,b) = \frac{c(a,b)}{c(a,b) + \mathrm{dis}(a,b)}.$$

This similarity function satisfies a number of desirable properties. For all individuals a, b,

- $0 \leq s(a,b) \leq 1$;
- $s(a,b) = 1$ if and only if $a = b$;
- $s(a,b) = s(b,a)$;

We may now define the notion of membership degree of an object o in a graded category.

Definition 16 *Given a graded category $\tilde{C}$ and an arbitrary individual name o, the degree of membership of o in $\tilde{C}$ is given by*

$$\mu_{\tilde{C}}(o) = \mathop{S}_{p \in \mathrm{Prot}(\tilde{C})} s(o,p).$$

Since the category of an item in the left-hand-side of a rule may be vague or imprecise, the degrees of truth of such an item with respect to the actual situation may be partial. This implies that a rule can be partially activated, i.e., the state of affairs to be reached thanks to the compliance to that rule can be uncertain.

Let us consider the following rule $r: b_1,\ldots,b_n \Rightarrow l$, where the clauses b_i have the form "o_i is $\tilde{C}_i$" and let $\tilde{C}_1,\ldots,\tilde{C}_n$ be the categories of $b_1,\ldots,b_n$, respectively. A clause b_i of a norm involving a graded category may thus be true only to a degree. The premise of the norm may be partially true and a norm may thus apply only to some extent.

If the membership of an instance in a category depends on its similarity to the prototype of the category and also on the purpose of the norm, then we must conclude that both the prototype of a category and the similarity measure used to compute the membership might vary as a function of the purpose. While it may be hard to see how the similarity measure could change as a function of purpose, it is reasonable to assume that the legislators may have different prototypes in mind for a category with the same name when they write norms for different purposes.

This amounts to assuming that, given a graded category $\tilde{C}$, its set of prototypes may vary as a function of the purpose or goal G of the norm. We write $\text{Prot}(\tilde{C} \mid G)$ to denote the set of the prototypes of category $\tilde{C}$ when the purpose of a norm is G.

The degree of truth α_{iG} of clause b_i = "o_i is $\tilde{C}_i$", given that the purpose of the norm is G, may be computed as

$$(6) \qquad \alpha_{iG} = \mu_{\tilde{C}_i}(o_i \mid G) = \underset{p \in \text{Prot}(\tilde{C}\mid G)}{S} s(o_i, p).$$

Definition 17 *The degree to which the premise $b_1,\ldots,b_n$ of rule of the form $b_1,\ldots,b_n \Rightarrow l$ is satisfied, given that the purpose of r is G, is given by*

$$\text{Deg}(b_1,\ldots,b_n \Rightarrow l \mid G) = \underset{i=1,\ldots,n}{T} \alpha_{iG}.$$

The state of affairs which is reached thanks to the compliance of r will be associated with the truth degree of $\text{Deg}(r \mid G)$ — this is also the degree associated to l after the activation of r.

5 Fuzzy Argumentation and Fuzzy Labeling

In recent years, several research efforts have attempted to combine formal argumentation and fuzzy logic, in such a way that the uncertainty of arguments can be measured by their fuzzy degrees, while the conflicts between arguments can be properly handled by Dung's argumentation semantics. Among them, Tamani and Croitoru [2014] proposed a quantitative preference based argumentation system, called F-ASPIC. Based on ASPIC and fuzzy set theory, it can be used to model structured argumentation with fuzzy concepts. However, it is not clear how the status of a fuzzy argument is evaluated. Meanwhile, da Costa Perira et al. [2011] introduce a labeling-based approach to evaluate the status of fuzzy arguments. Therefore, these two approaches are combined to lay a foundation for legal interpretation.

5.1 Fuzzy Argumentation System

A fuzzy argumentation system based on Tamani and Croitoru's F-ASPIC is proposed, with some adaptations to make it fit our framework, and with the addition of the fuzzy labeling algorithm proposed by [da Costa Pereira et al., 2011].

The main differences between our framework and F-ASPIC [Tamani and Croitoru, 2014] are as follows.

In our framework, we do not need to represent rules with different degrees of importance, as Tamani and Croitoru do. Unlike in F-ASPIC, the antecedent of a rule may be partially satisfied, if it involves graded categories. As a consequence, the consequent of that rule will have a partial truth degree and an argument depending on that rule has a partial membership in the set $\mathscr{A}$ of "active" arguments in the senese of da Costa Pereira et al.. So, although from a semantical point of view these gradual notions of partial truth or satisfaction are quite different from Tamani and Croitoru's notion of *importance* and *strength*, they lead to a mathematical treatment which is formally identical. Our main adaptation of F-ASPIC is therefore to replace, in the wording and in the formalism, these notions.

For the sake of simplicity, we assume that every element of the language and every rule are fallible. Hence, we do not differentiate between strict rules and defeasible rules, as ASPIC+ does, but we assume that we only have defeasible rules. This assumption makes the rationality postulates [Amgoud, 2014] trivially satisfied. However, it does not make things technically simpler (partial truth is basically preserved via strict rules, since they encode indisputable inferences). As a matter of fact, since strict rules satisfy contraposition (i.e., $P \Rightarrow Q$ is equivalent to $\neg Q \Rightarrow \neg P$), while defeasible rules do not have to, such behavior, when required, has to be explicitly simulated.

Definition 18 (Fuzzy argumentation system) *A fuzzy argumentation system, denoted as FAS, is a tuple $(\mathscr{L}, cf, \mathscr{R}, n, \mathrm{Deg})$ where*

- $\mathscr{L}$ *is a logical language.*

- cf *is a contrariness function (in this chapter, we only consider the classical negation $\neg$),*

- $\mathscr{R}$ *is the set of (defeasible) inference rules of the form $\phi_1, \ldots, \phi_m \Rightarrow \phi$ (where $\phi_i, \phi \in \mathscr{L}$).*

- $n : \mathscr{R} \mapsto \mathscr{L}$ *is a naming convention for rules.*

- $\mathrm{Deg} : \mathscr{R} \to [0,1]$ *is a function returning the degree of activation of a rule, given a grounding of the formulas occurring in it. Intuitively, $\mathrm{Deg}(r)$ represents the degree of truth of the antecedent of r.*

In the original F-ASPIC system, fuzzy arguments are then constructed with respect to a fuzzy knowledge base $\mathscr{K}$, assigning a degree of importance $\mu_{\mathscr{K}}(p)$ to each proposition $p \in \mathscr{L}$. In our framework, however, we do not attach a degree of importance to propositions of formulas *per se*, but we need to evaluate a degree of truth of their grounding with respect to graded categories. To be more precise, the atomic propositions that are liable to have a partial degree of truth are those of the form "x is C", where C is a graded category. Given a substitution of variable x with an individual object o, the truth value of the grounding "o is C" will be given, as suggested in the previous section, by the similarity measure $s(o, p)$ of o to one of the prototypes p of C (i.e., one p in the set $\mathrm{Prot}(C)$). To this aim, we keep the same symbol $\mathscr{K}$, but we regard it as a fuzzy valuation function.

Definition 19 (Fuzzy Valuation Function) *A fuzzy valuation function in a FAS $= (\mathscr{L}, cf, \mathscr{R}, n, \mathrm{Deg})$ is a fuzzy set $\mathscr{K} : \mathscr{L}_{\mathrm{ground}} \to [0,1]$ such that:*

- *if $\phi \in \mathscr{L}_{\mathrm{ground}}$ is a ground atomic proposition of the form "o is C", with C a graded category,*

$$\text{(7)} \qquad \mathscr{K}(o \text{ is } C) = \mathop{S}_{p \in \mathrm{Prot}(C)} s(o, p);$$

- *if $\phi \in \mathscr{L}_{\mathrm{ground}}$ is a ground atomic proposition not involving graded categories, $\mathscr{K}(\phi) \in \{0, 1\}$;*

- *if $\phi, \psi \in \mathscr{L}_{\mathrm{ground}}$,*

$$\begin{aligned} \mathscr{K}(\neg \phi) &= 1 - \mathscr{K}(\phi), \\ \mathscr{K}(\phi \wedge \psi) &= T(\mathscr{K}(\phi), \mathscr{K}(\psi)) \\ \mathscr{K}(\phi \vee \psi) &= S(\mathscr{K}(\phi), \mathscr{K}(\psi)) \end{aligned}$$

where T represents a triangular norm and S an associated triangular co-norm.

Let $r : b_1, \ldots, b_n \Rightarrow l$ be a rule. In a very simple case, the degree of activation Deg of r simply corresponds to the value returned by the Fuzzy Valuation Function $\mathscr{K}(\bigwedge_{1 \leq k \leq n} b_k)$.

Definition 20 (Fuzzy argument) *A fuzzy argument A on the basis of an argumentation theory with fuzzy valuation function $\mathscr{K}$ and a fuzzy argumentation system is*

- *ϕ if $\phi \in \mathscr{L}$ with: $\mathrm{Prem}(A) = \{\phi\}$, $\mathrm{Conc}(A) = \phi$, $\mathrm{Sub}(A) = \{A\}$, $\mathrm{Rules}(A) = \emptyset$.*

- *$A_1, \ldots, A_m \Rightarrow \phi$ if $A_1, \ldots, A_m$ are arguments such that there exists a rule $\mathrm{Conc}(A_1)$, $\ldots$, $\mathrm{Conc}(A_m) \Rightarrow \psi$ in $\mathscr{R}$. In this case, $\mathrm{Prem}(A) = \mathrm{Prem}(A_1) \cup \cdots \cup \mathrm{Prem}(A_m)$, $\mathrm{Conc}(A) = \psi$, $\mathrm{Sub}(A) = \mathrm{Sub}(A_1) \cup \cdots \cup \mathrm{Sub}(A_m) \cup \{A\}$, $\mathrm{Rules}(A) = \mathrm{Rules}(A_1) \cup \cdots \cup \mathrm{Rules}(A_m) \cup \{\mathrm{Conc}(A_1), \ldots, \mathrm{Conc}(A_m) \Rightarrow \psi\}$.*

Given an argument A, $\mathrm{Conc}(A)$ denotes the conclusion of A, $\mathrm{Prem}(A)$ the set of the premises of A, $\mathrm{Sub}(A)$ the set of the sub-arguments of A (including A itself), and $\mathrm{Rules}(A)$ the set of rules involved in A.

Then, the degree of activation of each argument is measured by a fuzzy degree, called *strength of argument* in F-ASPIC, which can also be interpreted as a degree of membership in the set of active arguments, defined as follows.

Definition 21 (Strength of argument) *Given a fuzzy argument A, its strength, denoted $\mathscr{A}(A)$, is defined as follows:*

- *if A is of the form ϕ, then $\mathscr{A}(A) = \mathscr{K}(\phi)$;*

- *otherwise,*

$$\mathscr{A}(A) = \underset{r \in Rules(A)}{S} T\left(\mathrm{Deg}(r), \underset{\phi \in Prem(A)}{T} \mathscr{K}(\phi)\right). \qquad (8)$$

Then, with respect to the notions of rebut, undercut and defeat in ASPIC, the counterparts in the setting of fuzzy argumentation are defined as follows.

Unlike F-ASPIC, our framework does not require the definition of a fuzzy counterpart of the rebut, undercut, and defeat relation. We rely on the usual crisp relations, defined as follows.

Definition 22 (Attacks) *A attacks B iff A undercuts, rebuts or undermines B, where the function n is a naming convention for rules, which maps each rule to a well-formed formula in $\mathscr{L}$ [Modgil and Prakken, 2014a], and*

- *A undercuts B (on B') iff $Conc(A) = \neg n(r)$ for some $B' \in Sub(B)$.*

- *A rebuts B (on B') iff $Conc(A) = \neg \phi$ for some $\exists B' \in Sub(B)$ of the form $B''_1, \ldots, B''_m \Rightarrow \phi$.*

- *A undermines B (on B') iff $Conc(A) = \neg \phi$ for some $B' = \phi$, $\phi \in Prem(B)$.*

Definition 23 (Defeat) *A defeats B iff A undercuts B on B', or A rebuts (undermines) B on B' and $\mathscr{A}(A) \not< \mathscr{A}(B')$.*

We use $\mathscr{A}$ and $\mathscr{D}$ to denote, respectively, the fuzzy set of active arguments (whose membership is their strength) and the defeat relation between them. Then, a fuzzy argumentation framework is represented as $\mathscr{F} = (\mathscr{A}, \mathscr{D})$.

This fuzzification of $\mathscr{A}$ provides a natural way of associating strengths to arguments, and suggests rethinking the labeling of an argumentation framework in terms of fuzzy degrees of argument acceptability [da Costa Pereira et al., 2011]. The status of arguments can thus be evaluated by means of Fuzzy AF-labeling.

Definition 24 (Fuzzy AF-labeling) *Let $(\mathscr{A}, \mathscr{D})$ be a fuzzy argumentation framework. A fuzzy AF-labeling is a total function $\alpha: \mathscr{A} \mapsto [0,1]$.*

Definition 25 (Fuzzy Reinstatement labeling) *Let $(\mathscr{A}, \mathscr{D})$ be a fuzzy argumentation framework, and α be a fuzzy AF-labeling. We say that α is a fuzzy reinstatement labeling iff, for all argument A,*

$$\alpha(A) = \min\{\mathscr{A}(A), 1 - \max_{B:(B,A) \in \mathscr{D}} \alpha(B)\} \qquad (9)$$

Da Costa Periera *et al.* [2011] made clear that given a fuzzy argumentation framework, its fuzzy reinstatement labeling may be computed by solving a system of n non-linear equations, where $n = \|\text{supp}(\mathscr{A})\|$, i.e., the number of arguments belonging to some non-zero degree in the fuzzy argumentation framework, of the same form as Equation 9, in n unknown variables, namely, the labels $\alpha(A)$ for all $A \in \text{supp}(\mathscr{A})$.

This can be done quite efficiently using an iterative method as follows: we start with an all-in labeling (a labeling in which every argument is labeled with the degree it belongs to $\mathscr{A}$). We denote by $\alpha_0 = \mathscr{A}$ this initial labeling, and by α_t the labeling obtained after the tth iteration of the labeling algorithm.

Definition 26 *Let α_t be a fuzzy labeling. An iteration in α_t is carried out by computing a new labeling α_{t+1} for all arguments A as follows:*

$$(10) \qquad \alpha_{t+1}(A) = \frac{1}{2}\alpha_t(A) + \frac{1}{2}\min\{\mathscr{A}(A), 1 - \max_{B:(B,A)\in\mathscr{D}} \alpha_t(B)\}.$$

Note that Equation 10 guarantees that $\alpha_t(A) \leq \mathscr{A}(A)$ for all arguments A and for each step of the algorithm.

The above definition actually defines a sequence $\{\alpha_t\}_{t=0,1,\ldots}$ of labelings, whose convergence has been proven [da Costa Pereira *et al.*, 2011]. We may now define the fuzzy labeling of a fuzzy argumentation framework as the limit of $\{\alpha_t\}_{t=0,1,\ldots}$.

Definition 27 *Let $\langle \mathscr{A}, \mathscr{D} \rangle$ be a fuzzy argumentation framework. A fuzzy reinstatement labeling for such argumentation framework is, for all arguments A,*

$$(11) \qquad \alpha(A) = \lim_{t \to \infty} \alpha_t(A).$$

Once this fuzzy reinstatement labeling has been computed, $\alpha(A)$ gives the degree to which each argument A in the framework is accepted; this degree may be used to compute the corresponding degree to which the purpose of a norm is G:

$$(12) \qquad \alpha(G) = \max_{A:\text{Conc}(A)=G} \alpha(A).$$

As it is clear from the above definitions, an argument may be accepted partially and thus the purpose of a norm may be uncertain. Now, different strategies may be used to deal with such an uncertainty. One possibility is to consider the purpose G for which $\alpha(G)$ is maximal. Another is to evaluate the norm with respect to all purposes such that $\alpha(G) > 0$ and then combine the results weighted by ther corresponding $\alpha(G)$.

6 Interpreting a Norm with Flexibility

In addition to taking graded categories into account, any norm is always associated with a purpose: that is what is called the *purpose* of the norm. The idea is then to capture the fact that, when a legislator states a norm, she has in mind a state of affairs to be reached through compliance with that norm. With that in mind, the degree to which a concept in the rule belongs to a category would also depend on the purpose associated with the rule. In other words, given a norm like $b_1, \ldots, b_n \Rightarrow l$, the degree associated to l depends on the degrees of truth of conditions b_i. These degrees depend in turn on the purpose associated to the norm: for example, the greater the extent to which the prohibition to smoke in public spaces promotes the goal *public health*, the greater is the degree of applicability of a rule like Public_Space $\Rightarrow$ No_Smoking assuming the fuzziness of the concept Public_Space. However, the actual purpose of the legislator can be controversial [Liebwald, 2013]: for example, not enough evidence or factual information might be available which could help discover what the legislator was intending when writing a norm. Note that the historical purpose could be obsolete due to social, economic or political change, and the legislator has not reacted in a timely manner or at all. Here, as done in legal theory [Peczenik, 1989; Sartor, 2005], we adopt an objective teleological approach to interpretation, which means that the purpose of a norm is the one that any rational interpreter would assign to it. Hence, we use an argumentative system which will determine which purpose, with respect to the current knowledge, is the most plausible purpose of a norm.

The case study in our chapter is the application of the Italian Legislative Act n. 40/2004 on "Medically Assisted Reproduction." Before the declaration of uncostitutionality ruled by the Constitutional Court (opinion n. 96/2015), the statute included section 4, par. 1: "The recourse to medically assisted reproduction techniques is allowed only [...] in the cases of sterility or infertility [...]." The purpose of the discussion is to see whether this provision can be interpreted so that non-sterile or fertile couples, in which one or both spouses are immune carriers of a serious genetic anomaly, could access those techniques.

These couples are able to conceive and bear a child, though the probability that the baby will contract the disease is high. These diseases are normally severely disabling, provoke physical dysfunctions, often prevent the full psychological development of the baby, and can cause premature death. The mentioned medical techniques can detect the illness in advance and consequently let the parents take aware decisions about the pregnancy.

The legislative act does not explicitly define 'sterility' and 'infertility.' On the basis of art. 7 l. 40/2004, every three years, the Ministry of Health is required to promulgate a decree containing the updated guidelines for the application of the law. According to these guidelines, the terms 'sterility' and 'infertility' are considered synonyms and refer to the lack of conception, in addition to those cases of certified pathology, after 12/24 months of regular sexual relations in a heterosexual couple.

In civil law systems, when it comes to statutory interpretation, one option is teleological interpretation, according to which, when interpreting a provision, judges often take into account what explicit or implicit purposes can be ascribed to the norm [Peczenik, 1989; Liebwald, 2013].

As for the purposes, law n. 40/2004 states as follows:

Art. 1, on "Purposes". Par.1: In order to favour the solution of reproduction problems caused by human sterility or infertility, it is allowed the recourse to medically assisted reproduction techniques, according to the conditions and the modalities provided for by the present law, which guarantees the respect of the rights of all the subjects involved, included the conceived baby.

Let us also consider the following norm from art. 4 of L. n. 40/2004:

The use of techniques of medically assisted procreation is [...] confined to the cases with issue of infertility or [...] sterility certified by a medical procedure.

Law n. 40/2004 is connected to other statutes of the legal system. In particular, the Italian Legislative Act n.194/1978 on "Social Protection of Maternity and Abortion" provides for the possibility of a therapeutic abortion if, during pregnancy, a pathological condition is ascertained, including those relating to significant anomalies or malformations of the baby, that put at risk the physical or psychic health of the woman." Severe genetic diseases are thus included. Moreover, along law n. 194/1978, the chance of a serious danger for the life of the woman is seen as a reason to proceed to abortion. This second legislative act is thus meant to promote the right to health both of the mother and of the child.

In light of the previous remarks, we can outline a list of interpretive arguments supporting different interpretations. Our main target is to see what interpretation better promotes the purposes that can be ascribed to the norm, if a purpose can be considered prominent, and what attacks can occur.

In what follows we present a plausible set of rules representing norms and interpretive legal arguments about such norms [Rotolo et al., 2015]. In both cases, fuzzy argumentation is related to the promotion of legal purposes.

In particular, the following (defeasible) rules can identify the basic the interpretive arguments arg_1, arg_2, arg_3, respectively, at stake:

$$r_1: \neg Ste(x), Rsn\_Exp\_Life(x) \Rightarrow \neg Med\_Rpr(x)$$
$$r_2: Med\_Rpr(x), Genetic\_Dis(x), Well\_Being(x) \Rightarrow Sol\_Rep\_Prob(x)$$
$$r_3: \neg Sol\_Rep\_Prob(x), Genetic\_Dis(x) \Rightarrow \neg Rsn\_Exp\_Life(x)$$
$$r_4: Gener\_Child(x) \Rightarrow \neg Ste(x)$$

where

- $Ste(x) =$ "x is sterile",
- $Med\_Rpr(x) =$ "x can access to medically assisted reproduction techniques",
- $Rsn\_Exp\_Life(x) =$ "x grants a reasonably expected life",
- $Genetic\_Dis(x) =$ "x is affected by a serious genetic disease",
- $Well\_Being(x) =$ "x enjoys psychological well-being",

- Sol_Rep_Prob(x) = "legally solved for x the reproduction problems",
- Gener_Child(x) = "x can generate children".

Consider the case mentioned above: a couple is actually able to conceive and generate children (Gener_Child(CP)), but they are both carriers of a serious genetic disease (Genetic_Dis(CP)), which does not allow children to live for more than a few years. Then according to the above rules, we have the following arguments:

$$arg_1 = \neg\text{Sol\_Rep\_Prob}(CP), \text{Genetic\_Dis}(CP) \Rightarrow$$
$$\neg\text{Rsn\_Exp\_Life}(CP)$$
$$arg_2 = \text{Gener\_Child}(CP) \Rightarrow \neg\text{Ste}(CP) \Rightarrow$$
$$\text{Rsn\_Exp\_Life}(CP), \neg\text{Ste}(CP) \Rightarrow \neg\text{Med\_Rpr}(CP)$$
$$arg_3 = \text{Med\_Rpr}(CP), \text{Genetic\_Dis}(CP),$$
$$\text{Well\_Being}(CP) \Rightarrow \text{Sol\_Rep\_Prob}(CP).$$

The attack relation between arguments are: arg_1 attacks arg_2, arg_2 attacks arg_3, and arg_3 attacks arg_1. Then, we have the following argumentation framework:

$$arg_1 \rightleftarrows arg_2 \longrightarrow arg_3$$

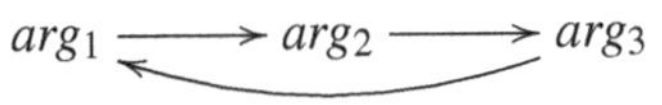

Figure 2. An argumentation framework

Let us consider these purposes:

- **Hlth_Of_MnC** = "purpose: the right to health both of the mother and the child"; this purpose is associated to rule r_2, i.e., we assume that r_2 promotes purpose **Hlth_Of_MnC**;

- **No_Eugenic** = "purpose: no eugenic selection"; this purpose is associated to rules r_1 and r_4, i.e., we assume that r_1 and r_4 promote purpose **No_Eugenic**.

For the sake of illustration, let us also assume that only two concepts are fuzzy: Gener_Child and Well_Being. Hence, if we consider, for example, r_4, this means that fuzziness depends only on the fact that rule r_4 makes the degree of $\neg\text{Ste}(CP)$ as dependent on the degree of capability of generating children by CP. No other source of vagueness are considered for r_4. Analogous considerations apply to rule r_2 in regard to Well_Being.

Given these purposes, we can measure the *degrees to which the premise* of rules r_2 and r_4 are satisfied by CP.

- **Rule r_4:** Let us assume that only one prototype p_1 is associated to Gener_Child and **No_Eugenic** (for example, a standard fertile couple statistically identified in the population of couples) in which, among others, the expected life of children is greater than 50 years and the incidence of genetic diseases is less than 20%. Clearly, these are distinctive features that differentiates p_1 with respect to CP:

suppose that the overall distinctive features are $d_1,\ldots,d_6$, while the common features are $c_1,\ldots,c_4$.

If we apply Definition 15, then $s(\text{CP}, p_1) = \frac{4}{4+6} = \frac{4}{10} = 0.4$. Since p is the unique prototype for Gener_Child with respect to **No_Eugenic** and that G for r_4 is {**No_Eugenic**}, then it is easy to check that (see, in particular, Definitions 16 and 19)

$$\mu_{\text{Gener\_Child}}(\text{CP}) = \text{Deg}(r_4 \mid G) = \mathscr{H}(\text{Gener\_Child}(\text{CP})) = 0.4.$$

- **Rule r_2:** Let us assume that only one prototype p_2 is associated to Well_Being and **Hlth_Of_MnC** and that the overall distinctive features are $d'_1,\ldots,d'_16$, while the common features are $c'_1,\ldots,c'_4$. For the same reason, given that $A(r_2)$ stands for Med_Rpr(CP) $\wedge$ Genetic_Dis(CP) $\wedge$ Well_Being(CP),

$$s(\text{CP}, p_2) = \mu_{\text{Well\_Being}}(\text{CP}) = \text{Deg}(r_2 \mid G') = \mathscr{H}(A(r_2)) = 0.2.$$

Given these degrees of activation of rules, the following table illustrates how to apply the machinery of fuzzy labeling to this scenario, given the above degrees of activation of the rules that determine the strength of arguments. As we noted, we defined the fuzzy labeling of a fuzzy argumentation framework as the limit of $\{\alpha_t\}_{t=0,1,\ldots}$. The convergence is obtained quickly: a small number of iterations is enough to get close to the limit.

t	$\alpha_t(\text{arg1})$	$\alpha_t(\text{arg2})$	$\alpha_t(\text{arg3})$
0	1	0.4	0.2
1	0.9	0.2	0.2
2	0.85	0.15	0.2
3	0.825	0.15	0.2
4	0.8125	0.1625	**0.2**
5	**0.8**	0.175	↓
6	↓	**0.2**	

Table 1. Fuzzy labeling

Therefore, arg_1 is accepted to degree 0.8 while arg_2 and arg_3 are given a much lower acceptance degree, namely 0.2. In other words, arg_1 is much more acceptable than arg_2 and arg_3. Its important to observe that these degrees just represent an order of plausibility, as if saying that arg_1 is four times as plausible as arg_2 or arg_3.

7 Related work

Young et al. [2016] endowed Brewka's prioritized default logic (PDL) with argumentation semantics using the ASPIC$^+$ framework for structured argumentation [Modgil and Prakken, 2014b]. More precisely, their goal is to define a preference ordering over arguments $\succsim$, based on the strict total order over defeasible rules defined to instantiate

$ASPIC^+$ to PDL, so as to ensure that an extension within PDL corresponds to the justified conclusions of its $ASPIC^+$ instantiation. Several options are investigated, and they demonstrate that the standard $ASPIC^+$ *elitist* ordering cannot be used to calculate $\succsim$ as there is no correspondence between the argumentation-defined inferences and PDL, and the same holds for a disjoint elitist preference ordering. The authors come up with a new argument preference ordering definition which captures both preferences over arguments and also *when* defeasible rules become applicable in the arguments' construction, leading to the definition of a strict total order on defeasible rules and corresponding non-strict arguments. Their representation theorem shows that a correspondence always exists between the inferences made in PDL and the conclusions of justified arguments in the $ASPIC^+$ instantiation under stable semantics.

Brewka and Eiter [1999] consider programs supplied with priority information, which is given by a supplementary strict partial ordering of the rules. This additional information is used to solve potential conflicts. Moreover, their idea is that conclusions should be only those literals that are contained in at least one answer set. They propose to use preferences on rules for selecting a subset of the answer sets, called the *preferred answer sets*. In their approach, a rule is applied unless it is defeated via its assumptions by rules of higher priorities.

Dung [2016] presents an approach to deal with contradictory conclusions in defeasible reasoning with priorities. More precisely, he starts from the observation that often, the proposed approaches to defeasible reasoning with priorities (e.g., [Brewka, 1989; Schaub and Wang, 2001; Modgil and Prakken, 2013]) sanction contradictory conclusions, as exemplified by $ASPIC^+$ using the weakest link principle together with the elitist ordering which returns contradictory conclusions with respect to its other three attack relations, and the conclusions reached with the well known approach of Brewka and Eiter [1999]. Dung shows then that the semantics for any complex interpretation of default preferences can be characterized by a subset of the set of stable extensions with respect to the normal attack relation assignments, i.e., a normal form for ordinary attack relation assignments. Dung's *normal attack relation* satisfies some desirable properties (Credulous cumulativity and Attack monotonicity) that cannot be satisfied by the $ASPIC^+$ semantics [Dung, 2016], i.e., the semantics of structured argumentation with respect to a given ordering of structured arguments (elitist or democratic pre-order) in $ASPIC^+$. In the setting of this paper, this notion could be defined as follows. Let $\alpha = (a_1, \ldots, a_n)$ and $\beta = (b_1, \ldots, b_m)$ be arguments constructed from a hierarchical abstract normative system. Since we have no Pollock style undercutting argument (as in $ASPIC^+$) and each norm is assumed to be defeasible, α is said to normally attack argument β if and only if β has a sub-argument β' such that $concl(\alpha) = \overline{concl(\beta')}$, and $r((a_{n-1}, a_n)) \geq r((b_{m-1}, b_m))$. According to the weakest link principle and Definition 23, the normal defeat relation is equivalent to the defeat relation using the last link principle in this paper.

Kakas *et al.* [2014] present a logic of arguments called *argumentation logic*, where the foundations of classical logical reasoning are represented from an argumentation perspective. More precisely, their goal is to integrate into the single argumentative representation framework both classical reasoning, as in propositional logic, and defeasible reasoning.

You *et al.* [2001] define a prioritized argumentative characterization of non-monotonic reasoning, by casting default reasoning as a form of prioritized argumentation. They illustrate how the parameterized formulation of priority may be used to allow various extensions and modifications to default reasoning.

We, and all these approaches, share the idea that an argumentative characterization of NMR formalisms, like prioritized default logic in Young's case and hierarchical abstract normative systems in our approach, contributes to make the inference process more transparent to humans. However, the targeted NMR formalism is different, leading to different challenges in the representation results. To the best of our knowledge, no other approach addressed the challenge of an argumentative characterization of prioritized normative reasoning.

Prakken and Sartor [2013] proposed to define a dynamic argumentation system as a tuple $S = \langle \mathcal{L}, -, \mathcal{R}, n \rangle$ where $\mathcal{L}$ is a logical language including symbols for predicates, functions, constants and variables, $=$ for equality, $\neg$ for negation and $\rightsquigarrow$ for normative conditionals, and the universal quantifier $\forall$, $\mathcal{R}$ is the set of inference rules, and n is the naming convention. A norm has the form $\forall (L_1 \wedge \ldots \wedge L_n \rightsquigarrow L)$, where $L_1, \ldots, L_n$ are literals. In particular, they define inference schemes for validity ($Valid(N(\phi)) \rightarrow \phi$), and applicability (i.e., undercutting, $\neg Applicable(w) \rightarrow \neg DMP(w)$). As future direction, the authors foster the extension of the framework by enriching the logical language with a formal account of modalities such as obligation. This is the issue we addressed in this chapter.

Van der Torre and Villata [2014] extend their dynamic legal argumentation framework with deontic modalities, and they propose an general framework for legal reasoning based on ASPIC-like argumentation and input/output logic. The framework allows to reason over normative concepts like factual and deontic detachment, and to assess norms' equivalence. The properties of our logical framework are proved. All new concepts are illustrated by a running example. Our main technical contribution is to give a formal analysis of legal argumentation, and a bridge to standard formalisms for normative systems like input/output logic. Compared to other input/output logics, van der Torre and Villata do not have weakening of the output or aggregation of obligations due to the clausal language. For a comparison with other deontic logics in the recent handbook on deontic logic and normative systems we can define the inference relation in terms of consequence sets as usual (e.g., $KB \models \phi$ iff $\phi \in Out(KB)$).

A framework for legal interpretation capable of taking graded, purpose-dependent institutional facts into account has been proposed by da Costa Periera *et al.* [2017]. Such a framework uses argumentation to handle conflicts between different interpretations of legal concepts. The originality of this proposal lies in the use of argumentation to identify the most likely purpose of a norm, which in turn circumscribes the interpretation of the categories (institutional facts, legal concepts) referred to by the norm. The idea of using many-valued logics in argumentation theory is not new. Just to name a few, [Cayrol and Lagasquie-Schiex, 2005] define a notion of gradual acceptability such that a numerical value is assigned to each argument on the basis of its attackers; Janssen *et al.* [2008] propose a fuzzy approach enriching the expressive power of classical argumentation, whose originality lies in the fact that the framework allows to represent the relative strength of the attacks; Grossi and Modgil [2015] propose a graded generalization of argumentation semantics in which the origin of the justification degrees is supposed to be exclusively endogenous, i.e., based exclusively on the topology of the attack relation. Qualitative approaches to arguments' acceptability have been proposed in preference-based argumentation frameworks (*PAF*) [Amgoud and Cayrol, 1997], value-based argumentation frameworks (*VAF*) [Bench-Capon, 2002b], and weighted argumentation frameworks (*WAF*) [Dunne *et al.*, 2011]. These approaches do not define graded semantics: *(i)* *PAF*s take into account preference orderings in the selection of acceptable conflicting arguments; *(ii)* *VAF*s are based on the assumption that some arguments can be stronger than others with respect to a certain value they advance, and this affects the success of an attack; and *(iii)* in *WAF*s, the weights are used for deciding which attacks can be ignored when computing the extensions. In these approaches, however, preference, values, and weights are provided only as input for the computation of extensions; they do not return an acceptability degree for arguments as output. Finally, Gabbay [2012] proposes an equational approach which returns multiple (graded) solutions, and thus several rankings for one argumentation framework.

Other frameworks for legal argumentation are listed below, but all of them concentrate on specific problems of reasoning with legal arguments, whilst the aim of our framework, as well as of Prakken and Sartor, is to integrate various aspects so far addressed separately towards a logic comprehensive model of dynamic legal argumentation. The combination of inferences establishing the validity of norms with inferences using valid norms has been proposed by Yoshino [1995]. The view that valid norms are defeasible reasons for legal conclusions was at the core of reason based logic by Hage [1997]. Arguments about applicability and inapplicability of norms are discussed by Gordon, Prakken and colleagues [1993; 1996]. Modeling reasoning with norms through argumentation schemes has been formalized by Verheij [2003]. Further connections between norms and argumentation include, among others, case based reasoning [Ashley, 1990], arguing in rule based systems [Prakken, 1993; Prakken and Sartor, 1996], dialogues and dialectics [Gordon, 1993], argument schemes [Gordon and Walton, 2009; Bex *et al.*, 2003].

Several works in the literature of AI & Law have considered the role of purposes in the legal interpretation. Indeed, this idea is standard in legal theory and the purpose of legal rules is recognised by jurists as decisive in clarifying the scope of the legal concepts that qualify the applicability conditions for those rules [Bench-Capon, 2002a; Prakken, 2002; Skalak and Rissland, 1992; Hage, 1997]. [Bench-Capon, 2002a; Prakken, 2002] use purposes/goals and values in frameworks of case based reasoning for modeling precedents mainly in a common law context. [Skalak and Rissland, 1992] analyse a number of legal arguments even in statutory law, which include cases close to the ones discussed here. Hage [1997] addresses, among others, the problem of reconstructing extensive and restrictive interpretation. This is done in Reason-Based Logic, a logical formalism that can deal with rules and reasons: the idea is that the satisfaction of rules' applicability conditions is usually a reason for application of these rules, but there can also be other (and possibly competing) reasons, among which we have the goals that led the legislator to make the rules. More recently, various work [Boella et al., 2009; Boella et al., 2010; Zurek and Araszkiewicz, 2013] proposed formal models for teleological interpretation in statutory law. All these approaches in AI & Law highlight the importance of rule purposes/goals. However, it seems that no work so far has attempted to couple this view with fuzzy logic and argumentation. In this perspective, we believe that this chapter may contribute to fill a gap in the literature.

8 Conclusions

In this chapter, we discuss three examples from the literature of handling norms by means of formal argumentation. First, we discuss how the so-called Greedy and Reduction approaches can be represented using the weakest and the last link principles respectively [Liao et al., 2016]. Based on such representation results, formal argumentation can be used to explain the detachment of obligations and permissions from hierarchical normative systems in a new way. Second, we discuss a dynamic ASPIC-based legal argumentation theory [Prakken and Sartor, 2013], and we discuss how existing logics of normative systems can be used to analyse such new argumentation systems [2014]. Third, we show how argumentation can be used to reason about other challenges in normative systems as well, by discussing a model for arguing about legal interpretation [da Costa Pereira et al., 2017]. In particular, we show how fuzzy logic combined with formal argumentation can be used to reason about the adoption of graded categories and thus address the problem of open texture in normative interpretation. We refer to the original papers for further details.

Our aim to discuss these three examples is to inspire new applications of formal argumentation to the challenges of normative reasoning in multiagent systems. We do not assume that the possible interactions between normative reasoning and formal argumentation is restricted to the three examples we discuss in this chapter. Besides resolving conflicting norms, norm compliance, norm dynamics and norm interpretation, it has been used also to argue about enforced obligations and permissions, and to establish norms' validity by deriving their conclusions. Moreover, other central challenge in normative multiagent system are discussed in the chapter of Pigozzi and van der Torre, and we believe that formal argumentation is also applicable to various other challenges. For example, agents can argue about the creation or emerging of norms from the mental states of individual agents, or how normative systems can be merged.

BIBLIOGRAPHY

[Alchourron and Makinson, 1981] Carlos E. Alchourron and David Makinson. Hierarchies of regulations and their logic. In Risto Hilpinen, editor, *New studies in deontic logic*, pages 125–148. Springer, 1981.

[Amgoud and Cayrol, 1997] Leila Amgoud and Claudette Cayrol. Integrating preference orderings into argument-based reasoning. In *ECSQARU-FAPR*, pages 159–170, 1997.

[Amgoud, 2014] Leila Amgoud. Postulates for logic-based argumentation systems. *Int. J. Approx. Reasoning*, 55(9):2028–2048, 2014.

[Araszkiewicz and Zurek, 2015] Michal Araszkiewicz and Tomasz Zurek. Comprehensive framework embracing the complexity of statutory interpretation. In *Legal Knowledge and Information Systems - JURIX 2015: The Twenty-Eighth Annual Conference, Braga, Portual, December 10-11, 2015*, pages 145–148, 2015.

[Araszkiewicz and Zurek, 2016] Michal Araszkiewicz and Tomasz Zurek. Interpreting agents. In *Legal Knowledge and Information Systems - JURIX 2016: The Twenty-Ninth Annual Conference*, pages 13–22, 2016.

[Ashley, 1990] Kevin D. Ashley. *Modeling legal argument - reasoning with cases and hypotheticals*. Artificial Intelligence and Legal Reasoning. MIT Press, 1990.

[Ashley, 1991] Kevin D. Ashley. Reasoning with cases and hypotheticals in HYPO. *International Journal of Man-Machine Studies*, 34(6):753–796, 1991.

[Bench-Capon et al., 2010] Trevor J. M. Bench-Capon, Henry Prakken, and Giovanni Sartor. *Argumentation in Legal Reasoning*. Argumentation in Artificial Intelligence. Springer, 2010.

[Bench-Capon, 2002a] Trevor J. M. Bench-Capon. The missing link revisited: The role of teleology in representing legal argument. *Artif. Intell. Law*, 10(1-3):79–94, 2002.

[Bench-Capon, 2002b] Trevor J. M. Bench-Capon. Value-based argumentation frameworks. In *NMR 2002*, pages 443–454, 2002.

[Bex et al., 2003] Floris Bex, Henry Prakken, Chris Reed, and Douglas Walton. Towards a formal account of reasoning about evidence: Argumentation schemes and generalisations. *Artif. Intell. Law*, 11(2-3):125–165, 2003.

[Boella et al., 2009] Guido Boella, Gguido Governatori, Antonino Rotolo, and Leendert W. N. van der Torre. Lex Minus Dixit Quam Voluit, Lex Magis Dixit Quam Voluit: A formal study on legal compliance and interpretation. In P. Casanovas, U. Pagallo, G. Sartor, and G. Ajani, editors, *AICOL-I/IVR-XXIV and AICOL-II/JURIX 2009 Revised Selected Papers*, pages 162–183, 2009.

[Boella et al., 2010] Guido Boella, Guido Governatori, Antonino Rotolo, and Leendert W. N. van der Torre. A logical understanding of legal interpretation. In F. Lin, U. Sattler, and M. Truszczynski, editors, *Proceedings of the Twelfth International Conference on the Principles of Knowledge Representation and Reasoning (KR 2010)*, 2010.

[Brewka and Eiter, 1999] Gerhard Brewka and Thomas Eiter. Preferred answer sets for extended logic programs. *Artificial Intelligence*, 109(1-2):297–356, June 1999.

[Brewka, 1989] Gerhard Brewka. Preferred subtheories: An extended logical framework for default reasoning. In N. S. Sridharan, editor, *Proceedings of the 11th International Joint Conference on Artificial Intelligence. Detroit, MI, USA, August 1989*, pages 1043–1048. Morgan Kaufmann, 1989.

[Cayrol and Lagasquie-Schiex, 2005] Claudette Cayrol and Marie-Christine Lagasquie-Schiex. Graduality in argumentation. *J. Artif. Intell. Res. (JAIR)*, 23:245–297, 2005.

[da Costa Pereira et al., 2011] Célia da Costa Pereira, Andrea G. B. Tettamanzi, and Serena Villata. Changing one's mind: Erase or rewind? possibilistic belief revision with fuzzy argumentation based on trust. In Toby Walsh, editor, *Proceedings of the Twenty-Second International Joint Conference on Artificial Intelligence (IJCAI'11), Barcelona, Catalonia, Spain, July 16–22, 2011*, pages 164–171. AAAI, 2011.

[da Costa Pereira et al., 2017] Célia da Costa Pereira, Andrea G. B. Tettamanzi, Beishui Liao, Alessandra Malerba, Antonino Rotolo, and Leendert van der Torre. Combining fuzzy logic and formal argumentation for legal interpretation. In *ICAIL*. ACM, 2017.

[D'Amato, 1983] Anthony D'Amato. Legal uncertainty. *California Law Review*, 71(1):1–55, 1983.

[Dung, 1995] Phan Minh Dung. On the acceptability of arguments and its fundamental role in nonmonotonic reasoning, logic programming and n-person games. *Artif. Intell.*, 77(2):321–358, 1995.

[Dung, 2016] Phan Minh Dung. An axiomatic analysis of structured argumentation with priorities. *Artif. Intell.*, 231:107–150, 2016.

[Dunne et al., 2011] Paul E. Dunne, Anthony Hunter, Peter McBurney, Simon Parsons, and Michael Wooldridge. Weighted argument systems: Basic definitions, algorithms, and complexity results. *Artif. Intell.*, 175(2):457–486, 2011.

[Gabbay, 2012] Dov M. Gabbay. Equational approach to argumentation networks. *Argument & Computation*, 3(2-3):87–142, 2012.

[Gordon and Walton, 2009] Thomas F. Gordon and Douglas Walton. Legal reasoning with argumentation schemes. In *Proceedings of the Twelfth International Conference on Artificial Intelligence and Law (ICAIL 2009)*, pages 137–146, 2009.

[Gordon, 1993] Thomas F. Gordon. The pleadings game. *Artif. Intell. Law*, 2(4):239–292, 1993.

[Grossi and Modgil, 2015] Davide Grossi and Sanjay Modgil. On the graded acceptability of arguments. In *IJCAI*, pages 868–874. AAAI Press, 2015.

[Hage, 1997] Jaap Hage. *Reasoning with Rules: An Essay on Legal Reasoning and Its Underlying Logic*. Kluwer, 1997.

[Hansen, 2008] Jörg Hansen. Prioritized conditional imperatives: problems and a new proposal. *Autonomous Agents and Multi-Agent Systems*, 17(1):11–35, 2008.

[Hart, 1994] Herbert L. A. Hart. *The Concept of Law*. Clarendon Press, Oxford, 1994.
[Heck, 1932] Philipp Heck. *Begriffsbildung und Interessensjurisprudenz*. Mohr Siebeck, Tübingen, 1932.
[Janssen et al., 2008] Jeroen Janssen, Martine De Cock, and Dirk Vermeir. Fuzzy argumentation frameworks. In *Procedings of the 12th International Conference on Information Processing and Management of Uncertainty in Knowledge-Based Systems (IPMU 2008)*, pages 513–520, 2008.
[Kakas et al., 2014] Antonis C. Kakas, Francesca Toni, and Paolo Mancarella. Argumentation for propositional logic and nonmonotonic reasoning. In Laura Giordano, Valentina Gliozzi, and Gian Luca Pozzato, editors, *Proceedings of the 29th Italian Conference on Computational Logic, Torino, Italy, June 16-18, 2014.*, volume 1195 of *CEUR Workshop Proceedings*, pages 272–286. CEUR-WS.org, 2014.
[Lakoff and Jonhson, 1980] George Lakoff and Mark Jonhson. *Metaphors We Live By*. University of Chicago Press, Chicago, 1980.
[Lakoff, 1987] George Lakoff. *Women, Fire, and Dangerous Things*. University of Chicago Press, Chicago, 1987.
[Liao et al., 2016] Beishui Liao, Nir Oren, Leendert van der Torre, and Serena Villata. Prioritized norms and defaults in formal argumentation. In *Proceedings of the 13th International Conference on Deontic Logic and Normative Systems (DEON2016)*, pages 139–154, 2016.
[Liebwald, 2013] Doris Liebwald. Law's capacity for vagueness. *Int J Semiot Law*, 26:391–423, 2013.
[MacCormick and Summers, 1991] D.N. MacCormick and R.S. Summers, editors. *Interpreting Statutes: A Comparative Study*. Ashgate, 1991.
[Makinson and van der Torre, 2000] David Makinson and Leendert van der Torre. Input/output logics. *J. Philosophical Logic*, 29(4):383–408, 2000.
[Malerba et al., 2016] Alessandra Malerba, Antonino Rotolo, and Guido Governatori. Interpretation across legal systems. In *Legal Knowledge and Information Systems - JURIX 2016: The Twenty-Ninth Annual Conference*, pages 83–92, 2016.
[Modgil and Prakken, 2013] Sanjay Modgil and Henry Prakken. A general account of argumentation with preferences. *Artif. Intell.*, 195:361–397, 2013.
[Modgil and Prakken, 2014a] Sanjay Modgil and Henry Prakken. The $ASPIC^+$ framework for structured argumentation: a tutorial. *Argument & Computation*, 5(1):31–62, 2014.
[Modgil and Prakken, 2014b] Sanjay Modgil and Henry Prakken. The $ASPIC^+$ framework for structured argumentation: a tutorial. *Argument & Computation*, 5(1):31–62, 2014.
[Navara, 2007] Mirko Navara. Triangular norms and conorms. *Scholarpedia*, 2(3):2398, 2007. revision #137537.
[Peczenik, 1989] Aleksander Peczenik. *On Law and Reason*. Kluwer, 1989.
[Pigozzi and van der Torre, to appear] Gabriella Pigozzi and Leendert van der Torre. Arguing about constitutive and regulative norms. *Journal of applied nonclassical logics*, to appear.
[Prakken and Sartor, 1996] Henry Prakken and Giovanni Sartor. A dialectical model of assessing conflicting arguments in legal reasoning. *Artif. Intell. Law*, 4(3-4):331–368, 1996.
[Prakken and Sartor, 2013] Henry Prakken and Giovanni Sartor. Formalising arguments about norms. In Kevin D. Ashley, editor, *JURIX*, volume 259 of *Frontiers in Artificial Intelligence and Applications*, pages 121–130. IOS Press, 2013.
[Prakken, 1993] Henry Prakken. A logical framework for modelling legal argument. In *Proceedings of the Fourth International Conference on Artificial intelligence and Law (ICAIL 1993)*, pages 1–9, 1993.
[Prakken, 2002] Henry Prakken. An exercise in formalising teleological case-based reasoning. *Artif. Intell. Law*, 10:113–133, 2002.
[Rotolo et al., 2015] Antonino Rotolo, Guido Governatori, and Giovanni Sartor. Deontic defeasible reasoning in legal interpretation: two options for modelling interpretive arguments. In Ted Sichelman and Katie Atkinson, editors, *ICAIL 2015*, pages 99–108, 2015.
[Sartor, 2005] Giovanni Sartor. *Legal reasoning: A cognitive approach to the law*, volume 5 of *A Treatise of Legal Philosophy and General Jurisprudence*. Springer, Berlin, 2005.
[Schaub and Wang, 2001] Torsten Schaub and Kewen Wang. A comparative study of logic programs with preference. In Bernhard Nebel, editor, *Proceedings of the Seventeenth International Joint Conference on Artificial Intelligence, IJCAI 2001, Seattle, Washington, USA, August 4-10, 2001*, pages 597–602. Morgan Kaufmann, 2001.
[Schweizer and Sklar, 1960] Berthold Schweizer and Abe Sklar. Statistical metric spaces. *Pacific J. Math.*, 10(1):313–334, 1960.
[Schweizer and Sklar, 1983] Berthold Schweizer and Abe Sklar. *Probabilistic metric spaces*. North Holland series in probability and applied mathematics. North Holland, 1983.
[Shapiro, 2011] Scott J. Shapiro. *Legality*. Harvard University Press, 2011.

[Skalak and Rissland, 1992] David B. Skalak and Edwina L. Rissland. Arguments and cases: An inevitable intertwining. *Artif. Intell. Law*, 1:3–44, 1992.

[Tamani and Croitoru, 2014] Nouredine Tamani and Madalina Croitoru. A quantitative preference-based structured argumentation system for decision support. In *Proceedings of the 2014 IEEE International Conference on Fuzzy Systems (FUZZ-IEEE)*, pages 1408–1415, 2014.

[Tosatto *et al.*, 2015] Silvano Colombo Tosatto, Pierre Kelsen, Qin Ma, Marwane El Kharbili, Guido Governatori, and Leendert W. N. van der Torre. Algorithms for tractable compliance problems. *Frontiers of Computer Science*, 9(1):55–74, 2015.

[Tversky, 1977] Amos Tversky. Features of similarity. *Psychological Review*, 84(4):327–352, 1977.

[van der Torre and Villata, 2014] Leendert W. N. van der Torre and Serena Villata. An aspic-based legal argumentation framework for deontic reasoning. In Simon Parsons, Nir Oren, Chris Reed, and Federico Cerutti, editors, *Computational Models of Argument - Proceedings of COMMA 2014, Atholl Palace Hotel, Scottish Highlands, UK, September 9-12, 2014*, volume 266 of *Frontiers in Artificial Intelligence and Applications*, pages 421–432. IOS Press, 2014.

[Vanpaemel *et al.*, 2005] Wolf Vanpaemel, Gerrit Storms, and Bart Ons. A varying abstraction model for categorization. In Bruno G. Bara, Lawrence Barsalou, and Monica Bucciarelli, editors, *Proceedings of the 27th Annual Conference of the Cognitive Science Society*, pages 2277–2282, Mahwah, NJ, 2005. Lawrence Erlbaum.

[Verheij, 2003] Bart Verheij. Dialectical argumentation with argumentation schemes: An approach to legal logic. *Artif. Intell. Law*, 11(2-3):167–195, 2003.

[Yoshino, 1995] Hajime Yoshino. The systematization of legal meta-inference. In L. Thorne McCarty, editor, *Proceedings of the Fifth International Conference on Artificial Intelligence and Law, ICAIL '95, College Park, Maryland, USA, May 21-24, 1995*, pages 266–275. ACM, 1995.

[You *et al.*, 2001] Jia-Huai You, Xianchang Wang, and Li-Yan Yuan. Nonmonotonic reasoning as prioritized argumentation. *IEEE Trans. Knowl. Data Eng.*, 13(6):968–979, 2001.

[Young *et al.*, 2016] Anthony P. Young, Sanjay Modgil, and Odinaldo Rodrigues. Prioritised default logic as rational argumentation. In Catholijn M. Jonker, Stacy Marsella, John Thangarajah, and Karl Tuyls, editors, *Proceedings of the 2016 International Conference on Autonomous Agents & Multiagent Systems, Singapore, May 9-13, 2016*, pages 626–634. ACM, 2016.

[Zadeh, 1965] Lotfi A. Zadeh. Fuzzy sets. *Information and Control*, 8:338–353, 1965.

[Zurek and Araszkiewicz, 2013] Tomasz Zurek and Michal Araszkiewicz. Modeling teleological interpretation. In *International Conference on Artificial Intelligence and Law, ICAIL '13, Rome, Italy, June 10-14, 2013*, pages 160–168, 2013.

12
Logics for Games, Emotions and Institutions
EMILIANO LORINI

1 Introduction

Agents in the societies can be either human agents or artificial agents. The focus of this paper is both on: (i) the present society in which human agents interact with the support of ICT through social networks and media, and (ii) the future society with mixed interactions between human agents and artificial systems such as autonomous agents and robots. Indeed, new technologies will come for future society in which such artificial systems will play a major role, so that humans will necessarily interact with them in their daily lives. This includes autonomous cars and other vehicles, robotic assistants for rehabilitation and for the elderly, robotic companions for learning support.

There are two main general observations underlying the present paper. The first is that interaction plays a fundamental role in existing information and communication technologies (ICT) and applications (e.g., Facebook, Ebay, peer-to-peer systems) and will become even more fundamental in future ICT. The second is that the cognitive aspect is crucial for the design of intelligent systems that are expected to interact with human agents (e.g., embodied conversational agents, robotic assistants, etc.). The system must be endowed with a psychologically plausible model of reasoning and cognition in order to be able (i) to understand the human agent's needs and to predict her behaviour, and (ii) to behave in a believable way thereby meeting the human agent's expectations.

Formal methods have been widely used in artificial intelligence (AI) and in the area of multiagent systems (MAS) for modelling intelligent systems as well as different aspects of social interaction between artificial and/or human agents. The aim of the present paper is to offer a general overview of the way logic and game theory have been and can be used in AI in order to build formal models of socio-cognitive, normative and institutional phenomena.

We take a bottom-up perspective to the analysis of normative and institutional facts that is in line with some classical analysis in organization theory such as the one presented in March & Simon's famous book "Organizations" [1958], described as a book in which they:

> "...surveyed the literature on organization theory, starting with those theories that viewed the employee as an instrument and physiological automaton, proceeding through theories that were centrally concerned with the motivational and affective aspects of human behavior, and concluding with theories that placed particular emphasis on cognitive processes"

[March and Simon, 1958, p. 5].

The present paper is organized in two main sections. Section 2 is devoted to cognitive aspects, while Section 3 is devoted to institutional ones. Section 2 starts from the assumption that cognitive agents are, by definition, endowed with a variety of mental attitudes such as beliefs, desires, preferences and intentions that provide input for practical reasoning and decision-making, trigger action execution, and generate emotional responses. We first present a conceptual framework that:

- clarifies the relationship between intention and action and the role of intention in practical reasoning;

- explains how moral attitudes such as standards, ideals and moral values influence decision-making;

- explains how preferences are formed on the basis of desires and moral values;

- clarifies the distinction between the concept of goal and the concept of preference;

- elucidates how mental attitudes including beliefs, desires and intentions cause emotional responses, and how emotions retroactively influence decision-making and mental attitudes by triggering belief revision, desire change and intention reconsideration.

Then, we explain how game theory and logic have been used in order to develop formal models of such cognitive phenomena. We put special emphasis on a specific branch of game theory, called epistemic game theory, and on a specific family of logics, so-called agent logics. The aim of epistemic game theory is to extend the classical game-theoretic framework with epistemic notions such as the concepts of belief and knowledge, while agent logics are devoted to explain how different types of mental attitudes (e.g., belief, desires, intentions) are related, how they influence decision and action, and how they trigger emotional responses.

Section 3 builds the connection between mental attitudes and institutions passing by the concept of collective attitude. Collectives attitudes such as joint intention, group belief, group goal, collective acceptance and joint commitment have been widely explored in the area of collective intentionality, the domain of social philosophy that studies how agents function and act at the group level and how institutional facts relate with physical (brute) facts (cf. [Ludwig and Jankovic, 2016; Tollefsen, 2002] for a general introduction of the research in this area). Section 3 is devoted to explain (i) how collective attitudes such as collective acceptance or common belief are formed either through aggregation of individual attitudes or through a process of joint perception, (ii) how institutional facts are grounded on collective attitudes and, in particular, how the existence and modification of institutional facts depend on the collective acceptance of these facts by the agent in the society and on the evolution of this collective acceptance. We also discuss existing logics for institutions that formalize the connection between collective attitudes and institutional facts.

In Section 4 we conclude by briefly considering the opposite path leading from norms and institutions to minds. In particular, we explain how institutions and norms, whose existence depends on their acceptance by the agents in the society, retroactively influence the agents' mental attitudes, decisions and actions.

2 Mental attitudes and emotions

In this section, we start with a discussion of two issues related with the representation of mental attitudes and emotions: (i) the cognitive processing leading from goal generation to action (Section 2.1), and (ii) the representation of the cognitive structure of emotions and of their influence on behaviour (Section 2.2). Then, we briefly explain how these cognitive aspects have been incorporated into game theory (Section 2.3). Finally, we consider how mental attitudes and emotion are formalized in logic and the connection between the representation of mental attitudes in logic and the representation of mental attitudes in game theory (Section 2.4).

2.1 A cognitive architecture

The conceptual background underlying our view of mental attitudes is summarized in Figure 1. (Cf. [Lorini, 2016a] for a logical formalization of some aspects of this view.) The cognitive architecture represents the process leading from generation of desires and moral values and formation of beliefs via sensing to action performance.

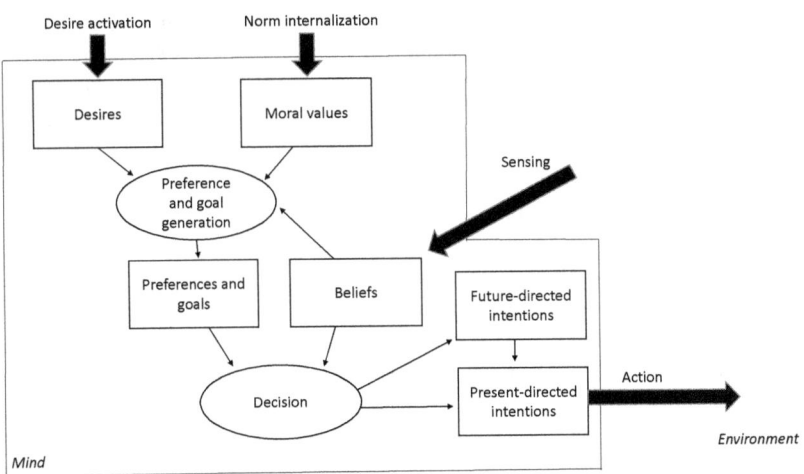

Figure 1. Cognitive architecture

The origin of beliefs, desires and moral values An important and general distinction in philosophy of mind is between epistemic attitudes and motivational attitudes. This distinction is in terms of the *direction of fit* of mental attitudes to the world. While epistemic attitudes aim at being true and their being true is their fitting the world, motivational attitudes aim at realization and their realization is the world fitting them [Platts, 1979; Anscombe, 1957; Humberstone, 1992]. Searle [Searle, 1979] calls "mind-to-world" the first kind of *direction of fit* and "world-to-mind" the second one.

There are different kinds of epistemic and motivational attitudes with different functions and properties. Examples of epistemic attitudes are beliefs, knowledge and opinions, while examples of motivational attitudes are desires, preferences, moral values and intentions. However, the most primitive and basic forms of epistemic and motivational attitudes are beliefs, desires and moral values.

Beliefs are mental representations aimed at representing how the physical, mental and social worlds are. Indeed, there are beliefs about natural facts and physical events (e.g., I believe that tomorrow will be a sunny day), introspective beliefs (e.g., I believe that I strongly wish that tomorrow will be a sunny day), and beliefs about mental attitudes of other agents (e.g., I believe that you believe that tomorrow will be a sunny day).

Following the Humean conception, a desire can be viewed as an agent's attitude consisting in an anticipatory mental representation of a pleasant state of affairs (representational dimension of desires) that motivates the agent to achieve it (motivational dimension of desires). The motivational dimension of an agent's desire is realized through its representational dimension, in the sense that, a desire motivates an agent to achieve it *because* the agent's anticipatory representation of the desire's content gives her pleasure so that the agent is "attracted" by it. For example when an agent desires to eat sushi, she is pleased to imagine herself eating sushi. This pleasant representation motivates her to go to the "The Japoyaki" restaurant in order to eat sushi. This view of desires unifies the standard theory of desire (STD) — focused on the motivational dimension — and the hedonic theory of desire (HTD) — focused on the hedonic dimension —. A third theory of desire has been advanced in the philosophical literature (see [Schroeder, 2004]), the so-called reward theory of desire (RTD). According to RTD what qualifies a mental attitude as a desire is the exercise of a capacity to represent a certain fact as a reward.[1]

Another fundamental aspect of desire is the *longing aspect*. The idea is that for an agent to desire something, the agent should be in a situation in which she does not have what she desires and she yearns for it. In other words, a state of affairs is desired by an agent only if the agent conceives it as *absent*. The following quotation from Locke [Locke, 1989, Book II, Chap. XXI] makes this point clear:

> To return then to the inquiry, what is it that determines the will in regard to our actions? And that...is not, as is generally supposed, the greater good in view: but some (and for the most part the most pressing) uneasiness a man is at present under. This that which successively determines the will, and sets us upon those actions, we perform. This uneasiness we may call, as it is, desire; which is uneasiness of the mind for want of some *absent good*...

This quotation seems in contradiction with what we claimed above, namely, that desire is based on the anticipatory representation of a pleasant state of affairs. However, the stronger the anticipated pleasure associated with a desire, the more painful is its

---

[1] According to [Dretske, 1988], desire is also a necessary condition for reward. In particular, desire determines what counts as a reward for an agent. For example, a person can be rewarded with with water only if she is thirsty and she desires to drink.

current lack of fulfillment — the term "uneasiness" in the previous quotation —, as in the case of longing for a drink when thirsty, for instance. So the contradiction is only apparent. This aspect of uneasiness described by Locke should not be confused with the concept of aversion which is traditionally opposed to the concept of desire (see [Schroeder, 2004, Chap. 5]). As emphasized above, if an agent desires a certain fact to be true, then she possesses an anticipatory mental representation of a *pleasant* fact motivating her to make the fact true. On the contrary, if an agent is averse to something, then she possesses an anticipatory mental representation of an *unpleasant* fact motivating her to prevent the fact from being true.

Moral values, and more generally moral attitudes (ideals, standards, etc.), originate from an agent's capability of discerning what is (morally) good from what is (morally) bad. If an agent has a certain ideal φ, then she thinks that the realization of the state of affairs φ ought to be promoted because φ is good in itself. Differently from desires, moral values do not necessarily have a hedonic and somatic component: their fulfillment does not necessarily give pleasure and their transgression does not necessarily give displeasure 'felt' from the body.

There are different ways to explain the origin of beliefs, desires, moral values. Beliefs are formed either via direct sensing from the external environment (e.g., I believe that there is a fire in the house since I can see it), communication (e.g., I believe that there is a fire in the house since you told me this and I trust what you say) and inference (e.g., I believe that there is a fire in the house since I already believe that smoke comes out from the house and if there is smoke coming out from the house then there is fire). One might argue that belief formation via direct sensing is more primitive than belief formation via communication and that the latter can be reduced to the former. Indeed, in the context of communication, the hearer first *perceives* the speaker's utterance, which is nothing but the performance of a physical action (e.g., uttering a certain sound, performing a certain gesture, emitting a certain light signal, etc.) and forms a belief about what the speaker has uttered. Then, she infers the meaning of the speaker's utterance (i.e., what the speaker wants to express by uttering a certain sound, by performing a certain gesture, by emitting a certain light signal, etc.). Although this is true for communication between humans and between artificial systems situated in the physical environment such as robots, it is not necessarily true for communication in an artificial domain in which there is no precise distinction between an utterance and its meaning. In the latter situation, the speaker may transmit to the hearer a message (e.g., a propositional formula) with a precise and non-ambiguous meaning.

The concept of trust plays a fundamental role in belief formation via direct sensing and via communication. Indeed, the hearer will not believe what the speaker says unless she believes that the speaker is a reliable source of information, thereby trusting the speaker's judgment. Similarly, for belief formation via direct sensing, an agent will not believe what she sees unless she believes that her perceptual apparatus works properly, thereby trusting it. The issue whether trust is reducible to other mental attitudes is relevant here. A justifiable approach consists in conceiving *communication-based trust* as a belief about the reliability of a source of information, where "reliable" means that, in the normal conditions, what the source says about a given issue is true.

The explanation about the origin of desires adopted in Figure 1 is that they are activated under certain conditions. For instance, according to Maslow's seminal theory of human motivation, "...everyday conscious desires are to be regarded as symptoms, as surface indicators of more basic needs" [Maslow, 1943, p. 392]. Maslow identified a set of basic (most of the time unconscious) needs of human agents including physiological needs,[2] need for safety, need for love and belonging, need for self-esteem and need for self-actualization. For example, a human agent's desire of drinking a glass of water could be activated by her basic physiological need for bodily balance including a constant body temperature, constant salt levels in the body, and so on. If certain variables of the agent's body are unbalanced and this unbalance is detected,[3] the agent receives a negative unpleasant signal from her body thereby entering in a state of felt displeasure and uneasiness — in the Lockean sense —. Consequently, she becomes intrinsically motivated to restore bodily balance. The connection between the agent's basic need for bodily balance and the agent's desire of drinking a glass of water may rely on the agent's previous experiences and be the product of operant conditioning (also called instrumental learning). Specifically, the agent may have learnt that, under certain conditions, drinking a glass a water is "a suitable means for" restoring balance of certain variables of the body. Indeed, every time the agent drunk water when she was feeling thirsty, she got a reward by making her basic need for bodily balance satisfied.[4]

In the case of artificial agents, conditions of desire activation should be specified by the system's designer. For example, a robotic assistant who has to take care of an old person could be designed in such a way that, every day at 4 pm, the desire of giving a medicine to the old person is activated in its mind.

As for the origin of moral values, social scientists (e.g., [6]) have defended the idea that there exist innate moral principles in humans such as fairness which are the product of biological evolution. Other moral values, as highlighted in Figure 1, have a cultural and social origin, as they are the product of the internalization of some external norm. A possible explanation is based on the hypothesis that moral judgments are true or false only in relation to and with reference to one or another agreement between people forming a group or a community. More precisely, an agent's moral values are simply norms of the group or community to which the agent belongs that have been internalized by the agent. This is the essence of the philosophical doctrine of moral relativism (see, e.g., [20]). For example, suppose that an agent believes that in a certain group or community there exists a norm (e.g., an obligation) prescribing that a given state of affairs should be true. Moreover, assume that the agent identifies herself as a member of this group or community. In this case, the agent will internalize the norm, that is, the external norm will become a moral value of the agent and will affect

---

[2]Maslow referred to the concept of homeostasis, as the living system's automatic efforts to maintain a constant, normal state of the blood stream, body temperature, and so on.

[3]Converging empirical evidences from neuroscience show that the hypothalamus is responsible for monitoring these bodily conditions.

[4]Following [Schroeder, 2004], one might argue that most conscious desires (including the desire to eat at a particular time and the desire to drink water) are instrumental, as they are activated *in order to* satisfy more basic needs of the individual.

the agent's decisions. For example, suppose that a certain person is (and identifies herself as) citizen of a given country. As in every civil country, it is prescribed that citizens should pay taxes. Her sense of national identity will lead the person to adopt the obligation by imposing the imperative to pay taxes to herself. When having to decide whether to pay taxes, she will decide to do it, not simply in order to avoid being sanctioned and being exposed to punishment, but also because she is motivated by the moral obligation to paying taxes.

From desires and moral values to preferences According to contemporary theories of human motivation both in philosophy and in economics (e.g., [Searle, 2001; Harsanyi, 1982]), preferences of a rational agent may originate either (i) from somatically-marked motivations such as desires or physiological needs and drives (e.g., the goal of drinking a glass of water originated from the physiological drive of thirst), or (ii) from moral considerations and values (e.g., the goal of helping a poor person originated from the moral value of taking care of needy people). More generally, there exists desire-dependent preferences and desire-independent ones originated from moral values. This distinction allows us to identify two different kinds of moral dilemmas. The first kind of moral dilemma is the one which is determined by the logical conflict between two moral values. The paradigmatic example is the situation of a soldier during a war. As a member of the army, the soldier feels obliged to kill his enemies, if this is the only way to defend his country. But, as a catholic, he thinks that human life should be respected. Therefore, he feels morally obliged not to kill other people. The other kind of moral dilemma is the one which is determined by the logical conflict between desires and moral values. The paradigmatic example is that of Adam and Eve in the garden of Eden. They are tempted by the desire to eat the forbidden fruit and, at the same time, they have a moral obligation not to do it.

According to the cognitive architecture represented in Figure 1, desires and moral attitudes of an agent are two different parameters affecting the agent's preferences. This allows us to draw the distinction between *hedonistic* agents and *moral* agents. A purely hedonistic agent is an agent who acts in order to maximize the satisfaction of her own desires, while a purely moral agent is an agent who acts in order to maximize the fulfillment of her own moral values. In other words, if an agent is purely hedonistic, the utility of an action for her coincides with the personal good the agent will obtain by performing this action, where the agent's personal good coincides with the satisfaction of the agent's own desires. If an agent is purely moral, the utility of an action for her coincides with the moral good the agent will promote by performing this action, where the agent's promotion of the moral good coincides with the accomplishment of her own moral values. Utility is just the quantitative counterpart of the concept of preference, that is, the more an agent prefers something, the higher its utility. Of course, purely hedonistic agents and purely moral agents are just extreme cases. An agent is more or less moral depending on whether the utility of a given option for her is more or less affected by her moral values. More precisely, the higher the influence of the agent's moral values on evaluating the utility of a given decision option, the more moral the agent. The extent to which an agent's utility is affected by her moral values can be called *degree of moral sensitivity*.[5]

Goals The reason why, in Figure 1, preferences and goals are included in the same box is that we conceive goals as intimately related with preferences. In particular, we assume that an agent has φ as a goal (or wants to achieve φ) if and only if: (i) the agent prefers φ to be true to φ to be false, and (ii) the agent considers φ a possible state of affairs (φ is compatible with what the agent believes). The second property is called *realism* of goals by philosophers (cf. [Bratman, 1987; Davidson, 1980; McCann, 1991]). It is based on the idea that an agent cannot reasonably pursue a goal unless she thinks that she can *possibly* achieve it, i.e., there exists at least one possible evolution of the world (a history) that the agent considers possible along which φ is true. Indeed, an agent's goal should not be incompatible with the agent's beliefs. This explains the influence of beliefs on the goal generation process, as depicted in Figure 1.[6] The first property is about the motivational aspect of goals. For φ to be a goal, the agent should not be indifferent between φ and $\neg\varphi$, in the sense that, the agent prefers a situation in which φ is true to a situation in which φ is false, *all other things being equal*. In other words, the utility of a situation increases in the direction by the formula φ *ceteris paribus* ("all else being equal") [Wellman and Doyle, 2001]. This property also defines Von Wright's concept of "preference of φ over $\neg\varphi$" [Von Wright, 1963].[7] According to this interpretation, a goal is conceived as a *realistic ceteris paribus preference for* φ.

However not all goals have the same status. Certain goals have a motivating force while others do not have it. Indeed, the fact that the agent prefers φ being true to φ being false does not necessarily imply that the agent is motivated to achieve a state in which φ is true and that she decides to perform a certain action *in order to* achieve it. For φ to be a motivating goal, for every possible situation that the agent envisages in which φ is true and for every possible situation that the agent envisages in which φ is false, the agent has to prefer the former to the latter. In other words, there is no way for the agent to be satisfied without achieving φ.[8]

An example better clarifies this point. Suppose Mary wants to buy a reflex camera Nikon and, at the same time, she would like to spend no more than 300 euros. In other words, Mary has two goals in her mind:

- G1: the goal of buying a reflex camera Nikon, and

- G2: the goal of spending no more than 300 euros.

She goes to the shop and it turns out that all reflex cameras Nikon cost more than 300 euros. This implies that Mary believes that she cannot achieve the two goals at the

---

[5]This degree can be conceived as a personality trait. In the case of human agents, it is either culturally acquired or genetically determined. In the case of artificial agents, it is configured by the system designer.

[6]The idea that beliefs form an essential ingredient of the goal generation process is also suggested by [Broersen *et al.*, 2002].

[7]Von Wright presents a more general concept of "preference of φ over ψ" which has been recently formalized in a modal logic setting by [van Benthem *et al.*, 2009]. See also [Rescher, 1967] for an interpretation of this *ceteris paribus* condition based on the concept of logical independence between formulas.

[8]The term 'satisfied' just means that the agent achieves what she prefers.

same time, as she envisages four situations in her mind but only three are considered possible by her: the situation in which only the goal G1 is achieved, the situation in which only the goal G2 is achieved and the situation in which no goal is achieved. The situation in which both goals are achieved is considered impossible by Mary. This is not inconsistent with the previous definition of goal since Mary still believes that it is possible to achieve each goal separately from the other. Figure 2 clearly illustrates this: the full rectangle includes all worlds that Mary envisages, so-called *information set*, while the dotted rectangle includes all worlds that Mary considers actually possible, so-called *belief set*.[9] (Cf. [Kraus and Lehmann, 1988; Lorini, 2016b] for a logical account of the distinction between information set and belief set.)

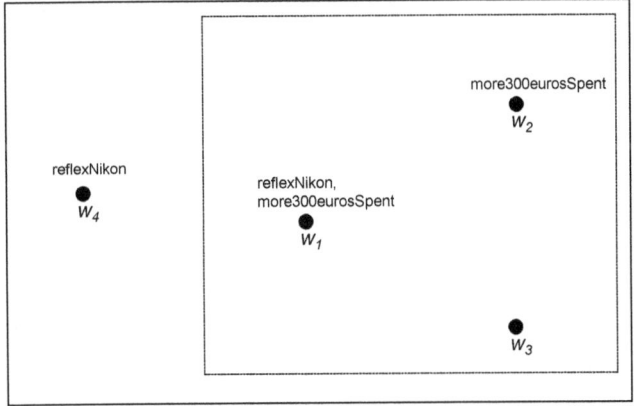

Figure 2. Example for goals

Mary decides to save her money since the goal G2 is a motivating one, while the goal G1 is not. To see that G1 is a goal, it is sufficient to observe that, *all other things being equal*, Mary prefers a situation in which she buys a Nikon to a situation in which she does not buy it. In fact, w_4 is preferred to w_3 and w_1 is preferred to w_2. Moreover, w_4 and w_3 are equal in everything except at w_4 Mary buys a Nikon while at w_3 she does not. Similarly, w_1 and w_2 are equal in everything except at w_1 Mary buys a Nikon while at w_2 she does not. To see that G1 is not motivating, it is sufficient to observe that there exists a situation in which Mary does not buy a Nikon (w_3) which is preferred to a situation in which she buys it (w_1). Finally, to see that G2 is a motivating goal, we just need to observe that every situation in which she spends no more than 300 euros (w_3 and w_4) is preferred to every situation in which this is not the case (w_1 and w_2). Thus, on the basis of what she believes, Mary concludes that

---

[9] Mary's information set includes all worlds that, according to Mary, are compatible with the laws of nature. For instance, Mary can perfectly envisage a world in which she is the president of French republic even though she considers this actually impossible.

she can only achieve her goal G2 by saving her money and by buying nothing in the shop.

From preferences and beliefs to actions As the cognitive architecture in Figure 1 highlights, beliefs and preferences are those mental attitudes which determine the agent's choices and are responsible for the formation of new intentions about present actions (present-directed intentions) and future actions (future-directed intentions). As emphasized in the literature in philosophy [Bratman, 1987; Mele, 1992] and AI [Bratman *et al.*, 1988], a future-directed intention is the element of a partial or a complete plan of the agent: an agent may have the intention to perform a sequence of actions later (e.g., the action of going to the train station in two hours followed by the action of taking the train from Paris to Bruxelles at 10 am) in order to achieve a certain goal (e.g., the goal of being in Bruxelles at the European Commission at 2 pm). A present-directed intention is a direct motivation to perform an action now.

In particular, decision is determined by beliefs, preferences and a general rationality criterion stating what an agent should do on the basis of what she believes and what she prefers. Different kinds of rationality criteria have been studied in the areas of decision theory and game theory ranging from expected utility maximization, maxmin and maxmax to satisficing [Simon, 1956]. Once the choice has been made by the agent and the corresponding intention has been formed, the action is performed right afterwards or later. Specifically, an agent forms the intention to perform a certain action at a given point in time and, once the time of the planned action execution is attained, the agent performs the action unless before attaining it, she has reconsidered her prior intention.

2.2 A cognitive view of emotion

In the recent years, emotion has become a central topic in AI. The main motivation of this line of research lies in the possibility of developing computational and formal models of artificial agents who are expected to interact with humans. To ensure the accuracy of a such formal models, it is important to consider how emotions have been defined in the psychological literature. Indeed, in order to build artificial agents with the capability of recognizing the emotions of a human user, of behaving in a believable way, of affecting the user's emotions by the performance of actions directed to her emotions (e.g. actions aimed at reducing the human's stress due to her negative emotions, actions aimed at inducing positive emotions in the human), such agents must be endowed with an adequate model of human emotions.

Appraisal theory The most popular psychological theory of emotion in AI is the so-called appraisal theory (cf. [Scherer *et al.*, 2001] for a broad introduction to this theory). It has emphasized the strong relationship between emotion and cognition, by stating that each emotion can be related to specific patterns of evaluations and interpretations of events, situations or objects (appraisal patterns) based on a number of dimensions or criteria called *appraisal variables* (e.g. goal relevance, desirability, likelihood, causal attribution). Appraisal variables are directly related to the mental attitudes of the individual (e.g. beliefs, predictions, desires, goals, intentions). For instance, when prospecting the possibility of winning a lottery and considering 'I win the lottery' as a desirable event, an agent might feel an intense hope. When prospecting the possibility of catching a disease and considering 'I catch a disease' as an undesirable event, an agent might feel an intense fear.

Most appraisal models of emotions assume that explicit evaluations based on evaluative beliefs (i.e. the belief that a certain event is good or bad, pleasant or unpleasant, dangerous or frustrating) are a necessary constituent of emotional experience. On the other hand, there are some appraisal models mostly promoted by philosophers [Searle, 1983; Gordon, 1987] in which emotions are reduced to specific combinations of beliefs and desires, and in which the link between cognition and emotion is not necessarily mediated by evaluative beliefs. Reisenzein [Reisenzein, 2009] calls *cognitive-evaluative* the former and *cognitive-motivational* the latter kind of models. For example, according to cognitive-motivational models of emotions, a person's happiness about a certain fact φ can be reduced to the person's belief that φ obtains and the person's desire that φ obtains. On the contrary, according to cognitive-evaluative models, a person feels happy about a certain fact φ if she believes that φ obtains and she evaluates φ to be good (desirable) for her. The distinction between cognitive-evaluative models and cognitive-motivational models is reminiscent of the opposition between the Humean view and the anti-Humean view of desire in philosophy of mind. According to the Humean view, belief and desires are distinct mental attitudes that are not reducible one to the other. Moreover, according to this view, there are no necessary connections between beliefs and desires, i.e., beliefs do not necessarily require corresponding desires and, viceversa, desires do not necessarily require corresponding beliefs. On the contrary, the anti-Humean view defends the idea that beliefs and desires are necessarily connected. A specific version of anti-Humeanism is the so-called "Desire-as-Belief Thesis" criticized by the philosopher David Lewis in [Lewis, 1988] (see also [Lewis, 1996; Hájek and Pettit, 2004]). In line with cognitive-evaluative models, this thesis states that an agent *desires* something to the extent that she *believes* it to be good.

The popularity of appraisal theory in logic and AI is easily explained by the fact that it perfectly fits with the concepts and level of abstraction of existing logical and computational models of cognitive agents developed in these areas. Especially cognitive-motivational models use folk-psychology concepts such as belief, knowledge, desire and intention that are traditionally used in logic and AI for modelling cognitive agents.

The conceptual background underlying our view of appraisal theory is depicted in Figure 3 which is nothing but the cognitive architecture of Figure 1 extended with an emotion component.

Figure 3 highlights the role of mental attitudes in emotion. In particular, it highlights the fact that mental attitudes of different kinds such as belief, desires, preferences, goals, moral values and (present-directed or future-directed) intentions determine emotional responses. For example, as emphasized above, the emotional response of happiness is triggered by a *goal* and the *certain belief* that the content of one's goal is true. On the contrary, the emotional response of sadness is triggered by a *goal* and the *certain belief* that the content of one's goal is false. The emotional response of hope is triggered by a *goal* and the *uncertain belief* that the content of one's goal is

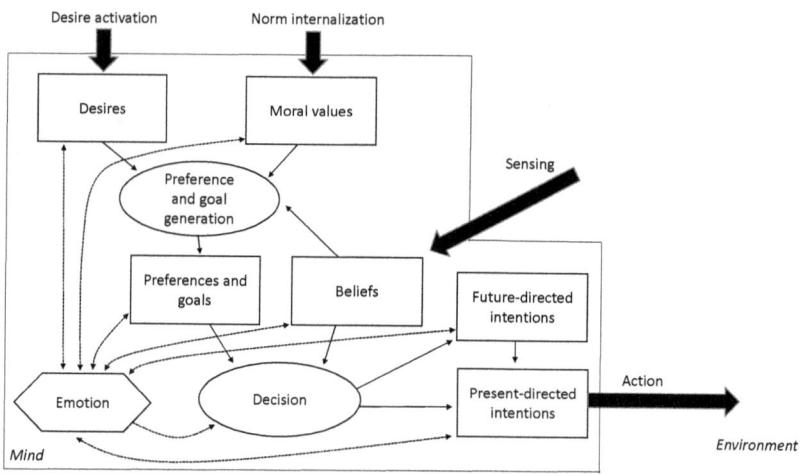

Figure 3. Cognitive architecture extended with emotions

true. On the contrary, the emotional response of fear is triggered by a *goal* and the *uncertain belief* that the content of one's goal is false. This view is consistent with a famous appraisal model, the so-called OCC psychological model of emotions [Ortony et al., 1988], according to which, while joy and distress are triggered by *actual consequences*, hope and fear are triggered by *prospective consequences* (or *prospects*). [Gratch and Marsella, 2004] interpret the term 'prospect' as synonymous of 'uncertain consequence' (in contrast with 'actual consequence' as synonymous of 'certain consequence').

Moral guilt and reproach are examples of emotions that are triggered by moral values [Haidt, 2003]. While moral guilt is triggered by the *belief* of being responsible for the violation of a *moral value* or *of being responsible for having behaved in a morally reprehensible way*, reproach is triggered by the *belief* of someone else being responsible for the violation of a *moral value* or *of someone else being responsible for having behaved in a morally reprehensible way*. In other words, guilt is triggered by self-attribution of responsibility for the violation of a moral value, while reproach is triggered by attribution to others of responsibility for the violation of a moral value.

Intentions as well might be responsible for triggering certain kinds of emotional response. For instance, as emphasized by psychological theories of anger (e.g., [Lazarus, 1991; Ortony et al., 1988; Roseman et al., 1996]), a necessary condition for an agent 1 to be angry towards another agent 2 is agent 1's belief that agent 2 has performed an action that has damaged her, that is, 1 believes that she has been kept from attaining an important goal by an improper action of agent 2. Anger becomes more intense when agent 1 believes that agent 2 has *intentionally* caused her damage. In this sense, an agent 1's belief about another agent 2's intention may have implications on the intensity of agent 1's anger.

Figure 3 also represents how emotions retroactively influence mental states and decision either (i) through coping or (ii) through anticipation and prospective thinking (i.e., the act of mentally simulating the future) in the decision-making phase.

Coping is the process of dealing with emotion, either externally by forming an intention to act in the world (problem-focused coping) or internally by changing the agent's interpretation of the situation and the mental attitudes that triggered and sustained the emotional response (emotion-focused coping) [Lazarus, 1991]. For example, when feeling an intense fear due to an unexpected and scaring stimulus, an agent starts to reconsider her beliefs and intentions in order to update her knowledge in the light of the new scaring information and to avoid running into danger (emotion-focused coping). Then, the agent forms an intention to go out of danger (problem-focused coping). Another agent can try to discharge her feeling of guilt for having damaged someone either by forming the intention to repair the damage (problem-focused coping) or by reconsidering the belief about her responsibility for the damage (emotion-focused coping). The coping process as well as its relation with appraisal is illustrated in Figure 4.

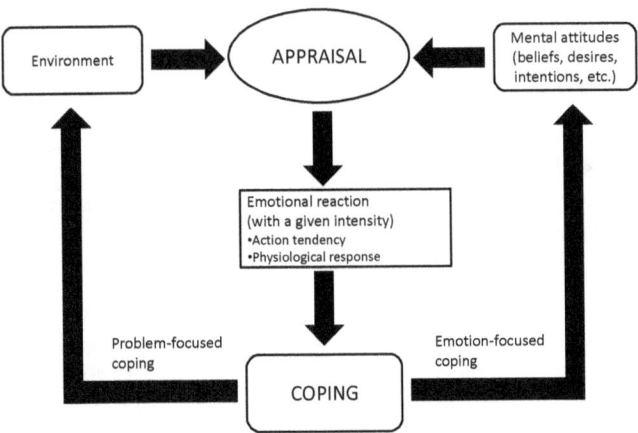

Figure 4. Appraisal and coping cycle

Influence of emotion on decision The influence of emotion on decision-making has been widely studied both in psychology and in economics. Rick & Loewenstein [Rick and Loewenstein, 2008] distinguish the following three forms of influence:

- **Immediate emotions**: real emotions experienced at the time of decision-making:
  - **Integral influences**: influences from immediate emotions that arise from contemplating the consequences of the decision itself,
  - **Incidental influences**: influences from immediate emotions that arise from

factors unrelated to the decision at hand (e.g., the agent's current mood or chronic dispositional affect);

- **Anticipated emotions**: predictions about the emotional consequences of decision outcomes (they are not experienced as emotions per se at the time of decision-making).

An example of integral influence of an immediate emotion is given by the following example.

Example 1 *Paul would like to eat some candies but her mother Mary has forbidden him to eat candies without her permission. Paul's fear of the sanction influences Paul's decision not to eat candies without asking permission.*

The following example illustrates incidental influence of an immediate emotion.

Example 2 *Mary has quarreled with her colleague Paul. At the end of the day she goes back home after work and on the metro a beggar asks her for money. Few hours after the quarrel with Paul, Mary is still in a bad mood and because of her current disposition she refuses the beggar's request.*

The following example illustrates the influence of anticipated emotions on decision.

Example 3 *Peter has to decide whether to leave his job as a researcher at the university of Paris and to accept a job offer as a professor at a university in the U.S. He decides to accept the job offer because he thinks that, if he refuses it, he will likely regret his decision.*

One of the most prominent theory of the integral influence of emotion on decision is Damasio's theory of the somatic marker [Damasio, 1994]. According to this theory, decision between different courses of actions leads to potentially advantageous (positive) or harmful (negative) outcomes. These outcomes induce a somatic response used to mark them and to signal their danger or advantage. In particular, a negative somatic marker 'signals' to the agent the fact that a certain course of action should be avoided, while a positive somatic marker provides an incentive to choose a specific course of action. According to Damasio's theory, somatic markers depend on past experiences. Specifically, pain or pleasure experienced as a consequence of an outcome are stored in memory and are felt again when the outcome is envisaged in the decision-making process. The following example clearly illustrates this.[10]

Example 4 *Mary lives in Toulouse and has to decide whether to go to Paris by plane or by train. Last time she traveled by plane she had a painful experience because of*

---

[10]Positive and negative somatic markers can operate either at a conscious level or at a unconscious/automatic level. This corresponds to Ledoux's distinction between explicit memory and implicit memory and between two possible elaborations of a stimulus inducing an emotional response [LeDoux, 1996]: conscious elaboration vs. automatic elaboration.

turbulence. Mary envisages the possibility of incurring again in a turbulence and gets frightened, thereby deciding to travel by train.

There have been several attempts to extend the classical expected utility model in order to incorporate anticipated emotions that are related to our uncertainty about the future, such as hopefulness, anxiety, and suspense [Caplin and Leahy, 2001]. Some economic models of decision-making consider how the anticipation of a future regret might affect a person's current decision [Loomes and Sugden, 1987]. In particular, according to these models, if a person believes that after choosing a certain action she will likely regret for having made this choice, she will be less willing to choose the action (than in the case in which she does not believe this). These models agree in defining regret as the emotion that stems from the comparison between the actual outcome deriving from a given choice and a counterfactual better outcome that might have been had one chosen a different action [Frijda *et al.*, 1989; Kahneman and Miller, 1986; Zeelenberg *et al.*, 2000]. More recently, some economists have studied the influence of strategic emotions such as interpersonal guilt and anger on decision [Battigalli and Dufwenberg, 2007; Charness and Dufwenberg, 2009; Hopfensitz and Reuben, 2009]. Following psychological theories of interpersonal guilt [Baumeister *et al.*, 1994; Tangney, 1995], models developed in this area assume that the prototypical cause of guilt is the infliction of harm, loss, or distress on a relationship partner. Moreover, they assume that if people feel guilty for hurting their partners and for failing to live up to their expectations, they will alter their behavior (to avoid guilt) in ways that seem likely to maintain and strengthen the relationship. This is different from the concept of moral guilt formalized by [Lorini and Muehlenbernd, 2015] according to which a person feels (morally) guilty if she believes that she is responsible for having behaved in a morally reprehensible way (see Section 2.4 for more details).

2.3 Interacting minds: from game theory to epistemic game theory

The idea highlighted in Section 2.1 of describing rational agents in terms of their epistemic and motivational attitudes, is also adopted by classical decision theory and game theory. In particular, classical decision theory accounts for the criteria and principles (e.g., expected utility maximization) that a rational agent should apply in order to decide what to do on the basis of her beliefs and preferences. Game theory generalizes decision theory to the multiagent case in which agents' decisions are interdependent and agents' actions might interfere between them so that: (i) the possibility for an agent to achieve her goals may depend on what the other agents decide to do, and (ii) agents form beliefs about the future choices of the other players and, consequently, their current decisions are influenced by what they believe the others will do. More generally, game theory involves a strategic component that is not considered by classical decision theory whose object of analysis is a single agent who makes decisions and acts in an environment she does not share with other agents.

Classical decision theory and game theory provide a quantitative account of individual and strategic decision-making by assuming that agents' beliefs and preferences can be respectively modeled by subjective probabilities and utilities. In particular, while subjective probability captures the extent to which a fact is *believed* by a certain agent, utility captures how much a certain state of affairs is *preferred* by the agent. In other words, subjective probability is the quantitative counterpart of the concept of belief, while utility is the quantitative counterpart of the concept of preference.[11]

One of the fundamental concepts of game theory is the concept of solution which is, at the same time, a prescriptive notion, in the sense that it prescribes how rational agents in a given interaction *should* play, and a predictive one, in the sense that it allows us to predict how the agents *will* play. There exist many different solution concepts both for games in normal form and for games in extensive form (e.g., Nash Equilibrium, iterated deletion of strongly dominated strategies, iterated deletion of weakly dominated strategies, correlated equilibrium, backward induction, forward induction, etc.) and new ones have been proposed in the recent years (see, e.g., [Halpern, 2011]). A major issue we face when we want to use a solution concept in order either to predict human behavior or to build some practical applications (e.g., for computer security or for multiagent systems) is to evaluate its significance. Some of the questions that arise in these situations are, for instance: given certain assumptions about the agents such as the assumption that they are rational (e.g., utility maximizers), under which conditions will the agents converge to equilibrium? Are these conditions realistic? Are they too strong for the domain of application under consideration? There is a branch of game theory, called epistemic game theory, which can help to answer these questions (cf. [Perea, 2012] for a general introduction to the research in this area). Indeed, the aim of epistemic game theory is to provide an analysis of the necessary and/or sufficient epistemic conditions of the different solution concepts, that is, the assumptions about the epistemic states of the players that are necessary and/or sufficient to ensure that they will play according to the prescription of the solution concept. Typical epistemic conditions which have been considered are, for example, the assumption that players have common belief (or common knowledge) about the rationality of every player,[12] the assumption that every player knows the choices of the others,[13] or the assumption that players are logically omniscient.[14]

Epistemic game theory shares concepts and methods with what Aumann calls interactive epistemology [Aumann, 1999]. The latter is the research area in logic and philosophy which deals with formal models of knowledge and belief when there is more than one rational agent or "player" in the context of interaction having not only knowledge and beliefs about substantive matters, but also knowledge and beliefs about

---

[11] Qualitative approaches to individual and strategic decision-making have been proposed in AI [Boutilier, 1994; J. and R., 1999] to characterize criteria that a rational agent should adopt for making decisions when she cannot build a probability distribution over the set of possible events and her preference over the set of possible outcomes cannot be expressed by a utility function but only by a qualitative ordering over the outcomes. For example, going beyond expected utility maximization, qualitative criteria such as the maxmin principle (choose the action that will minimize potential loss) and the maxmax principle (choose the action that will maximize potential gain) have been studied and axiomatically characterized [Brafman and Tennenholtz, ; Brafman and Tennenholtz, 1996].

[12] This is the typical condition of iterated deletion of strongly dominated strategies (also called iterated strong dominance).

[13] This condition is required in order to ensure that the agents will converge to a Nash equilibrium.

[14] See [Zvesper, 2010] for an analysis of iterated strong dominance after relaxing the assumption of logical omniscience.

the others' knowledge and beliefs. The concept of rationality corresponds either to the optimality criterion according to which an agent should choose an action which guarantees the highest utility, given what she believes the other agents will do, or the prudential criterion according to which an agent should not choose an action which ensures the lowest utility, given what she believes the other agents will do. An example of the former is expected utility maximization, while an example of the latter is weak rationality in the sense of [van Benthem, 2007] (cf. also [Myerson, 1991; Binmore, 1991]), according to which an agent should not choose an action which is strongly dominated by another action, given what the agent believes the other agents will do.

Epistemic game theory provides a useful framework for clarifying how agents' mental attitudes influence behaviours of agents in a social setting. In particular, it allows us to understand the subtle connection between beliefs, preferences and decision, as represented in Figure 1 given in Section 2.1, under the assumption that the agents' decisions are interdependent, in the sense that they are affected by what the agents believe the others will choose.[15]

2.4 Logics for mental attitudes, emotion and games

This section is devoted to discuss existing logics for mental attitudes and emotion proposed in AI as well as the connection between the representation of mental attitudes and emotion in logic and the representation of mental attitudes and emotion in game theory.

Logics for mental attitudes Since the seminal work of [Cohen and Levesque, 1990] aimed at implementing Bratman's philosophical theory of intention [Bratman, 1987], many formal logics for reasoning about mental attitudes of agents such as beliefs, desires and intentions have been developed. Among them we should mention the logics developed by [Lorini and Herzig, 2008; Lorini, 2016a; Herzig and Longin, 2004; Konolige and Pollack, 1993; Meyer et al., 1999; Miller and Sandu, 1997; Rao and Georgeff, 1991; Shoham, 1993; Singh and Asher, 1993; van Linder et al., 1998; Wooldridge, 2000].

The general term used to refer to this family of logics is *agent logics*. A subfamily is the family of BDI logics whose most representative example is the modal logic by [Rao and Georgeff, 1991] whose primitive constituents are the the concepts of belief (B), desire (D) and intention (I) which are expressed by corresponding modal operators. Another well-known agent logic is the so-called KARO framework developed by [Meyer et al., 1999]. KARO is a multi-modal logic framework based on a blend of dynamic logic with epistemic logic, enriched with modal operators for modeling mental attitudes such as beliefs, desires, wishes, goals and intentions.

---

[15]Although epistemic game theory and, more generally, game theory share with Figure 1 the concepts of belief and preference, they do not provide an account of the origin of beliefs, desires and moral values and of the connection between desires, moral values and preferences. Moreover, the concept of future-directed intention is not included in the conceptual apparatus of game theory and epistemic game theory. The same can be said for goals: the concept of goal is somehow implicit in the utility function but is not explicitly modeled.

Generally speaking, agent logics are nothing but formal models of rational agency whose aim is to explain how an agent endowed with mental attitudes makes decisions on the basis of both what she believes and what she wants or prefers. In this sense, the decisions of the agent are determined by both the agent's beliefs (the agent's epistemic states) and the agent's preferences (the agent's motivational states). As discussed in Section 2.1, the output of the agent's decision-making process is either a choice about what to do in the present, also called present-directed intention, or a choice about what to do in the future, also called future-directed intention. The idea that the behavior of an agent can be explained by attributing mental states to the agent and by having a sophisticated account of the relationship between her epistemic states and her motivational states and of the influence of these on the agent's decision-making process is something agent logics share with other disciplines including philosophy of mind [Dennett, 1987], cognitive sciences [Pylyshyn, 1984], psychology [Reisenzein, 2009] and artificial intelligence [Castelfranchi, 1998].

Logics for emotion More recently, agent logics have been used to formalize the cognitive structure and the coping strategies of different types of emotion. For instance, a logical formalization of emotion in the context of the KARO framework has been proposed. In particular, in the KARO framework each emotion type is represented with a special predicate, or fluent, in the jargon of reasoning about action and change, to indicate that these predicates change over time. For every fluent a set of effects of the corresponding emotions on the agent's planning strategies are specified, as well as the preconditions for triggering the emotion in terms of mental attitudes of agents. The latter correspond to generation rules for emotions. For instance, in [Meyer, 2006] generation rules for four basic emotions are given: joy, sadness, anger and fear, depending on the agent's plans. In [Turrini et al., 2010] generation rules for guilt and shame have been proposed.

A logical formalization of the OCC psychological model of emotions [Ortony et al., 1988] has been proposed in [Adam et al., 2009].

Surprise is the simplest emotion that is triggered by the mismatch between an expectation that an event will possibly occur and an incoming input (i.e., what an agent perceives). In [Lorini and Castelfranchi, 2007] a logical theory of surprise is proposed. The theory clarifies two important aspects of this cognitive phenomenon. First, it addresses the distinction between surprise and astonishment, the latter being the emotion triggered by something an agent could not reasonably expect. The crucial difference between surprise and astonishment is that the former necessarily requires an explicit expectation in the agent's mind, while the latter does not. One can be astonished by something since, at the moment she perceives it, she realizes that it was totally unpredictable, without having formulated an expectation in advance. For example, suppose Mary is working in her office. Suddenly, someone knocks the door and enters into Mary's office. Mary sees that the person is a policeman. She is astonished by this fact even though, before perceiving it, she did not have explicit in her mind the expectation that "a policeman will not enter into the office". Secondly, the theory clarifies the role of surprise in belief change by conceiving it as a basic mechanism which is responsible for triggering belief reconsideration.

In a more recent paper [Lorini and Schwarzentruber, 2011], a logical formalization of counterfactual emotions has been provided. Counterfactual emotions, whose prototypical example is regret, are those emotions that are based on counterfactual reasoning about agents' choices. Other examples are rejoicing, disappointment, and elation. The formalization is based on an epistemic extension of STIT logic (the logic of "seeing to it that") by Belnap et al. [Belnap *et al.*, 2001; Horty, 2001; Broersen, 2011; Lorini, 2013] and allows to capture the cognitive structure of regret and, in particular, the counterfactual belief which is responsible for triggering this emotion, namely the *belief that a counterfactual better outcome might have been, had the agent chosen a different action*. In [Lorini *et al.*, 2014], the STIT logical analysis of counterfactual emotions is extended to moral emotions. The latter involve counterfactual reasoning about responsibility for the transgression of moral values. In particular, the proposed formalization accounts for the attribution of responsibility for the violation of a moral value either to the self or to the other. This is a fundamental constituent of moral emotions such as guilt, reproach, moral pride and moral approval. For example, according to the proposed analysis, guilt is triggered by the *belief that one is responsible for having behaved in a morally reprehensible way*. A game-theoretic account of moral guilt, which parallels the STIT logical analysis, has been given in [Lorini and Muehlenbernd, 2015].

The problem of emotion intensity has also been adressed by logicians. Following existing psychological models of emotion based on appraisal theory, intensity of these emotions is defined as a function of two cognitive parameters, the strength of the expectation and the strength of the desire which are responsible for triggering the emotional response. For instance, the intensity of hope that a certain event will occur is a monotonically increasing function of both the strength of the expectation and the strength of the desire that the event will occur. The logical theory of appraisal and coping presented in [Dastani and Lorini,] also considers the behavioral aspects of such emotions: how the execution of a certain coping strategy depends on the intensity of the emotion generating it. Specifically, it is assumed that: (i) an agent is identified with a numerical value which defines her tolerance to the negative emotion, and (ii) if the intensity of the negative emotion (e.g., fear) exceeds this value then the agent will execute a coping strategy aimed at discharging the negative emotion.

Logics for games The relationship between logic and game theory has been explored in both directions: *games for logic* and *logic for games*. On the one hand, methods and techniques from game theory have been applied to formal semantics, proof theory and model checking for different kinds of logic [Hintikka and Sandu, 1997; Gradel, 2002; Keiff, 2011]. On the other hand, logical representation languages have been proposed in computer science and AI to represent game-theoretic concepts such as the concepts of strategy, capability, winning strategy as well as solution concepts such as Nash equilibrium and backward induction. This includes logics such as Coalition Logic [Pauly, 2002], Alternating-time Temporal Logic (ATL) [Alur *et al.*, 2002] and STIT (the logic of "seeing to it that") [Belnap *et al.*, 2001; Horty, 2001].

More recently, logics for epistemic game theory have been proposed by incorporating epistemic components in existing logics for games and developing new logical formalisms that can represent, at the same time, the structure of the game and the mental attitudes and rationality of the players involved in the game.

Much of the work in the field of epistemic game theory is based on a *quantitative* representation of uncertainty and epistemic attitudes. Notable examples are the analysis of the epistemic foundations for forward induction and for iterated admissibility based on Bayesian probabilities [Stalnaker, 1998; Halpern and Pass, 2009], conditional probabilities [Battigalli and Siniscalchi, 2002] or lexicographic probabilities [Brandenburger *et al.*, 2008]. The distinction between quantitative and qualitative approaches to uncertainty has been widely discussed in the AI literature (cf. [Goldszmidt and Pearl, 1996]). While in quantitative approaches belief states are characterized by classical probabilistic measures or by alternative numerical accounts, such as lexicographic probabilities or conditional probabilities [Battigalli and Siniscalchi, 2002], qualitative approaches do not use any numerical representation of uncertainty but simply a plausibility ordering on possible worlds structures inducing an epistemic-entrenchment-like ordering on propositions.

Both logics for epistemic game theory based on a qualitative representation of epistemic attitudes [Baltag *et al.*, 2009; Lorini, 2016b; Lorini and Schwarzentruber, 2010] and logics for epistemic game theory based on probability theory [Halpern and Pass, 2009; Bjorndahl *et al.*, 2014] have been proposed in the recent years. The main motivation for the latter is to exploit logical methods in order to provide sound and complete axiomatics for important concepts studied in epistemic game theory such as rationality and common knowledge of rationality. The main motivation for the former is to show that interesting results about the epistemic foundation for solution concepts in game theory can be proved in a qualitative setting, without necessarily exploiting the complex machinery of probability theory.

The connection between logical models of epistemic states based on Kripke semantics and formal models of epistemic states based on the concept of type space has also been explored [Galeazzi and Lorini, 2016; Klein and Pacuit, 2014]. While the former have been mainly proposed by logicians in AI [Fagin *et al.*, 1995] and philosophy [Stalnaker, 2006], the latter have been proposed by game theorists in economics [Harsanyi, 1967]. The main motivation for this research lies in the possibility of building a bridge between two research communities that study the same concepts and phenomena from different perspectives.

3 From mental attitudes to institutions via collective attitudes

In this section we gradually move from minds to institutions. The connection between the former and the latter is built via the concept of collective attitude. Specifically, we discuss a particular view of institutions: the idea that institutional facts are grounded on the agents' collective attitudes that, in turn, originate from the agents' mental attitudes.

Section 3.1 starts with a discussion about the different functions and origins of collective attitudes, while Section 3.2 clarifies the connection between collective attitudes and institutions. Finally, Section 3.3 explains how this connection has been formalized in logic.

3.1 Collective attitudes

Collectives such as groups, teams, coorporations, organizations, etc. do not have minds. However, we frequently ascribe intentional attitudes to them in the same way as we ascribe intentional attitudes to individuals. For example, we may speak of what our family prefers, of what the goal of a coorporation or organization is, of what the scientific community think about a certain issue, and so on.

Aggregate vs. common attitudes An important distinction in the theory of collective attitudes is between aggregate attitudes and common attitudes. As emphasized by [List, 2014] "...an aggregate attitude (of a collective) is an aggregate or summary of the attitudes of the individual members of the collective, produced by some aggregation rule or statistical criterion...". A typical example of aggregate attitude produced by a statistical criterion is shared belief, namely the fact that all agents (or most of the agents) in a set of agents believe that a certain proposition p is true. An example of aggregate attitude produced by an aggregation rule is the collective acceptance of a jury about a given proposition p obtained by majority voting: the jury believes that the proposition p is true if and only if the majority of the members of the jury has expressed the individual opinion that p is true. Aggregate attitudes produced by aggregation rules are the objects of analysis of judgement aggregation, an important research area in social sciences and AI (see [Grossi and Pigozzi, 2014; List, 2012] for an introduction to judgement aggregation). Differently from common attitudes, aggregate attitudes do not require a level of common awareness by the members of the group. That is, a group can hold an aggregate attitude even though the members of the group do not necessarily believe so. For example, the fact that two agents share the belief that p is true does not necessarily imply that they individually believe that they share this belief. As emphasized by [List, 2014] "...a common attitude (of a collective) is an attitude held by all individual members of the collective, where their holding it is a matter of common awareness", where the term "common awareness" refers to the fact that every member of the group believes that the group has the common attitude, that every member of the group believes that every member of the group believes that the group has the common attitude, and so on. A typical example of common attitude is common belief: every agent in the group believes that p is true, every agent in the group believes that every agent in the group believes that p is true, and so on ad infinitum.

Functions of collective attitudes Collective attitudes play a crucial role in the society as: (i) they provide the basis of our common understanding through communication, (ii) they ensure coordination between agents, (iii) they are fundamental constituents of collaborative activities between agents acting as members of the same team.

In linguistic, the concept of common ground in a conversation is typically conceived as the common knowledge (or common belief) that the speaker and the hearer have about the rules of the language they use and about the meaning of the expressions uttered by the speaker [Stalnaker, 2002]. Indeed, language use in conversation is a form of social activity that requires a certain level of coordination between what the speaker means and what the addressee understands the speaker to mean. Any utterance of the speaker is in principle ambiguous because the speaker could use it to express a variety of possible meanings. Common ground — as a mass of information and facts mutually believed by the speaker and the addressee — ensures coordination by disambiguating the meaning of the speaker's utterance. For example, suppose two different operas, "Don Giovanni" by Mozart and "Il Barbiere di Siviglia" by Rossini, are performed in the same evening at two different theaters. Mike goes to see Don Giovanni and the next morning sees Mary and asks "Did you enjoy the opera yesterday night?", identifying the referent of the word "opera" as Don Giovanni. In order to ensure that Mary will take "opera" as referring to Don Giovanni, it has to be the case that the night before Mary too went to see Don Giovanni, that Mary believes that Mike too went to see Don Giovanni, that Mary believes that Mike believes that Mary too went to see Don Giovanni, and so on.

Moreover, since the seminal work by David Lewis [Lewis, 1969], the concept of common belief has been show to play a central role in the formation and emergence of social conventions.

Finally, collective attitudes such as common goal and joint intention are traditionally used in in the philosophical area and in AI to account for the concept of collaborative activity [Bratman, 1992; Grosz and Kraus, 1996; Dunin-Keplicz and Verbrugge, 2002; Dunin-Keplicz and Verbrugge, 2010]. Notable examples of collaborative activity are the activities of painting a house together, dancing together a tango, or moving a heavy object together. Two or more agents acting together in a collaborative way need to have a common goal and need to form a shared plan aimed at achieving the common goal. In order to make collaboration effective, each agent has to commit to her part in the shared plan and form the corresponding intention to perform her part of the plan. Moreover, she has to monitor the behaviors of the others and, eventually, to reconsider her plan and adapt her behavior to the new circumstances.

The origin of collective attitudes Where do collective attitudes come from? How are they formed? There is no single answer to these questions, as collective attitudes can originate in many different ways.

As explained above, aggregate attitudes are the product of aggregation procedures like majority voting or unanimity (cf. [List and Pettit, 2011]). The agents in a certain group decide to use a certain aggregation rule. Then, every agent expresses her opinion about a certain issue p and the aggregation rule is used to determine what the group believes or what the group accepts. Examples of collective attitudes originating from the aggregation of individual attitudes are group belief and collective acceptance.

Collective attitudes, such as shared belief and common belief, can also be formed through communication or joint perception. A source of information announces to all agents in a group that a certain proposition p is true. Under the assumption that every agent perceives what the information source says and that every agent in the group trusts the information source's jugement about p, the agents will share the belief that p is true as a result of the announcement. Creation of common belief through communication requires satisfaction of certain conditions that are implicit in the concept of public announcement, as defined in the context of public announcement logic (PAL) [Plaza, 1989], the simplest logic in the family of dynamic epistemic logics (DEL) [van

Ditmarsch *et al.*, 2007]. Specifically, to ensure that an announcement will determine a common belief that the announced fact is true, every agent in the group has to perceive what the information source says, every agent in the group has to perceive that every agent in the group perceives what the information source says, and so on. The latter is called *co-presence* condition in the linguistic literature [Clark and Marshall, 1981].

The concept of co-presence becomes particularly relevant in the perspective of designing artificial systems situated in a physical environment that need to acquire common belief of certain facts in order to achieve coordination and to make collaboration effective. For example, imagine two robots moving in the physical environment. A source of information signals to them that there is a danger. It does this by emitting a red light. The robots will be able to form different levels of mutual belief about this fact depending on: (i) their spatial positions and the orientation of their sensors with respect to the source of information, and (ii) the perception of the other robots' spatial positions and of the orientations of the other robots' sensors with respect to the source of information. The concept of co-presence applies not only to agents interacting in a physical environment but also to agents interacting in a virtual environment (e.g., virtual characters of a videogame).

A side note: collective acceptance vs. common belief A property that clearly distinguishes collective acceptance from common belief is that common belief implies shared belief, while collective acceptance does not: when there is a common belief in a group of agents C that a certain proposition p is true then each agent in C individually believes that p is true, while it might be the case that there is a collective acceptance in C that p is true, and at the same time one or several agents in C do not individually believe that p is true. For example, the members of a Parliament might collectively accept (*qua* members of the Parliament) that launching a military action against another country is legitimate because by majority voting the Parliament decided so, even though some of them — who voted against the military intervention — individually believe the contrary. This difference is due to the fact that collective acceptance is a kind of aggregate attitude which can be formed through aggregation procedures others than unanimity.

Another important difference between collective acceptance and common belief is the irreducibility of collective acceptance to the individual level. In particular, it has been emphasized that, while common belief is strongly linked to individual beliefs and can be reduced to them, collective attitudes such as collective acceptance cannot be reduced to a composition of individual attitudes. This aspect is particularly emphasized by Gilbert [Gilbert, 1987] who follows Durkheim's non-reductionist view of collective attitudes [Durkheim, 1982]. According to Gilbert, any proper group attitude cannot be defined only as a label on a particular configuration of individual attitudes, as common belief is. In [Gilbert, 1989; Tuomela, 2007] it is suggested that a collective acceptance of a set of agents C is based on the fact that the agents in C identify themselves as members of a certain group, institution, team, organization, etc. and recognize each other as members of the same group, institution, team, organization, etc. Common belief and common knowledge, as traditionally defined in epistemic logic [Fagin *et al.*, 1995], do not entail this aspect of mutual recognition and identification with respect to the same group, institution, team, organization, etc.

3.2 Grounding institutions and norms on collective attitudes

In the previous section we have explained how collective attitudes are generated from mental attitudes through aggregation procedures, communication or joint perception.

The next step in our analysis is to explain how institutions and norms are grounded on collective attitudes of different types including collective acceptance and common belief. The term "grounded" means that the existence and the evolution of an institution or a norm depend on the existence and the evolution of the collective attitudes of the agents who are members of the institution and who are subject to the norm.

We focus here on two forms of grounding that have been considered in the literature: the grounding of institutions on collective acceptance and the grounding of conventions on common belief.

Collective acceptance and institutions The problem of understanding what institutions are and how they function has been addressed both in social sciences, in philosophy and in legal theory. Computer scientists working in the area of multiagent systems have been interested in devising artificial institutions, modeling their dynamics and the different kinds of rules and norms of an institution that agents have to deal with. Following [North, 1990, p. 3], artificial institutions can be conceived as "the rules of the game in a society or the humanly devised constraints that structure agents' interaction". In some models of artificial institutions norms are conceived as means to achieve coordination among agents and agents are supposed to comply with them and to obey the authorities of the system [Esteva et al., 2001]. More sophisticated models of institutions leave to the agents' autonomy the decision whether to comply or not with the specified rules and norms of the institution [Ågotnes et al., 2007; Lopez y Lopez et al., 2004]. However, all previous models abstract away from the legislative source of the norms of an institution, and from how institutions are created, maintained and changed by their members.

What these models of artificial institutions neglect is the fundamental relationship between institutions and the collective attitudes of their members and, in particular, the fact that the existence and the dynamics of an institution (norms, rules, institutional facts, etc.) are determined by the collective attitudes of the agents which identify themselves as members of the institution. This aspect is emphasized in the following quote from [Mantzavinos et al., 2004, p. 77]:

> "only because institutions are anchored in peoples minds do they ever become behaviorally relevant. The *elucidation of the internal aspect is the crucial step* in adequately explaining the emergence, evolution, and effects of institutions." [Emphasis added].

Prominent philosophical theories of institutional reality conceive collective acceptance as the collective attitude on which institutions are grounded [Searle, 1995; Tuomela, 2002]. The relationship between acceptance and institutions has also been emphasized in the philosophical doctrine of Legal Positivism [Hart, 1992]. According to Hart, the foundations of an institution consist of adherence to, or acceptance of, an ultimate rule of recognition by which the validity of any rule of the institution may be evaluated.[16]

Common belief and conventions Convention is a concept that has been widely studied in economics [Sugden, 2004], philosophy [Binmore, 2005; Tummolini et al., 2013] and computer science [Walker and Wooldridge, 1995; Villatoro et al., 2011; Shoham and Tennenholtz, 1997; Sen and Airiau, 2007], given the fundamental role it plays in the regulation of both human and artificial societies.

Eating manners, the kind of clothes we wear in office, and the side of the road on which we drive are mundane examples of convention. Roughly, a social convention is a customary, arbitrary and self-enforcing rule of behavior that is generally followed and expected to be followed in a group or in a society at large [Lewis, 1969]. When a social convention is established, everybody behaves in an agreed-upon way even if they did not in fact explicitly agree to behave in this way. A social convention can thus be seen as a kind of tacit agreement that has evolved out of a history of previous interactions [Sugden, 2004; Tummolini et al., 2013].

Since the seminal contribution by David Lewis [Lewis, 1969], the modern approach to conventions is rooted both in epistemic logic and in evolutionary game theory. The *epistemic approach* to the study of conventions has focused on the characterization of the kind of mutual beliefs and expectations that are required for a group to adopt a certain convention [Cubitt and Sugden, 2003; Sillari, 2005; Vanderschraaf, 1995] and on the distinction between the epistemic conditions of conventions in contrast with the epistemic conditions of social norms [Bicchieri, 2006]. The epistemic approach clearly highlights the fact that conventions are grounded on collective attitudes. Indeed, according to the well-known definition of convention by David Lewis [Lewis, 1969, pp. 76], a given regularity of behavior R is a convention for a population of agents P at a recurrent situation S, only if the agents in the population P *mutually expect* everyone in P to conform to the regularity R in the situation S (and commonly believe so). In other words, for a convention to exist, the agents in the population have to form a mutual expectation about each other's behavior (and a common belief about this). Consider the example of driving on the left-hand side in the UK. This is a convention as every person in the UK expects other people in the UK to drive on the left-hand side of the road. Moreover, every person in the UK expects other people to drive on the left-hand side of the road *because and as long as* she expects other people to expect everyone to drive on the left-hand side of the road.

The *evolutionary approach* to the study of conventions has focused on the conditions under which a certain convention can emerge in a given population of agents depending on the agents' learning capabilities. Notable examples of this approach are the models by Kandori et al. [Kandori et al., 1993] and Young [Young, 1993] which make predictions about the conditions under which agents converge to equilibrium in a certain coordination game by learning the others' play and adjusting their strategies over time. For instance, Kandori et al.'s model investigates the dynamic process that leads the agents to converge to the risk dominant equilibrium in a repeated 2×2 coordination game.

---

[16]In Hart's theory, the rule of recognition is the rule that specifies the ultimate criteria of validity in a legal system.

It is worth noting that the epistemic approach and the evolutionary approach to the study of conventions have not yet been reconciled. Indeed, none of the existing evolutionary models of conventions deals with the epistemic aspect of conventions, as they do not assume agents to be cognitive and only consider a simplified notion of convention as a mere regularity of behavior.

3.3 Logics for institutions

In [Lorini *et al.*, 2009] a modal logic of collective acceptance is proposed, in accordance with the philosophical theories of this notion discussed in Section 3.2. In the logic, collective acceptance is conceived as the collective attitude that some agents have *qua* members of the same institution. In particular, a collective acceptance held by a set of agents C *qua* members of a certain institution x is the kind of acceptance the agents in C are committed to when they are "functioning together as members of the institution x", that is, when the agents in C identify and recognize each other as members of the institution x. For example, in the context of the institution Greenpeace agents (collectively) accept that their mission is to protect the Earth *qua* members of Greenpeace. The state of acceptance *qua* members of Greenpeace is the kind of acceptance these agents are committed to when they are functioning together as members of Greenpeace, that is, when they identify and recognize each other as members of Greenpeace. The logic accounts for different kinds of aggregation procedures that the members of an institution may adopt in order to build a collective acceptance of a given fact. This includes unanimity, majority and a criterion based on leadership according to which what the members of an institution collectively accept coincides with the acceptance of the legislator of the institution. Moreover, the logic clearly distinguishes collective acceptance from common belief, by emphasizing the fact that, while common belief is reducible to individual beliefs, collective acceptance cannot be reduced to individual attitudes of the members of an institution. The fact that collective acceptance is not reducible to individual attitudes is reflected in the formal semantics of the logic. While in epistemic logic common belief is commonly represented by means of the transitive closure of the union of the accessibility relations for the individual beliefs, the accessibility relation for collective acceptance is not definable in terms of the accessibility relations for individual beliefs or individual acceptances. Moreover, collective acceptance entails the notion of "group identification" that is not reducible to the individual level.

Following the idea of some prominent philosophical theories of institutions [Searle, 1995; Tuomela, 2002] according to which institutional reality only exists in relation with the collective acceptance of institutional facts by the members of the institution, a systematic analysis of institutional concepts in the context of this logic is given. This includes the concepts of weak permission, strong permission, obligation and constitutive rule.

The relationship between the logic of collective of acceptance and existing logics of institutions has also been investigated. This includes the comparison between the logic of collective acceptance and the logic of institutional facts proposed by [Jones and Sergot, 1996] and refined more recently by [Grossi *et al.*, 2006]. According to [Jones and Sergot, 1996; Grossi *et al.*, 2006], the primary aspect of institutional facts is their being true in the context of an institution x.

In [Lorini et al., 2009], the bridge between collective acceptance and informal institutions is built by assuming that:

> a certain fact φ is true in the context of an informal institution x if only if the members of the informal institution x collectively accept that φ is true (in the context of x).

Differently from formal or legal institutions, informal institutions have no official of the law who is in charge of promulgating new norms and who is the guarantor of their validity. An example of informal institution is a language whose rule specifying the relationship between a certain utterance and its meaning is shared by a group of people: in the context of this group, the utterance has a certain meaning since the language speakers collectively accept this.

In [Lorini and Longin, 2008], the analysis is extended to formal and legal institutions in which legislators and officials of the law exist who are in charge of either creating new norms or suppressing existing ones out of collective deliberation and who are guarantors of the norms' validity. Specifically, it is assumed that:

> a certain fact φ is true in the context of a formal institution x if only if the legislators of the institution x collectively accept that φ is true (in the context of x).

For example, according to the French law, the legal drinking age is 18 since this fact is accepted by the French legal authority. As emphasized in Section 3.2, this is close to Hart's idea that a legal norm exists because it adheres to the standards of validity specified by the ultimate rule of recognition that has to be accepted by the legal authority. For example, the Italian legal authority accepts that a norm is valid as far as it has been promulgated by the Italian parliament and published in the "Gazzetta Ufficiale della Repubblica Italiana" (Official Gazette of the Italian Republic).

4 Conclusion: closing the circle

In the previous sections we have explained: (i) the role of mental attitudes in decision-making and in action performance as well as the relationship between mental attitudes and emotion (Section 2), (ii) how collective attitudes are generated from mental attitudes as well as the relationship between institutions and norms, on the one hand, and collective attitudes, on the other hand (Section 3). More generally, we have moved from the mental level to the collective level and, then, from the collective level to the institutional-normative level. It is now time to close the circle by going back to mind.

The relevant question here is the following: how do institutions and norms, that are grounded on agents' collective attitudes retroactively influence decision-making and action?

First of all, for a norm or convention to affect an agent's decision, it has to be recognized by the agent, that is, the agent has to believe that the norm or convention exists and that if she does not conform to it, she will incur a violation. The latter is called *normative belief* by [Conte and Castelfranchi, 1999] (see also [Andrighetto et al., 2010]). Recognition of a convention is guaranteed, if the agent belongs to the group of agents in which the convention holds. Indeed, as emphasized in Section 3.2, according to Lewis' definition, a certain regularity of behavior R is a convention for a population of agents P if and only if the agents in P mutually expect everyone in P to conform to the regularity R and commonly believe so. Thus, if agent i is a member of P and R is convention for P, then i has to believe that R is convention for P. The latter follows from the fact that if the agents in P have a common belief that some proposition p holds, then every agent in P has to believe so.[17]

Once the norm or convention with its associated costs and sanction for violation has be recognized by an agent, the agent will take it into consideration in her decision-making process. For the sake of clarity, we here distinguish *norm compliance* from mere *norm following*. Norm compliance requires the *goal* to conform to the content of the norm. In other words, for an agent to comply with a norm, she has to be motivated by the goal of conforming to what the norm prescribes. For example, an agent complies with the norm of paying taxes if she wants to pay taxes, after having recognized the corresponding norm that she ought to pay taxes. Norm following just requires that the agent chooses an action *knowing that* this choice will lead her to conform to what the norm prescribes. To sum up, while norm compliance requires *purposively* (or *intentionally*) conforming to what the norm prescribes, norm following only requires *knowingly* conforming to what the norm prescribes. Under the assumption that "purposively doing" implies "knowingly doing", norm compliance can be seen as a special case of norm following.

Two different forms of norm compliance exist. As we have emphasized in Section 2.1, some norms are internalized by the agent and give rise to moral values. If the agent decides to comply with them, she does it for ethical or moral reasons. In these cases, the agent's goal of conforming to what the norm prescribes is mainly originated from moral considerations. This is *ethical or moral compliance*. For example, an agent may comply with the legal obligation to pay taxes for ethical or moral reasons: the agent wants to pay taxes because she is motivated by the moral value to behave honestly. More generally, ethical compliance requires that the agent's goal of conforming to what the norm prescribes does not depend on the agent's actual desires[18] but only on the agents' actual moral values.[19]

In other cases, the agent complies with the norm because she desires to avoid the sanction or the social cost as a consequence of the violation and because she fears punishment. This is *opportunistic compliance* which is typical for conventions such as the following one:

> Except for pizza, sandwiches and other "finger foods", don't eat with your fingers.

This is a convention in Europe, as every person in Europe expects other people in

---

[17]This property can be formally proved in the logic of common belief [Fagin et al., 1995].

[18]This means that if the agent did have different desires in her mind, he would have had still the goal to follow the norm.

[19]This means that it is possible for the agent to reconsider her actual moral values in such a way that her goal to follow the norm is also reconsidered.

Europe to follow it and every group of European people has a common belief that each of them expects the others to follow the convention. An European person believes that the convention exists and wants to follow it because she desires to avoid the social cost associated with the violation (e.g., the cost of being publicly blamed if she eats the food with her fingers).

In the case of opportunistic compliance, the agent wants to conform with what the norm prescribes because the consequences of norm violation (e.g., sanction, social cost, punishment) are undesirable for her, while the consequences of norm fulfillment (e.g., reward, social approval) are desirable for her. More generally, opportunistic compliance requires that the agent's goal of conforming to what the norm prescribes does not depend on the agent's actual moral values but only on the agents' actual desires.

We conclude the paper with the general observation that, although norm compliance has been extensively studied in the area of multiagent systems, with an emphasis on both its logical aspects [Ågotnes *et al.*, 2009; Rotolo, 2011; Knobbout and Dastani, 2012], and computational aspects [Criado Pacheco *et al.*, 2013; Alechina *et al.*, 2012; Lopez y Lopez *et al.*, 2004; van Riemsdijk *et al.*, 2013; Lee *et al.*, 2014], there is still no formal model which captures the distinctions between norm following and norm compliance, and between ethical compliance and opportunistic compliance. We believe this is an important issue. Its understanding would allow to complement a bottom-up approach to institutions, grounding them on the mental level via the collective level, with a top-down approach, explaining how institutions and norms influence the agents' cognition.

BIBLIOGRAPHY

[Adam *et al.*, 2009] C. Adam, A. Herzig, and D. Longin. A logical formalization of the OCC theory of emotions. *Synthese*, 168(2):201–248, 2009.

[Ågotnes *et al.*, 2007] T. Ågotnes, W. van der Hoek, and M. Wooldridge. Quantified coalition logic. In *Proceedings of the Twentieth International Joint Conference on Artificial Intelligence (IJCAI'07)*, pages 1181–1186. AAAI Press, 2007.

[Ågotnes *et al.*, 2009] T. Ågotnes, W. van der Hoek, and M. Wooldridge. Robust normative systems and a logic of norm compliance. *Logic Journal of the IGPL*, 18(1):4–30, 2009.

[Alechina *et al.*, 2012] N. Alechina, M. Dastani, and B. Logan. Programming norm-aware agents. In *Proceedings of the 11th International Conference on Autonomous Agents and Multiagent Systems (AAMAS 2012)*, pages 1057–1064. ACM Press, 2012.

[Alur *et al.*, 2002] R. Alur, T. Henzinger, and O. Kupferman. Alternating-time temporal logic. *Journal of the ACM*, 49:672–713, 2002.

[Andrighetto *et al.*, 2010] G. Andrighetto, M. Campennì, F. Cecconi, and R. Conte. The complex loop of norm emergence: a simulation model. In K. Takadama, C. C. Revilla, and G. Deffuant, editors, *The Second World Congress on Social Simulation*, LNAI. Springer-Verlag, 2010.

[Anscombe, 1957] G. E. M. Anscombe. *Intention*. Basil Blackwell, 1957.

[Aumann, 1999] R. Aumann. Interactive epistemology I: Knowledge. *International Journal of Game Theory*, 28(3):263–300, 1999.

[Baltag *et al.*, 2009] A. Baltag, S. Smets, and J. A. Zvesper. Keep "hoping" for rationality: a solution to the backward induction paradox. *Synthese*, 169(2):301–333, 2009.

[Battigalli and Dufwenberg, 2007] P. Battigalli and M. Dufwenberg. Guilt in games. *The American Economic Review*, 97(2):170–176, 2007.

[Battigalli and Siniscalchi, 2002] P. Battigalli and M. Siniscalchi. Strong belief and forward induction reasoning. *J. of Economic Theory*, 106(2):356–391, 2002.

[Baumeister *et al.*, 1994] R. F. Baumeister, A. M. Stillwell, and T. F. Heatherton. Guilt: an interpersonal approach. *Psychological Bullettin*, 115(2):243–267, 1994.

[Belnap et al., 2001] N. Belnap, M. Perloff, and M. Xu. *Facing the future: agents and choices in our indeterminist world*. Oxford University Press, New York, 2001.

[Bicchieri, 2006] C. Bicchieri. *The grammar of society: the nature and dynamics of social norms*. Cambridge University Press, 2006.

[Binmore, 1991] K. Binmore. *Fun and Games: A Text on Game Theory*. D. C. Heath and Company, 1991.

[Binmore, 2005] K. Binmore. *Natural Justice*. Oxford University Press, 2005.

[Bjorndahl et al., 2014] A. Bjorndahl, J. Y. Halpern, and R. Pass. Axiomatizing rationality. In *Proceedings of the Fourteenth International Conference on Principles of Knowledge Representation and Reasoning: (KR 2014)*. AAAI Press, 2014.

[Boutilier, 1994] C. Boutilier. Towards a logic for qualitative decision theory. In *Proceedings of International Conference on Principles of Knowledge Representation and Reasoning (KR'94)*, pages 75–86. AAAI Press, 1994.

[Brafman and Tennenholtz,] R. I. Brafman and Moshe Tennenholtz. An axiomatic treatment of three qualitative decision criteria. *Journal of the ACM*, 47(3):452–482.

[Brafman and Tennenholtz, 1996] R. I. Brafman and Moshe Tennenholtz. On the foundations of qualitative decision theory. In *Proceedings of the Thirteenth National Conference on Artificial Intelligence (AAAI'96)*, pages 1291–1296. AAAI Press, 1996.

[Brandenburger et al., 2008] A. Brandenburger, A. Friedenberg, and J. Keisler. Admissibility in games. *Econometrica*, 76:307–352, 2008.

[Bratman et al., 1988] M. Bratman, D. J. Israel, and M. E. Pollack. Plans and resource-bounded practical reasoning. *Computational Intelligence*, 4:349–355, 1988.

[Bratman, 1987] M. Bratman. *Intentions, plans, and practical reason*. Harvard University Press, Cambridge, 1987.

[Bratman, 1992] Michael E. Bratman. Shared cooperative activity. *The Philosophical Review*, 101(2):327–41, 1992.

[Broersen et al., 2002] J. Broersen, M. Dastani, J. Hulstijn, and L. van der Torre. Goal generation in the boid architecture. *Cognitive Science Quarterly*, 2(3-4):428–447, 2002.

[Broersen, 2011] J. Broersen. Deontic epistemic stit logic distinguishing modes of mens rea. *Journal of Applied Logic*, 9(2):137–152, 2011.

[Caplin and Leahy, 2001] A. Caplin and J Leahy. Psycological expected utility theory and anticipatory feelings. *Quarterly Journal of Economics*, 116(1):55–79, 2001.

[Castelfranchi, 1998] Cristiano Castelfranchi. Modelling social action for AI agents. *Artificial Intelligence*, 103:157–182, 1998.

[Charness and Dufwenberg, 2009] G. Charness and M. Dufwenberg. Guilt in games. *Econometrica*, 74(6):1579–1601, 2009.

[Clark and Marshall, 1981] H. Clark and C. Marshall. Definite reference and mutual knowledge. In A. K. Joshi, B. L. Webber, and I. A. Sag, editors, *Elements of discourse understanding*. 1981.

[Cohen and Levesque, 1990] P. R. Cohen and H. J. Levesque. Reasons: Belief support and goal dynamics. *Artificial Intelligence*, 42:213–61, 1990.

[Conte and Castelfranchi, 1999] R. Conte and C. Castelfranchi. From conventions to prescriptions. towards an integrated view of norms. *Artificial Intelligence and Law*, 7:323–340, 1999.

[Criado Pacheco et al., 2013] N. Criado Pacheco, E. Argente, P. Noriega, and V. Botti. Human-inspired model for norm compliance decision making. *Information Sciences*, 245:218–239, 2013.

[Cubitt and Sugden, 2003] R. P. Cubitt and R. Sugden. Common knowledge, salience and convention: a reconstruction of david lewis' game theory. *Economics and Philosophy*, 19:175–210, 2003.

[Damasio, 1994] A. Damasio. *Descartes Error: Emotion, Reason and the Human Brain*. Putnam Publishing, New York, 1994.

[Dastani and Lorini,] M. Dastani and E. Lorini. A logic of emotions: from appraisal to coping. In *Proceedings of the Eleventh International Joint Conference on Autonomous Agents and Multiagent Systems (AAMAS 2012)*, ACM Press, pages 1133-1140.

[Davidson, 1980] D. Davidson. Intending. In *Essays on Actions and Events*. Oxford University Press, New York, 1980.

[Dennett, 1987] D. C. Dennett. *The Intentional Stance*. MIT Press, Cambridge, Massachusetts, 1987.

[Dretske, 1988] F. Dretske. *Explaining behavior: reasons in a world of causes*. MIT Press, 1988.

[Dunin-Keplicz and Verbrugge, 2002] B. Dunin-Keplicz and R. Verbrugge. Collective intentions. *Fundamenta Informaticae*, 51(3):271–295, 2002.

[Dunin-Keplicz and Verbrugge, 2010] B. Dunin-Keplicz and R. Verbrugge. *Teamwork in Multi-Agent Systems: A Formal Approach*. Wiley, 2010.

[Durkheim, 1982] E. Durkheim. *The rules of Sociological Method*. Free Press, New York, 1982. first published in French in 1895.

[Esteva et al., 2001] M. Esteva, J. Padget, and C. Sierra. Formalizing a language for institutions and norms. In *Intelligent Agents VIII (ATAL'01)*, volume 2333 of *LNAI*, pages 348–366, Berlin, 2001. Springer Verlag.

[Fagin et al., 1995] R. Fagin, J. Halpern, Y. Moses, and M. Vardi. *Reasoning about Knowledge*. MIT Press, Cambridge, 1995.

[Frijda et al., 1989] N. H. Frijda, P. Kuipers, and E. Ter Schure. Relations among emotion, appraisal, and emotional action readiness. *Journal of Personality and Social Psychology*, 57(2):212–228, 1989.

[Galeazzi and Lorini, 2016] P. Galeazzi and E. Lorini. Epistemic logic meets epistemic game theory: a comparison between multi-agent Kripke models and type spaces. *Synthese*, 193(7):2097–2127, 2016.

[Gilbert, 1987] M. Gilbert. Modelling collective belief. *Synthese*, 73(1):185–204, 1987.

[Gilbert, 1989] M. Gilbert. *On Social Facts*. Routledge, London and New York, 1989.

[Goldszmidt and Pearl, 1996] M. Goldszmidt and J. Pearl. Qualitative probability for default reasoning, belief revision and causal modeling. *Artificial Intelligence*, 84:52–112, 1996.

[Gordon, 1987] R. M. Gordon. *The structure of emotions*. Cambridge University Press, Cambridge, 1987.

[Gradel, 2002] E. Gradel. Model checking games. *Electronic Notes in Theoretical Computer Science*, 67:15–34, 2002.

[Gratch and Marsella, 2004] J. Gratch and S. Marsella. A domain independent framework for modeling emotion. *Journal of Cognitive Systems Research*, 5(4):269–306, 2004.

[Grossi and Pigozzi, 2014] D. Grossi and G. Pigozzi. *Judgment Aggregation: A Primer*. Synthesis Lectures on Artificial Intelligence and Machine Learning. Morgan & Claypool Publishers, 2014.

[Grossi et al., 2006] D. Grossi, J.-J. Ch. Meyer, and F. Dignum. Classificatory aspects of counts-as: An analysis in modal logic. *Journal of Logic and Computation*, 16(5):613–643, 2006.

[Grosz and Kraus, 1996] Barbara Grosz and Sarit Kraus. Collaborative plans for complex group action. *Artificial Intelligence*, 86(2):269–357, 1996.

[Haidt, 2003] J. Haidt. The moral emotions. In R. J. Davidson, K. R. Scherer, and H. H. Goldsmith, editors, *Handbook of affective sciences*, pages 852–870. 2003.

[Hájek and Pettit, 2004] A. Hájek and P. Pettit. A theory of human motivation. *Australian Journal of Philosophy*, 82:77–92, 2004.

[Halpern and Pass, 2009] J. Y. Halpern and R. Pass. A logical characterization of iterated admissibility. In A. Heifetz, editor, *Proc. of TARK 2009*, pages 146–155, 2009.

[Halpern, 2011] J. Y. Halpern. Beyond nash equilibrium: Solution concepts for the 21st century. In K. R. Apt and E. Gradel, editors, *Lectures in Game Theory for Computer Scientists*, pages 264–289. 2011.

[Harsanyi, 1967] J. C. Harsanyi. Games with incomplete information played by 'bayesian' players. *Management Science*, 14:159–182, 1967.

[Harsanyi, 1982] J. Harsanyi. Morality and the theory of rational behaviour. In A.K. Sen and B. Williams, editors, *Utilitarianism and Beyond*. Cambridge University Press, Cambridge, 1982.

[Hart, 1992] H. L. A. Hart. *The concept of law*. Clarendon Press, Oxford, 1992. new edition.

[Herzig and Longin, 2004] Andreas Herzig and Dominique Longin. C&L intention revisited. In Didier Dubois, Chris Welty, and Mary-Anne Williams, editors, *Proceedings 9th Int. Conf. on Principles on Principles of Knowledge Representation and Reasoning(KR2004)*, pages 527–535. AAAI Press, 2004.

[Hintikka and Sandu, 1997] J. Hintikka and G. Sandu. Game-theoretical semantics. In J. van Benthem and A. ter Meulen, editors, *Handbook of Logic and Language*, pages 361–410. Elsevier, 1997.

[Hopfensitz and Reuben, 2009] A. Hopfensitz and E. Reuben. The importance of emotions for the effectiveness of social punishment. *The Economic Journal*, 119(540):1534–1559, 2009.

[Horty, 2001] J. F. Horty. *Agency and Deontic Logic*. Oxford University Press, Oxford, 2001.

[Humberstone, 1992] I. L. Humberstone. Direction of fit. *Mind*, 101(401):59–83, 1992.

[J. and R., 1999] Doyle J. and Thomason R. Background to qualitative decision theory. *The AI Magazine*, 20(2):55–68, 1999.

[Jones and Sergot, 1996] Andrew Jones and Marek J. Sergot. A formal characterization institutionalised power. *Journal of the IGPL*, 4:429–445, 1996.

[Kahneman and Miller, 1986] D. Kahneman and D. T. Miller. Norm theory: comparing reality to its alternatives. *Psychological Review*, 93(2):136–153, 1986.

[Kandori et al., 1993] M. Kandori, G. Mailath, and R. Rob. Learning, mutation, and long run equilibria in games. *Econometrica*, 61:29–56, 1993.

[Keiff, 2011] L. Keiff. Dialogical logic. In E. N. Zalta, editor, *The Stanford Encyclopedia of Philosophy*. 2011.

[Klein and Pacuit, 2014] D. Klein and E. Pacuit. Changing types: Information dynamics for qualitative type spaces. *Studia Logica*, 102:297–319, 2014.

[Knobbout and Dastani, 2012] M. Knobbout and M. Dastani. Reasoning under compliance assumptions in normative multiagent systems. In *Proceedings of the 11th International Conference on Autonomous Agents and Multiagent Systems (AAMAS 2012)*, pages 331–340. ACM Press, 2012.

[Konolige and Pollack, 1993] K. Konolige and M. E. Pollack. A representationalist theory of intention. In R. Bajcsy, editor, *Proceedings 13th International Joint Conference on Artificial Intelligence (IJCAI 93)*, pages 390–395, San Francisco, CA, 1993. Morgan Kaufmann Publishers.

[Kraus and Lehmann, 1988] S. Kraus and D. J. Lehmann. Knowledge, belief and time. *Theoretical Computer Science*, 58:155–174, 1988.

[Lazarus, 1991] R. S. Lazarus. *Emotion and adaptation*. Oxford University Press, New York, 1991.

[LeDoux, 1996] J. LeDoux. *The emotional Brain*. Simon and Schuster, New York, 1996.

[Lee et al., 2014] J. Lee, J. Padget, B. Logan, D. Dybalova, and N. Alechina. Run-time norm compliance in BDI agents. In *Proceedings of the Proceedings of the 2014 international conference on Autonomous agents and multi-agent systems (AAMAS 2014)*, pages 1581–1582. ACM Press, 2014.

[Lewis, 1969] D. K. Lewis. *Convention: a philosophical study*. Harvard University Press, Cambridge, 1969.

[Lewis, 1988] D. Lewis. Desire as belief. *Mind*, 97:323–332, 1988.

[Lewis, 1996] D. Lewis. Desire as belief ii. *Mind*, 105:303–313, 1996.

[List and Pettit, 2011] C. List and P. Pettit. *Group Agency: The Possibility, Design, and Status of Corporate Agents*. Oxford University Press, 2011.

[List, 2012] C. List. The theory of judgment aggregation: an introductory review. *Synthese*, 187(1):179–207, 2012.

[List, 2014] C. List. Three kinds of collective attitudes. *Erkenntnis*, 79(9):1601–1622, 2014.

[Locke, 1989] J. Locke. *An essay concerning human understanding*. Clarendon Press, Oxford, 1989.

[Loomes and Sugden, 1987] G. Loomes and R. Sugden. Testing for regret and disappointment in choice under uncertainty. *Economic J.*, 97:118–129, 1987.

[Lopez y Lopez et al., 2004] F. Lopez y Lopez, M. Luck, and M. d'Inverno. Normative agent reasoning in dynamic societies. In *Proceedings of the Third International Conference on Autonomous Agents and Multi-Agent Systems (AAMAS'04)*, pages 732–739. ACM Press, 2004.

[Lorini and Castelfranchi, 2007] E. Lorini and C. Castelfranchi. The cognitive structure of Surprise: looking for basic principles. *Topoi: An International Review of Philosophy*, 26((1)):133–149, 2007.

[Lorini and Herzig, 2008] E. Lorini and A. Herzig. A logic of intention and attempt. *Synthese*, 163(1):45–77, 2008.

[Lorini and Longin, 2008] E. Lorini and D. Longin. A logical account of institutions: from acceptances to norms via legislators. In *Proceedings of the International Conference on Principles of Knowledge Representation and Reasoning (KR 2008)*, pages 38–48. AAAI Press, 2008.

[Lorini and Muehlenbernd, 2015] E. Lorini and R. Muehlenbernd. The long-term benefits of following fairness norms: a game-theoretic analysis. In *Proceedings of the 18th Conference on Principles and Practice of Multi-Agent Systems (PRIMA 2015)*, pages 301–318, Berlin, 2015. Springer-Verlag.

[Lorini and Schwarzentruber, 2010] E. Lorini and F. Schwarzentruber. A Modal Logic of Epistemic Games. *Games, Epistemic Game Theory and Modal Logic*, 1(4):478–526, 2010.

[Lorini and Schwarzentruber, 2011] E. Lorini and F. Schwarzentruber. A logic for reasoning about counterfactual emotions. *Artificial Intelligence*, 175(3-4):814–847, 2011.

[Lorini et al., 2009] E. Lorini, D. Longin, B. Gaudou, and A. Herzig. The logic of acceptance: grounding institutions on agents' attitudes. *Journal of Logic and Computation*, 19(6):901–940, 2009.

[Lorini et al., 2014] E. Lorini, D. Longin, and E. Mayor. A logical analysis of responsibility attribution : emotions, individuals and collectives. *Journal of Logic and Computation*, 24(6):1313–1339, 2014.

[Lorini, 2013] E. Lorini. Temporal STIT logic and its application to normative reasoning. *Journal of Applied Non-Classical Logics*, 23(4):372–399, 2013.

[Lorini, 2016a] E. Lorini. A logic for reasoning about moral agents. *Logique et Analyse*, 58(230):177–218, 2016.

[Lorini, 2016b] E. Lorini. A minimal logic for interactive epistemology. *Synthese*, 193(3):725–755, 2016.

[Ludwig and Jankovic, 2016] K. Ludwig and M. Jankovic. Collective intentionality. In L. McIntyre and A. Rosenberg, editors, *The Routledge Companion to the Philosophy of Social Science*. Routledge, New York, 2016.

[Mantzavinos et al., 2004] C. Mantzavinos, D.C. North, and S. Shariq. Learning, institutions, and economic performance. *Perspectives on Politics*, 2:75–84, 2004.

[March and Simon, 1958] J. G. March and H. A. Simon. *Organizations*. Wiley, New York, 1958.

[Maslow, 1943] A. H. Maslow. A theory of human motivation. *Psychological Review*, 50:370–396, 1943.

[McCann, 1991] H. McCann. Settled objectives and rational constraints. *American Philosophical Quarterly*, 28:25–36, 1991.

[Mele, 1992] A. R. Mele. *Springs of Action: Understanding Intentional Behavior*. Oxford University Press, Oxford, 1992.

[Meyer et al., 1999] J. J. Ch. Meyer, W. van der Hoek, and B. van Linder. A logical approach to the dynamics of commitments. *Artificial Intelligence*, 113(1-2):1–40, 1999.

[Meyer, 2006] J.-J. Ch. Meyer. Reasoning about emotional agents. *International J. of Intelligent Systems*, 21(6):601–619, 2006.

[Miller and Sandu, 1997] K. Miller and G. Sandu. Weak commitments. In G. Holmstron-Hintikka and R. Tuomela, editors, *Contemporary Action Theory, vol.2: Social Action*. Kluwer Academic Publishers, Dordrecht, 1997.

[Myerson, 1991] R. Myerson. *Game Theory: Analysis of Conflict*. Harvard University Press, 1991.

[North, 1990] D.C. North. *Institutions, Institutional Change, and Economic Performance*. Cambridge University Press, Cambridge, 1990.

[Ortony et al., 1988] Andrew Ortony, G.L. Clore, and A. Collins. *The cognitive structure of emotions*. Cambridge University Press, Cambridge, MA, 1988.

[Pauly, 2002] M. Pauly. A modal logic for coalitional power in games. *Journal of Logic and Computation*, 12(1):149–166, 2002.

[Perea, 2012] A. Perea. *Epistemic game theory: reasoning and choice*. Cambridge University Press, 2012.

[Platts, 1979] M. Platts. *Ways of meaning*. Routledge and Kegan Paul, 1979.

[Plaza, 1989] J. A. Plaza. Logics of public communications. In M. Emrich, M. Pfeifer, M. Hadzikadic, and Z. Ras, editors, *Proceedings of the 4th International Symposium on Methodologies for Intelligent Systems*, 201-216, 1989.

[Pylyshyn, 1984] Z. Pylyshyn. *Computation and Cognition: Toward a Foundation for Cognitive Science*. MIT Press, Cambridge, Massachusetts, 1984.

[Rao and Georgeff, 1991] A. S. Rao and M. P. Georgeff. Modelling rational agents within a BDI architecture. In *Proceedings of KR'91*, San Francisco, CA, 1991. Morgan Kaufmann Publishers.

[Reisenzein, 2009] R. Reisenzein. Emotional experience in the computational belief-desire theory of emotion. *Emotion Review*, 1(3):214–222, 2009.

[Rescher, 1967] Nicholas Rescher. Semantic foundations for the logic of preference. In N. Rescher, editor, *The logic of decision and action*. University of Pittsburgh Press, 1967.

[Rick and Loewenstein, 2008] S. Rick and G. Loewenstein. The role of emotion in economic behavior. In M. Lewis, J. Haviland-Jones, and L. Feldman-Barrett, editors, *The Handbook of Emotion*. Guilford, New York, 2008.

[Roseman et al., 1996] I.J. Roseman, A.A. Antoniou, and P.E. Jose. Appraisal determinants of emotions: Constructing a more accurate and comprehensive theory. *Cognition and Emotion*, 10:241–277, 1996.

[Rotolo, 2011] A. Rotolo. Norm compliance of rule-based cognitive agents. In *Proceedings of the 22nd International Joint Conference on Artificial Intelligence (IJCAI 2011)*, pages 2716–2721. AAAI Press, 2011.

[Scherer et al., 2001] K. R. Scherer, A. Schorr, and T. Johnstone, editors. *Appraisal Processes in Emotion: Theory, Methods, Research*. Oxford University Press, Oxford, 2001.

[Schroeder, 2004] T. Schroeder. *Three faces of desires*. Oxford University Press, 2004.

[Searle, 1979] J. Searle. *Expression and meaning*. Cambridge University Press, 1979.

[Searle, 1983] J. Searle. *Intentionality: An Essay in the Philosophy of Mind*. Cambridge University Press, New York, 1983.

[Searle, 1995] John Searle. *The Construction of Social Reality*. The Free Press, New York, 1995.

[Searle, 2001] J. Searle. *Rationality in Action*. MIT Press, Cambridge, 2001.

[Sen and Airiau, 2007] S. Sen and S. Airiau. Emergence of norms through social learning. In *Proceedings of the 20th International Joint Conference on Artificial Intelligence (IJCAI 2007)*, pages 1507–1512. ACM Press, 2007.

[Shoham and Tennenholtz, 1997] Y. Shoham and M. Tennenholtz. On the emergence of social conventions: modeling, analysis, and simulations. *Artificial Intelligence*, 94(1-2):139–166, 1997.

[Shoham, 1993] Y. Shoham. Agent-oriented programming. *Artificial Intelligence*, 60:51–92, 1993.

[Sillari, 2005] G. Sillari. A logical framework for convention. *Synthese*, 147(2):379–400, 2005.

[Simon, 1956] H. A. Simon. Rational choice and the structure of the environment. *Psychological Review*, 63(2):129–138, 1956.

[Singh and Asher, 1993] M. Singh and N. Asher. A logic of intentions and beliefs. *Journal of Philosophical Logic*, 22:513–544, 1993.

[Stalnaker, 1998] R. Stalnaker. Belief revision in games: forward and backward induction. *Mathematical Social Sciences*, 36:31–56, 1998.

[Stalnaker, 2002] R. Stalnaker. Common ground. *Linguistics and Philosophy*, 25(5-6):701–721, 2002.

[Stalnaker, 2006] R. Stalnaker. On logics of knowledge and belief. *Philosophical Studies*, 128:169–199, 2006.

[Sugden, 2004] R. Sugden. *Economics of rights, co-operation and welfare (2nd Edition)*. Palgrave Macmillan, 2004.

[Tangney, 1995] J. P. Tangney. Recent advances in the empirical study of shame and guilt. *American Behavioral Scientist*, 38(8):1132–1145, 1995.

[Tollefsen, 2002] D. P. Tollefsen. Collective intentionality and the social sciences. *Philosophy of the Social Sciences*, 32(1):25–50, 2002.

[Tummolini et al., 2013] L. Tummolini, G. Andrighetto, C. Castelfranchi, and R. Conte. A convention or (tacit) agreement betwixt us: on reliance and its normative consequences. *Synthese*, 190(4):585–618, 2013.

[Tuomela, 2002] R. Tuomela. *The Philosophy of Social Practices: A Collective Acceptance View*. Cambridge University Press, Cambridge, 2002.

[Tuomela, 2007] R. Tuomela. *The Philosophy of Sociality*. Oxford University Press, Oxford, 2007.

[Turrini et al., 2010] P. Turrini, J.-J. Ch. Meyer, and C. Castelfranchi. Coping with shame and sense of guilt: a Dynamic Logic Account. *Journal Autonomous Agents and Multi-Agent Systems*, 20(3):401–420, 2010.

[van Benthem et al., 2009] J. van Benthem, P. Girard, and O. Roy. Everything else being equal: A modal logic for ceteris paribus preferences. *Journal of Philosophical Logic*, 38:83–125, 2009.

[van Benthem, 2007] J. van Benthem. Rational dynamics and epistemic logic in games. *International Game Theory Review*, 9(1):13–45, 2007.

[van Ditmarsch et al., 2007] H. P. van Ditmarsch, W. van der Hoek, and B. Kooi. *Dynamic Epistemic Logic*. Kluwer Academic Publishers, 2007.

[van Linder et al., 1998] B. van Linder, van der Hoek, and J.-J. Ch. W., Meyer. Formalising abilities and opportunities. *Fundamenta Informaticae*, 34:53–101, 1998.

[van Riemsdijk et al., 2013] M. B. van Riemsdijk, L. A. Dennis, M. Fisher, and K. V. Hindriks. Agent reasoning for norm compliance: a semantic approach. In *Proceedings of the 2013 international conference on Autonomous agents and multi-agent systems (AAMAS 2013)*, pages 499–506. ACM Press, 2013.

[Vanderschraaf, 1995] P. Vanderschraaf. Convention as correlated equilibrium. *Erkenntnis*, 42(1):65–87, 1995.

[Villatoro et al., 2011] D. Villatoro, S. Sen, and J. Sabater-Mir. Exploring the dimensions of convention emergence in multiagent systems. *Advances in Complex Systems*, 14(2):201–227, 2011.

[Von Wright, 1963] G. H. Von Wright. *The logic of preference*. Edinburgh University Press, 1963.

[Walker and Wooldridge, 1995] W. Walker and M. Wooldridge. Understanding the emergence of conventions in multi-agent systems. In *Proceedings of the First International Conference on Multi-Agent Systems (ICMAS-95)*, pages 384–389. AAAI Press, 1995.

[Wellman and Doyle, 2001] M. P. Wellman and J. Doyle. Preferential semantics for goals. In *Proceedings of the Ninth National conference on Artificial intelligence (AAAI'91)*, pages 698–703. AAAI Press, 2001.

[Wooldridge, 2000] M. Wooldridge. *Reasoning about Rational Agents*. MIT Press, Cambridge, 2000.

[Young, 1993] H. P. Young. The evolution of conventions. *Econometrica*, 61:57–84, 1993.

[Zeelenberg et al., 2000] M. Zeelenberg, W. van Dijk, A. S. R. Manstead, and J. van der Pligt. On bad decisions and disconfirmed expectancies: the psychology of regret and disappointement. *Cognition and Emotion*, 14(4):521–541, 2000.

[Zvesper, 2010] J. A. Zvesper. *Playing with Information*. PhD thesis, University of Amsterdam, The Netherlands, 2010.

www.ingramcontent.com/pod-product-compliance
Lightning Source LLC
Chambersburg PA
CBHW051032160426
43193CB00010B/910